Selenium Contamination in Water

Selenium Contamination in Water

Edited by

Pooja Devi
CSIR- CSIO, India

Pardeep Singh
PGDAV College, University of Delhi, India

Arindam Malakar
Nebraska Water Center, University of Nebraska, USA

Daniel Snow
Nebraska Water Center, University of Nebraska, USA

WILEY Blackwell

This edition first published 2021

Registered Offices
John Wiley & Sons, Inc., 111 River Street, Hoboken, NJ 07030, USA
John Wiley & Sons Ltd, The Atrium, Southern Gate, Chichester, West Sussex, PO19 8SQ, UK

Editorial Office
9600 Garsington Road, Oxford, OX4 2DQ, UK

For details of our global editorial offices, customer services, and more information about Wiley products visit us at www.wiley.com.

Wiley also publishes its books in a variety of electronic formats and by print-on-demand. Some content that appears in standard print versions of this book may not be available in other formats.

Library of Congress Cataloging-in-Publication Data

Names: Devi, Pooja, 1988– editor. | Singh, Pardeep, editor. | Malakar, Arindam, editor. | Snow, Daniel (Daniel D.), editor. | John Wiley & Sons, Inc., publisher.
Title: Selenium contamination in water / edited by Pooja Devi, Pardeep Singh, Arindam Malakar, Daniel Snow.
Description: Hoboken, NJ : Wiley-Blackwell, 2021. | Includes index.
Identifiers: LCCN 2020047563 (print) | LCCN 2020047564 (ebook) | ISBN 9781119693451 (cloth) | ISBN 9781119693536 (adobe pdf) | ISBN 9781119693543 (epub)
Subjects: LCSH: Water–Pollution. | Selenium. | Drinking water–Contamination. | Soils–Selenium content.
Classification: LCC TD427.S38 S44 2021 (print) | LCC TD427.S38 (ebook) | DDC 628.1/68–dc23
LC record available at https://lccn.loc.gov/2020047563
LC ebook record available at https://lccn.loc.gov/2020047564

Cover Design: Wiley
Cover Images: Reproduced from Y.V. Nancharaiah, P.N.L. Lens, Ecology and biotechnology of selenium-respiring bacteria, *MMBR* 79: 61–80. doi: 10.1128/MMBR.00037-14/ with permission of the publisher.

Set in 9.5/12.5pt STIXTwoText by SPi Global, Pondicherry, India

10 9 8 7 6 5 4 3 2 1

To My Beloved, Disciplined, Courteous, and Devoted Student,
Late Mr. Rishabh Jain,
May your Soul Rest in Peace Wherever You Are

LATE MR. RISHABH JAIN
30.03.1991 – 10.01.2021

Contents

List of Contributors

Pinki Rani Agrawal
CSIR- National Physical Laboratory
New Delhi
India

Academy of Scientific and Innovative Research (AcSIR)
Ghaziabad
India

Abolfazl Arjmandi
Department of Chemical Engineering
Babol Noshirvani University of Technology
Babol
Iran

Amit Bansiwal
Sophisticated Environmental Analytical Facility (SEAF)
CSIR-National Environmental Engineering Research Institute (CSIR-NEERI)
Nagpur
Maharashtra
India

Lavisha Bashambu
Dr. S. S. Bhatnagar University Institute of Chemical Engineering and Technology
Panjab University
Chandigarh
India

Vijay K. Bharti
DRDO-Defence Institute of High Altitude Research
Leh
Ladakh
India

Madhulika Bhati
Principal Scientist CSIR – National Institute of Science
Technology and Development Studies (CSIR-NISTADS)
New Delhi
India

Jennifer Cooper
Department of Agronomy and Horticulture
University of Nebraska
Lincoln
Nebraska
USA

Rashmi Dahake
CSIR-National Environmental Engineering Research Institute
Nagpur
Maharashtra
India

Academy of Scientific and Innovative Research (AcSIR)
Ghaziabad
India

Arup Giri
DRDO-Defence Institute of High Altitude Research
Leh
Ladakh
India

Rahul Harshwardhan
Office of Registrar General of India
Ministry of Home Affairs
New Delhi
India

Asmita Jadhav
CSIR-National Environmental Engineering Research Institute
Nagpur
Maharashtra
India

Rishabh Jain
Academy of Scientific and Innovative Research (AcSIR)
Ghaziabad
India

Central Scientific Instruments Organisation
Chandigarh
India

Rupali Jha
Department of Chemical Engineering
HBTU
Kanpur
Uttar Pradesh
India

Charu Jhamaria
Environmental Science
IIS (deemed to be University)
Jaipur
Rajasthan
India

Sushil Kumar Kansal
Dr. S.S. Bhatnagar University Institute of Chemical Engineering and Technology
Panjab University
Chandigarh
India

Paramdeep Kaur
Department of Biotechnology
Panjab University
Chandigarh
India

S. Kaviya
Department of Applied Chemistry
Cochin University of Science and Technology (CUST)
Kerala
India

Ketki Kulkarni
Sophisticated Environmental Analytical Facility (SEAF)
CSIR-National Environmental Engineering Research Institute (CSIR-NEERI)
Nagpur
Maharashtra
India

Pradeep Kumar
Department of Chemical Engineering & Technology
IIT (BHU)
Varanasi
Uttar Pradesh
India

Praveen Kumar
School of Materials Sciences
Indian Association for the Cultivation of Science
Kolkata
India

Anuradha Kumari
School of Environmental Science
Jawaharlal Nehru University
New Delhi
India

Nitin Labhsetwar
Sophisticated Environmental Analytical Facility (SEAF)
CSIR-National Environmental Engineering Research Institute (CSIR-NEERI)
Nagpur
Maharashtra
India

Dipa Mahato
Junior Research Fellow and Senior Research Fellow
CSIR – National Institute of Science Technology and Development studies (CSIR-NISTADS)
New Delhi
India

Vishal Mishra
School of Biochemical Engineering
Indian Institute of Technology (BHU)
Varanasi
Uttar Pradesh
India

Neerja Mittal
Academy of Scientific and Innovative Research (AcSIR)
Ghaziabad
India

Central Scientific Instruments Organisation
Chandigarh
India

Deepanshu Monga
Department of Biotechnology
Panjab University
Chandigarh
India

Ketki Nagpure
Sophisticated Environmental Analytical Facility (SEAF)
CSIR-National Environmental Engineering Research Institute (CSIR-NEERI)
Nagpur
Maharashtra
India

Ipsita Nandi
Institute of Environmental Science
Banaras Hindu University
Varanasi
Uttar Pradesh
India

Sakshi Narula
CSIR – National Institute of Science Technology and Development studies (CSIR-NISTADS)
New Delhi
India

Majid Peyravi
Department of Chemical Engineering
Babol Noshirvani University of Technology
Babol
Iran

Jayashree Rajesh Prasad
Department of Computer Science
Singhad College of Engineering
Pune
Maharashtra
India

Pooja Devi
Academy of Scientific and Innovative Research (AcSIR)
Ghaziabad
India

Central Scientific Instruments Organisation
Chandigarh
India

Puneet Ranjan
DRDO-Defence Institute of High Altitude Research
Leh
Ladakh
India

Awanish Kumar Sharma
Department of Physics
GEU
Dehradun
Uttarakhand
India

Rahul Sharma
CSIR- National Physical Laboratory
New Delhi
India

Academy of Scientific and innovative Research (AcSIR)
Ghaziabad
India

Sulaxna Sharma
THDC, IHET
Tehri
Uttarakhand
India

Baljinder Singh
Department of Biotechnology
Panjab University
Chandigarh
India

Pardeep Singh
Department of Environment Sciences
University of Delhi
New Delhi
India

Surinder Singh
Dr. S.S. Bhatnagar University Institute of Chemical Engineering and Technology
Panjab University
Chandigarh
India

Vishal Singh
School of Biochemical Engineering
Indian Institute of Technology (BHU)
Varanasi
Uttar Pradesh
India

Dhirendra Kumar Srivastava
Council of Science and Technology
Lucknow
Uttar Pradesh
India

Anupma Thakur
Academy of Scientific and Innovative Research (AcSIR)
Ghaziabad
India

Central Scientific Instruments Organisation
Chandigarh
India

Gyanendra Tripathi
Department of Bioengineering
Integral University
Lucknow
Uttar Pradesh
India

Kailas L. Wasewar
Advance Separation and Analytical Laboratory (ASAL)
Department of Chemical Engineering
Visvesvaraya National Institute of Technology (VNIT)
Nagpur
Maharashtra
India

Deepak Yadav
Department of Chemical Engineering
HBTU
Kanpur
Uttar Pradesh
India

1

Mapping of Selenium Toxicity and Technological Advances for its Removal

A Scentiometric Approach

Madhulika Bhati[1], *Jayashree Rajesh Prasad*[2], *Charu Jhamaria*[3], *Sakshi Narula*[1], *and Dipa Mahato*[1]

[1] *CSIR – National Institute of Science, Technology and Development Studies (CSIR-NISTADS), New Delhi, India*
[2] *Department of Computer Science, Singhad College of Engineering, Pune, Maharashtra, India*
[3] *Environmental Science, IIS (deemed to be University), Jaipur, Rajasthan, India*

1.1 Introduction

Selenium contamination is worldwide phenomenon. Contamination of selenium in surface and ground water in river basins is one of the critical problems nowadays.

As selenium shows a narrow tolerance limit (40–400 mg/day), the problem of both deficiency and toxicity of Se have being identified in many parts of India and the world (WHO 2009). Different countries are making efforts to remove selenium from water using different technologies. Some specific processing technologies include reduction, bioremediation, phytoremediation coagulation, electro-coagulation, co-precipitation, electric kinetics, adsorption, chemical precipitation, and membrane technology.

In periodic table, selenium (Se) lies between sulfur and tellurium in Group VIA and between arsenic and bromine in Period 4. Selenium originated from the Greek word "Selene," which means "moon." It was discovered in 1817 by Jöns Jacob Berzelius, "Father of Swedish Chemistry," during the manufacturing of sulfuric acid, when he observed and analyzed a red deposit on the wall of lead chambers (Tinggi 2003). Selenium, mostly in combination with minerals that contain sulfur, is found in the Earth's crust. It generally exists in four oxidation stages, i.e. elemental selenium, selenites, and selenates. The oxidation states of the selenium found in nature are −2, 0, +4, or +6. Selenium is an essential micronutrient for its development and reproduction for humans, insects, fish, and many other species, but is known often to be toxic in amounts beyond the acceptable limits (Stadtman and TC 1978). Selenium's Acceptable Limit (AL) for groundwater is fixed at 0.01 mg/l (ppm) by the Bureau of Indian Standards (BIS). The spectrum of selenium consumption for optimum human and animal safety is very narrow, meaning that low selenium consumption is correlated with developmental defects and disease and a high amount of selenium contribute to toxicity (Reilly 2006; Kurokawa and Berry 2013). It is required for

Selenium Contamination in Water, First Edition. Edited by Pooja Devi, Pardeep Singh, Arindam Malakar, and Daniel Snow.

healthy joints, heart, and eyes. Its role in DNA synthesis, the immune system, and the reproductive system is of critical value. A well-known example of selenium toxicity is Keshan disease in Hubei Province, China, where several deaths have been reported (Yang and Xia 1995; Kipp 2015). Selenium is primarily found in water as selenate (SeO_4^{2-}) and selenite (SeO_3^{2-}). Among the two compounds, selenate is the more stable and thus comparatively difficult to remove from water solutions. Different physiochemical qualities, including oxidation reduction, pH, their chemical types, and sorbing surfaces, regulate selenium in soils or irrigation waters (Mondal et al. 2004). While being widely used in the production of photocells, rectifiers, photocopy machines, paints, crystal glass, and pesticides, it is extremely hazardous if found in drinking water exceeding 10 μg/l where the extent of selenium use for industrial applications comprising electronics, photocopying machinery, glass, rubber, etc. varies from 1 to 7000 μg/l. Proclamation of selenium adulteration to the surroundings is frequently connected with copper casting accomplishments. Selenium can enter water supplies through interaction with selenium carrying minerals or through contact with polluted soil or from emancipation from excavations, earthen deposits, discharge from factories, or from agricultural overflow escaping natural selenium amalgams from desiccated, undeveloped terrestrial areas. Selenium (Se) compounds in groundwater have attracted the attention of scientists due to the increased contamination as result of increasing industrialization and activities like mining, combustion of coal, and use of selenium-contaminated water for irrigation (Luoma and Presser 2009; Gibson et al. 2012). Selenium generally enters the food chain through both surface and underground waters which are used for irrigation and drinking purposes (Dhillon and Dhillon 2003; Zhang et al. 2014).

The objective of this chapter is to discover the noteworthy research going on in this domain using a scientometric approach and visualization tools. This study explores the most-cited investigations, authors, institutions, and countries since 2000. It also presents insights about the leading technological developments in the removal of selenium.

This chapter systematically reviews high-impact literature to identify, evaluate, and interpret the work of researchers, scholars, and practitioners so as to develop insights into various removal methods, such as sedimentation, filtration, activated carbon adsorption, ion exchange, reverse osmosis, and biological techniques for removing selenium from water and wastewater.

1.1.1 Contamination Status of Selenium

The authors have searched relevant literature related to the contamination status of selenium. Results indicate that selenium contamination is a global problem that affects a wide variety of human actions, from the most traditional farming methods to the most modern production processes. The USA, Canada, Republic of China, India, Japan, and Brazil are continuously making efforts to find out the technological solutions to remove the selenium.

In India, many areas are contaminated with selenium (Table 1.1). Punjab is the most affected region in Northwest India, with over 1000 ha of polluted fields. The main factors for the mobilization of Se and its bioavailability are alkaline soil pH, production of Se bio-accumulators, and inadequate industrial effluents/emissions treatment (Paikaray 2016). In the Majha belt of Punjab, which includes the Amritsar, Gurdaspur, and Tarn Taran

Table 1.1 Selenium contamination status in Indian States.

Author	Year	State	Districts	Contamination level
Virk	2018	**Punjab** (Majha Belt)	Tarn Taran	0.010–0.076
			Gurdaspur	0.010–0.094
			Amritsar	0.010–0.039
Virk	2019	**Punjab** (Doaba belt)	Jalandhar (Most contaminated)	0.016–0.022
			Kapurthala	0.01–0.082
			Hoshiarpur	0.011–0.029
Virk	2019	**Punjab** (Malwa belt)	Ludhiana	0.016–0.14
			Roop Nagar	0.01–0.20
			Ferozpur	0.011–0.025
			Fatehgarh Sahib	0.011–0.028
Yadav et al.	2005	**Rajasthan**	Jaipur	Lower selenium level in Soil
			Alwar	
Ghosh et al.	2008	**Maharashtra**	Mumbai	0.08–1.10 mg/l
K. Chandrasekaran*, Manjusha Ranjit, and J. Arunachalam	2009	**Punjab**		0.012–0.093
Sharma and Kumar	2006	**Bihar**	Bihar	
Kumar and Riyazuddin	2011	**Tamilnadu**	Chennai	0.15–0.43 μg/l and 0.16–4.73 μg/l
Singh and Kumar	1976	**Uttar Pradesh**	Muzaffarnagar	
Vediya and Shrivastava	2008	**Gujarat**	Ahmadabad	84.7–503.7 μg/l

districts, underground water is mainly contaminated with selenium and many other heavy metals (Virk 2018). Another belt which has groundwater quality-affected habitations is the Doaba belt of Punjab, which includes the Jalandhar, Kapurthala, and Hoshiarpur districts (Virk 2019). As per Punjab Water Supply and Sanitation Department (PWSSD) data, selenium contamination of groundwater is greatest in the Jalandhar district in this belt. Selenium was detected in 105 habitations in Jalandhar, 30 in Kapurthala, and 19 in Nawanshahar. High-level contamination of selenium was reported in the Malwa region of Ludhiana, with 90 habitations having higher than permissible limit of the metal with up to 0.140 mg/l content of selenium against the permissible limit of 0.01 mg/l. Another belt in Punjab, the "Malwa belt" of Ludhiana, Ferozepur, Roop Nagar, and Fatehgarh Sahib districts, also has selenium content in water sampled from tube wells (Virk 2019). The Doaba belt is the most selenium-contaminated belt of Punjab. Also, in Punjab, surface soils are three to five times richer in selenium content compared to sub-surface soils (Dhillon and Dhillon 1991).

Overall, some areas are contaminated with selenium, others suffer from a deficiency of selenium. It has been inferred that deficiency-related problems are common in arsenic (As)-polluted sites like West Bengal (Das et al. 1995; Roychowdhury et al. 2003; Spallholz et al. 2004). In Western India, elevated soil selenium levels can be seen in Jaipur and Alwar, Rajasthan and selenium enrichment in brassica plants sown there has been shown (Yadav et al. 2005). In Maharashtra, up to 3 mg/kg and more than 1100 mg/l Se has been reported in soil and water, respectively, around Thane creek, Mumbai (Ghosh et al. 2008). In Central India, Jharkhand shows elevated levels of Se as it is a coal-fired region (Finkelman and Stracher 2011) and Bihar shows elevated levels because of industrial effluents and emissions from the Barauni industrial area (Sharma and Kumar 2006). Lowlands and ponds have been contaminated with a high level (>10 mg/l) of Se due to disposal of untreated industrial wastes degrading the quality of underlying groundwater around Chennai City, Tami Nadu' (Kumar and Riyazuddin 2011). In Northern and North-western India, Se content shows high variability. In Uttar Pradesh, in the Western part of Muzaffarnagar, there is a high selenium content in ground water, and different parts of Haryana show Se content ranging from 1 to 10.5 mg/kg (Singh and Kumar 1976). In Haryana, surface soils are mostly enriched with selenium other than areas receiving high rainfall like Ambala, Kurukshetra, Rohtak, and Sonipat.

1.1.2 Mapping Selenium Research Dynamics Advances

Keeping in view the main objective of the chapter – to identify, evaluate, and interpret the work of researchers, scholars, and practitioners for developing insights into the technologies used for the removal of selenium from water – the following methodologies have been implemented:

1) To begin with, some free searches were done on Google Scholar to find "remediation, treatment technologies for selenium-contaminated waters," and by studying suitable references, a string was developed. The Web of Science database was used for extraction of data. The string applied was "selenium" with further search in "treatment OR remediation OR contamination OR removal" AND "Water." The number of publications which appeared was 3840, which were analyzed. Extraction of data based on countries revealed that the USA leads in publishing papers in this domain, followed by China and Canada, India and Japan.
2) The authors selected papers with significant influence on technologies for selenium removal in the form of published articles, research papers, white papers and review reports on the said domain. The literature comprises material, notions, data, and evidence inscribed from a precise perspective to accomplish confident objectives or express convinced interpretations on the landscape of the technologies to be developed and the way they are to be investigated. The authors further wish to evaluate effectively this literature in the context of the research being proposed.
3) Thereafter, authors began by mapping a number of key strategies for selenium removal from waste and drinking waters. This research suggests mapping of the concepts, opinions, and thoughts from a volume of quality literature. The authors followed a reputable technique for expressing facts and intelligent processes. The mapping presented in this chapter is a description as a "visual outline," a "pictorial depiction," and a "topographical allegory" of the research in the domain of water purification. Mapping in this research presents a "perceptible indication" of a researcher's thoughts and clarification of the research domain which can be delivered to both policy makers and scientists.

The present study evaluates the performance of selenium treatment technologies based on several parameters such as, country, city of publication, and Web of Science Core Collection Times Cited Count (which measures the technological impact of literature). The most influential literature is studied for the Multidisciplinary Sciences research. The authorship outline, cooperative catalogue, joint factor, adapted collaborative quantity and research profile. The study also examines international publications output and impact in terms of citations per paper and usage counts in various journals.

This study analyzes the evolution outline of literature related to selenium removal in water and soil in India during 1987–2020 (33 years). The Web of Science is a multidisciplinary international bibliographic archive. The Web of Science international multidisciplinary bibliographical database was used in order to classify the publications reported internationally on the basis of specific scientific field.

1.1.3 Bibliometric Analysis

The Web of Science (WoS) international multidisciplinary bibliographical database used to identify the contributions published worldwide in the field of various technological domains suggested the contribution of various fields of science, as shown in Figure 1.1.

As is evident from Figure 1.1, the research is dominated by the Environmental Sciences, Metrology and Atmospheric Sciences, and the Water Resources domain collectively, which makes up 20% of the total research. This one fifth share of the contribution is followed by Chemistry/Multidisciplinary with 10%, Biotechnology and Applied Microbiology with 10%, and Soil Science with another share of 10% contribution. Adding on these three chunks of domains forms 50% of the total contribution, which is further strengthened by

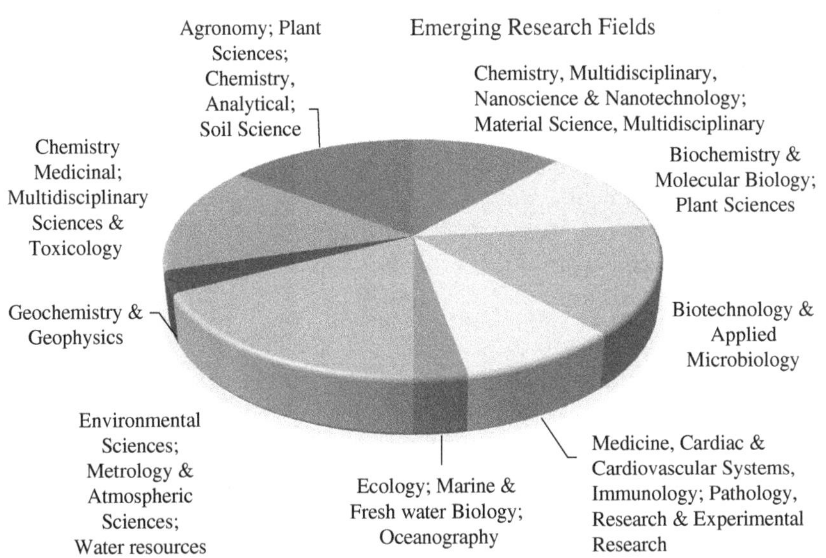

Figure 1.1 Emerging research field for selenium removal technologies published worldwide.

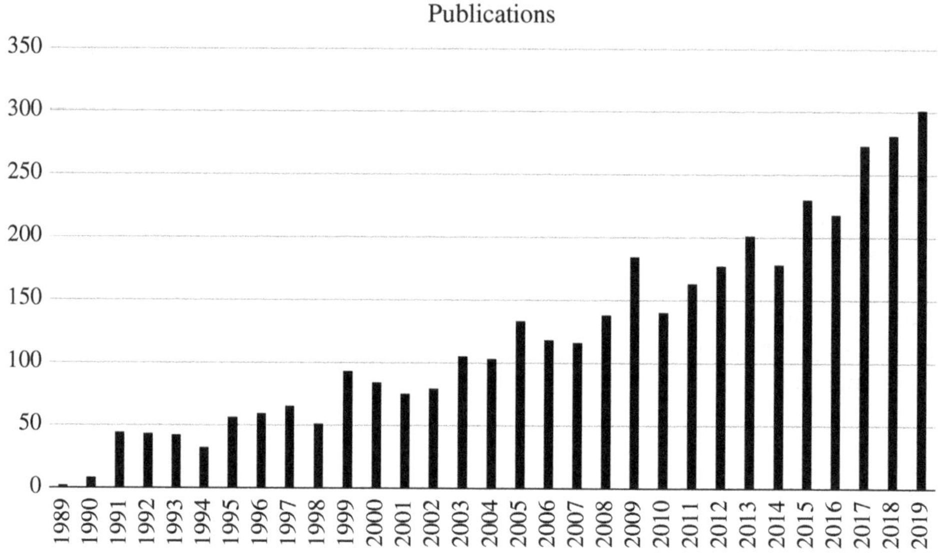

Figure 1.2 Selenium research publication trends.

research from: Medicine/Immunology/Pathology; Biochemistry and Molecular Biology/Plant Sciences; Nanoscience and Nanotechnology/Material Science; Ecology/Marine and Fresh Water Biology; Geochemistry and Geophysics; and Agronomy and Plant Sciences.

The most influential research that spans over the domains is from Chemistry, Multidisciplinary Sciences, Nanoscience and Nanotechnology, and Material Science. Salt et al. (1995), Matoba et al. (2002), Grätzel (2009), and Elmolla and Chaudhuri (2010) have made significant contributions with the following components:

- judiciously eradicating problematic contaminants from water discharge
- investigating tarn size restrictions
- sewage flow management
- complying with nitrate, sulfate, chloride discharge constraints
- recuperating heavy metal content
- freshwater scarcity/confines
- groundwater treatment technologies
- mine dewatering
- acid mine waste water.

Quantitative assessment of the research has been presented in these publications considering year-wise research outcome, geographical distribution, and type of collaboration, characteristics of highly productive institutions and the method of communication used by the scientists. In a similar manner, the authors explored the second category of domain shown in Figure 1.1 denoted by Biochemistry & Molecular Biology and Plant Sciences (Mandal et al. 2001; Zhu et al. 2009). The representative research outcomes highlight issues related to polluted soils and waters with key environmental and human-related health issues, which may be moderately resolved by the emergent phytoremediation expertise.

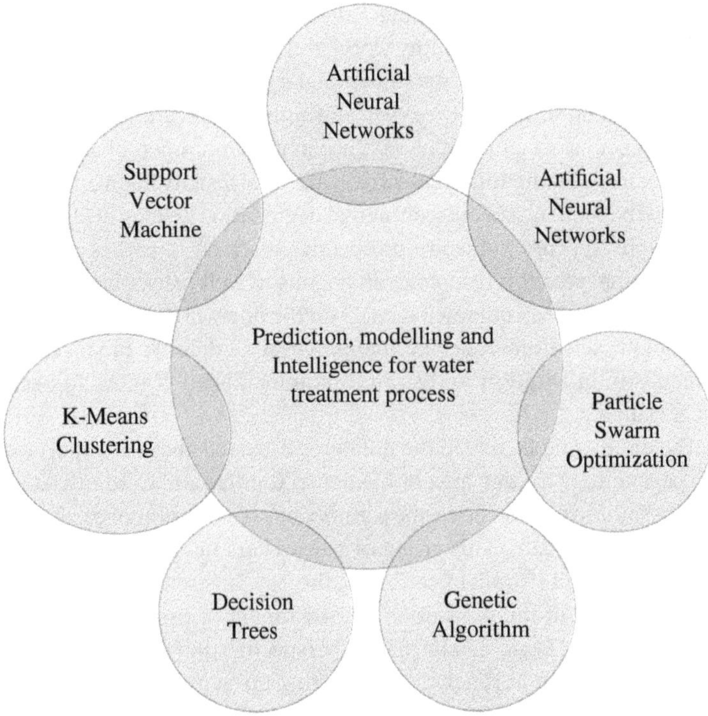

Figure 1.3 Artificial Intelligence, Machine Learning, and Deep Learning mechanisms in modeling and prediction of water treatment parameters.

In situ phyto-extraction and phyto-stabilization have proved to be significantly effective in the removal of toxic metals from the soil (Khan et al. 2020a). This lucrative plant-based tactic for remediation benefits from the amazing capability of plants to essence rudiments and mixtures from the atmosphere and to metabolize various molecules in their materials. Toxic heavy metals and organic contaminants are the chief goals for phytoremediation. Lately, facts of the biological and molecular apparatuses of phytoremediation commenced to arise collectively with organic and technological approaches intended to heighten and expand phytoremediation. Organic amendments and hydrologic regimes have also been found to be effective for Se removal by using constructed wetlands microcosms (Zhao et al. 2020) Furthermore, numerous arena prosecutions confirm the possibility of ecological cleaning by means of plants. This survey quintessence on the utmost established subsections of phytoremediation expertise and on the biotic apparatuses that brands phytoremediation exertion. Although plants alone show the ability to remove the toxic agricultural pollutants using different strategies, integrated approaches such as microbes and plant associations (rhizoremediation) are also proving to be effective options for metal removal (Khan et al. 2020b).

The authors' attention was captured by eminent research in the field of Biotechnology and Applied Microbiology (Reeves 1997; Salt et al. 1998; Vara and de Oliveira Freitas 2003; Eapen and D'Souza 2005; Tang et al. 2015). Significantly, active selenium occurs in oxic and anoxic environments, and this presence plays an important role in carbon and nitrogen mineralization by bacteriological anaerobic breathing. Selenium-breathing bacteria (SeRB) come from a geologically isolated, primeval or filthy world and play a significant role in the selenium process. The chalcogen selenium and its microbial cycle have aroused few concerns, with comparable operational similarity to oxygen and sulfur. Extracellular polymeric substances (EPS), a high-molecular-weight biopolymer originated from microbial metabolism, have been found to reduce selenite into non-soluble and low-toxicity elemental selenium, which would prevent the sequence of environmental degradation (Zhang et al. 2020). This collection of publications appears to inspire prospective work on microbes that use selenate and selenite as terminal electron receptors, analogous to well-researched sulfate-reducing bacteria. Summaries have been completed lately on the noteworthy developments in the role of SeRB in the biotic selenium cycle and their environmental role, phylogenetic classification, and metabolism, in addition to selenium biomineralization tools and eco-friendly biotechnological claims.

While the presence of selenium and its role in the domains discussed above is important, there is a remarkable experimental research area in Medicine, Cardiac and Cardiovascular Systems, Immunology and Pathology that shows the significance of selenium in cardiovascular health (Weschenfelder et al. 2020). This group of researchers made a praiseworthy linkage of selenium metabolism and its role in cardiac pathology. It is remarkable to note that the consumption of selenium in the prevention and treatment of cardiovascular ailments remains an indefinable area. From an alternative perspective, the chief purpose of selenium here is antioxidant defense through its amalgamation as selenocysteine into enzyme groups, for example glutathione peroxidases and thioredoxinreductases. Moreover, selenium compounds are various and have multifaceted metabolic effects, and thus there is partial dependence on selenoprotein expression. However, apart from the valuable properties of selenium, proved in clinical observations, selenium certainly may be destructive. It is thought-provoking that the biological activities of selenium concurrently have consequences that may influence gene expression, the causing negative sequelae typically seen. Selenium nanoparticles (SeNPs) could substantially reduce hyperlipidemia and vascular damage in mice, most likely by controlling the metabolism of cholesterol and reducing oxidative stress by antioxidant selenoenzymes/selenoproteins (Guo et al. 2020). Removal of unsafe inorganic species of selenium, including selenite and selanate, has recently been tested by drinking water researchers for efficacy (Meher et al. 2020).

This review now comes to the most impacted domain: Environmental Sciences, Metrology, Atmospheric Sciences, and Water Resources (Zayed et al. 1998a, 1998b; Amthor 2001; Jiang et al. 2013). Selenium (Se) is one of the vital essentials in food harvests as an outcome of rigorous plant production in several nations. Next, Se has emerged as the focus of research into various parts of the biosphere. This domain presents the contemporary understanding of Se in the agroecosystem. The existence of selenium in the atmosphere from soil to food systems is considered. The most hopeful and pressing potential nanotechnology developments (Yang et al. 2005; Liu et al. 2015) are in agriculture; and the manufacture nano-selenium elements, and agronomic nanotechnology and its practice and justifiable expansion are also emphasized.

In other research (Smedley et al. 2002), the bioconcentration factor for plants, which plays a significant role in geochemical prospecting and animal nutrition, is studied for a detailed plant/soil system. This paper reveals in-depth geochemical annotations for an improved understanding of the ecological qualities of Se, such as its resemblance to sulfur and tellurium.

Approximately 20% of the most influential research under discussion comes under the domain of Chemistry, Medicine Science, Multidisciplinary Sciences and Toxicology (Brown et al. 1999; Elmolla and Chaudhuri 2010). The extracts of this research domain consider the average selenium absorptions in forms of water from approximately 0.4 to 16 μg/l. In marine creatures there is a robust association between the Se deposits in the water and those in the body tissues. Many creatures bioaccumulate Se by factors as high as 1000 to 4000, making selenium poisonous to humans and other creatures. Yang et al. (2005) and Lenz and Lens (2009) claim existence of micronutrient for several creatures in trivial quantities; and they claim further that occupational selenium exterminating is frequently unintentional and occasional.

During the publication analysis, Artificial Intelligence (AI), Machine Learning (ML), and Deep Learning (DL) have emerged as a powerful tool to actuate water treatment processes. The most important role of AI, ML, and DL is that it enables efficient prediction and modeling in the conventional water treatment, distribution, and selenium removal processes.

AI methods, comprising Artificial Neural Network (ANN), Genetic Algorithm (GA), and Particle Swarm Optimization (PSO), are deployed (Gupta and Gupta 1998) for the assessment of Se (IV) deletion from aqueous solutions.

Combination of GA and PSO optimizes the specifications of ANN. The authors claimed to reduce the error to less than 3% with this arrangement of AI tools. It is found that ANN-PSO and ANN-GA models have proven to be a perfect choice for demonstrating and enhancing Se (IV) removal by the adsorption and reduction apparatuses.

The authors also studied the Publication Agencies mapping for all the most influential research papers discussed here. Figure 1.4 shows Representative Highest Influential Publications for selenium removal technological published worldwide. It can be clearly seen that the highest number of highly cited research is published by Springer, making up 25% of all publications. The next 20% of contributions comes from Elsevier Science BV in Amsterdam. The other well-regarded publications range from 1 to 10% of total contributions.

The major objective of this section is to present a modality that helps researchers to visualize "outlines" in the said research field to "perceive effects" that may then be unexplored, by recognizing "gaps" in the research field and "limitations" to issues under examination. The authors emphasize that the fundamental outcome of this study is identifying probable innovative areas of study and the constraints.

To add to the discussion discussed so far, some technologies indicating comprehensive treatments, such as desalination and brine management solutions, are observed. A robust ultra-high recovery reverse osmosis system has been proposed by several researchers for distant commercial set-ups enabling reverse osmosis to confiscate scaling ions and thereby causing maximum recovery and smooth operations. There are suggestions that low-slung temperature evaporator-crystallizers be used. Analysis results indicate that countries are using different treatment technologies, including reduction techniques, phytoremediation, bioremediation, coagulation-flocculation, electro-coagulation (EC), electrochemical methods, adsorption, co-precipitation, electro kinetics, membrane technology, and chemical

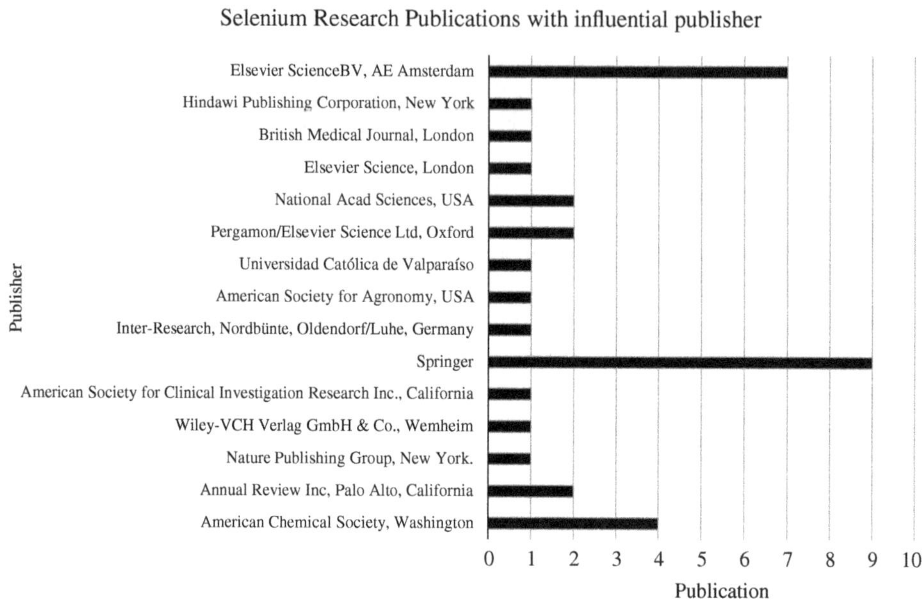

Figure 1.4 Representative most influential publications for selenium removal technology published worldwide.

precipitation. However, the order of use of particular methods varies from country to country.

Word dynamics analysis was carried out by key words, title, and abstract on the papers published by Asian countries. India and China emerged as the top five Asian countries. The pattern of word dynamics indicated that Se removal using phytoremediation is emerging as main technology, followed by bioremediation and UASB reactor. Adsorption, reduction, and sorption emerge as the dominant methods for Se removal technology in China.

1.2 Selenium Reduction Technologies Used in India

Phytoremediation is a technology in which selenium get accumulated in plant parts and then plants can be harvested and incinerated. Phytovolatilization, which is also a type of phytoremediation, is more effective, in which a few plant species which can tolerate selenium can be grown on selenium-contaminated sites and then can volatilize less toxic forms or non-toxic forms of it in the environment (Dhillon and Dhillon 2003). This technique efficiently remediates soils contaminated with selenium as well as being an eco-friendly technology and using fewer resources.

In Punjab, mustard is found to be the best accumulator of selenium. It is may be because of the high concentration of sulfur in mustard and the resemblance of selenium to sulfur. In addition to mustard, other plants, including onion, garlic, broccoli, Brazilian nuts, and mushroom, also show a strong affinity to Se (Sharma et al. 2015). In soil systems, many remediation methods can be applied for selenium management, such as gypsum

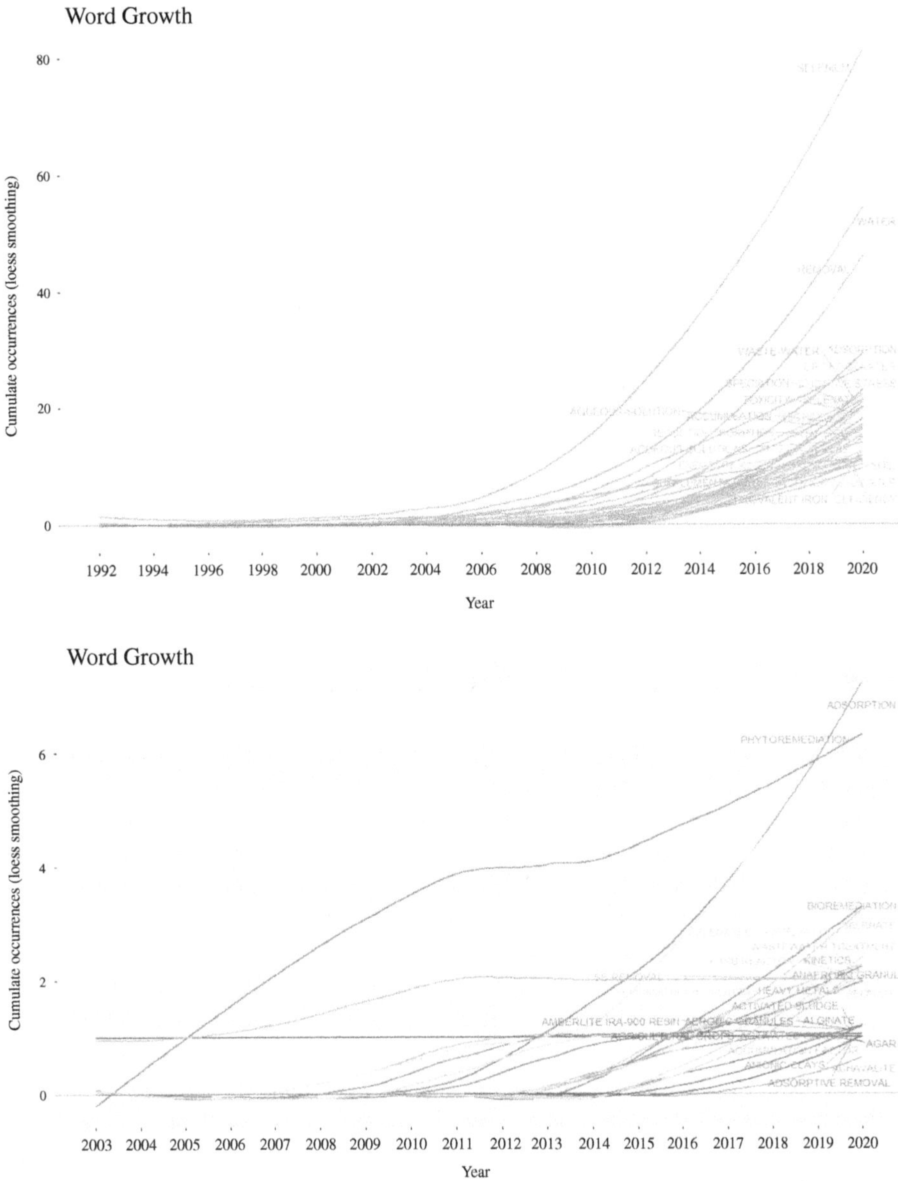

Figure 1.5 Upper panel: word dynamics for India. Lower panel: word dynamics for China.

application in selenium-contaminated farms, which has been found to be efficient in its reduction levels (Dhillon and Dhillon 2000). Application of organic matter such as press mud and poultry manure has shown to reduce Se levels in rice, wheat, and maize up to 97% in Se-contaminated farms (Dhillon et al. 2010). In situ bioremediation can be achieved by bacterial cultures belonging to b-Proteobacteria and Bacilli class, which are fairly selenium-tolerant microorganisms (Ghosh et al. 2008; Prakash et al. 2010).

Biofortification, in which harvested plant parts are decomposed in agricultural soil which can be used further for the enrichment of food products with Se (Bañuelos et al. 2015), is another method.

Among the other technologies being used for the remediation of selenium-contaminated waters are: ion exchange, reverse osmosis, nanofiltration, solar ponds, chemical reduction with iron, microalgal–bacterial treatment, alumina adsorption, Fe^{+3} coagulation/filtration, lime softening, and ferrihydrite adsorption (El-Shafey 2007; Luo et al. 2008). Use of waste wheat bran can be an eco-friendly technology in a continuous up-flow fixed-bed column system for biosorption of selenium species in aqueous solution (Hasan et al. 2010). One of the efficient methods of selenium removal from waste water can be use of double-layered synthetic hydroxide materials (Zn/Al, Mg/Al, and Zn/Fe) as an adsorbent (Mandal et al. 2009).

Another technology demonstrates photoreductive removal of selenium (IV) using spherical binary oxide photo catalysts under visible light. As a range of scavengers, EDTA (ethylene diamine tetra acetic acid) and formic acid are found to be the most suitable for the reduction reaction, and of these two, formic acid is found best for reduction of selenium – the catalyst used for the process is TiO_2, which is non-corrosive, non-toxic, and has high photoactivity, high photostability, and an economical nature. It has been reported that catalyst can be used repeatedly at least five times with marginal change in the activity (Aman et al. 2011).

1.3 Selenium Reduction Technologies Used in China

Nanoscale zero-valent iron (nZVI) is the most widely applied nanomaterial as an adsorbent for groundwater and hazardous waste treatment. It is also effective for selenium treatment and removal. Batch experiments have been conducted and show that nano-ZVI has approximately a removal rate three times or higher than those of micro-scale iron, nanoscale iron oxides, $Fe(OH)_3$, nanoscale TiO_2, and activated alumina for selenium removal (Ling et al. 2015). ZVI or nano-ZVI is effective for Se^{VI} removal from wastewater by reducing to more adsorptive Se^{IV} and/or to insoluble Se species (i.e. Se^0, Se^I, and Se^{II}). A key role is known to be played by dissolved Fe^{2+} in the reductive removal of selenate by ZVI. There are two major roles for Fe^{2+}: (i) it participates in selenate reduction directly as partial electron donor with a Fe^{2+}:Se stoichiometry of ~1:1, and (ii) helping in the transformation of the passive layer on iron grain and corrosion products to magnetite, favoring electron transfer and thus enhancing selenate removal. ZVI was the main electron donor. In a ZVI-Se^{VI}-Fe^{2+} system, sequential reduction of selenate is reported where elemental Se was the dominant reductive product. Selenate reduction by ZVI assisted by Fe^{2+} has been identified as a sustainable treatment method for wastewater contaminated with selenate (Tang et al. 2014).

It was observed in the trials that the removal efficiency of Se (VI) by ZVI was only 4.8% within 120 minutes, although a much higher removal efficiency (62.1%) was obtained by nZVI. Owing to its smaller particle size (approximately 100 nm), nZVI has a higher capacity for removal of Se (VI) than ZVI (approximately 150 m). The smaller size of the iron particles suggests a greater surface area that gives more stable Se (VI) reaction sites. In comparison, the removal efficiency of Se (VI) by nZVI/Al-bent was 95.7%, which was far higher than that of nZVI (62.1%). It suggests that there is a synergistic impact on the elimination of Se (VI) in the nZVI/Al-bent composite method (Li et al. 2015). A study demonstrated that iron oxides and aluminum oxide are effective adsorbents for Se removal (Kuan et al. 1998; Sheha and El-Shazly 2010; Sharrad et al. 2012). Coagulation was found to be an effective process for removal of metal from water (Hu et al. 2015). They found that the Fe-based coagulant was much more efficient than the Al-based coagulants. It was further suggested that researchers' pre-reduction of Se (VI) to Se (IV) seems to be necessary to achieve effective Se removal. A synergistic effect of using ZVI was demonstrated by Liang et al. (2015) with magnetic effect and it was found that removal efficiency increased, with a reduction in removal time at a pH of 6.0. Graphene nano sheets, as a novel nano adsorbent, can also be used for removing various pollutants from water by certain modifications. To overcome the structural limits of graphene (aggregation) and graphene oxide (hydrophilic surface) in water, sulfonated graphene (GS) is prepared by diazotization reaction using sulfanilic acid (Shen and Chen 2015). Another technology demonstrated for removal of selenium from water is by sequestration by Cationic Layered Rare Earth Hydroxide $Y_2(OH)_5Cl{\cdot}1.5H_2O$. One study confirmed that Se (IV) and Se (VI) are successfully sorbed onto $Y_2(OH)_5Cl{\cdot}1.5H_2O$, which is also further supported by Raman and EDS measurements (Zhu et al. 2017). Another study demonstrated a promotional effect of the use of Mn^{2+} and Co^{2+} selenate removal by ZVI. It was found that Selenite (SeIV) was the predominant reductive product in the presence of Co^{2+}; however, selenite and elemental Se (Se^0) were the main reductive products in the presence of Mn^{2+}. But this process shows promotional effect in anaerobic conditions (Tang et al. 2014).

Taking the role of AI further, there is another powerful area of predictive modeling which is a method used in predictive analytics to generate a statistical visualization of future behavior. Predictive analytics is the domain of data mining related to anticipating likelihoods and inclinations. Alternatively, AI deals with intelligent acts, i.e. the activities that describe them as intelligent. Subsequent to the thought process, the sole purpose is to evaluate the influence of AI algorithms for the implementation of intelligent predictive models. On piece of promising research (Pinto et al. 2009) answers many crucial issues by construction predictive models. These models stimulate prediction of manganese and turbidity echelons in treated water, to guarantee that the water supply does not distress community healthiness in a undesirable mode and observes the existing regulation. Additionally, popular supervised classification algorithms such as decision trees and the unsupervised k-means algorithm build clustering models.

Recently, it has been interesting to note that the presence of selenium in plants has been modeled to show a tight borderline limit between nutritious prerequisite and toxic supplement in plants (Soil Science Society of America 2008). The AI algorithms beautifully model how the steep dose response curve caused by bioaccumulation properties have led to the description of selenium as a "tinderbox" modeled through anthropogenic events.

1.4 Selenium Research Dynamics Using AI Techniques

AI and Deep learning methods have spread their capabilities in depicting contests for water-sanitation amenities and research forums. Deep learning presents an outstanding substitute to countless studies in optimization (Dentel 1995). Compared to out-of-date machine learning algorithms, deep learning has a robust learning capability to efficiently utilize data sets for data mining and knowledge mining. The objective of this investigation is to assess the prevailing unconventional methods. This paper further explores the boundaries and predictions of deep learning.

Furthermore, a novel placement of a machine learning ensemble in the field of water distribution networks was studied by Camarinha-Matos and Martinelli (1998). This was an innovative application to govern and implement water distribution networks by a supervision classification, a distributed information management framework for water quality monitoring and recreation.

An alternate implementation of ANN along with SVM (Haghiabi et al. 2018) investigates water quality prediction. These authors reach a valuable outcome from their research, that "tansig" and "RBF", which are transfer and kernel functions, demonstrate significant performance compared to other functions. SVM proves to be the most accurate model compared to other machine learning algorithms.

Finally, it is time to end up the discussion by looking at the most important issue of all, i.e. cost. It is extremely important to give a keen thought to the cost issue for wastewater treatment. Machine learning has been uniquely deployed (Torregrossa et al. 2018) for efficient energy cost modeling for wastewater treatment plants. The researchers have innovatively proposed cost as a parameter to evaluate the performance of the system.

Thus, these technologies ensure that performance is accurately predicted and assists in ensuring that efforts are made to deal with issues in advance. Machine learning generates innovative visions that can be used as evidence for future research on scheduling the distribution of the water resources

1.5 Conclusion

The presence of selenium in plants has been modeled to show a tight borderline limit between nutritious prerequisite and toxic supplement. The steep dose response curve caused by the bioaccumulation properties of selenium have led to the description of this element as a "tinderbox." Water treatment for selenium removal is a component of successful selenium management strategy. Several technologies have been used by countries for selenium removal. There is a noticeable evolutionary role played by machine learning and artificial intelligence techniques in modeling and estimating the parameters contributing to efficient performance of systems.

At this stage, benchmarking plays a significant role in assessing the performance of technologies in terms of their value proposition, environmental impacts (following the principle of clean technology with proper treatment of sludge as product obtained) or, in other words, satisfying all the components of ASSURED analysis: A (Affordable), S (Scalable), S (Sustainable), U (Universal), R (Rapid), E (Excellent), D (Distinctive). A credible benchmarking by assessing the technologies based on the ASSURED parameters will help to screen technology which is more capable of being replicable, non-disruptive, and scalable.

Acknowledgment

The authors are thankful to the Director, CSIR-NISTADS, Management of Sinhgad Technical Education Society, Pune, and Management of IIS (deemed to be University), Jaipur for their continuous support and guidance in carrying out this research work.

Conflict of Interest

The authors do not have a conflict of interest.

References

Aman, N., Mishra, T., Hait, J., and Jana, R. (2011). Simultaneous photoreductive removal of copper (II) and selenium (IV) under visible light over spherical binary oxide photocatalyst. *Journal of Hazardous Materials* 186 (1): 360–366.

Amthor, J. (2001). Effects of atmospheric CO2 concentration on wheat yield: review of results from experiments using various approaches to control CO2 concentration. *Field Crops Research* 73 (1): 1–34.

Bañuelos, G., Arroyo, I., Pickering, I. et al. (2015). Selenium biofortification of broccoli and carrots grown in soil amended with Se-enriched hyperaccumulator Stanleyapinnata. *Food Chemistry* 166: 603–608.

Brown, G.E., Foster, A.L., and Ostergren, J.D. (1999). Mineral surfaces and bioavailability of heavy metals: a molecular-scale perspective. *Proceedings of the National Academy of Sciences* 96 (7): 3388–3395.

Camarinha-Matos, L. and Martinelli, F. (1998). Application of machine learning in water distribution networks. *Intelligent Data Analysis* 2 (4): 311–332.

Das, D., Chatterjee, A., Mandal, B. et al. (1995). Arsenic in ground water in six districts of West Bengal, India: the biggest arsenic calamity in the world. Part 2. Arsenic concentration in drinking water, hair, nails, urine, skin-scale and liver tissue (biopsy) of the affected people. *The Analyst* 120 (3): 917.

Dentel, S.K. (1995). Use of the streaming current detector in coagulation monitoring and control. *Aqua-Journal of Water Supply: Research and Technology* 44 (2): 70–79.

Dhillon, K. and Dhillon, S. (1991). Selenium toxicity in soils, plants and animals in some parts of Punjab, India. *International Journal of Environmental Studies* 37 (1–2): 15–24.

Dhillon, K. and Dhillon, S. (2000). Selenium accumulation by sequentially grown wheat and rice as influenced by gypsum application in a seleniferous soil. *Plant and Soil* 227 (1–2): 243–248.

Dhillon, S. and Dhillon, K. (2003). Quality of underground water and its contribution towards selenium enrichment of the soil – plant system for a Seleniferous region of Northwest India. *Journal of Hydrology* 272 (1–4): 120–130.

Dhillon, K., Dhillon, S., and Dogra, R. (2010). Selenium accumulation by forage and grain crops and volatilization from seleniferous soils amended with different organic materials. *Chemosphere* 78 (5): 548–556.

Eapen, S. and D'Souza, S.F. (2005). Prospects of genetic engineering of plants for phytoremediation of toxic metals. *Biotechnology Advances* 23 (2): 97–114.

Elmolla, E. and Chaudhuri, M. (2010). Photocatalytic degradation of amoxicillin, ampicillin and cloxacillin antibiotics in aqueous solution using UV/TiO_2 and UV/H_2O_2/TiO_2 photocatalysis. *Desalination* 252 (1–3): 46–52.

El-Shafey, E. (2007). Removal of Se(IV) from aqueous solution using sulphuric acid-treated peanut shell. *Journal of Environmental Management* 84 (4): 620–627.

Finkelman, R.B. and Stracher, G.B. (2011). Environmental and health impacts of coal fires. *Coal and Peat Fires: A Global Perspective: Coal – Geology and Combustion* 1: 115–125.

Ghosh, A., Mohod, A.M., Paknikar, K.M., and Jain, R.K. (2008). Isolation and characterization of selenite-and selenate-tolerant microorganisms from selenium-contaminated sites. *World Journal of Microbiology and Biotechnology* 24 (8): 1607–1611.

Gibson, B., Blowes, D., Lindsay, M., and Ptacek, C. (2012). Mechanistic investigations of Se(VI) treatment in anoxic groundwater using granular iron and organic carbon: an EXAFS study. *Journal of Hazardous Materials* 241: 92–100.

Grätzel, M. (2009). Recent advances in sensitized mesoscopic solar cells. *Accounts of Chemical Research* 42 (11): 1788–1798.

Guo, L., Xiao, J., Liu, H., and Liu, H. (2020). Selenium nanoparticles alleviate hyperlipidemia and vascular injury in ApoE-deficient mice by regulating cholesterol metabolism and reducing oxidative stress. *Metallomics* 12 (2): 204–217.

Gupta, U.C. and Gupta, S.C. (1998). Trace element toxicity relationships to crop production and livestock and human health: implications for management. *Communications in Soil Science and Plant Analysis* 29 (11–14): 1491–1522.

Haghiabi, A., Nasrolahi, A., and Parsaie, A. (2018). Water quality prediction using machine learning methods. *Water Quality Research Journal* 53 (1): 3–13.

Hasan, S., Ranjan, D., and Talat, M. (2010). Agro-industrial waste 'wheat bran' for the biosorptive remediation of selenium through continuous up-flow fixed-bed column. *Journal of Hazardous Materials* 181 (1–3): 1134–1142.

Hu, C., Chen, Q., Chen, G. et al. (2015). Removal of Se(IV) and Se(VI) from drinking water by coagulation. *Separation and Purification Technology* 142: 65–70.

Jiang, K., Zhou, K., Yang, Y., and Du, H. (2013). A pilot-scale study of cryolite precipitation from high fluoride-containing wastewater in a reaction-separation integrated reactor. *Journal of Environmental Sciences* 25 (7): (1331–13)37.

Khan, M.I., Cheema, S.A., Anum, S. et al. (2020a). Phytoremediation of agricultural pollutants. In: *Phytoremediation* (ed. B.R. Shmaefsky), 27–81. Cham: Springer.

Khan, M., Shaheen, S., Ali, S. et al. (2020b). In situ phytoremediation of metals. In: *Phytoremediation* (ed. B.R. Shmaefsky), 103–121. Cham: Springer.

Kipp, A.P., Strohm, D., Brigelius-Flohé, R. et al. (2015). Revised reference values for selenium intake. *Journal of Trace Elements in Medicine and Biology* 32: 195–199.

Kuan, H., Lo, L., Wang, M.K., and Lin, C.F. (1998). Removal of Se (IV) and Se (VI) from water by aluminum-oxide-coated sand. *Water Research* 32 (3): 915–923.

Kumar, A. and Riyazuddin, P. (2011). Speciation of selenium in groundwater: seasonal variations and redox transformations. *Journal of Hazardous Materials* 192 (1): 263–269.

Kurokawa, S. and Berry, M.J. (2013). Selenium. Role of the essential metalloid in health. In: *Interrelations Between Essential Metal Ions and Human Diseases* (eds. A. Sigel, H. Sigel and A. Sigel), 499–534. Dordrecht: Springer.

Lenz, M. and Lens, P. (2009). The essential toxin: the changing perception of selenium in environmental sciences. *Science of the Total Environment* 407 (12): 3620–3633.

Li, Y., Cheng, W., Sheng, G. et al. (2015). Synergetic effect of a pillared bentonite support on SE(VI) removal by nanoscale zero valent iron. *Applied Catalysis B: Environmental* 174–175: 329–335.

Liang, L., Guan, X., Huang, Y. et al. (2015). Efficient selenate removal by zero-valent iron in the presence of weak magnetic field. *Separation and Purification Technology* 156: 1064–1072.

Ling, L., Pan, B., and Zhang, W. (2015). Removal of selenium from water with nanoscale zero-valent iron: mechanisms of intraparticle reduction of Se(IV). *Water Research* 71: 274–281.

Liu, X., Wang, S., Zhou, Y. et al. (2015). Regionalization and spatiotemporal variation of drought in China based on standardized precipitation evapotranspiration index (1961–2013). *Advances in Meteorology* 2015: 950262.

Luo, Q., Tsukamoto, T., Zamzow, K., and Miller, G. (2008). Arsenic, selenium, and sulfate removal using an ethanol-enhanced sulfate-reducing bioreactor. *Mine Water and the Environment* 27 (2): 100–108.

Luoma, S.N. and Presser, T.S. (2009). Emerging opportunities in management of selenium contamination. *Environmental Science and Technology* 43 (22): 8483–8487.

Mandal, B., Ogra, Y., and Suzuki, K. (2001). Identification of dimethylarsinous and monomethylarsonous acids in human u(rin)e of the arsenic-affected areas in West Bengal, India. *Chemical Research in Toxicology* 14 (4): 371–378.

Mandal, S., Mayadevi, S., and Kulkarni, B. (2009). Adsorption of aqueous selenite (Se(IV)) species on synthetic layered double hydroxide materials. *Industrial & Engineering Chemistry Research* 48 (17): 7893–7898.

Matoba, T., Shimokawa, H., Kubota, H. et al. (2002). Hydrogen peroxide is an endothelium-derived hyperpolarizing factor in human mesenteric arteries. *Biochemical and Biophysical Research Communications* 290 (3): 909–913.

Meher, A.K., Jadhav, A., Labhsetwar, N., and Bansiwal, A. (2020). Simultaneous removal of selenite and selenate from drinking water using mesoporous activated alumina. *Applied Water Science* 10 (1): 10.

Mondal, K., Jegadeesan, G., and Lalvani, S. (2004). Removal of selenate by Fe and NiFe nanosized particles. *Industrial & Engineering Chemistry Research* 43 (16): 4922–4934.

Paikaray, S. (2016). Origin, mobilization and distribution of selenium in a soil/water/air system: a global perspective with special reference to the Indian scenario. *CLEAN–Soil, Air, Water* 44 (5): 474–487.

Pinto, A., Fernandes, A., Vicente, H., and Neves, J. (2009). Optimizing water treatment systems using artificial intelligence-based tools. *WIT Transactions on Ecology and the Environment* 125: 185–194.

Prakash, N.T., Sharma, N., Prakash, R., and Acharya, R. (2010). Removal of selenium from Se enriched natural soils by a consortium of bacillus isolates. *Bulletin of Environmental Contamination and Toxicology* 85 (2): 214–218.

Reeves, D.W. (1997). The role of soil organic matter in maintaining soil quality in continuous cropping systems. *Soil and Tillage Research* 43 (1–2): 131–167.

Reilly, C. (2006). Selenium in foods. In: *Selenium in Food and Health* (ed. C. Reilly), 158–172. Dordrecht: Springer.

Roychowdhury, T., Tokunaga, H., and Ando, M. (2003). Survey of arsenic and other heavy metals in food composites and drinking water and estimation of dietary intake by the villagers from an arsenic-affected area of West Bengal, India. *Science of the Total Environment* 308 (1–3): 15–35.

Salt, D.E., Blaylock, M., Kumar, N.P. et al. (1995). Phytoremediation: a novel strategy for the removal of toxic metals from the environment using plants. *Bio/Technology* 13 (5): 468–474.

Salt, D.E., Smith, R.D., and Raskin, I. (1998). Phytoremediation. *Annual Review of Plant Biology* 49 (1): 643–668.

Sharma, M.C. and Kumar, P. (2006). Impact of pollution on livestocks of adjoining areas of Barauni industrial area in northern India. *Toxicology Letters* 164: S178–S179.

Sharma, V.K., McDonald, T.J., Sohn, M. et al. (2015). Biogeochemistry of selenium. A review. *Environmental Chemistry Letters* 13 (1): 49–58.

Sharrad, M.O.M., Liu, H., and Fan, M. (2012). Evaluation of FeOOH performance on selenium reduction. *Separation and Purification Technology* 84: 29–34.

Sheha, R.R. and El-Shazly, E.A. (2010). Kinetics and equilibrium modeling of Se (IV) removal from aqueous solutions using metal oxides. *Chemical Engineering Journal* 160 (1): 63–71.

Shen, Y. and Chen, B. (2015). Sulfonated graphene nanosheets as a superb adsorbent for various environmental pollutants in water. *Environmental Science & Technology* 49 (12): 7364–7372.

Singh, M. and Kumar, P. (1976). Selenium distribution in soils of bio-climatic zones of Haryana. *Journal of the Indian Society of Soil Science* 24 (1): 62–67.

Smedley, P.L., Nicolli, H.B., Macdonald, D.M.J. et al. (2002). Hydrogeochemistry of arsenic and other inorganic constituents in groundwaters from La Pampa, Argentina. *Applied Geochemistry* 17 (3): 259–284.

Soil Science Society of America. (2008). Biological selenium removal: solution to pollution? *ScienceDaily* (16 September), http://www.sciencedaily.com/releases/2008/09/080915121321.htm.

Spallholz, J.E., Boylan, L.M., and Rhaman, M.M. (2004). Environmental hypothesis: is poor dietary selenium intake an underlying factor for arsenicosis and cancer in Bangladesh and West Bengal, India? *Science of the Total Environment* 323 (1–3): 21–32.

Stadtman, T.C. and TC, S. (1978). Some selenium-dependent biochemical processes. *Advances in Enzymology* 48: 1–28.

Tang, C., Huang, Y.H., Zeng, H., and Zhang, Z. (2014). Reductive removal of selenate by zero-valent iron: the roles of aqueous Fe2+ and corrosion products, and selenate removal mechanisms. *Water Research* 67: 166–174.

Tang, C., Cheng, N., Pu, Z. et al. (2015). NiSe nanowire film supported on nickel foam: an efficient and stable 3D bifunctional electrode for full water splitting. *Angewandte Chemie International Edition* 54 (32): 9351–9355.

Tinggi, U. (2003). Essentiality and toxicity of selenium and its status in Australia: a review. *Toxicology Letters* 137 (1–2): 103–110.

Torregrossa, D., Leopold, U., Hernández-Sancho, F., and Hansen, J. (2018). Machine learning for energy cost modelling in wastewater treatment plants. *Journal of Environmental Management* 223: 1061–1067.

Vediya, S.D. and Shrivastava, A.K. (2008). Physico-chemical studies of sediments collected from Sabarmati river bank (Gujarat), India. *Plant Archives* 8 (1): 347–349.

Vara Prasad, M.N. and de Oliveira Freitas, H.M. (2003). Metal hyperaccumulation in plants: biodiversity prospecting for phytoremediation technology. *Electronic Journal of Biotechnology* 6 (3): 285–321.

Virk, H.S. (2018). Selenium contamination of groundwater of Majha Belt of Punjab (India). *Research and Reviews: A Journal of Toxicology* 8 (2): 1–7.

Virk, H.S. (2019). Selenium contamination of groundwater of Doaba Belt of Punjab, India. *Research & Reviews: A Journal of Toxicology* 9 (1): 1–8.

Weschenfelder, C., Schaan de Quadros, A., Lorenzon dos Santos, J. et al. (2020). Adipokines and adipose tissue-related metabolites, nuts and cardiovascular disease. *Metabolites* 10 (1): 32.

World Health Organization. (2009). *Global Health Risks: Mortality and Burden of Disease Attributable to Selected Major Risks.*

Yadav, S.K., Singh, I., Singh, D., and Han, S.D. (2005). Selenium status in soils of northern districts of India. *Journal of Environmental Management* 75 (2): 129–132.

Yang, G.Q. and Xia, Y.M. (1995). Studies on human dietary requirements and safe range of dietary intakes of selenium in China and their application in the prevention of related endemic diseases. *Biomedical and Environmental Sciences: BES* 8 (3): 187–201.

Yang, L., Shahrivari, Z., Liu, P.K.T. et al. (2005). Removal of trace levels of arsenic and selenium from aqueous solutions by calcined and uncalcined layered double hydroxides (LDH). *Industrial & Engineering Chemistry Research* 44 (17): 6804–6815.

Zayed, A., Gowthaman, S., and Terry, N. (1998a). Phytoaccumulation of trace elements by wetland plants: I. Duckweed. *Journal of Environmental Quality* 27 (3): 715–721.

Zayed, A., Lytle, C.M., and Terry, N. (1998b). Accumulation and volatilization of different chemical species of selenium by plants. *Planta* 206 (2): 284–292.

Zhang, Y., Li, Y., Li, J. et al. (2011). Enhanced removal of nitrate by a novel composite: nanoscale zero valent iron supported on pillared clay. *Chemical Engineering Journal* 171 (2): 526–531.

Zhang, M., Tang, S., Huang, X. et al. (2014). Selenium uptake, dynamic changes in selenium content and its influence on photosynthesis and chlorophyll fluorescence in rice (Oryza sativa L.). *Environmental and Experimental Botany* 107: 39–45.

Zhang, X., Fan, W.Y., Yao, M.C. et al. (2020). Redox state of microbial extracellular polymeric substances regulates reduction of selenite to elemental selenium accompanying with enhancing microbial detoxification in aquatic environments. *Water Research* 172: 115538.

Zhao, Q., Huang, J.C., He, S., and Zhou, W. (2020). Enhancement of a constructed wetland water treatment system for selenium removal. *Science of the Total Environment* 714: 136741.

Zhu, Y.G., Pilon-Smits, E.A., Zhao, F.J. et al. (2009). Selenium in higher plants: understanding mechanisms for biofortification and phytoremediation. *Trends in Plant Science* 14 (8): 436–442.

Zhu, L., Zhang, L., Li, J. et al. (2017). Selenium sequestration in a cationic layered rare earth hydroxide: a combined batch experiments and EXAFS investigation. *Environmental Science & Technology* 51 (15): 8606–8615.

2

Selenium Distribution and Chemistry in Water and Soil

Sulaxna Sharma[1] *and Awanish Kumar Sharma*[2]

[1] *THDC, IHET, Tehri, Uttarakhand, India*
[2] *Department of Physics, GEU, Dehradun, Uttarakhand, India*

2.1 Introduction

Selenium is a rarer element on earth and ranked as 70th element among naturally occurring elements in order of its natural abundance. It was first discovered by Swedish chemist Jöns Jacob Berzelius in 1818 when preparing sulfuric acid. He named it after the Greek word *selene* (meaning "moon"). Selenium after its discovery was almost unnoticed for decades (Hamilton 2004), until the identification of its nutritional essentiality in animals in 1957 (Schwarz and Foltz 1957). The interest of researchers in the trace element has grown substantially with the recognition of its nutritional role as antioxidant and toxicity among human beings (Rotruck et al. 1973). It is dispersed widely in low concentration in the Earth's crust (Berrow and Ure 1989; Russell 2011) and its concentrations in the Earth's crust and sea water are 0.004 and 0.09 μg/g, respectively. However, the average global concentration of Se in soil is 0.4 μg/g (i.e. ranging from 0.01–2 μg/g) (Dungan et al. 2002) and 3.1 μg/g in soil of Northwest India (Bajaj et al. 2011). However, only bioavailable Se is the key to determine Se in plant tissues ranging from 0.02–4000 μg/g, depending upon low to high seleniferous soil, which enters into food chain. Hamilton (2004) has reported that the trace to moderate amount, i.e. 40–400 μg/day, of Se is necessary for development and normal growth, and to maintain homeostatic functions, respectively. The deficiency of Se in diet (<3 μg/day, reported by Yang et al. (1983), in China) can weaken immune system and can cause Keshan disease and Kashin–Beck disease, (Coppinger and Diamond 2001). On the other hand, a higher concentration (>400 μg/day) of it may cause health hazard to living beings, including gastrointestinal, respiratory, and cardiovascular problems. It may also result in fragile nails, excessive tooth decay, hair loss, garlic-smelling breath, discoloration and mental problems (Pedrero and Madrid 2009). The intake of Se recommended by World Health Organization (WHO) and different recommended Se intakes followed in different countries are summarized in Table 2.1.

Selenium Contamination in Water, First Edition. Edited by Pooja Devi, Pardeep Singh, Arindam Malakar, and Daniel Snow.

Table 2.1 Recommended daily nutrient requirement of selenium (μg/d) as per WHO and NAS and followed by different countries and their daily Se intakes.

			Unique recommended Se intakes (μg/day) followed by countries[c,d]						
Group	**WHO[a]**	**NAS[b]**	**USA**	**Canada**	**UK**	**Europe**	**Australia**	**China**	**India**
Men	40	55	55	55	75	55	85	[e]	[e]
Women	30	55	55	55	60	55	70	[e]	[e]
Daily Se intakes (μg/day)									
Men			90[a]	98–224[a]	60[c]	38–48[a]	[e]	3–11,[a,e] 1338[a,f]; 9–11[c,e], 5000[c,f]	48[a]
Women			74[a]		60[c]		[e]		

[a] WHO (2005).
[b] National Academy of Science (NAS) (2000).
[c] Thomson (2003).
[d] Thomson (2004).
[e] Keshan disease area.
[f] Seleniferous area.
[g] Data not available.

For adults, the recommended dietary allowance (RDA) is 55 μg/day (see Table 2.1) and tolerable upper intake is 400 μg/day, set by the Institute of Medicine of the National Academy of Sciences (NAS 2000) in the US, while more than 5 mg/day can be fatal (Mayer (2010). Amweg et al. (2003) has reported that excessive Se replaces the sulfur from amino acids that subsequently changes the 3D structure of proteins and impairs the enzymatic functions. Levander and Burk (2006) have discussed the biological properties of organoselenium compounds in metabolism and have reported that organoselenium compounds, i.e. di-/trimethylselenides and selenoamino acids, viz. selenomethionine, selenocystine, and selenohomocystine compounds, are of interest in nutrition and health. Ralston et al. (2009) have reported that dimethylselenide is volatile neutral molecule and readily excreted through breathing and thus provides a natural way to export excess of Se from body. The established "no observed adverse effect level" (NOAEL) in China for human population is 15.5 μg Se/kg body weight as compared to the "recommended daily amount," i.e. 0.9 μg Se/kg body weight in US (Ralston et al. 2009). It has also been reported by the group that for humans there is about ~20-fold factor between nutrient dietary intake and the threshold intake that result in a toxic level of Se. On considering its narrow tolerance range of nutritional deficiency (<40 μg/day) and toxic level (>400 μg/day) (Levander and Burk 2006) for humans, Se is being termed as "essential toxin" (Stolz et al. 2002) or "double-edged sword" (Fernández-Martínez and Charlet 2009) element.

Apart from essential dietary value of Se, it has several industrial uses including ceramic, pigment, pharmaceuticals, organic catalyst, as photocopying material, in glass manufacturing industries as dopants, photocell devices, electronic equipment, batteries, antidandruff products, etc. (Mayland et al. 1989; Russell 2011). For commercial applications it is produced worldwide from mud deposited on the anode during electrolytic refining of copper (Russell 2011). Literature reports state that the annual production of Se is around 1500 tones and recycled amount from industrial waste is approximately 150 tones. Therefore, in addition to natural sources, emission of Se-using industries mentioned above and combustion of fossil fuel including coal burns are other main sources of Se in environment that results Se as an emerging pollutant. Sharma and Kumar (2006) have reported that in Baurani industrial area of Bihar, India, groundwater area up to 30 km away has polluted by the effluents from oil refineries, chemical fertilizers, and thermal power plants. A study conducted by the National Irrigation Water Quality Program (NIWQP), in various irrigated regions, in different states of western US between 1986 and 1993 (Seiler 1995, 1998). The study demonstrated that when irrigated aquifers are underlain by marine shales (as mentioned in Eq. 2.1), the agricultural drainage waters reaches to toxic Se level. Bioavailability of Se in many areas of the countries like USA, Australia, Ireland, China, and India is high enough to cause toxicity in soil and water (Presser and Piper 1998). In low land areas and ponds the deposition of solid wastes is the major source of elevated Se concentration in ground water. Except certain seleniferous areas concentration of Se in drinking water is usually less than 10 μg/l. Vegetables and crops are responsible for a Se-rich diet; the reason may be the adsorption of particulate Se onto them. Moreover, non-vegetarian food is more Se-rich than vegetarian food (Paikaray 2016). Se intake through inhalation is much lesser (<0.1%) in comparison to drinking water and food. Even in highly seleniferous soils, the contribution of Se by drinking water is relatively low in comparison to food stuffs. In the polluted areas water and soil are highly enriched with Se and are a concern for human and animal health. This attracts the attention of scientific community worldwide, in recent years. Therefore it is essential to know the chemistry of Se including its oxidation states, isotopes and allotropic forms, chemical, and physical properties of Se. Such properties are helpful to understand the nature of Se compounds and their structure. Additionally it would be easy to illustrate a relationship between environmental conditions responsible for distribution of different chemical compounds of Se in biogeological samples at different sites.

2.2 Environmental Distribution and Forms

Selenium is the 34th element and is placed in the VIA (or 16th) group of periodic table following the order O, S, **Se**, Te, Po, and Lv, within the group. Chemical properties of Se are analogous to those of sulfur and it mainly shows following oxidation states selenide (−2), elemental selenium (0), selenite (+4), and selenate (+6). Se occurs in minerals, for example pyrites, where it replaces sulfur partially and also form compounds with other elements (Butterman and Brown 2004).

In nature, elemental selenium (Se^0) exists in six stable isotopic forms, ^{74}Se, ^{76}Se, ^{77}Se, ^{78}Se, ^{80}Se, and ^{82}Se, whereas 24 unstable isotopes of Se has also been reported so far, with half-life 20 ms to 295 000 years (Audi et al. 2003). The details of unstable isotopes, their atomic mass, half-life, and mode of decay are reported in detail by Perrone et al. (2015). The isotopes ^{78}Se and ^{80}Se are more common due to their high natural abundance, i.e. 24 and 50%, respectively. Of the radio-nucleotides, ^{75}Se ($t_{1/2}$ = 120 days), owing to its reasonably high half-life value, is suitable for determination of biological traces (Irons et al. 2006), whereas ^{75}Se, ^{77m}Se, and ^{81}Se may possibly be used for quantitative determination of Se through radiologic diagnostics (Li and Zheng 1990).

Se exists in allotropic forms either in amorphous form or any of the three crystalline forms viz., α-monoclinic, β-monoclinic, and hexagonal forms (National Research Council 1983). Amorphous Se is red in color (Jovari et al. 2003) and its viscosity is highly dependent on temperature. At 230 °C it is a free-flowing liquid and on reducing the temperature up to 80 °C its viscosity increases and it forms polymeric chains. However, on further reducing the temperature there is a decrease in its viscosity and it forms ring-shaped aggregates. It forms Se_8 rings and has deep red color in its monoclinic crystalline form. The shape of α-monoclinic Se is flat hexagonal and polygonal crystals while it has needle like shape in β-monoclinic crystalline form. Hexagonal crystalline form with spiral Se chains is the most stable form of Se and is gray in color. Monoclinic crystalline and amorphous forms are transformed into hexagonal form at temperature >110 and 70–210 °C, respectively. The variation of its physical properties with its allotropic form has been reviewed in detail by Crystal (1972) and Chizhikov and Shchastlivyi (1968).

Since lithosphere, hydrosphere, atmosphere, and biosphere are the integrated components of environment, as a result it is indispensable to understand the cycling of Se in soil, water, and air. During volcanic eruption, at high temperature Se and S vaporize to gaseous form. On cooling, Se condenses and forms a layer over ionized micro-particulate of atmosphere that eventually precipitates with rainfall and appends to the rocks and/or enters to the water bodies. In several research papers the abundance/concentration of Se is reported in different terms that includes: weight by weight: microgram per gram (μg/g) or (mg/kg) is equivalent to parts per million (ppm); and weight by volume: microgram per liter (μg/l) = 1 ppb Se), here in this chapter μg/g and μg/l units are used throughout. In the earth's crust the average abundance of Se is ~0.09 μg/g (Lakin 1972) and barely found in its native state. Owing to the chemical resemblance of Se to its analogue S, it is widely distributed in environment as a major and minor constituent of most of the sulfide ores (Cooper et al. 1970) or as selenides of nickel (Ni), copper (Cu), silver (Ag), lead (Pb), and Mercury (Hg). Uranium ore contain highest (~600 μg/g) of Se content (Ralston et al. 2009). Rocks contain around 40% of the Se of the total of Earth crust (Wang and Gao 2001), values reported for igneous rocks (0.35 μg/g) (Fordyce 2005), sedimentary rocks (0.0881 μg/g) (Tamari et al. 1990), shales (0.24–277 μg/g) (Lakin and Davison 1967), phosphatic rocks (1.4–178 μg/g) (Robbins and Carter 1970), coal (1–5 μg/g) (Cooper et al. 1970), limestone (0.03–0.08 μg/g) (Fordyce 2005), and sandstone (0–112 μg/g) (Lakin and Davison 1967). Leaching from these Se-rich sources can elevate the Se concentration in environment up to 1200 μg/g (Paikaray 2016). Rosenfeld and Beath (1964) have compiled the data of Se concentrations in rocks and seleniferous soils. These seleniferous rocks are the major source of Se in soil, ground water, and atmosphere.

The distribution of seleniferous rocks and deposits in geologic and topographic map indicate the areas of Se-rich soil and water. Soil and water at distant places can become seleniferous when soluble and/or suspended Se species are imported to such areas through surface water. In the environment, concentration of Se varies from place to place; this uneven distribution of Se is mainly governed by the processes including weathering, interaction of rocks and water, and microbial activities. These processes control the transportation of Se from rocks to soil, water, and air. In soil the amount of Se is primarily influenced by parent material and possible leaching from rocks during soil formation. The confined environmental surroundings such as properties of soil, aeration, organic matter, pH, and microbial activity are also the responsible factors for Se distribution in soil. The average range and mean global concentration of Se in soil is 0.01–2 and 0.4 μg/g, respectively (Dungan et al. 2002). Further, the reactivity and bioavailability of Se, in addition to total concentration, also depends on different chemical forms available in soil and water. Here bioavailability means the part of substance that becomes soluble and accessible for adsorption by means of membrane (Reeder et al. 2006; Ruby et al. 1996) of living organisms.

The toxic, tolerable, and deficient areas of Se level exist alongside and for these different Se levels local environmental conditions are responsible. Depending upon the sampling done for the available Se concentration in vegetation and plants grown in the soil, seleniferous soil has been categorized as toxic, moderate, and low level of Se. Soil that provides an adequate amount of Se to make toxic plants is referred to as toxic seleniferous soil. Contrary to this, the soil may have high Se level as exhibited by toxic Se soil but provide less Se to the plants, known as non-toxic seleniferous soil. From deficient to most-seleniferous soil, the concentration of Se reported is 0.01 and 1200 μg/g (Fleming 1980; Jacobs 1989; Neal 1995). Many countries have elevated level of Se including USA (Presser 1994), India (Dhillon and Dhillon 2003), Ireland (Seby et al. 1997), and China (Wang and Gao 2001). Central region and Great Plains of North USA and Prairie region of Canada (Ihnat 1989) is formed from Cretaceous shale (2 μg/g) and exhibit relatively high concentration. In Australia, Ireland, and various other countries with toxic Se level, shales are the parent material (Johnson 1975). Florida, South Carolina, and Tennessee are ranging lower (0.8–9 μg/g) in Se due to phosphatic rocks of that region (Rader and Hill 1935). A low level of Se has been documented in Finland and New Zealand. Se content in Hawaiian and Japanese volcanic sulfur ranged from 1026 to 2000 and 67–206 μg/g, respectively (Lakin and Davison 1967). Many parts of Africa were recognized with low Se; however, in Asia both high and low Se concentrations have been reported (National Research Council 1983). Most of the parts of the world are characterized as moderate to low bioavailability as compared to high Se soil content. Among most studied Se-contaminated water bodies, Kesterson Reservoir of San Joaquin Valley, California USA is one of them. The main Se source is Se-rich marine sedimentary rocks (mean values = 8.9 μg/g) of the coastal range, which raise the Se content to the reservoir by weathering and other beneath mechanisms (*Milne* 1998; Presser and Piper 1998). Human and industrial activities are also responsible for the discharge of Se in rivers and lakes. Dhillon and Dhillon (2003) have compiled a comprehensive review on seleniferous soils.

Water applied for irrigation purpose to soil can mobilize the soluble Se species from upper soil surface and create a hydraulic gradient, and as a result Se is discharged to surface water. Apart from natural Se sources, this agriculture drain water and other anthropogenic activities can also discharge Se to water bodies and is available to aquatic life. The aquatic bodies have also played an important role in the transportation of Se from one place to other. The estimated value for the fluxes of Se is ~14 000 tons per year that are relocated from continents to oceans along aquatic pathway (Nriagu 1989, 1991). The mode of Se transfer is either via selenium-bearing sediments (~85%) or through soluble species (Cutter and Bruland 1984). Leaching from soil to groundwater is another segment of hydrological transport of selenium species. However, contribution through leaching processes is not very clear because of the heterogeneity of the interfaces between water-rocks, -soils and -sediment (Dhillon et al. 2008). Ihnat (1989) has reported that the average residence period of selenium in the deep ocean is ~1100 years, almost of the order of residence time of water. The soluble selenium species from industrial waste move downward through infiltration and elevate the Se level in ground water during monsoon season (Kumar and Riyazuddin 2011). Se concentration in water is ranged between 1 and 10 μg/l; due to the emission of geogenic and/or anthropogenic Se, this may exceed 100 μg/l occasionally in outstanding cases. In industrial effluents Se concentration is in 10–100 μg/l range, under rare conditions it can exceeds up to 1000 μg/l. Typical Se concentrations in ambient water are <1 μg/l in the dearth of direct Se sources. The existing Se concentration of marine waters is 0.02–0.04 μg/l. However, for fresh water the background Se concentration is considered to be comparable to marine water. Ghosh et al. (2008) reported that the elevated concentration of Se in soil (1100 μg/l) and water (3 μg/g) in localities of Mumbai and Punjab of India is due to the alkaline pH (pH 8–9.2) of soil in industrially polluted areas. The concentration of Se is high in the saturated water aquifers which have fluctuations in their water table. Se concentration in aquifers also depends on mixing ratio of discharge stream into receiving water. In addition, the probable interaction of discharge to Se removing sediments is relatively more important than its dispersion in aqueous phase. The water quality criterion for Se in water is primarily determined by ecotoxicological considerations for aquatic living organism, e.g. predatory fish and waterfowl, which are at the top of aquatic food chains. Countries like the US and Canada provide regulatory guidelines that include 5 and 1 μg/l (with varying guidelines for individual regions up to 100 μg/l), respectively (US Environmental Protection Agency 2008). The relationship between Se concentration in water and environmental toxicology is not fully understood because of the existing complexity of the biogeochemical Se cycle. However, several efforts are in progress to establish water quality measures on considering different ecosystems that are subjected to emission of Se in particular areas specific for sites.

In addition to this, adsorption and desorption of elements, precipitation of minerals, and incineration of municipal wastes (Plant et al. 2004) have also contributed to the insertion of Se into atmosphere. Organometallic compounds of Se are introduced partly to the atmosphere by chemical or microbial redox reactions and to soil and water by metabolic uptake and release by animals and plants (McNeal and Balistrieri 1989). Consequently, it enters into the food chain through crops, plant, and aquatic lives (Paikaray 2016). Frost (1967) has reported that sea water, earth crust, animals, and plants contain 0.004, 0.09, 1–20, and 0.02–4000 μg/g, respectively, which indicates that plants and animals have ability to concentrate Se from earth crust. It again enters into environment through the decomposition of these species and has excreted from human body (~50–80% through urine) to environment. For this in-and-out pathway of Se into environment, several cycles have been proposed including geological cycling of Se where animals and plants had a role, proposed by Moxon et al. (1939) and Lakin and Davidson (1967). Shrift (1964) and Frost (1973) have proposed a biological cycle of Se in which involvement of reduction–oxidation (redox) reactions of Se by plants, bacteria, and fungi have been incorporated. Allaway et al. (1967) and Olson (1967) have reviewed the cycling of low and high levels, respectively, of Se in soils, plants, and animals.

Se forms more than 170 solid compounds, whereas very limited compounds of Se are known that are in liquid (SeF_4, Se_2Cl_2, and CSe_2) and gaseous (H_2Se and SeF_4) state (Butterman and Brown 2004). Owing to its varied oxidation states it forms different type of compounds with sulfur (selenium sulfide (Se_2S_2) and polysulfides), oxygen (selenate, SeO_4^{--} and selenite SeO_3^{--}), hydrogen selenide (H_2Se), and some organometalic compounds with methyl such as dimethylselenide ($(CH_3)_2Se$), dimethyl selenone ($(CH_3)_2SeO_2$), and dimethyl diselenide ($(CH_3)_2Se_2$) that are widely distributed in biogeologic forms. On substitution of sulfur it forms selenocysteine, selenocystine, and selenomethionine with amino acids (Chabroullet (2007); Losi and Frankenberger (1997); Haygarth (1994); Fernández-Martínez and Charlet 2009). Therefore, it is clear that a range of inorganic and organic molecular forms of Se are present in environmental material such as rocks, soils, the aquatic system, and in air.

2.3 Species of Selenium

Bioavailability of Se species is mainly governed by the distribution of Se in particle (insoluble Se species) and dissolved phases. Selenate (SeO_4^{2-}) and selenite (SeO_3^{2-}) are the soluble inorganic forms of Se, among which selenate (SeO_4^{2-}) is considered the most soluble Se form (~90–95% of soluble se in arid agriculture water) (Masscheleyn et al. 1990). However, Se(0) and Se(-II) are usually insoluble until available in suspended form. A quantitative relation between particulate and soluble Se species can be depicted in terms of distribution coefficient (K_d = Se per unit mass particulate material/Se per unit volume water, in units of equivalent) (IAEA 1994). This is frequently used to calculate the sorption behavior of Se in soil. Thus, accumulation of Se in animals is mainly in the particulate form as compared with that of the amount taken up from water (Luoma et al. 1992). Biogeochemical transformation reactions govern the Se concentration based on interaction between soluble and particle Se species. Concentration of Se in the food web is dependent on efficiency of transformation from dissolved to particulate Se by microorganism and determined in terms of μg of Se/g. Nakamaru et al. (2005) have reported that K_d of Se is 600–800 l/kg for Japanese agricultural andosols.

Existence of Se species, along physiological pH and within permissible redox potential range of water, is shown in the Pourbaix diagram for Se (Figure 2.1; Mayland et al. 1989). In the Figure 2.1 the highlighted part shows the natural soil conditions. Se(0) is insoluble and on oxidation can be converted to higher oxidation state Se(IV) and Se (VI) species. In natural soil conditions, selenate (SeO_4^{2-}; the most toxic form of Se) and selenite (SeO_3^{2-}) are the soluble oxyanions that can be transformed into elemental Se(0) where organic matter and microbial species participate in this process. Selenate (SeO_4^{2-}) is most soluble form of Se, with low adsorption and precipitation capacity, which predominates in alkaline and high redox potential soil conditions (Fordyce 2007). On the other hand, selenite (SeO_3^{2-}) occurs in neutral and acidic soil conditions (Kabata-Pendias 2011) with moderate redox potential range (*Milne* 1998). Se(-I) and Se(-II) species are stable in the form of metallic selenides and organometallic compounds (e.g. DMSe dimethylselenide) under strong reducing conditions. These species are insoluble and immobilize unless available in suspended form. Most of the Se(-II) species are toxic in nature; however, it has been stated that the toxicity of gaseous DMSe is lesser (~500–700 times) than SeO_4^{2-} (Calderone et al. 1990; Stork et al. 1999). Figure 2.1, suggest that HSe^- doesn't exist in natural conditions; however, in water it gets converted to H_2SeO_3, a highly oxidizing selenic acid. Besides these inorganic Se species, various organic selenides in the presence of microbial species present in natural conditions in soil and water bodies (Cutter 1982; Lenz et al. 2006).

For transportation of selenium aquatic bodies play important role (Cutter and Bruland 1984), therefore it is important to know the species present in the aqueous medium.

The scheme suggests that selenite is thermodynamically unstable and converted to selenate under well-aerated alkaline water, but the process is slow, i.e. kinetically this transformation is not accomplished in ambient surface water until microbes are not involved. In water with high redox potential, selenate is the major species, as redox potential decreases in sediment–water interface and in sediment pore water selenite is the major species. Selenite shows high adsorption capacity over mineral sites such as Fe and Mn hydroxide minerals, etc. as compared to selenate. As a result Se concentration in aqueous phase decreases dramatically irrespective of the redox potential of the system. Other than these soluble inorganic Se species occur in highly reducing waters by reaction of Se compounds with reduced S moieties. In nature, selenate is the major species in irrigation drain water that inserted to wetland (Presser and Ohlendorf 1987; Zhang and Moore 1996). Selenite is widespread species in industrial waste associated particularly with fossil fuel (Cutter and San Diego-McGlone 1990). Organic Se is available wherever it is recycled (Takayanagi and Wong 1984). In the presence of microbial activity soluble selenite and

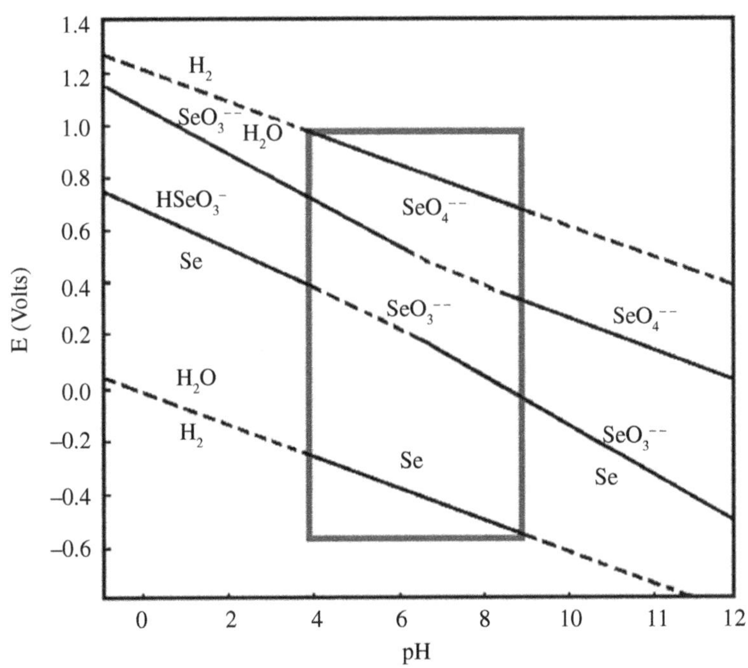

Figure 2.1 E-pH diagram of selenium in soils (*Source:* adapted from Mayland et al. 1989 with permission from John Wiley and Sons).

selenate can be reduced to undissolved elemental Se, so in highly reducing waters, selenite and selenate are not encountered. Insoluble Se species are very less reactive and thus are less significant from a quantitative point of view in aerated water. Besides inorganic Se species various other organic Se species have been assumed to present in small fractions of the entire Se concentration in well aerated surface water. However, their contribution is considered to be negligible for the overall Se cycling, but may have possible implications for uptake by aquatic organisms and may have ecotoxicity effect on them.

2.4 Interaction of Selenium with Organic Matters and Microorganisms

In combination with other naturally occurring ecological conditions, organic matters (OM) have also play a vital role. It has been reported that Se interacts with OM in soil and aquatic environments (Abrams et al. 1990; Zhang and Moore 1996). The distribution coefficient (K_d) for Se-OM is 1800l/kg which is twice that for Se to soil clay. This association of Se-OM governs the mobility of Se species by complexation process (Filella and Town 2003). In soil and water the high heterogeneity of OM (Sutton and Sposito 2005) makes it more complex to study the properties of OM because of the occurrence of higher OM content in top soil; Se concentration is high in surface soil as compared to deeper soil layers (Zhang and Moore 1996).

Ferri and Sangiorgio (2001) have studied the interaction of selenite with natural polysaccharides of different molecular weight. The studies revealed that the charge on organic species doesn't affect the formation constants but its strength is dependent on the structure of the complex. It has been reported that Se binds well with OM in acidic soils (Christensen et al. 1989; Gustafsson and Johnson 1992). Interaction with humic and fulvic acids has been studied and correlated with animal (Wang et al. 1996) and plant uptake (Kang et al. 1991; Zhang and Moore 1996). The interaction of Se species with fulvic acid (soluble) is more bioavailable than humic acid (suspended particulate) (Qin et al. 2012). This can be explained by the retention of negatively charged oxyanions by the OM that immobilizes the Se species in anoxic conditions. The study of interaction of selenium oxyanions and humic acid has been carried out in by Bruggeman et al. (2005, 2007). The interaction of inorganic Se with humic acid as well as with plants and microbial processes may immobilize Se (Tolu et al. 2014) and make them less bioavailable. In the presence of reducing phase (e.g. FeS_2), selenate is expected to reduce to selenite which is removed by precipitation (Bruggeman et al. 2005). This elemental Se can be further altered to ferroselite ($FeSe_2$) inside the sediments by precipitation reactions or by assimilation into solid phases (e.g. pyrite) (Velinsky and Cutter 1991). $FeSe_2$ on oxidation can be converted to selenate according to the reaction (Eq. 2.1) (Bailey et al. 2012).

$$2FeSe_2 + 7O_2 + 2H_2O \rightarrow 2Fe^{2+} + 4SeO_4{}^{2-} + 4H^+ \tag{2.1}$$

Gustafsson and Johnson (1992) have reported that considerable amount of Se fractionation is present in OM rich Bs horizon of Swedish podzols due to the high adsorption capacity of anions by OM. Wang and Gao (2001) have reported that despite having sufficient Se concentration (0.1–0.3 μg/g) in soils, Keshan and Kashin –Beck diseases (Se deficiency) are widespread in the northeastern regions of China. This is because the soil is rich in OM. This result in immobilization of Se and makes it less bioavailable. Se concentration in wetland soils (Cooke and Bruland 1987; Thompson-Eagle and Frankenburger 1992) and evaporation ponds (Fan and Higashi 1998) also depends on biogeochemical volatilization of Se.

In the environment, reduction of dissolved Se oxyanions to particulate Se in oxygenated water occurs via plant uptake and by microorganisms and affects Se bioavailability. Through a dissimilatory reduction (energy metabolism) process microorganisms can use soluble selenate (SeO_4^{2-}) and selenite (SeO_3^{2-}) as electron acceptors and reduce to insoluble and less bioavailable Se(0) or Se(-II) species. In this process SeO_3^{2-} acts as an intermediate. These insoluble Se(0) and Se(-II) are further transformed to volatile organic-Se species and escape to the atmosphere. Hamdy and Gissel-Nielsen (1976) have studied under laboratory conditions and reported that the volatilization of reduced Se is higher in sandy soil than a humic soil which is about 8 and 5% of added Se, respectively. The selenate owing to its kinetically inert nature does not get precipitated and remains in equilibrium with other selenium species formed in soil or ground water that are rich in humic acid. The following reactions (Eqs. 2.2 and 2.3) for the reduction of Se oxianions to elemental Se have been reported by Bailey et al. (2012) with the generic carbon species:

$$CH_2O + 2SeO_4^{2-} \rightarrow CO_2 + 2SeO_3^{2-} + H_2O \tag{2.2}$$

$$CH_2O + SeO_3^{2-} + 2H^+ \rightarrow CO_2 + Se^0 + 2H_2O \tag{2.3}$$

(where CH_2O represents generic organic carbon compound).

This process provides an efficient way for detoxification of environment as insoluble Se and organic-Se compounds are less toxic than soluble Se oxyanions (Bailey et al. 2012). Appropriate electron (e^-) donor substances such as sugars, alcohols, humic substances, organic acids and hydrogen (Bailey et al. 2012; Korom 1992) have been used by Se-reducing microorganisms. Se oxyanions are behaved as electron (e^-) acceptor in dissimilatory reduction reactions. Dissimilatory reduction process contributes significantly to detoxification of a seleniferous environment as compared to the assimilatory reduction process. Dissimilatory reduction offers a cost-effective way for environmental remediation and therefore attracts the attention of researchers. Eswayah et al. (2016) have reviewed in detail the microbial transformations of unlike forms of Se, assimilation, dissimilation reduction process, anaerobic respiration of several microorganisms supported by dissimilatory reduction, formation of nanoparticulate elemental selenium, etc. Retention of Se by cattle manure has also been reported by Ogaard et al. (2006). In the batch studies the group observed that on addition of cattle manure there is a decrease in adsorption of both selenite and selenate to a loam soil, since manure contains organic acids of low molecular weight such as acetic, formic, and oxalic acids which contend with Se for available sorption sites (Ogaard 1996).

2.5 Interaction of Selenium with Clay Mineral

In soil the presence of clay and minerals plays a vital role in sorption and desorption process of Se species and thus affect their bioavailability. The content of Se is better in soil with elevated clay content than in sandy soil (Gissel-Nielsen 1975). Gissel-Nielsen (1971) has studied the effect of texture and pH value on Se uptake from sandy and loamy soils. Caolinite ($Al_2Si_2O_5(OH)_4$), a clay mineral, is formed by octahedral sheets of alumina having a positive charge at their edges, that serve as binding sites for Se (Bar-Yosef and Meek 1987). Several other minerals such as Fe/Al oxyhydroxides, manganese oxide, and hydroxyapatite (Monteil-Rivera et al. 2000) are also present in the soils (Eich-Greatorex et al. 2010); the presence of such minerals varies from place to place. The available binding sites for these minerals, i.e. positively or negatively charged, are dependent on reaction conditions, e.g. in alkaline and acidic reaction conditions the charge on aluminum hydroxide minerals is negative and positive, respectively. Over the surfaces of these minerals selenate and selenite get sorbed under acidic soil conditions (Balistrieri and Chao 1990). However, selenate has lesser adsorption capacity and more bioavailable than selenite. The presence of citric and oxalic acid affect the sorption of selenate and selenite. The presence of acetic and formic acids facilitates the adsorption of selenate but does not affect the adsorption of selenite (Dynes and Huang 1997; Wijnja and Schulthess 2000). This kind of behavior of selenate and selenite can be elucidated by the formation of an outer-sphere complex of selenate which bounds with non-specific anion exchange sites with minerals, whereas selenite forms an inner-sphere complex by ligand exchange method (i.e. form covalent bonds) that is irreversible in nature (Peak and Sparks 2002). Similar to selenate and selenite, acetic and formic acids form outer-sphere complexes, whereas citric and oxalic acids form inner-sphere complexes at the mineral-water interface (Dynes and Huang 1997; Wijnja and Schulthess 2000). Besides this, sorption of selenite in the presence of fulvic acids has no effect on aluminum oxides (Zuyi et al. 2000), but promotes the adsorption on amorphous iron oxides (Tam et al. 1995).

The adsorption process is affected by the existence of contending ions such as phosphate (PO_4^{3-}) and sulfate (SO_4^{2-}), and the nature of bonding formed between the competing ions has also governed the mobility of Se species by affecting their sorption in soil and water bodies. In soil, these species are mainly added through agricultural products including gypsum- and phosphate-rich fertilizers. For Japanese soils it has been reported that sorption of Se decreases in the presence of PO_4^{3-} ion; for selenite this effect is more pronounced than for selenate. Therefore, in phosphate-rich areas selenite is more bioavailable (Nakamaru et al. 2006; Nakamaru and Sekine 2008) whereas for sulfate-rich soil selenate is more available for biota. This kind of behavior of selenite and selenate is due to their chemical similarities to phosphate (Eich-Greatorex et al. 2010) and sulfate, respectively. This supports the study of Nakamaru and Sekine (2008), i.e. an increase in concentration of sulfate ions does not change selenite sorption. To reduce the selanate sorption for plants it has been reported that $BaCl_2$ is more effective than sulfate ions as it forms an insoluble $BaSeO_4$ compound (Mikkelsen et al. 1989). Further, in neutral medium Se uptake decreased in the presence of sulfate (Williams and Thoronton 1972), whereas increased Se uptake has been observed in an alkaline medium (Carter et al. 1972).

Singh (1979) has studied the consequences of supplementary Se and P on arid matters and also examined the chemical composition of *Brassicajuncea Cos*, in a greenhouse test. The study reveals that when Se is added without and with P there is a manifold improvement in the concentration of Se in plants with respect to control. Se uptake by *Trifolium alexandrinum* has increased effectively for all variants of Se on addition of 50 and 100 ppm P. The effect is also dependent on the presence of auxiliary ions in fertilizers (Khattak et al. 1991) such as superphosphate ($Ca(H_2PO_4)_2$) (Fleming 1962), which reduces the Se uptake that may be due to the existence of competing SO_4^{2-} ions (component of superphosphate). Carter et al. (1972) have proposed that phosphate can work in two ways: in the presence of Se phosphate competes with Se at sorption sites, or promotes the cell division of plant roots and thus increases Se uptake.

2.6 Conclusion

Selenium is an essentio-toxin element with chemistry similar to that of sulfur. Several Se species are available in nature and are distributed widely in soil and water. Physico-chemical properties including originating rocks, pH, redox potential, organic matters, microorganisms, mineral, properties of clay, competing ions, etc. are responsible for the incidence of the different Se species in soils. In soil the main sources of Se are seleniferous rocks, burning of fossil fuels, disposal of Se-containing industrial waste and addition of agricultural products, etc. These sources are responsible for the accumulation of different chemical species of Se in soil. Depending upon Se concentration in soil can be divided into toxic, moderate, and deficient levels which may exist side by side. Bioavailability of Se in the plants and crops supported by soil depends on local environmental factors, e.g. in acidic pH range (4.5–6.5), rich in OM, presence of strongly reducing species and also in the region of low rainfall, non-toxic seleniferous soil predominates, i.e. insoluble selenide (-II) and elemental Se(0) forms are present majorly. In a solution of well-aerated alkaline soil selenate and selenite are the most soluble species but the presence of such species may change. In humid conditions the occurrence of accumulated zones of minerals renders the bioavailability of Se to plants. The interaction of Se species with OM including fulvic acid is more bioavailable than humic acid because interaction of Se species with humic acid immobilizes Se and therefore becomes less bioavailable. Sorption and desorption process in the presence of clay content, OM and minerals in soil play a crucial role in bioavailability of Se. Sorption of Se is at its maximum between pH 3 and 5 and therefore it is less available for plants in loamy than in sandy soil. Besides this, OM facilitates the simulation of microbial activity that changes the oxidation state of Se (VI to -II). This may also result in methylation, which facilitates strong binding of Se to OM. The reduced form of Se is sorbed on OM and less mobile. Fungal growth in the presence of OMs affects the mobilization of Se and its availability hindered dramatically from soil to plant. Se concentration in soil also varies along the depth profile of soil. Presence of free calcium carbonate and increased pH (alkaline) reveals toxic seleniferous soils. To evaluate Se contamination understanding related to particulate Se is important, because every species has different bioavailability. The bioavailability of Se in soil follows the following order: selenate ($Se^{6\backslash +}$) > selenomethionine > selenocysteine > selenite (Se^{4+}) > elemental selenium (Se^0) > selenide (Se^{2-}). Remedial treatments are highly desired to overcome the problems related to Se pollution by industrial waste. To this end it has been proposed by the researchers that treatment of industrial discharge is necessary before disposal. On the other hand it can be minimized by reducing the concentration of Se level in the soil by encouraging use of gypsum, OM such as mud, and poultry manure. Besides this, there is also more-advanced in-situ bioremediation by different bacterial cultures to reduce or tolerate Se concentrations in soil.

References

Abrams, M.M., Burau, R.G., and Zasoski, R.J. (1990). Organic selenium distribution in selected California soils. *Soil Sci. Soc. Am. J.* 54: 979–982.

Allaway, W.H., Cary, E.E., and Ehlig, C.F. (1967). The cycling of low levels of selenium in soils, plants and animals. In: *Selenium in Biomedicine: A Symposium* (ed. O.H. Muth). Westport, CT: AVI.

Amweg, E.L., Stuart, D.L., and Weston, D.P. (2003). Comparative bioavailability of selenium to aquatic organisms after biological treatment of agricultural drainage water. *Aquat. Toxicol.* 63: 13–25.

Audi, G., Bersillon, O., Blachot, J., and Wapstra, A.H. (2003). The NUBASE evaluation of nuclear and decay properties. *Nucl. Phys. A* 729: 3–128.

Bailey, R.T., Hunter, W.J., and Gates, T.K. (2012). The influence of nitrate on selenium in irrigated agricultural groundwater systems. *J. Environ. Qual.* 41: 783–792.

Bajaj, M., Eiche, E., Neumann, T. et al. (2011). Hazardous concentrations of selenium in soil and groundwater in north-west India. *J. Hazard. Mater.* 189: 640–646.

Balistrieri, L.S. and Chao, T.T. (1990). Adsorption of selenium by amorphous iron oxhydroxide and manganese dioxide. *Geochim. Acta.* 54: 739–751.

Bar-Yosef, B. and Meek, D. (1987). Selenium desorption from Ca-kaolinite. *Commun. Soil Sci. Plant Anal.* 18 (7): 771–779.

Berrow, M.L. and Ure, A.M. (1989). Geologic material and soil. In: *Occurrence and Distribution of Selenium* (ed. M. Inat), 213–242. Boca Raton, FL: CRC Press, Inc.

Berzelius, J.J. (1818). Recherches sur un nouveau corps mineral trouve dans le souffre fabrique a Fahlun. *Annales de Chimie et de Physique* 9: 160–180. (225–267, 337–365).

Bruggeman, C., Maes, A., Vancluysen, J., and Vandenmussele, P. (2005). Selenite reduction in boom clay: effect of FeS_2, clay minerals and dissolved organic matter. *Environ. Pollut.* 137: 209–221.

Bruggeman, C., Maes, A., and Vancluysen, J. (2007). The interaction of dissolved boom clay and gorleben humic substances with selenium oxyanions (selenite and selenate). *Appl. Geochem.* 22: 1371–1379.

Butterman, W.C., Jr. and Brown, R.D. (2004). Mineral Commodity Profiles: Selenium, U.S. Geological Survey Open-File Report 03-018, Reston, VA: US Government, pp. 1–19.

Calderone, S.J., Frankenberger, W.T., Parker, D.R., and Karlson, U. (1990). Influence of temperature and organic amendments on the mobilization of selenium in sediments. *Soil Biol. Biochem.* 22: 615–620.

Carter, D.L., Robbinson, C.W., and Brown, M.J. (1972). Effect of phosphorus fertilization on the selenium concentration in alfalfa. *Soil Sci. Soc. Am. Proc.* 36: 624–628.

Chabroullet, C. (2007). Etude de la remobilisation d'elements traces a partir d'un sol de surface contamine: influence du vieillissement des composes organiques du sol sur la remobilization du selenium. Thesis. University of Grenoble, 234.

Chizhikov, D.M. and Shchastlivyi, V.P. (1968). *Selenium and Selenides*. London: Collet's.

Christensen, B.T., Bertelsen, F., and Gissel-Nielsen, G. (1989). Selenite fixation by soil particle-size separates. *J. Soil Sci.* 40: 641–647.

Cooke, T.D. and Bruland, K.W. (1987). Aquatic chemistry of selenium: evidence of biomethylation. *Environ. Sci. Tech.* 21: 1214–1219.

Cooper, W.C., Bennett, K.G., and Croxton, F. (1970). The history, occurrence, and properties of selenium. In: *Selenium* (eds. R.A. Zingaro and W.C. Cooper), 3–27. New York;: Van Nostrand Reinhold.

Coppinger, R.J. and Diamond, A.M. (2001). Selenium deficiency and human disease. In: *Selenium: Its Molecular Biology and Role in Human Health* (ed. D.L. Hatfield), 219–331. Netherlands: Kluwer Academic Press.

Crystal, R.G. (1972). Elemental selenium: structure and properties. In: *Organic Selenium Compounds: Their Chemistry and Biology* (eds. D.L. Klayman and W.H.H. Gunther), 1188. New York: Wiley.

Cutter, G.A. (1982). Selenium in reducing waters. *Science* 217: 829–831.

Cutter, G.A. and Bruland, K.W. (1984). The marine biogeochemistry of selenium – a re-evaluation. *Limnol. Oceanogr.* 29: 1179–1192.

Cutter, G.A. and San Diego-McGlone, M.L.C. (1990). Temporal variability of selenium fluxes in San Francisco Bay. *Sci. Total Environ.* 97/98: 235–250.

Dhillon, K.S. and Dhillon, S.K. (2003). Distribution and management of seleniferous soils. In: *Advances in Agronomy*, 1ee (ed. D. Sparks), 119–184. Amsterdam: Elsevier.

Dhillon, S.K., Dhillon, K.S., Kohli, A., and Khera, K.L. (2008). Evaluation of leaching and runoff losses of selenium from seleniferous soils through simulated rainfall. *J. Plant Nutr. Soil Sci-Zeitschrift Fur Pflanzenernahrung Und Bodenkunde* 171: 187–192.

Dungan, R.S., Yates, S.R., and Frankenberger, W.T. (2002). Volatilization and degradation of soil-applied dimethylselenide. *J. Environ. Qual.* 31: 2045–2050.

Dynes, J.J. and Huang, P.M. (1997). Influence of organic acids on selenite sorption by poorly ordered aluminum hydroxides. *Soil Sci. Soc. Am. J.* 61: 772–783.

Eich-Greatorex, S., Krogstad, T., and Sogn, T.A. (2010). Effect of phosphorus status of the soil on selenium availability. *J. Plant Nutr. Soil Sci.* 173: 337–344.

Eswayah, A.S., Smith, T.J., and Gardiner, P. (2016). Microbial transformations of selenium species of relevance to bioremediation. *Appl. Environ. Microbiol.* 82 (16): AEM.00877-16.

Fan, T.W.-M. and Higashi, R.M. (1998). Biochemical fate of selenium in microphytes: natural bioremediation by volatilization and sedimentation in aquatic environments. In: *Environmental Chemistry of Selenium* (eds. W.T. Frankenberger Jr. and R.A. Engberg), 545–564. New York: Marcel Dekker Inc.

Fernández-Martínez, A. and Charlet, L. (2009). Selenium environmental cycling and bioavailability: a structural chemist point of view. *Rev. Environ. Sci. Biotechnol.* 8: 81–110.

Ferri, T. and Sangiorgio, P. (2001). Selenium speciation in waters: role of dissolved polysaccharides on the mobilization process. *Ann. Chim.* 91: 229–238.

Filella, M. and Town, R.M. (2003). Implications of natural organic matter binding heterogeneity on understanding trace metal complexation in aquatic systems. *J. Phys. IV France* 107: 479–482.

Fleming, G.A. (1962). Selenium in Irish soils and plant. *Soil Sci.* 91 (1): 28–35.

Fleming, G.A. (1980). Essential micronutrients II: iodine and selenium. In: *Applied Soils Trace Elements*, 1ee (ed. B.E. Davis), 199–234. New York: Wiley.

Fordyce, F. (2005). Selenium deficiency and toxicity in the environment. In: *Essentials of Medical Geology*, 1ee (ed. O. Selinus), 373–415. Uppsala: Elsevier.

Fordyce, F. (2007). Selenium geochemistry and health. *Ambio* 36: 94–97.

Frost, D.V. (1967). Significance of the symposium. In: *Symposium on Selenium and Bio-Medicine* (ed. O.H. Muth), 7–26. Westport, CT: Avi Publishing Co.

Frost, D.V. (1973). The selenium cycle and health. Paper delivered at the 66th Annual Meeting of the Air Pollution Control Association, Chicago.

Ghosh, A., Mohod, A.M., Paknikar, K.M., and Jain, R.K. (2008). Isolation and characterization of selenite- and selenate tolerant microorganisms from selenium-contaminated sites. *World J. Microbiol. Biotechnol.* 24: 1607–1611.

Gissel-Nielsen, G. (1971). Influence of pH and texture of the soil on plant uptake of added selenium. *J. Agric. Food Chem.* 19 (6): 1165–1167.

Gissel-Nielsen, G. (1975). Foliar application and pre-sowing treatment of cereals with selenite. *Z. Pflanzenern. Bodenk.* 1: 97–105.

Gustafsson, J.P. and Johnson, L. (1992). Selenium retention in the organic matter of Swedish forest soils. *J. Soil Sci.* 43: 461–472.

Hamdy, A.A. and Gissel-Nielsen, G. (1976). Fractionation of soil selenium. *Z. Pflanzenern. Bodenkd.* 6: 697–703.

Hamilton, S.J. (2004). Review of selenium toxicity in the aquatic food chain. *Sci. Total Environ.* 326: 1–31.

Haygarth, P.M. (1994). Global importance and global cycling of selenium. In: *Selenium in the Environment* (eds. W.T. Frankenberger Jr. and S. Benson), 1–28. Hong Kong: Marcel Dekker, Inc.

IAEA. (1994). *Handbook of Parameter Values for the Prediction of Radionuclides Transfer in Temperate Environments*. Vienne: International Atomic Energy Agency.

Ihnat, M. (1989). *Occurence and Distribution of Selenium*. Boca Raton, FL: CRC Press.

Irons, R., Carlson, B.A., Hatfield, D.L., and Davis, C.D. (2006). Both selenoproteins and low molecular weight selenocompounds reduce colon cancer risk in mice with genetically impaired selenoprotein expression. *J. Nutr.* 136: 1311–1317.

Jacobs, L.W. (1989). Selenium in agriculture and the environment. Soil Science of America Special Publication, Wisconsin.

Johnson, C.M. (1975). Selenium in soils and plants: contrast in conditions providing safe but adequate amounts of selenium in the food chain. In: *Trace Elements in Soil-Plant-Animals Systems* (eds. D.J.D. Nicholas and A.R. Egan), 137–158. New York: Academic Press.

Jovari, P., Delaplane, R.G., and Pusztai, L. (2003). Structural models of amorphous selenium. *Phys. Rev. B* 67: 172201.

Kabata-Pendias, A. (2011). *Trace Elements in Soils and Plants*, 4ee. Boca Raton, FL: Taylor and Francis Group.

Kang, Y., Yamada, H., Kyuma, K. et al. (1991). Selenium in soil humi-acid. *Soil Sci. Plant Nutr.* 37: 241–248.

Khattak, R.A., Page, A.L., Parker, D.R., and Bakhtar, D. (1991). Accumulation and interactions of arsenic, selenium, molybdenum and phosphorus in alfalfa. *J. Environ. Qual.* 20: 165–168.

Korom, S.F. (1992). Natural denitrifi cation in the saturated zone: a review. *Water Resour. Res.* 28: 1657–1668. https://doi.org/10.1029/92WR00252.

Kumar, A.R. and Riyazuddin, P. (2011). Speciation of selenium in groundwater: seasonal variations and redox transformations. *J. Hazard. Mater.* 192: 263–269.

Lakin, H.W. (1972). Selenium accumulation in soils and its absorption by plants and animals. *Geol. Soc. Am. Bull.* 83: 181–189.

Lakin, H.W., and Davidson, D.F. (1967). The relation of the geochemistry of selenium to its occurrence in soils. In: *Selenium in Biomedicine: A Symposium* (ed. O.H. Muth), 27. Westport, CT: AVI.

Lenz, M., Gmerek, A., and Lens, P.N.L. (2006). Selenium speciation in anaerobic granular sludge. *Int. J. Environ. Anal. Chem.* 86: 615–627.

Levander, O.A. and Burk, R.F. (2006). Update of human dietary standards for selenium. In: *Selenium its Molecular Biology and Role in Human Health* (eds. D.L. Hatfield, M.J. Berry and V.N. Gladyshev), 399–410. New York;: Springer.

Li, J. and Zheng, C.J. (1990). *The Handbook of the Environmental Background Values*. Beijing: China Environmental Science Press.

Losi, M.E. and Frankenberger, W.T. (1997). Reduction of selenium oxyanions by Enterobacter cloacae SLD1a-1: isolation and growth of the bacterium and its expulsion of selenium particles. *Appl. Environ. Microbiol.* 63: 3079–3084.

Luoma, S.N., Johns, C., Fisher, N. et al. (1992). Determination of selenium bioavailability to a benthic bivalve from particulate and solute pathways. *Environ. Sci. Tech.* 26: 485–491.

Masscheleyn, P.H., Delaune, R.D., and Patrick, W.H. (1990). Transformations of selenium as affected by sediment oxidation-reduction potential and pH. *Environ. Sci. Technol.* 24: 91–96.

Mayer, M.A. (2010). Band structure engineering of ZnO1−xSex alloys. *Appl. Phys. Lett.* 97: 022104.

Mayland, H.F., James, L.J., Panter, K.E. and Sonderegger, J.L. (1989). Selenium in seleniferous environment. In: *Selenium in Agriculture and the Environment*. SSSA Special Publication 23 (ed. L.W. Jacobs), 15–50. Madison, WI: ASA and SSSA.

McNeal, J.M. and Balistrieri, L.S. (1989). Geochemistry and occurrence of Se: an overview. Se in agriculture and the environment. *Soil Sci. Soc. Am. J.* 23: 1–13.

Mikkelsen, R.L., Page, A.L., and Bingham, F.T. (1989). Factors affecting selenium accumulation by agricultural crops. In: *Selenium in Agriculture and the Environment*. SSSA Special Publication 23 (ed. L.W. Jacobs), 65–94. Madison, WI: ASA and SSSA.

Milne, J.B. (1998). The uptake and metabolism of inorganic selenium species. In: *Environmental Chemistry of Selenium*, 1ee (eds. W.T. Frankenberger and R.A. Engberg), 315. New York: Marcel Dekker.

Monteil-Rivera, F., Fedoroff, M., Jeanjean, J. et al. (2000). Sorption of selenite (SeO_3^{2-}) on hydroxyapatite: an exchange process. *J. Colloid Interface Sci.* 221: 291–300.

Moxon, A.L., Olson, O.E. and Searight, W.V. (1939). Selenium in rocks, soils and plants. South Dakota Agric. Exp. Stn. Tech. Bull. 2:94.

Nakamaru, Y.M. and Sekine, K. (2008). Sorption behavior of selenium and antimony in soils as a function of phosphate ion concentration. *Soil Sci. Plant Nutr.* 54 (3): 332–341.

Nakamaru, Y., Tagami, K., and Uchida, S. (2005). Distribution coefficient of selenium in Japanese agricultural soils. *Chemosphere* 58: 1347–1354.

Nakamaru, Y., Tagami, K., and Uchida, S. (2006). Effect of phosphate addition on the sorption-desorption reaction of selenium in Japanese agricultural soils. *Chemosphere* 63: 109–115.

National Academy of Science (NAS) (2000). *Selenium, in Dietary Reference Intakes for Vitamin C, Vitamin E, Selenium, and Carotenoids*, 284–324. Washington, DC: National Academy of Sciences, National Academy Press.

National Research Council (NRC). (1983). Distribution. In: *Selenium in Nutrition*. 2nd rev. ed, 10–39. Washington, DC: National Academy of Sciences.

Neal, R.H. (1995). *Selenium*. London: Blackie Academic & Professional.

Nriagu, J.O. (1989). *Occurrence and Distribution of Selenium*. Florida: CRC Press.

Nriagu, J.O. (1991). *Heavy Metals in the Environment*. Edinburgh: CEP Consultants.

Ogaard, A.F. (1996). Effect of fresh and composted cattle manure on phosphate retention in soil. *Acta Agric. Scand. B. Soil Plant Sci.* 46: 98–105.

Ogaard, A.F., Sogn, T.A., and Eich-Greatorex, S. (2006). Effect of cattle manure on selenate and selenite retention in soil. *Nutr. Cycl. Agroecosyst.* 76: 39–48.

Olson, O.E. (1967). Soil, plant, animal cycling of excessive levels of selenium. In: *Selenium in Biomedicine: A Symposium* (ed. O.H. Muth), 297. Westport, CT: AVI.

Paikaray, S. (2016). Origin, mobilization and distribution of selenium in a soil/water/air system: a global perspective with special reference to the Indian scenario. *CSAWAC* 44(5):451–586.

Peak, D. and Sparks, D.L. (2002). Mechanisms of selenate adsorption on iron oxides and hydroxides. *Environ. Sci. Technol.* 36: 1460–1466.

Pedrero, Z. and Madrid, Y. (2009). Novel approaches for selenium speciation in foodstuffs and biological specimens: a review. *Anal. Chim. Acta.* 634: 135–152.

Perrone, D., Monteiro, M., and Nunes, J.C. (2015). The chemistry of selenium. In: *Selinium: Chemistry, Analysis, Function and Effects* (ed. V.R. Preedy), 3–18. London: The Royal Society of Chemistry.

Plant, J.A., Kinniburgh, D., Smedley, P. et al. (2004). Arsenic and selenium. In: *Treatise on Geochemistry Series*, vol. 9 (eds. H.D. Holland and K.K. Turekain), 17–66. Amsterdam: Elsevier.

Presser, T.S. (1994). The Kesterson effect. *Environ. Manag.* 18: 437–454.

Presser, T.S. and Ohlendorf, H.M. (1987). Biogeochemical cycling of selenium in the San Joaquin Valley, California, USA. *Environ. Manage.* 11: 805–821.

Presser, T.S. and Piper, D.Z. (1998). Mass balance approach to selenium cycling through the San Joaquin valley: from source to river to bay. In: *Environmental Chemistry of Selenium*, 1ee (eds. W.T. Frankenberger and R.A. Engberg), 153–182. New York: CRC Press.

Qin, H.B., Zhu, J.M., and Su, H. (2012). Selenium fractions in organic matter from Se-rich soils and weathered stone coal in selenosis areas of China. *Chemosphere* 86 (6): 626–633.

Rader, L.F. Jr. and Hill, W.L. (1935). Occurance of selenium in natural phosphates, superphosphates and phosphoric acid. *J. Agric. Res.* 51: 1071.

Ralston, N.V.C., Unrine, J. and Wallschläger, D. (2009). *Biogeochemistry and Analysis of Selinium and its Species*. North American Metals Council.

Reeder, R.J., Schoonen, M.A.A., and Lanzirotti, A. (2006). Metal speciation and its role in bioaccessibility and bioavailability. In: *The Emergent Field of Medical Mineralogy and Geochemistry*, 1ee (ed. J.J. Rosso), 59–113. Mineralogical Society of America and the Geochemical Society, USA.

Robbins, C.W. and Carter, D.L. (1970). Selenium concentrations in phosphorus fertilizer materials and associated uptake by plants. *Soil Sci. Soc. Am. Proc.* 34: 506.

Rosenfeld, I. and Beath, O.A. (1964). *Selenium: Geobotany, Biochemistry Toxicity and Nutrition*. New York;: Academic Press.

Rotruck, J.T., Pope, A.L., Ganther, H.E., and Hoekstra, W.G. (1973). Selenium: biochemical role as a component of glutathione peroxidase. *Science* 179: 588–590.

Ruby, M.V., Davis, A., Schoof, R. et al. (1996). Estimation of lead and arsenic bioavailability using a physiologically based extraction test. *Environ. Sci. Technol.* 30: 422–430.

Russell, B. (2011). Selenium stories. *Nat. Chem.* 3: 570.

Schwarz, K. and Foltz, C.M. (1957). Selenium as an integral part of factor 3 against dietary necrotic liver degeneration. *J. Am. Chem. Soc.* 79: 3292–3293.

Seby, F., Gautier, M.P., Lespes, G., and Astruc, M. (1997). Selenium speciation in soils after alkaline extraction. *Sci. Total Environ.* 207: 81–90.

Seiler, R.L. (1995). Prediction of areas where irrigation drainage may induce selenium contamination of water. *J. Environ. Qual.* 24: 973–979.

Seiler, R.L. (1998). Prediction of lands susceptible to irrigation induced contamination of water. In: *Environmental Chemistry of Selenium* (eds. W.T. Frankenburger and R.A. Engberg), 397–418. New York, NY: Marcel Dekker.

Sharma, M.C. and Kumar, P. (2006). Impact of pollution on livestocks of adjoining areas of Barauni industrial area in northern India. *Toxicol. Lett.* 164: S178–S179.

Shrift, A. (1964). A selenium cycle in nature. *Nature* 201: 1304.

Singh, M. (1979). Effect of selenium and phosphorus on the growth and chemical composition of raya (Brassica juncea). *Plant Soil* 51: 485–490.

Stolz, J.F., Basu, P., and Oremland, R.S. (2002). Microbial transformation of elements: the case of arsenic and selenium. *Int. Microbiol.* 5: 201–207.

Stork, A., Jury, W.A., and Frankenberger, W.T. (1999). Accelerated volatilization rates of selenium from diff erent soils. *Biol. Trace Elem. Res.* 69: 217–234.

Sutton, R. and Sposito, G. (2005). Molecular structure in soil humic substances: the new view. *Environ. Sci. Technol.* 39: 9009–9015.

Takayanagi, K. and Wong, G.T.F. (1984). Total selenium and selenium (IV) in the James River estuary and southern Chesapeake Bay. *Estuarine, Coastal Shelf Sci.* 18: 113–119.

Tam, S.C., Chow, A., and Hadley, D. (1995). Effects of organic-component on the immobilization of selenium on iron oxyhydroxide. *Sci. Total Environ.* 164: 1–7.

Tamari, Y., Ogawa, H., Fukumoto, Y. et al. (1990). Selenium content and its oxidation-state in igneous rocks, rock-forming minerals, and a reservoir sediment. *Bull. Chem. Soc. Jpn.* 63: 2631–2638.

Thompson-Eagle, E.T. and Frankenburger, W.T. Jr. (1992). Bioremediation of soils contaminated with selenium. *Adv. Soil Sci.* 17: 261–310.

Thomson, C.D. (2003). Selenium: physiology. In: *Encyclopaedia of Food Science, Food Technology and Nutrition* (eds. B. Caballero, L.C. Trugo and P.M. Finglas), 5117–5124. London: Academic Press.

Thomson, C.D. (2004). Assessment of requirements for selenium and adequacy of selenium status: a review. *Eur. J. Clin. Nutr.* 58: 391–402.

Tolu, J., Thiry, J., Bueno, M. et al. (2014). Distribution and speciation of ambient selenium in contrasted soils, from mineral to organic rich. *Sci. Total Environ.* 479–480: 93–101.

US Environmental Protection Agency. (2008). Washington, DC, EPA/600/R-08/130. . . . American Chemical Society, Washington, DC, 68(4):647–652, (1996).

Velinsky, D.J. and Cutter, G.A. (1991). Geochemistry of selenium in a coastal salt marsh. *Geochim. Cosmochim. Acta* 55: 179–191.

Wang, Z.J. and Gao, Y.X. (2001). Biogeochemical cycling of selenium in Chinese environments. *Appl. Geochem.* 16: 1345–1351.

Wang, Z.J., Xu, Y., and Peng, A. (1996). Influences of fulvic acid on bioavailability and toxicity. *Biol. Trace Elem. Res.* 55: 147–162.

WHO (2005). Selenium. 194–216. Geneva: World Health Organization.

Wijnja, H. and Schulthess, C.P. (2000). Interaction of carbonate and organic anions with sulfate and selenate adsorption on an aluminum oxide. *Soil Sci. Soc. Am. J.* 64: 898–908.

Williams, C. and Thoronton, I. (1972). The effect of soil additives on the uptake of molibdenum and selenium from soils from different environments. *Plant Soil* 39: 395–406.

Yang, G., Wang, S., and Zhon, R. (1983). Endemic selenium intoxication of humans in China. *Am. J. Clin. Nutr.* 37: 872–881.

Zhang, Y.Q. and Moore, J.N. (1996). Selenium fractionation and speciation in a wetland system. *Environ. Sci. Technol.* 30: 2613–2619.

Zuyi, T., Taiwei, C., Jinzhou, D. et al. (2000). Effect of fulvic acids on sorption of U(VI), Zn, Yb, I and Se(IV) onto oxides of aluminum, iron and silicon. *Appl. Geochem.* 15: 133–139.

3

Occurrence and Sources of Selenium Contamination in Soil and Water and its Impacts on Environment

Rashmi Dahake[1,2], *Amit Bansiwal*[1], *and Asmita Jadhav*[1]

[1] *CSIR-National Environmental Engineering Research Institute (CSIR-NEERI), Nagpur, Maharashtra, India*
[2] *Academy of Scientific and Innovative Research (AcSIR), Ghaziabad, India*

3.1 Introduction

Selenium (Se) is considered to be a vital element for living organisms. In 1817 a Swedish chemist, Jöns Jakob Berzelius, discovered this element during sulfuric acid production. The name given was complementary to its sister element tellurium. The meaning of tellurium in Greek is "earth" and selenium is named after the Greek moon goddess, Selene. As Se is an essential micronutrient, its importance has been grown widely in recent years. The global estimates state that approximately 3000 million population of the world suffer from Se deficiency and are prone to the various health impacts. The Se-deficient countries reported so far include USA, Canada, Japan, China, Finland, Scotland, New Zealand, and Spain (Fordyce 2013). Se requirements in human and animals can be met by consuming food which contains Se. Consumption of crops such as cereals, wheat, and vegetables which contain Se in small concentrations can fulfill the Se requirement of a whole population (Awika 2011). The presence of Se in human populations is majorly dependent upon the level in soil and food that is being produced in such soil. The addition of Se in the food chain is desirable and acceptable when it is in low concentration; however, the increased level of Se leads to an adverse effect on the human population and environment (Rayman 2008).

Dietary Se in human and animals has its own advantages, but overconsumption or exposure can lead to harmful effects. Concentration and speciation of the Se is the deciding factor for the impact of Se on environment. There is a great amount of literature available on Se which majorly focuses on nutrition and its occurrence in food chain. In recent years Se has emerged as a pollutant which may contaminate and cause severe toxicity to the environment. Researchers have started evaluating the occurrence, sources, and harmful effect of Se and are working on different mitigating, removal, and remediation technologies. Brief insights into Se as an emerging contaminant in water and soil are covered in this chapter, along with its natural and anthropogenic sources and uses.

Selenium Contamination in Water, First Edition. Edited by Pooja Devi, Pardeep Singh, Arindam Malakar, and Daniel Snow.

3.1.1 Chemistry of Se

Se is a metalloid with atomic number 34 and molecular mass of 78.971 g/mol, which means it is heavier than Fe, Ni, and Mn, and lighter than Ag, Pb, and Au. In the periodic table it occurs in fourth period and belongs to the XVI (chalcogen) group. Se occurs in different allotropic forms such as amorphous, crystalline, and metallic which varies its physical properties. The two distinct allotropes of amorphous Se are finely divided brick red form and vitreous form. The only difference between these two allotropes is the route of production; otherwise at microscopic level they appear alike. The crystalline allotropes appear in various monoclinic varieties such as red to brown in color and metallic gray or black Se. Both monoclinic and amorphous forms of Se converts spontaneously at 70–120 °C, whereas the hexagonal crystalline allotrope is thermodynamically most stable at temperatures above 110 °C. Se exists in different oxidation states due to its strong metallic properties. It can be reduced to the selenide (Se^{-2}) oxidation state or oxidized to the +4, i.e. selenite (SeO_3^{2-}) or +6, i.e. selenate (SeO_4^{2-}) oxidation states (Fernández-Martínez and Charlet 2009).

It is a trace element with five stable isotopes ^{74}Se, ^{76}Se, ^{77}Se, ^{78}Se, ^{80}Se reported till date, based on half-life ranging from 20 ms to 295 000 years whereas 24 unstable Se isotopes are present in nature (Aston 1922). The nuclear waste disposal research has drawn attention toward the fission product of ^{235}U which has the isotope of Se as ^{79}Se. ^{82}Se has a very long half-life of approximately 10^{20} years, hence it is considered as the stable isotope which yields ^{80}Kr through double beta decay.

Se shows similar chemical behavior to sulfur and similar physiological properties to arsenic. The valance electronic configuration of Se ([Ar] $3d^{10}$ $4s^2$ $4p^4$) is similar to that of sulfur and shows that the properties such as size of atom both in ionic and covalent states, electronegativity, ionization potential, and bond energies are similar to that of sulfur, but in biological systems where sulfur tends to reduce, Se tends to oxidize (Wessjohann et al. 2007).

3.1.2 Forms of Se

The occurrence of Se in the environment is mostly in two forms, i.e. organic and inorganic Se. The organic form includes selenoaminoacids, selenopeptides, and selenoproteins whereas different minerals contain it in the form of selenites and selenates, which are inorganic forms of Se. The inorganic forms are more toxic than organoSe compounds and have less bioavailability. The main classes of organoSe compounds include selenols, diselenides, selenoxides, selenones, Se acids, selenides, Se halides, and selenaheterocyclic compounds (Back 2006).

3.1.3 Chemical and Physical Properties

Se is a fairly reactive element. It is 16 Group element with a melting point of 317.3 K, a boiling point of 553.7 K, and density of 1.823 g/cm^3. Se is placed in the oxygen group of the periodic table. It usually occurs in combination with metals like mercury, copper, silver, or lead (Guoxiong Hua 2011).

It combines easily with hydrogen, fluorine, chlorine, and bromine. It reacts with nitric and sulfuric acids. It also combines with a number of metals to form compounds called selenides such as magnesium selenide (MgSe). Se dioxide has a characteristic odor of rotten horseradish.

The Pauling scale electronegativity of Se is 2.55, which is slightly lower than oxygen due to the size of atom indicated by van der Waals and ionic radius as 0.14 and 0.198 nm respectively. The energy required to remove the first electron from the valence shell is 940.7 kJ/mol, which is quite lower than that required for the second and third electrons, as removal of one electron leads to stability of the atom (Brown 1914). Se and tellurium are often associated with each other. They tend to occur together in the Earth and have somewhat similar properties.

3.2 Sources and Occurrence of Se in the Environment

Se is widely present in nature. The distribution of Se in soil, water, and air is either by natural routes or by anthropogenic activities. Plant–soil environments undergo different Se functions, speciation, metabolism, and biological transformation which have different consequences in human and animal health (El-Ramady et al. 2014a). Within the environment, transformation of Se is continuously happening from one chemical form to another. Some of the sources are presented below.

3.2.1 Natural Sources

A variety of elements in the environment are combined with Se due to its complex chemical behavior and these Se compounds are widespread in the rocks, soils, water, and air. It is distributed in the atmosphere by processes such as volcanic activity, hot springs, weathering of soils and rocks, forest wildfires, movement of groundwater, chemical and biological redox reactions beneath the soil, and mineral formations. The occurrence of Se in the soil or in water sources depends mainly on the climate, topography, and parental material (El-Ramady et al. 2014b). Minerals such as sulfide minerals are associated with selenite- and selenide-containing rocks. The erosion of these rocks is responsible for the occurrence of Se in soils in the form of elemental Se, for instance selenite salts, ferric selenite, or in its organic form (Malisa 2001). The anionic forms of Se, i.e. selenite and selenate, are highly soluble, bio-available, mobile, and toxic above certain limits. Decomposition of plant which accumulate Se are the sources of organic forms of Se.

The available level of Se in soil is mainly dependent on the texture of soil, rainfall, the kind of soil, and the concentration of organic matter. The lowest concentration of Se is obtained in sandy soils while the highest is in organic and calcareous soil. The major Se-controlling factors in the soil are Eh and pH whereas other parameters like organic ligands, clay, and hydroxides also play a significant role. The worldwide occurrence of Se is in very broad range from 0.005 to 3.5 mg/kg. The concentration of Se in plants depends on the surrounding soil's Se levels (Sakizadeh et al. 2016). A greater amount of selenate is taken up by plants than selenite and it is metabolized in chloroplasts by pathways similar to sulfur.

Se originates in water from the soil leaching and atmospheric deposits. The Se can form a stable complex with particulate or colloidal matter or sediments or can be dissolved in water. The movement of Se from the top layer is governed by the transport phenomenon and hence Se can occur in water as suspended or dissolved fraction (El-Ramady et al. 2014b).

The concentration of Se in sea water varies between 0.1 to 0.35 μg/l (Gaillardet et al. 2003). Natural waters have a concentration less than 1 μg/l (Conde and Sanz Alaejos 1997). Global Se average concentration in river waters is given in the range of 0.02–0.5 μg/l while in groundwater and surface water it ranges from 0.06 to 400 μg/l (World Health Organization 2011). The concentration of Se in water is dependent upon pH. The compounds which are soluble in water are converted at high and low pH and this results in an increase in concentration.

Se can be transported to the air by natural processes such as volcanic eruptions, soil erosion, forest fires, and evaporation from ocean and sea. Mosses and peats in marine regions with increased Se levels when volatilized release Se into the air in the form of hydrogen selenide and elemental Se, selenites, and selenates in particulate form. Hydrogen selenide and Se dioxide are unstable in the air and are converted by oxidation into Se and H_2O and into selenicious acid in moist conditions (Belcher et al. 1980). The ambient air Se concentration is generally very low and varies from 0.1 to 10 ng/m^3 (Lee and Duffield 1979; Gilbert and Fornes 1980).

The presence of Se in grains and vegetables is mainly accountable for the Se contents of soil in which they grow. The natural concentration of Se is very low, i.e. approximately 6 mg/g in vegetables like carrot, tomatoes, potatoes, cucumber, etc., even if they are grown in seleniferous soils. Some vegetables effectively accumulate Se from seleniferous soil, for instance onions and asparagus accumulate up to 17 μg/g. Fruits accumulate very low Se which is less than 10 μg/kg (Whanger 2004).

3.2.2 Anthropogenic Sources of Contamination in Environment

Some industrial waste and processes may contribute for the addition of high levels of Se in the environment. The major source of Se contamination in air is the combustion of fossil fuels (Shah et al. 2007). The elemental Se burns to form Se dioxide in the air. The main atmospheric Se emission includes oil refining factories, mining, and milling end product manufacturing. In 1978 Harr has reported that the fossil fuel burning, emissions from industries, and municipal wastes release almost 1500, 2700, and 360 tons of Se annually, respectively (Fishbein 1983). In Canada nonferrous smelters emitted 3.02 tons of Se in 1993 as reported by Skeaff and Dubreuil. Different kind of paper also contains Se (Skeaff and Dubreuil 1997).

Incineration of papers, municipal waste, and rubber tyres are additional sources of Se in the atmosphere. Few of the sources of Se in the atmosphere have been quantified yet. Fugitive dust emissions from the fly ash settling ponds and hazardous waste sites where Se compounds were disposed of are the potential sources of Se in air (Santhanam et al. 1979). In addition to the burning of fossil fuels, smelting, mining, the disposal of industrial waste, and the processing of nuclear fuel waste also release high levels of Se into the environment. Fossil fuels and coal contain significant amounts of Se which will be added to the atmosphere during its processing in industries. The Se is present in the crude oil in the form of seleniferous marine shale which contains a fairly good amount of Se which is sent into the atmosphere through waste water generated during refining process. Levels of release of Se through different refining processes are different. The desalting process releases less Se, and atmospheric distillation causes elemental Se to be formed and partitioned with suspended matter, and processed stream water discharges high levels of Se from hydrotreatment and catalytic cracking units. In the end step, the processed oil still contained increased levels of Se, which was examined during sour water treatment (Willig 2014).

Many electronics industrial waste also contains Se as Se is an essential component in the electronics accessories such as capacitors, printers/toners, and photocopiers etc. It is also used in semiconductors, photoelectric cells, glass, and pharmaceuticals production, the waste of which goes into the landfills and incineration and is added to environment. Leachates obtained from landfilling units of such industries wastes show increased levels of Se.

The radioactive isotope of Se, i.e. $^{79}Se_{34}$ may also be a component of nuclear waste. The release of this isotope contributes Se content in soil, plants, and water when accidentally released from nuclear waste repositories or power plants. Smelting of ores such as nickel, zinc, and copper also releases Se in atmosphere through volatization. The Se concentration in the environment is increasing day by day and may lead to environmental pollution which is very harmful for human beings (Duce 1987).

3.3 Drinking Water Standards and Criteria

The permissible limit of Se in the drinking water was fixed by different countries based on the dietary intake through food and water. The permissible limit for drinking water is 10 times more than that for surface water as aquatic animals require less Se than mammals (Barron et al. 2009). World Health Organization (WHO) has set the standard for drinking water at 40 μg/l (WHO 2012). The permissible value set by the United States Environment Protection Agency (USEPA) is 50 μg/l (EPA 2020). Different countries have set their standards based on the availability of different factors; for example, the Bureau of Indian Standard has fixed the permissible limit to 10 μg/l (BIS 2012). The water quality standards for drinking water of different countries are tabulated in Table 3.1.

3.4 Effect of Se in Human, Terrestrial, and Aquatic Life

Se is a vital necessity to human, aquatic, and terrestrial organisms. The presence of Se in the diet has nutritional benefits to the organisms on the Earth while its deficiency or excess consumption can lead to an adverse effect in humas as well as in aquatic and terrestrial life.

3.4.1 Human Population

Se is an essential micronutrient required for the human body. It is an important part of selenoproteins which will be required for various biological functions such as antioxidants

Table 3.1 Drinking water and surface water limits of different countries (EPA 2020; WHO 2012).

	Limit (μg/l)	
Country	**Drinking water**	**Surface water**
United States	50	5
Canada (British Columbia, Ontario, Quebec)	10	–
Canada (Alberta, Nova Scotia, Prince Edward Island)	50	–
South Africa	20	–
Germany	10	3
Ireland	10	–
Japan	10	–
India	10	–
Malaysia	10	–
Russia	10	–
Abu Dhabi	40	–
New Zealand	–	10
Australia	–	10

defense, thyroid hormone formation, DNA synthesis, fertility, and reproduction. In organisms Se is converted into different metabolites. Se plays an important role in slowing down the aging process and improvement of endurance of muscles besides vitamin E. Methylselenol also plays a part in cancer prevention (Papp et al. 2007).

The deficiency of Se causes biochemical changes which may lead to additional stress in humans and causes certain type of illness (Du et al. 2012). A combination of Se deficiency and another stress such as viral infection can cause Keshan disease. A disease, namely cardiomyopathy, (also termed as Keshan disease) occurred in China during 1970 due to deficiency of Se. It was recommended that a daily intake of 20 mcg can protect the adults from this disease. Kashin–Beck disease is a type of osteoarthritis which is associated with male infertility due to Se deficiency (Sanmartin et al. 2012).

Se supplementation has an antiviral effect in the human body which lowers the risk of autoimmune thyroid disease. Excessive Se can generate reactive oxygen species which may leads to carcinogenesis. A few pieces of research also specify that high Se concentration may develop tumorigenesis by uplifting TR 1 expression and to diabetes. Besides a close relation with carcinogenesis, Se is related to development of chronic degenerative diseases such as cardiovascular disease and amyotrophic lateral sclerosis (Ralston et al. 2008). Se is 40 times more neurotoxic in inorganic form than in organic form.

3.4.2 Aquatic and Terrestrial Life

Se concentrations characteristically are higher in oceanic organisms than in those from river ecosystems, and they are greater in areas having seleniferous soils or deposits along with the areas with industrial, agricultural, or municipal wastes dump sites. It is observed that the egg-laying vertebrates, namely fish and water birds, are the most sensitive organisms and could be at greatest risk in Se-contaminated environments. These organisms at the base of the food web have been detected to concentrate Se 10^2 to 10^6 times exceeding that present in the water column (Sharma et al. 2017). There are some reports of high levels of Se in fish, primarily from seleniferous areas of the western United States and from reservoirs contaminated by fly ash from combustion of coal. The transfer of organic Se to eggs in egg-laying vertebrates results in the production of reactive species that can cause not only oxidative stress but cellular dysfunction, which results in embryo mortality or teratogenic malformations observed in fish larvae and its population reduction (Chapman 2009).

Industries such as mining, agriculture, and petrochemicals discharge the waste generated by different operations into the water bodies, which causes contamination of Se in water. Generally the Se background concentrations in terrestrial invertebrates is found to be 0.1–2.5 mg/kg. Earthworms were found to bioaccumulate elevated concentrations of Se from selenite-enriched soil (up to 7.5 mg Se/kg). There are fewer reports regarding terrestrial habitat, Se bioaccumulation, and toxicity.

The USEPA recently recommended strict guidelines for water quality criterion for the protection of aquatic life: 1.2 μg/l in lentic ecosystems (still freshwater) and 3.1 μg/l in lotic ecosystems (Holmes and Gu 2016). These new stringent guidelines may be useful for the protection of aquatic and terrestrial life from Se contamination.

3.4.3 Biological Role

Se forms unusual amino acids: selenocysteine and selenomethionine. In humans Se is a nutrient that functions as a co-factor for reduction of antioxidant enzymes such as thioredoxin reductase and glutathione peroxidises found in some animal and plants (not all plants require Se). Dietary Se also reduces the effects of mercury toxicity, but it will be effective only at low doses of mercury. The excessive poisoning of Se can be measured through determination of Se in body fluids namely blood plasma, urine, and serum. There are some reports regarding cancer patients consuming selenomethionine through medicine that may achieve very high concentration of Se in plasma and urine of a person (Stadtman 1974).

3.5 Industrial Applications

Se is widely used in many industrial applications. More than 80% of global Se is produced as a by-product of the copper refining process. It is collected from the anode residue (Hoffmann 1989). The production of Se through refinery processes has increased to 2800 tons (in the year 2018) due to its huge demand in industrial applications (USGS 2020). Refinery production by country is depicted in Figure 3.1. Ammonium selenite is used in glass manufacturing processes to neutralize coloration that has occurred due to iron oxide. Sodium selenate is also used as the soil additive in Se-deficient areas. Ruby red glass, a variety of glass utilized in tableware, traffic, and other signal lenses and infrared equipment, was manufactured using 1–50 pounds of Se per ton of glass (Weyl 1981). Se in its high purity forms is also employed in semiconductors, specialty transformers, thermo elements, etc.

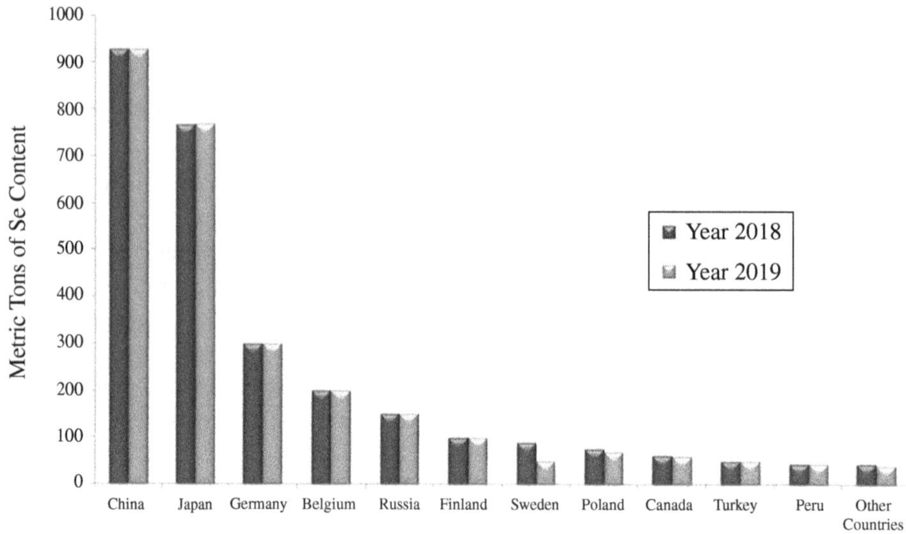

Figure 3.1 Global refinery production for 2018 and 2019 (USGS 2020).

About 23% of the total Se demand is for duplicating machine use. The Se-coated metal cylinders are employed in the dry photographic copying process in which electricity transfers the photographic image (Joannopoulos 1979). For the each 100 sq. feet of copying surface around 4 pounds of Se is needed (Bureau of Mines 1970). The inorganic pigments which are used in plastics, paints, enamels, inks, rubbers, and ceramics are manufactured using cadmium sulfoselenide (Waitkins et al. 1942).

To increase the resilience of rubber and to promote heat, oxidation, and abrasion resistance, Se compounds are used as accelerators and vulcanizing agents in rubber manufacturing (Dudley 1938). Se is added to the stainless steel during manufacturing, which will improve its characteristic properties such as forging, casting, and machining without sacrificing corrosion resistance. Se compounds such Se dioxides catalyze many organic reactions such as hydrogenation, oxidation, isomerization, and polymerization. Se is extensively utilized to help control micro cracking in electroplated chromium as reducing or oxidizing agent (Rashkov et al. 1983). Due to the antioxidant and antigalling properties of Se it is utilized in insecticides, parasiticides, bactericides and herbicides, mercury vapor detectors, insect repellents, etc. (Mechora 2019). Since 1974 the Food and Drug Administration has removed the ban on Se and established it as an essential micronutrient. Therefore, Se in small amounts is used as a food additive for livestock. Utilization of Se in industrial applications has been proven for a long time and the wastes generated from these industrial applications are polluting the environment day by day.

Se is also used as catalyst in some chemical reactions. In X-ray crystallography, incorporation of one or more Se atoms in place of sulfur helps with multiple or single wavelength anomalous dispersion. Se is widely used for toning of photographic prints and it intensifies tonal range of black and white photographic images, resulting in improved permanence of prints (Wall 1924). The worldwide consumption of Se in different applications is presented in Figure 3.2.

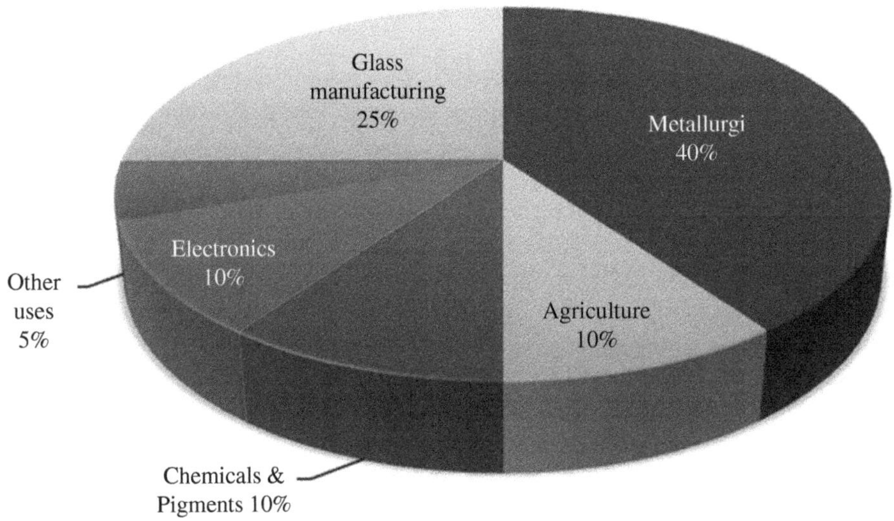

Figure 3.2 Global consumption of Se (USGS 2020).

3.6 Conclusions

This chapter summarized the occurrence, sources, and importance of Se in environment. The major source of Se is from natural activities and nowadays anthropogenic sources are also releasing a huge amount of Se into the environment. The Se from the soil and water enters into the food chain where it is bioaccumulated and transferred into humans and animals through diet.

Earlier, the deficiency of Se was of major concern as it could be one of the important factors in various diseases such as Keshan and Kashin–Beck diseases, infertility in female, etc. However, augmented consumption can also lead to the adverse health impacts.

Demand for Se has increased due to the growth of industrialization and the use of Se in different manufacturing processes. Major Se production is from refineries and copper smelting processes. In the United States fish and wildlife poisoning occurred due to elevated concentration of Se in water. Such episodes have grabbed the attention of researchers for data generation and possible implications of excessive Se and its hazards to human health. Se is an important trace element due to its Janus nature. It is an essential nutrient to human and animal while it can also act as potent toxin if consumed beyond certain limits. Timely dealing with this issue in connection with an environmental Se database can be useful to avoid any adverse impact on humans and the environment.

References

Aston, F.W. (1922). The isotopes of selenium and some other elements. *Nature*: 664. https://doi.org/10.1038/110664a0.

Awika, J.M. (2011). Major cereal grains production and use around the world. *ACS Symposium Series* 1089: 1–13. https://doi.org/10.1021/bk-2011-1089.ch001.

Back, T.G. (2006). Selenium: organoselenium chemistry. *Encyclopedia of Inorganic Chemistry* https://doi.org/10.1002/0470862106.ia214.

Barron, E., Belcher, R., Bogdanski, S.L. et al. (2009). The case for re-evaluating the upper limit value for selenium in drinking water in Europe. *Journal of Water and Health* 7 (4): 630–641. https://doi.org/10.2166/wh.2009.097.

Belcher, R. et al. (1980). Molecular emission cavity analysis. Part 14. Determination of selenium utilizing hydrogen selenide generation. *Analytica Chimica Acta* 113 (1): 13–20. https://doi.org/10.1016/S0003-2670(01)85109-6.

BIS. (2012). Indian Standards Drinking Water Specifications IS 10500: 2012. *Bureau of Indian Standard, Indian Standards Drinking Water Specifications*, 2(May): 11. http://cgwb.gov.in/Documents/WQ-standards.pdf.

Brown, F.C. (1914). The crystal forms of metallic selenium and some of their physical properties. *Physical Review* 4 (2): 85–98. https://doi.org/10.1103/PhysRev.4.85.

Bureau of Mines. (1970). *Mineral Facts and Problems*. Washington, DC: U.S. Dept. of the Interior, Bureau of Mines.

Chapman, P.M. (2009). Is selenium a global contaminant of potential concern? *Integrated Environmental Assessment and Management* https://doi.org/10.1897/1551-3793-5.3.353.

Conde, J.E. and Sanz Alaejos, M. (1997). Selenium concentrations in natural and environmental waters. *Chemical Reviews* 97 (6): 1979–2003. https://doi.org/10.1021/cr960100g.

Du, X.H., Dai, X., Song, R.X. et al. (2012). SNP and mRNA expression for glutathione peroxidase 4 in Kashin-Beck disease. *British Journal of Nutrition* 107 (2): 164–169. https://doi.org/10.1017/S0007114511002704.

Duce, A. (1987). A global atmospheric selenium budget. 92 (D11): 13289–13298. https://doi.org/10.1029/JD092iD11p13289.

Dudley, H.C. (1938). Selenium as a potential industrial Hazard. *Public Health Reports* 53 (8): 281–292.

El-Ramady, H. et al. (2014a). Selenium in soils under climate change, implication for human health. *Environmental Chemistry Letters* 13 (1): 1–19. https://doi.org/10.1007/s10311-014-0480-4.

El-Ramady, H.R. et al. (2014b). Selenium and nano-selenium in agroecosystems. *Environmental Chemistry Letters* 12 (4): 495–510. https://doi.org/10.1007/s10311-014-0476-0.

EPA (2020) *National Primary Drinking Water Regulations (NPDWRs)*. Available at: https://www.epa.gov/ground-water-and-drinking-water/national-primary-drinking-water-regulations#Inorganic.

Fernández-Martínez, A. and Charlet, L. (2009). Selenium environmental cycling and bioavailability: a structural chemist point of view. *Reviews in Environmental Science and Biotechnology* 3: 81–110. https://doi.org/10.1007/s11157-009-9145-3.

Fishbein, L. (1983). Environmental selenium and its significance. *Toxicological Sciences* 3 (5): 411–419. https://doi.org/10.1093/toxsci/3.5.411.

Fordyce, F.M. (2013). Selenium deficiency and toxicity in the environment. In: *Essentials of Medical Geology*. Rev. Ed. (eds. O. Selinus, B. Alloway, J.A. Centeno, et al.), 375–416. Dordrecht: Springer Netherlands https://doi.org/10.1007/978-94-007-4375-5_16.

Gaillardet, J., Viers, J., and Dupré, B. (2003). Trace elements in river waters. *Treatise on Geochemistry* 3: 225–272. https://doi.org/10.1016/B0-08-043751-6/05165-3.

Gilbert, R.D. and Fornes, R.E. (1980). Exchange of comments on neutron activation analyses for simultaneous determination of trace elements in ambient air collected on glass-fiber filters. *Analytical Chemistry* 52 (7): 1153. https://doi.org/10.1021/ac50057a037.

Guoxiong Hua, J.D.W. (2011). Organic phosphorus-selenium chemistry. In: *Selenium and Tellurium Chemistry from Small Molecules to Biomolecules and Materials* (ed. J.D., R. Woollins, Laitinen), 1–39. Springer Nature https://doi.org/10.1007/978-3-642-20699-3.

Hoffmann, J.E. (1989). Recovering selenium and tellurium from copper refinery slimes. *JOM* 41 (7): 33–38. https://doi.org/10.1007/BF03220269.

Holmes, A.B. and Gu, F.X. (2016). Emerging nanomaterials for the application of selenium removal for wastewater treatment. *Environmental Science: Nano.* 3 (5): 982–996. https://doi.org/10.1039/c6en00144k.

Joannopoulos, J.D. (1979). International conference on the physics of selenium and tellurium. In: *The Electronic Structure of Crystalline Phases of Se and Te* (eds. E. Gerlach and P. Grosse), 2–11. Dordrecht: Springer https://doi.org/10.1007/978-3-642-81398-6.

Lee, R.E. and Duffield, F.V. (1979). Sources of environmentally important metals in the atmosphere. In: *Ultratrace Metal Analysis in Biological Sciences and Environment* (ed. R.E. Lee, Jr. and F.V. Duffield), 146–171. Dordrecht: Springer. doi: 10.1021/ba-1979-0172.ch011.

Malisa, E.P. (2001). The behaviour of selenium in geological processes. *Environmental Geochemistry and Health* 23 (2): 137–158. https://doi.org/10.1023/A:1010908615486.

Mechora, Š. (2019). Selenium as a protective agent against pests: a review. *Plants* 8 (8): 262. https://doi.org/10.3390/plants8080262.

Papp, L.V., Lu, J., Holmgren, A., and Khanna, K.K. (2007). From selenium to selenoproteins: synthesis, identity, and their role in human health. *Antioxidants and Redox Signaling* 9 (7): 775–806. https://doi.org/10.1089/ars.2007.1528.

Ralston, N.V.C., Ralston, C.R., Blackwell, J.L. III, and Raymond, L.R. (2008). Dietary and tissue selenium in relation to methylmercury toxicity. *NeuroToxicology* 29 (5): 802–811. https://doi.org/10.1016/j.neuro.2008.07.007.

Rashkov, S., Monev, M., and Atanassov, N. (1983). Some possibilities for the use of stressed cathodically hydrogenated electrolytic nickel coatings for the deposition of microporous and microcracked chromium plates. *Surface Technology* https://doi.org/10.1016/0376-4583(83)90051-1.

Rayman, M.P. (2008). Food-chain selenium and human health: emphasis on intake. *British +Journal of Nutrition* https://doi.org/10.1017/S0007114508939830.

Sakizadeh, M., Sharafabadi, F.M., Shayegan, E., and Ghorbani, H. (2016). Concentrations and soil-to-plant transfer factor of selenium in soil and plant species from an arid area. *IOP Conference Series: Earth and Environmental Science* 44 (5) https://doi.org/10.1088/1755-1315/44/5/052027.

Sanmartin, C. and Plano, D. (2012). Selenium and clinical trials: new therapeutic evidence for multiple diseases. *Current Medicinal Chemistry* 18 (30): 4635–4650. https://doi.org/10.2174/092986711797379249.

Santhanam, C.J., Lunt, R.R., Johnson, S.L. et al. (1979). Health and environmental impacts of increased generation of coal ash and FGD sludges. *Environmental Health Perspectives* 33: 131–157. https://www.ncbi.nlm.nih.gov/pmc/articles/PMC1638118/pdf/envhper00478-0130.pdf.

Shah, P., Strezov, V., Stevanov, C., and Nelson, P.F. (2007). Speciation of arsenic and selenium in coal combustion products. *Energy and Fuels* 21 (2): 506–512. https://doi.org/10.1021/ef0604083.

Sharma, V.K., Sharma, V.K., McDonald, T.J., and Sohn, M. (2017). Assessment of toxicity of selenium and cadmium selenium quantum dots: a review. *Chemosphere* https://doi.org/10.1016/j.chemosphere.2017.08.130.

Skeaff, J.M. and Dubreuil, A.A. (1997). Calculated 1993 emission factors of trace metals for Canadian non-ferrous smelters. *Atmospheric Environment* 31 (10): 1449–1457. https://doi.org/10.1016/S1352-2310(96)00319-6.

Stadtman, T.C. (1974). Selenium biochemistry. *Science* 183 (4128): 915–922. https://doi.org/10.1126/science.183.4128.915.

USGS (2020) *Mineral Commodity Summaries 2020, United States Geological Survey*. Available at: https://pubs.usgs.gov/periodicals/mcs2020/mcs2020.pdf.

Waitkins, G.R., Bearse, A.E., and Shutt, R. (1942). Industrial utilization of selenium and tellurium. *Industrial & Engineering Chemistry* https://doi.org/10.1021/ie50392a002.

Wall, E.J. (1924). *Photographic Facts and Formulas*. Boston, MA: Springer.

Wessjohann, L.A., Schneider, A., Abbas, M., and Brandt, W. (2007). Selenium in chemistry and biochemistry in comparison to sulfur. *Biological Chemistry* https://doi.org/10.1515/BC.2007.138.

Weyl, W.A. (1981). *Coloured Glasses*. Sheffield: Society of Glass Technology.

Whanger, P.D. (2004). Selenium and its relationship to cancer: an update. *British Journal of Nutrition* 91 (1): 11–28. https://doi.org/10.1079/bjn20031015.

WHO (2012) Potable water specification, 2e. Rwanda Bureau of Standards, Kigali.

Willig, F.K. (2014). Selenium Recovery in Precious Metal Technology. Thesis for a Degree Project in Engineering Chemistry. (February). Available at: http://umu.diva-portal.org/smash/get/diva2:703367/FULLTEXT01.pdf.

World Health Organization (2011). *Selenium in Drinking-water Background*, vol. 2, 7. WHO Press. doi: WHO/HSE/WSH/10.01/14.

4

Selenium Toxicity in Domestic Animals

Sources, Toxicopathology, and Control Measure

Arup Giri, Puneet Ranjan, and Vijay K. Bharti

DRDO-Defence Institute of High Altitude Research, Leh, Ladakh, India

4.1 Introduction

Selenium (Se), a semi-metal or metalloid was discovered by Swedish chemist Berzelius in 1818 during the production of sulfuric acid. It was named after the Greek moon goddess, Selene. Se is a semi-metal as it is present together with oxygen and sulfur and has similar chemical characteristics to S (Lenz and Lens 2009). The atomic number of Se is 34. It produces rotting horseradish odor during heating. In fact, it is not present in nature in its pure form. It is found in water in the form of selenites and selenates. This is a trace element because it constitutes just 0.00008% of the Earth's crust (Bennett 1983; Bednarek and Bik 1994).

Selenium (Se) is being an essential micronutrient (Lu and Holmgren 2009) required for livestock animals for better cardiac and skeletal muscle growth, fertility, immune status, anti-carcinogenic, antiviral, anti-inflammatory, chemopreventive, antiparasitic, etc. (Leibovitz et al. 1990; Rayman 2000; Guisbiers et al. 2016; Peng et al. 2016; Stolzoff and Webster 2016; El-Ghazaly et al. 2017; Guisbiers et al. 2017; Maiyo and Singh 2017). It has now become toxic due to poisoning, selenium injection, fodders grown on selenium-toxic land, and indiscriminate minerals supplementation feeding. As early as the 13th century, the toxicity of Se was reported for the first time by Marco Polo, who found hoof disease and hair loss in horses due to toxically high levels of Se accumulated from plant consumption. In addition, Se is the most important constituent of the body's antioxidant defense mechanism through selenoproteins such as GPx (Hefnawy and Tórtora-Pérez 2010; Pascual and Aranda 2013). Despite all these useful qualities of Se, it causes several disorders in domestic animals after consuming high doses. The main sources of Se for livestock animals are soil, water, forages, and feed supplements. Coal-mining regions mainly contain a higher level of Se due to the presence of pyrite oxidation (Dreher and Finkelman 1992). The level of Se varies from state to state and from country to country (Dhillon and Dhillon 2003; Lenz and Lens 2009).

The presence of Se in groundwater and surface water mainly depends upon the characteristics of soil, and anthropogenic factors relating to that area (Giri 2019). The underground water of a seleniferous region showed 2.54–69.53 μg/l of Se, whereas a

Selenium Contamination in Water, First Edition. Edited by Pooja Devi, Pardeep Singh, Arindam Malakar, and Daniel Snow.

non-seleniferous region was characterized by 0.25–8.63 μg/l (Dhillon and Dhillon 1991). Soil is the main source for the higher accumulation of Se in forage (Lakin 1972; Wu et al. 1988; Harada et al. 1989; Stephen et al. 1989; McQuinn et al. 1991). Feed supplements also caused selenosis in livestock animals. Wahlstrom and Olson (1959) and Wilson et al. (1982) reported a focal symmetrical poliomyelomalacia after administration of a Se supplement in swine but chronic selenosis in cattle, sheep, and horse was rare after a Se supplement in feeds. Therefore, diagnosis of Se in soil, forages, and in livestock animals should be executed properly. The primary source of Se for plants is soil followed by water. Se uptake by plants mostly depends upon the soil's physico-chemical properties and basically the pH of the soil determines the Se uptake by plants. If a plant has the capability to uptake Se (i.e. seleniferous plants), then it will accumulate higher Se. The nature of the soil defines the Se availability to plants. It has been found that Se uptake is higher in heavier-textured soils that in clay-type soil. The heavier-textured soils promote Se uptake by plants. The pH of the soil is also important for Se uptake by plants. A study reported that high soil pH will increase the leaching of Se toward the sub-soil from the top soil. Therefore, when examining Se availability in the soil, it has been well documented that soil should be tested for pH level and the soil samples should be taken at different depths according to the pH (Davis et al. 2000).

Higher levels of Se exposure triggered alkali disease and the effect of selenium toxicity was blind staggers (National Research Council [NRC] 1983). The higher Se level triggered three forms of toxicity primarily: acute, sub-acute, and chronic toxicity. Animal poisoning is described as being acute, sub-chronic, or chronic (National Research Council [NRC] 1983). A lot of factors control the selenosis in an animal's body, such as the properties of Se and animal health properties. It was also found that selenosis varies from species to species of livestock animals (horses > cattle) (Ehlig et al. 1968; Rosenfeld and Beath 2013). Sub-chronic and chronic Se toxicity causes pathological changes mainly in skeletal muscle, heart, liver, spleen, and kidney as Se is mostly accumulated at a higher level in these organs (Rosenfeld and Beath 2013). Therefore, some preventive measurement like agricultural field management, the feeding of high protein-rich feed, dietary sulfate, methionine supplement, linseed meal, spraying sulfur-containing materials on fodders grown in selenium-rich soil, clean water, phyto-remediation etc. are being implemented to maintain the health and higher productivity of domestic animals (Dhillon and Dhillon 1991). Therefore, this book chapter discusses in detail the sources and mechanism of Se toxicity in the bodies of livestock animal, as well as the probable preventive measures to attenuate Se toxicity leading to better health and higher production of the livestock animals and reduction of selenium in the human food chain through milk and meat intake.

4.2 Sources of Selenium to Domestic Animals

Following various studies, it has been found that the sources of Se toxicity in domestic animals are soil, water, feed supplements, fodders grown on selenium-rich soil, etc. The Se levels in different food sources of different country are shown in Table 4.1.

Table 4.1 Presence of Se (mg/kg) in different foods of different countries.

	Food			
Country	**Cereals (mg/kg)**	**Meat (mg/kg)**	**Dairy food (mg/kg)**	**Vegetables (mg/kg)**
India	0.23–1.64[a]	–	0.01–0.09[b]	0.92–1.34[a]
Australia[c]	0.01–/0.31	0.06–0.34	</0.001–0.11	</0.001–0.022
U.K.[d]	0.11	0.12–0.6	0.01–0.085	0.005–0.01
U.S.[d]	0.3–0.56	0.06–1.33	0.006–0.3	0.004–0.07
Canada[d]	0.01	0.06–1.22	0.005–0.01	0.005–0.01
Finland[d]	0.02	0.05–0.48	0.002–0.025	0.002
New Zealand[d]	0.035	0.03–0.38	0.004–0.025	0.003

[a] Dhillon and Dhillon (1991).
[b] Giri (2019).
[c] Australia (Tinggi et al. 1992; Tinggi 1999; Tinggi and Conor Reilly 2001).
[d] Combs (1988).

4.2.1 Soil

Weathering is the primary natural process to release Se in the soil. In different forms, Se present in the soil like elemental selenium, calcium selenate, basic ferric selenite, and organic selenium compounds after the decomposing of animal and plant materials. The coal-mining region mainly contains a higher level of Se due to the presence of pyrite oxidation (Dreher and Finkelman 1992). The level of Se varies from states to states, country to country (Dhillon and Dhillon 2003; Lenz and Lens 2009). Usually, the soil contains 0.1–2 mg Se/kg (Fishbein 1983). It was found that soil in the US, India, and Ireland has a higher level of Se (100 mg Se/kg), while soil in Brazil and Argentina has a relatively lower level of Se (< 0.1 mg/kg) (Dhillon and Dhillon 2003; Lenz and Lens 2009). Depending on the higher and lower levels, one region may recognize each as a region that is seleniferous and nonseleniferous. Soil which contains 5 mg/kg Se is very toxic (Rogers et al. 1990).

4.2.2 Water

Se's presence in groundwater and surface water depends largely on soil characteristics, and anthropogenic factors relating to that area (Giri 2019). Underground water of a seleniferous region showed 2.54–69.53 μg/l of Se, whereas a nonseleniferous region was characterized by 0.25–8.63 μg/l (Dhillon and Dhillon 1991). Freshwater showed a higher level (0.06–400 mg Se/l) of Se than saltwater (0.04–0.12 mg Se/l) (Schutz and Turekian 1965; Höghberg 1986). At high altitudes, groundwater contains 0.03–0.12 and 0.00–0.11 mg/l Se in summer and winter, respectively. During summer and winter, river water contains 0.06–0.32 and 0.28–0.46 mg/l Se (Giri 2019).

4.2.3 Forage

A positive correlation was found between the presence of Se in the soil and Se level in forage. In plants, the amount of Se in the seeds is higher than in any other part of the crop. Specific levels of Se in different plants are due in particular to the plants' genetic composition (Harada et al. 1989; Stephen et al. 1989; McQuinn et al. 1991). Se in forage varies between 0.02–0.05 mg Se/kg DM in poisonous areas for Se (Whelan et al. 1994).

4.2.4 Feed Supplements

An enriched feed supplement is the second way to in which domestic animals can develop toxic levels, after Se-enriched forage is consumed. Following different cases of Se toxicity in domestic animals in the US, the FDA has specified that selenosis causing all forms of Se should be within 0.3 mgSe/kgDM (FDA 1987).

Wahlstrom and Olson (1959) and Wilson et al. (1982) reported focal symmetric poliomyelomalacia following administration of a Se supplement in swine but chronic selenosis in cattle, goats, and horses was rare after a Se supplement in feeds. Knott et al. (1958) reported the severe hair loss caused by the administration of 13.3 g Se for 82 days in pigs. Dietary Se dose in horses between 17 and 24 mg/l induced hair loss, inappetence, emaciation, and hoof lesions (Miller and Williams 1940). Oral application of Na_2SeO_3 along with a diet of 0.25–0.5 mg Se/kg BW grain-hay molasses induced animal mortality with many other characteristics such as depression, polioencephalomalacia, vomiting, gastrointestinal hazards, etc. (Maag et al. 1960). Many studies have shown that Se feed supplement within the seleniferous range has hindered growth rate, caused broken hooves, and decreased calving in cattle (Dinkel et al. 1963; Olson et al. 1973).

4.2.5 Diagnosis of Se in Soil, Forages, and in Livestock Animals

Soil followed by water are the primary sources of Se for plants. The amount of Se plant uptake not only depends upon the amount of Se present in the soil but also upon the soil's physico-chemical properties. Na_2SeO_4 is the most available form of Se for plant uptake (Hawrylak-Nowak 2013). Soil nature defines plant availability of Se. Se uptake has been found to be higher in the more heavily-textured soils than in softer soil. The heavier-textured soils are promoting plants to take up Se. Soil pH also has the value of plants taking up Se.

A study reported that high soil pH will increase the leaching of Se toward the sub-soil from the top soil. Therefore, when examining Se availability in the soil, it has been well documented that soil should be tested for pH level and the soil samples should be taken at different depths according to the pH. Seleniferous soil, water, and plants respectively showed >2 ppm, >50 ppb, and >1.0 ppm. Planting should be scheduled taking into account the presence of Se in top soil or sub-soil (Davis et al. 2000).

The overall allowable amounts for Se in human drinking water, livestock drinking water, and irrigation water are 0.01, 0.05, and 0.02 ppm, Therefore it is mandatory to diagnose the Se level in all water sources before using the water for drinking and cultivation (WHO 1987).

Plants with 5 ppm of accumulated Se caused livestock health hazards. A study reported that seleniferous plants may accumulate 15 000 ppm of Se (Rosenfeld 1964). However, seleniferous plants are not always responsible for selenosis in livestock animals. Plants grown in seleniferous soil may or may not accumulate the Se in their tissue. Plants which accumulate Se in their tissue are called obligatory seleniferous plants. Milk-vetches, poison-vetches and prince's plume are the some of the more common obligatory seleniferous plants (Davis et al. 2000). Livestock animals do not consume these plants because of their non-palatability but sometimes sheep consume these plants. Other plants accumulate Se in their tissue by normal processes. These types of plant are called facultative selenium accumulators. These plants are unpalatable to the livestock animals and emit a garlic-sulfur odor. Saltbushes, curlycup gumweed, broom snakeweed, and asters are plants in this group. If they are forage-deprived, cattle and horses consume these plants (Davis et al. 2000). The non-accumulator plants are also responsible for evoking selenosis in animals. Therefore, the presence of such plant species may categorize the grazing land (Davis et al. 2000).

Se is a necessary element for normal growth and fertility in livestock animals. Where there is Se deficiency or toxic Se levels, normal growth has reportedly been stunted in livestock animals, even causing death. The Se-deficient level in grazing cattle is 0.10 ppm, and 0.03 ppm in sheep. The toxic level is 1–5 ppm in cattle, sheep, and horses. This minute gap between Se deficiency and toxicosis has to be maintained by a livestock feeding and management system. Selenosis in livestock animals produces three types of toxicosis: chronic, acute, and sub-acute toxicity. In livestock animals there are some general diagnosed signs such as hunger and thirst, abdominal discomfort, excessive salivation, respiratory failure, dental grating, blindness, and paralysis, etc. In livestock selenosis results in hematological changes, alteration of blood biochemical parameters, reduced antioxidant enzymatic activity, etc. (Rosenfeld and Beath 1964; Davis et al. 2000).

4.3 Toxicopathology of Selenium in Different Domestic Animals

It is well-known that all animals are susceptible to selenium poisoning. Grazing activity of sheep, cattle, horses that increases the rate of selenosis might be due to the consumption of forage of seleniferous grasses or shrubs, whereas poisoning that occurs in poultry or swine and other domestic animals is as a result of being given seleniferous grain or some other seleniferous food. Subtypes of toxicity to selenium are:

i Acute intoxication:

This results from the ingestion of a lethal amount of primary-indicator or other plants which contains high levels of selenium (several hundred ppm Se). As plants containing high levels of selenium are unpalatable, acute selenium intoxication is uncommon. However, it is usually associated with hungry animals that might eat these seleniferous plants.

ii Chronic intoxication:

Chronic selenium poisoning is a consequence of the consumption of toxic levels of selenium (5–40 ppm) in plants such as grass or grain where selenium is bound in plant protein and is fairly insoluble in water. However, it has been shown that this condition can be caused by feeding the element's inorganic salts. The seleniferous feed must be eaten for weeks or months to have its effect. The key symptoms of poisoning in cattle and horses include hair loss, emaciation, hoof malformations, and lameness. Perhaps the worst economic impact arising from seleniferous forage intake is lower reproduction.

Blind stagger type chronic selenium poisoning is said to be induced by organic selenium compounds. To show evidence of poisoning the animals will need to ingest highly toxic amounts of these plants over a significant period of time. The condition of blind staggers is described in three stages. The animal has some vision loss in the first stage, and wanders in circles, disregarding objects along its way. In the second stage, the signs in stage one reach extreme severity and the forelegs of the animal are weak and give way. In the third stage, oral activity is arrested. The mechanism of the tongue and swallowing becomes paralyzed, respiration is staggered and quickened, abdominal pain is evident, and the cornea becomes cloudy. The third stage occurs unexpectedly and death frequently follows within a couple of hours, usually due to respiratory arrest.

Toxicopathology relies on the concentration of selenium in blood and organs ranging from 1 to 5 ppm across the entire livestock in the seleniferous zone (O'Toole and Raisbeck 1995). The sub-chronic and chronic toxicity of the Se causes pathological changes in the skeletal muscle, heart, liver, spleen, and kidney, as they are mostly accumulated at a higher level in these organs (Rosenfeld and Beath 2013). There are reports of histopathological changes such as disrupted endocardium layer, diffuse lung hemorrhages, parenchymal layer degeneration with focal necrosis in hepatic tissues, and nephritis (Faye and Seboussi 2008). Table 4.2 provides the recommended normal intake of selenium in animals per day.

4.3.1 Cattle

Toxicity in any animal depends on the different parameters or factors, such as forage organisms, animal feeding, and dietary components etc. Alkali disease or chronic cow toxicity occurs when 5–40 mg Se/kg BW diets are fed for several months. There is acute toxicity when eating feed, which contains 10–20 mg Se/kg BW (National Research Council [NRC] 1983). One research found that the feed conversion efficiency in young calves was affected by 10 mg Se/kg DM. The winter season is the most vulnerable to Se toxicity in dairy cattle, though handling Se-enriched stall feed (Jenkins and Hidiroglou 1986; Ropstad et al. 1988). Se toxicity in animals has caused many symptoms, such as hair loss, coat roughness, dullness and lack of strength, emaciation, joint bone deterioration, higher respiration levels, weakness and lameness, hooves detachments, etc. (Ghosh et al. 1993; Rosenfeld and Beath 2013).

4.3.2 Sheep

Oral selenite exposure with 50 mg Se/day in sheep caused death after exposure for 72 days. Postmortem reports determined that lesions were observed in heart and lungs (Glenn et al. 1964a, 1964b). The seleniferous range caused severe eye deformation in growing lambs (Rosenfeld and Beath 1947). The erythrocytic lysis process was also reported after selenite exposure to sheep and this might explain the hematological changes in sheep due to chronic selenosis (Young et al. 1981).

Table 4.2 RDI of selenium ingestion for different animals.

Animal species	RDI of Se	References
Beef cattle	100 μg/kg of FDM	National Research Council (NRC) (2001), Suttle (2010)
	0.20 mg/kg	McDowell (1992)
Camel	97–112 ng/ml	Zong-Ping et al. (1994)
	25–53 ng/ml	Rahim (2005)
Dairy cattle	300 μg/kg of FDM	National Research Council (NRC) (2001), Suttle (2010)
	0.30 mg/kg	McDowell (1992)
Donkeys	2 mg/day	Geor et al. (2013)
	0.1–0.15 mg/100 kg BW	National Research Council (NRC) (2007)
Goat	0.1 mg/kg of FDM	Papazafeiriou et al. (2016)
Horses	0.1 ppm of FDM	National Research Council (NRC) (2007), Pagan et al. (1999)
	0.10 mg/kg	McDowell (1992)
Pigs	0.15–0.30 mg/kg	Surai and Fisinin (2015)
Swine	0.10–0.30 mg/kg	McDowell (1992)
Sheep	0.1–0.2 mg/kg of FDM	National Research Council (NRC) (1985)
	0.10–0.20 mg/kg	McDowell (1992)

BW – body weight; FDM – food dry matter; RDI – recommended daily intake.
Source: Adapted from Hosnedlova et al. (2017).

Most of the studies indicated that toxicity in sheep and lambs was caused by 0.45–8 mg Se/kg LW (Caravaggi et al. 1970; Yaeger et al. 1998; Tiwary et al. 2006). Inorganic forms of Se caused toxicity in sheep and lambs. Giese (1984) found that most of the Se excreted via urine and feces after intravenous injection of high Se dose. The half-life of Se in sheep is 14.7 days (Blodgett and Bevill 1987). Se toxicity in sheep and lamb caused respiratory problems, hair loss, cracks on hooves and horn, ataxia, and reduced feed intake. Myocardial necrosis and pulmonary alveolar vasculitis were found after exposing lambs with 8 mg Se/kg LW (Gupta et al. 1982; Tiwary et al. 2006).

4.3.3 Pigs

Miller and Williams (1940) reported that oral selenite intoxication with swine body weight of 13–23 mg Se/kg caused paresis, tremor, vomiting, anorexia, diarrhea, and depression (Figure 4.1). 10 ppm of sowing exposure results in toxicity, shown by reduced design rate, litter size, and piglet weight (Wahlstrom and Olson 1959). A study by Diehl et al. (1975) reported that 2.0 and 1.2 mg/kg body weight administration caused muscle tetany, increased plasma glutamic oxaloacetic transaminase (GOT) activity, and clinical signs of tremor, ataxia, etc. As observed in sheep and calves, acute and chronic Se toxicity has also emerged in young pigs. Acute Se toxicity in young lambs following selenosis manifested similar

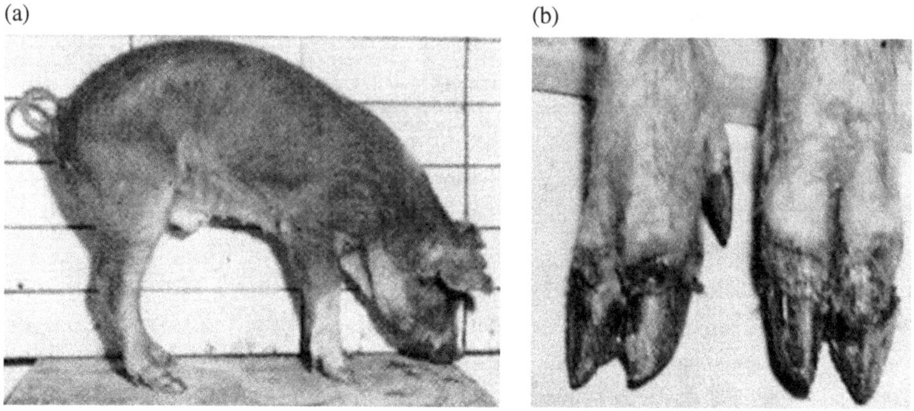

Figure 4.1 (a and b) Symptoms of selenosis in pigs (unthriftiness (a) and separation of the hoofs from the skin (b)). (*Source:* Adapted from Wahlstrom et al. (1956)).

symptoms (Shortridge et al. 1971). Chronic Se toxicity induced sluggishness, lack of energy, emaciation, hair roughness, hair loss, sluggishness of hooves, damage to the liver, heart failure, etc. (Underwood 1979).

Toxicity varies according to the pigs' genetic difference. Some of the studies have reported that pigs' susceptibility to Se toxicity varied depending on pigs' hair color (Wahlstrom and Olson 1959; Wahlstrom et al. 1984). Red-haired pigs were more prone to Se toxicity than black- or white-haired pigs (Wahlstrom et al. 1984). 4–8 mg Se/kg DM exerts Se toxicity in pigs in food grains; that also depends on the composition of food and other factors such as duration of exposure (Palmer et al. 1983; Goehring et al. 1984a; Goehring et al. 1984b). One study of a maize-soybean diet with 8 mg Se/kg DM impaired appetite and growth within five weeks of exposure. However, for the same time duration this concentration in wheat and oats had no effect on the pigs (Goehring et al. 1984a; Goehring et al. 1984b). 12 mg Se/kg DM of feed caused hoof lesions in pigs, as in dairy cattle and horses (Goehring et al. 1984b). The concentration in pigs with 5, 10, and 25 mg Se/kg DM for 120 days resulted in hoof lesion, edema, hyperemia, extreme spinal cord lesions, paralysis of the body, emaciation, postnecrotic atrophic liver cirrhosis, and lumbal poliomyelomalacia, etc. and inflammation (Goehring et al. 1984b; O'Toole and Raisbeck 1995). Feeding meal diets of corn-soybean along with 5 ppm of Se in swine results reduced growth and feed intake. The most influential measure of toxicity to Se is decreased growth rate. Up to 8.3 ppm Se addition in swine feed did not cause any health effects but after 12 ppm Se meal exposure caused hoof lesions (Mahan and Moxon 1980; Wahlstrom et al. 1984; Goehring et al. 1984b). The same diagnosis of hoof lesions was also published for Harrison et al. (1983) after exposure to Se. Poulsen et al. (1989) reported that after suckling the growth rate was stunted by exposure to Se, because the milk produced lower rates of Se than the feed received. They also reported that 16 mg of inorganic Se/kg feed along with diets based on barley produced no health hazards such as sow conception, number of piglets born, mortality rate, etc. But after Se supplementation, there was a tendency to lower piglet body weight as the colostrum and milk contained Se.

A higher dose of Se in feed also decreased feed intake, circular dark band on hoofs, and rate of gain following piglet weaning (Wahlstrom and Olson 1959; Goehring et al. 1984b; Mahan and Moxon 1984; Poulsen et al. 1989). Kim and Mahan (2001a) reported Se supplementation in swine muscle ad internal organs such as heart, kidney, liver etc. caused the higher Se accumulation. They also reported that inorganic Se caused serious swine problems. They came to this conclusion after feeding 0.3–10 ppm Se to the swine with organic and inorganic form Se. Results showed that exposure of 7–10 ppm Se for prolonged periods caused selenosis due to organic Se exposure after alteration of reproductive performance. On the other hand, inorganic Se during lactation period caused detrimental effects in the progeny (Kim and Mahan 2001b).

4.3.4 Poultry

Seleniferous region has impacted on the poultry industry after poor hatchability, embryo deformities, malformed toes, legs, wings, etc. (Franke 1934; Franke et al. 1936; Franke and Tully 1936; Moxon 1937). However, it has been also reported that the toxicity of Se in poultry or any other animals depends upon the other several factors which are intimately associated with the nature of diet, gender of animals, any co-exposure of other elements to the animals, etc. (Levander 1972). Selenate injection with 0.6–0.8 ppm in the air sac of poultry eggs before incubation caused the teratogenic effects on the developing embryo, such as malformed upper beak, missing eyes and legs, and stunted growth (Franke et al. 1936). The same type of results were reported by Sukra et al. (1976) and Khan and Gilani (1980). Poley et al. (1941) indicated that 4.0 and 10 ppm Se application by dietary grains caused lower growth rates for chicks. 15 ppm dietary application culminated in decreasing hatchability to zero (Poley et al. 1937). Trelease and Beath (1949) recorded Se toxicity embryo deformities, the toxicity caused by various chemicals exposure growing turkeys administered with 20 ppm Se in regular rations. 9.0 ppm Se application on laying turkey hens caused few embryonic malformations but this increased at 15.0 ppm Se application (Carlson et al. 1951).

One study reported that insertion of Se as selenious acid with 0.4 ppm into the yolk before the incubation caused toxicity and produced malformed embryos (Ridgway and Karnofsky 1952). Layed eggs of poultry reared after dieting with 8.0 ppm Se from wheat caused induction for necrosis in growing embryo, especially in the brain, spinal cord, and upper beak, and also decreased the growth of the embryo (Gruenwald 1958). This study was also supported by the findings of Rosenfeld and Beath (2013). Latshaw (1975) studied unequal distribution of injected Se in eggs, such as higher concentration in yolk than the albumen part of the eggs. Selenite injection of 18 pg Se/egg into the yolk caused abnormal changes in hematological profile after 19 days and a propensity toward malformed embryo formation was reported (Landauer 1940; Kury et al. 1967). One study reported that 2.0 ppm dietary Se did not produced any defective effects in the laying hens in terms of growth and egg laying, but administration of 8.0 ppm caused a reduction of egg laying and stunted body growth (Thapar et al. 1969). Ort and Latshaw (1978) administered 5, 7, and 9 ppm Se in laying hens. Results reported that a 5 ppm dose caused the reduce hatchability, a 7 ppm dose caused reduced egg weights, and a 9 ppm caused zero egg production. Selenite in a *Torula* yeast diet to chicks increased the chicks' mortality rate (El-Begearmi and Combs Jr 1982).

Mertz and Underwood (1986) found that application of 8 ppm Se in diet caused higher bioavailability of Se in feathers. Supplementing emus with 1.4 ppm Se caused reduced hatchability from 52% to 40% with fatal egg deformities (Kinder et al. 1995). Another study reported that 2.0 ppm Se from the administration of selenite by drinking water has no effect on chick growth patterns (Cantor et al. 1984). Drinking Se-containing 300 μgSe/l agricultural drainage water caused embryonic death and death of adult birds (Ohlendorf 1989). Kumar et al. (2018) conducted a study on the effects of sodium selenite on adult White Leghorn broiler birds. For 42 days they administered sodium selenite with 15 and 30 ppm. Results indicated that histopathological changes occurred in the liver, kidney, lungs, heart, and intestine (Figure 4.2a–d). There were diminished amounts of hemoglobin, packed volume of cells, ALT, and AST.

In one study, the following concentrations of inorganic selenium and selenized yeast were given to chickens: 0, 2, 5, 10, 15, 20, or 30 mg Se/kg for six months. Concentration with 10, 15, 20, or 30 mg Se/kg feed induced some pathological changes, such as renal edema; higher

(a) (b) (c) (d)

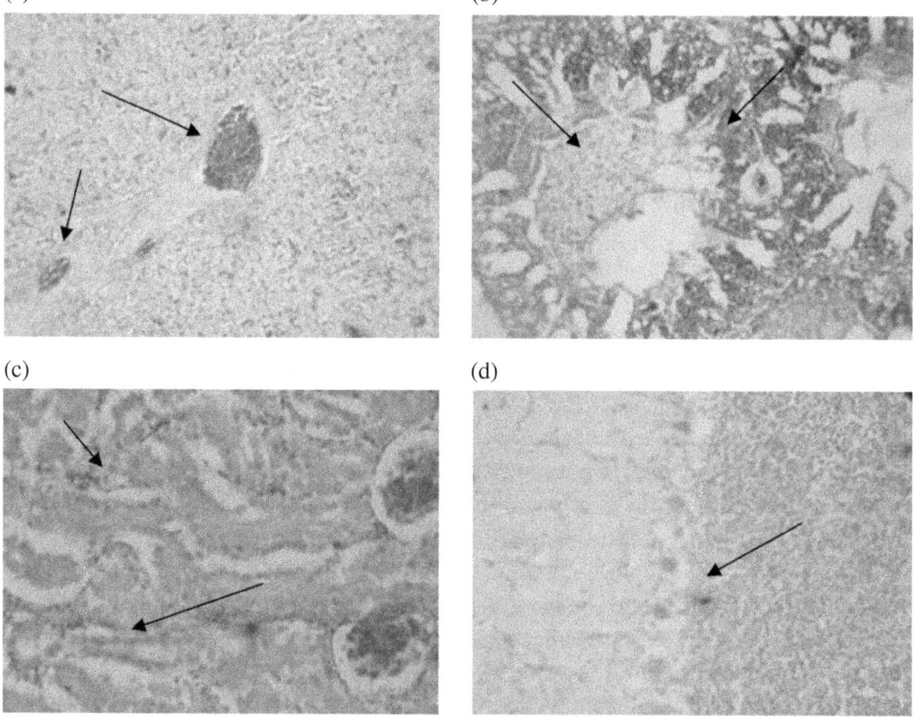

Figure 4.2 (a) The liver section of poultry birds that had ingested selenium showing congestion in blood vessels, congestion in sinusoids, necrosis in the parenchyma, and connective tissue proliferation between the lobules (*Source:* Adapted from Kumar et al. (2018)). (b) Section of birds' lungs showing congestion in the alveoli and oedema in the bronchi (*Source:* Adapted from Kumar et al. (2018)). (c) Kidney section of poultry birds that had ingested selenium showing degeneration of tubules, infiltrations of mononuclear cells (*Source:* Adapted from Kumar et al. (2018)). (d) Section of cerebellum of birds that had ingested selenium showing degeneration of Purkinje cells (*Source:* Adapted from Kumar et al. (2018)).

Table 4.3 Se toxicity in poultry birds.

Chemical form	Age	Dose (ppm)	Toxic effects	References
Na_2SeO_3 + Se in wheat	Immature	10	Growth ↓	Carlson and Leitis (1957)
Se in wheat	Laying hens	10	Hatchability ↓	Moxon and Wilson (1944)
Se in corn, barley, wheat	Adult	5	Mortality ↑	Moxon (1937)
Na_2SeO_3	Adult	6.5	Deformed embryos	Moxon (1937)
Na_2SeO_3	One day	8	Body weight ↓	Moxon (1937)
Selenous acid	One day	8	Mortality ↑	Thapar et al. (1969)
SeO_2	one day	5	Mortality ↑	Hill (1979)
Na_2SeO_3	Adult	8	Dead embryo	Gruenwald (1958)
Selenite	Adult	10	Growth ↓	Jensen and Chang (1976)
Selenite	One day	5	Growth ↓	Jensen (1986)

↑ – Increased; ↓ – Decreased; Na_2SeO_3 – Sodium selenite, SeO_2 – Selenium dioxide; ppm – parts per million.

rates of dystrophy cells were found in renal tubules (Todorović et al. 2004). Ort and Latshaw (1978) found that 5 mg Se/kg administration reduced hatchability and deformed embryos (absence of eyes, beaks, wing and feet irregularities). Another research suggested that 4 mg Se decreases the appetite of deformed eggs in chickens for seven days (Poley and Moxon 1938; Landauer 1940; Cantor et al. 1984). Table 4.3 lists more research on the toxicity caused by Se.

4.3.5 Camel and Horse

To date, only experimental selenosis in camels has been reported. One study by McDonald et al. (2011) reported that Se doses with 1 and 2 mg/kg body weight caused lassitude, inappetence, inability to raise, mild pulmonary congestion, alveolar edema, acute myositis, and edema in a different portion of the brain. Administration of 0.051–0.095 mg Se/kg LW to camels caused selenosis (Seboussi et al. 2008). Faye and Seboussi et al. (2008) reported that 8 and 12 mg Se/day administration caused several clinical symptoms such as hair discoloration, followed by alopecia, dark watery diarrhea, inferior cervical lymph increase, dyspneic respiration, and difficulty in walking. Pathological alterations in kidney, heart, and liver were also observed, such as congestions in Bowman's space and convoluted tubules, the proliferation of Purkinje fibers, degenerative changes in myofibers, degenerative changes hepatic cells of the hepatic lobules, edema in intercostal and diaphragm muscles, perivascular edema in brain, etc. (Figure 4.3a–l). Aitken (2001) reported that selenosis showed mainly hoof lesions in horses, along with decreased body growth.

4.3.6 Antelope

Raisbeck et al. (1995) studied the impact of Se on antelope treatment. For a month, antelopes were given water with <50 ppb Se. Following the acclimation one group of antelopes

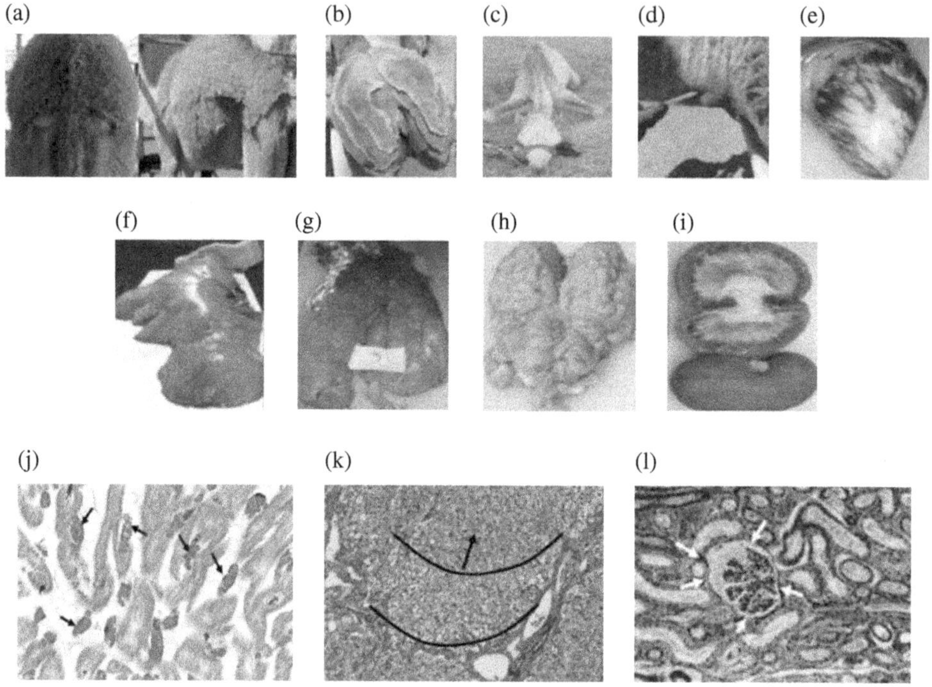

Figure 4.3 (a) Alopecia; (b) pad lesions; (c) sternal position; (d) hypertrophy of cervial lymphnoid; (e) heart congestion, soft discoloration; (f) liver congestion; (g) pulmunary congestion; (h) brain edema; (i) muscle discoloration; (j) camel heart congestion in capillary vessels; (k) degeneracy of liver in camel; (l) toxic effects in kidney of camel (*Source:* Adapted from Seboussi et al. (2012)).

was fed for 168 days with 25 ppm Se native grass hay along with alfalfa hay (0.3 ppm Se). The final exposed Se concentration was of 15 ppm. The control group was given feed with approximately 0.3 ppm Se. After the experiment it was found that the higher exposure to Se caused a reduction in antelopes' feed intake and a reduction in body weight. However, there were no lung, liver, or kidney abnormalities.

4.4 Control Measures of Selenium Toxicity

All of the previous reports have mentioned that Se toxicity could reduce livestock farming industry production in any country. Mapping of agricultural land based on the presence of compulsory and optional Se accumulator plants and non-accumulator plants is the ultimate control measure method (de Souza et al. 1999; Barillas et al. 2011; White 2016; Gupta and Gupta 2017; Schiavon and Pilon-Smits 2017). It will help to decide the measure of grazing land and actively manage the toxicity of Se in the soil. Before applying forage plants to animals of livestock or any other supplements these should be subjected to determination of the level of Se. Both of these preventive steps would potentially minimize the intake of Se in livestock (Tiwary et al. 2006). Transgenic plants have the potential to remove Se from soil (Pilon-Smits and LeDuc 2009). Apart from these etiological preventive steps, some studies stated that the amount of animal blood Se can be a great predictor for earlier determination of signs linked to Se toxicity. The presence of >1.5 μg/ml in cattle blood has been reported to be indicative of the onset of Se toxicity (Deore et al. 2002).

Preventive measures such as soil treatment, high protein-rich feed feeding, dietary sulfate, methionine supplement, linseed meal, spraying sulfur-containing materials on fodders grown in selenium-rich soil, clean water, etc. may reduce selenosis (Prakash et al. 2010; He et al. 2018; Wadgaonkar et al. 2018). To date, no particular studies have ever reported absolute attenuation of Se toxicity in animals as a result of the procedure. The study on intravenous administration of reduced glutathione (GSH) @ 5 mg/kg of BW was useful in arresting the Se toxicity after decreasing the activity of glutathione peroxidase (GSH-Px) (Deore et al. 2005). The study of Arora et al. (1975) and Arora (1985) reported that trace mineral mixture which includes 1 kg of magnesium sulfate ($MgSO_4$), 166 g of ferrous sulfate ($FeSO_4$), 24 g of copper sulfate ($CuSO_4$), 75 g of zinc sulfate ($ZnSO_4$), and 15 g of cobalt sulfate ($CoSO_4$) was a great help in recovering the Se toxicity in buffaloes after 21 days of exposure and within 50 days. One research recorded that both arsanilic acid (0.02%) and 3-nitro (0.005%) were similarly successful in preventing selenosis in pigs. These chemicals also had positive effects on weight gain as the control pigs developed between 67 and 100 pounds per pig (Wahlstrom et al. 1956). Se toxicity was also minimized by a protein-rich diet and mineral balance mixture in buffaloes (Fessler et al. 2003). Increasing horse grazing time to 50% on pasture significantly reduced Se concentration compared to those horses that had less pasture time (Crisman et al. 1994). The same pattern emerged in water buffaloes (Khanal and Knight 2010).

Seleniferous soil is the primary component in the food chain which transfers the higher levels of Se. Adding sulfur decreases the plant supply Se in soil. One piece of research suggested that the introduction of 1000 kg gypsum/ha in sugar cane increased the S:Se ratio and decreased the Se amount (Dhillon and Dhillon 1991). Continuous application of S can result in plant selenium deficiency, and eventually there will be trouble for domestic animals (Andrews et al. 1968; McDonald 1975; Fessler et al. 2003). One phytoremediation technique was reported to reduce the level of Se in soil, where it was claimed that kenaf and canola planting in seleniferous soil had scavenged the Se from soil and water (Wood 2000).

Carbohydrate-rich diets, protein, sulfate, and calcium in cattle decreased Se levels (Kessler 1993). Concentration above 2.4 g/kg DM of sulfur decreased Se absorption substantially (Spears and Weiss 2008). Administration of 0.8% calcium DM in feed significantly decreases Se in pregnant cows' blood and muscle (Harrison et al. 1983).

Vitamin E administration can reduce the toxicity to Se in cattle (NRC 1996). One study suggested that the use of vitamin E in dairy cattle with 15–60 foreign units (IU) decreases Se (NRC 1984, 1996).

4.5 Conclusion

Selenium (Se) is a vital mineral in domestic animals' nutrition which helps up-regulation of body antioxidants status and mitigation of free radicals. However, selenosis occurs in domestic animals due to the consumption of water, plants, and feed supplements containing higher levels of Se. Hence, feeding management plays an important role in mitigating Se toxicity and modulating Se-induced body metabolism. However, based on current understanding and ameliorative measures, selenosis cannot be prevented at advanced stage. Therefore, further studies should be carried out on the removal of cellular excess selenium, inhibition of selenium-induced metabolism using metal chelator, antagonist, competititive inhibition of enzyme co-factor, etc. as future preventive measures of selenosis in domestic animals.

References

Aitken, P. (2001). Selenium toxicity. *Practice* 23 (5): 286–289.

Andrews, E.D., Hartley, W.J., and Grant, A.B. (1968). Selenium-responsive diseases of animals in New Zealand. *New Zealand Veterinary Journal* 16 (1–2): 3–17.

Arora, S.P. (1985). Livestock problems related to geochemistry in India including selenium toxicity and goitre. *Monograph Series: Environmental Geochemistry and Health London: Science Reviews Limited* 164: 180.

Arora, S.P., Kaur, P., Khirwar, S.S. et al. (1975). Selenium levels in fodders and its relationship with Degnala disease. *Indian Journal of Dairy Science* 28: 249–253.

Barillas, J.R.V., Quinn, C.F., and Pilon-Smits, E.A. (2011). Selenium accumulation in plants – phytotechnological applications and ecological implications. *International Journal of Phytoremediation* 13 (supl): 166–178.

Bednarek, D. and Bik, D. (1994). Influence of selenium on animals' health. Part II. Result of deficiency. *Życie Wet* 7: 269–272.

Bennett, B.G. (1983). Exposure of man to environmental PCBs – an exposure commitment assessment. *Science of the Total Environment* 29 (1–2): 101–111.

Blodgett, D.J. and Bevill, R.F. (1987). Acute selenium toxicosis in sheep. *Veterinary and Human Toxicology* 29 (3): 233–236.

Cantor, A.H., Nash, D.M., and Johnson, T.H. (1984). Toxicity of selenium in drinking water of poultry. *Nutrition Reports International* 29 (3): 683–688.

Caravaggi, C., Clark, F.L., and Jackson, A.R.B. (1970). Acute selenium toxicity in iambs following intramuscular injection of sodium selenite. *Research in Veterinary Science* 11: 146–149.

Carlson, C.W. and Leitis, E. (1957). Methionine, betaine and choline as counteractants of selenium toxicity. *Poultry Science* 36 (5): 1108–1108.

Carlson, C.W., Kohlmeyer, W., and Moxon, A.L. (1951). Arsenic fails to control selenium poisoning in turkeys. *S. Dak. Farm Home Research* 3 (1): 20–22.

Combs, G.F. (1988). Selenium in foods. In: *Advances in Food Research*, vol. 32 (eds. C.O. Chichester and B.S. Schwegert), 85–113. San Diego, CA: Academic Press.

Crisman, M.V., Carmel, D.K., Lessard, P., and Ley, W.B. (1994). A survey of whole blood selenium concentrations of horses in Virginia and Maryland. *Journal of Equine Veterinary Science* 14 (5): 256–261.

Davis, J.G., Steffens, T.J., Engle, T.E. et al. (2000). Diagnosing selenium toxicity. *Natural Resources Series* 6: 109.

Deore, M.D., Srivastava, K.A., and Sharma, S.K. (2002). Blood selenium levels during different stages of selenosis in buffaloes and its evaluation as a diagnostic tool. *Veterinary and Human Toxicology* 44 (5): 260–263.

Deore, M.D., Srivastava, A.K., and Sharma, S.K. (2005). Effect of reduced glutathione treatment on selenosis, blood selenium concentration and glutathione peroxidase activity after repeated short-term selenium exposure in buffalo calves. *Toxicology* 213 (1–2): 169–174.

Dhillon, K.S. and Dhillon, S.K. (1991). Accumulation of selenium in sugarcane (*Sachharum officinarum* Linn.) in seleniferous areas of Punjab, India. *Environmental Geochemistry and Health* 13 (3): 165–170.

Dhillon, K.S. and Dhillon, S.K. (2003). Distribution and management of seleniferous soils. *Advances in Agronomy* 79 (1): 119–184.

Diehl, J.S., Mahan, D.C., and Moxon, A.L. (1975). Effects of single intramuscular injections of selenium at various levels to young swine. *Journal of Animal Science* 40 (5): 844–850.

Dinkel, C.A., Minyard, J.A., and Ray, D.E. (1963). Effects of season of breeding on reproductive and weaning performance of beef cattle grazing seleniferous range. *Journal of Animal Science* 22 (4): 1043–1045.

Dreher, G.B. and Finkelman, R.B. (1992). Selenium mobilization in a surface coal mine, Powder River Basin, Wyoming, USA. *Environmental Geology and Water Sciences* 19 (3): 155–167.

Ehlig, C.F., Allaway, W.H., Cary, E.E., and Kubota, J. (1968). Differences among plant species in selenium accumulation from soils low in available selenium 1. *Agronomy Journal* 60 (1): 43–47.

El-Begearmi, M.M. and Combs, G.F. Jr. (1982). Dietary effects on selenite toxicity in the chick. *Poultry Science* 61 (4): 770–776.

El-Ghazaly, M.A., Fadel, N., Rashed, E. et al. (2017). Anti-inflammatory effect of selenium nanoparticles on the inflammation induced in irradiated rats. *Canadian Journal of Physiology and Pharmacology* 95 (2): 101–110.

Faye, B. and Seboussi, R. (2008). Experimental selenium intoxication in camel. *Veterinariya* 3: 18–29.

Fessler, A.J., Moller, G., Talcott, P.A., and Exon, J.H. (2003). Selenium toxicity in sheep grazing reclaimed phosphate mining sites. *Veterinary and Human Toxicology* 45 (6): 294–298.

Fishbein, L. (1983). Environmental selenium and its significance. *Fundamental and Applied Toxicology* 3 (5): 411–419.

Food and Drug Administration (FDA) (1987). Food additives permitted in feed and drinking water of animals: selenium. *Federal Register* 52 (65): 10887–10888.

Franke, K.W. (1934). A new toxicant occurring naturally in certain samples of plant foodstuffs. I. Results obtained in preliminary feeding trials. *Journal of Nutrition* 8: 597–607.

Franke, K.W. and Tully, W.C. (1936). A new toxicant occurring naturally in certain samples of plant foodstuffs: VII. Low hatchability due to deformities in chicks produced from eggs obtained from chickens of known history. *Poultry Science* 15 (4): 316–318.

Franke, K.W., Moxon, A.L., Poley, W.E., and Tully, W.C. (1936). Monstrosities produced by the injection of selenium salts into hens' eggs. *Anatomical Record* 66: 15–22.

Geor, R.J., Coenen, M., and Harris, P. (2013). *Equine Applied and Clinical Nutrition E-Book: Health, Welfare and Performance*. Elsevier Health Sciences.

Ghosh, A., Sarkar, S., Pramanik, A.K. et al. (1993). Selenium toxicosis in grazing buffaloes and its relationship with soil and plant of West Bengal. *Indian Journal of Animal Sciences (India)* 63 (5): 557–560.

Giese, W.W. (1984). Isotope techniques in studies of selenium deficiency and toxicity syndromes in farm animals. In *Nuclear Techniques in Tropical Animal Diseases and Nutritional Disorders*, 97–112. Vienna: International Atomic Energy Agency.

Giri, A. (2019). Studies on trace minerals and heavy metals status in different sources of water and their bioavailability in milking cattle at high altitude. Ph.D. Thesis, Bharathiar University, Coimbatore, Tamilnadu, India.

Glenn, M.W., Jensen, R., and Griner, L.A. (1964a). Sodium selenate toxicosis: the effects of extended oral administration of sodium selenate on mortality, clinical signs, fertility, and early embryonic development in sheep. *American Journal of Veterinary Research* 25: 1479–1485.

Glenn, M.W., Jensen, R., and Griner, L.A. (1964b). Sodium selenate toxicosis: pathology and pathogenesis of sodium selenate toxicosis in sheep. *American Journal of Veterinary Research* 25: 1486–1494.

Goehring, T.B., Palmer, I.S., Olson, O.E. et al. (1984a). Effects of seleniferous grains and inorganic selenium on tissue and blood composition and growth performance of rats and swine. *Journal of Animal Science* 59 (3): 725–732.

Goehring, T.B., Palmer, I.S., Olson, O.E. et al. (1984b). Toxic effects of selenium on growing swine fed corn-soybean meal diets. *Journal of Animal Science* 59 (3): 733–737.

Gruenwald, P. (1958). Malformations caused by necrosis in the embryo: illustrated by the effect of selenium compounds in Chick embryos. *The American Journal of Pathology* 34 (1): 77.

Guisbiers, G., Wang, Q., Khachatryan, E. et al. (2016). Inhibition of *E. coli* and *S. aureus* with selenium nanoparticles synthesized by pulsed laser ablation in deionized water. *International Journal of Nanomedicine* 11: 3731.

Guisbiers, G., Lara, H.H., Mendoza-Cruz, R. et al. (2017). Inhibition of Candida albicans biofilm by pure selenium nanoparticles synthesized by pulsed laser ablation in liquids. *Nanomedicine: Nanotechnology, Biology and Medicine* 13 (3): 1095–1103.

Gupta, M. and Gupta, S. (2017). An overview of selenium uptake, metabolism, and toxicity in plants. *Frontiers in Plant Science* 7: 2074.

Gupta, R.C., Kwatra, M.S., and Singh, N. (1982). Chronic selenium toxicity as a cause of hoof and horn deformities in buffalo, cattle and goat. *Indian Veterinary Journal (India)* 59 (9): 738–740.

Harada, I., Shinohara, I., and Sato, S. (1989). The absorption characteristics of selenious acid applied to corn (*Zea mays* L.). *Journal of Rakuno Gakuen University. Natural Science (Japan)* 14: 49–55.

Harrison, J.H. and Conrad, H.R. (1984). Effect of dietary calcium on selenium absorption in the nonlactating dairy cow. *Journal of Dairy Science* 67: 1860.

Harrison, L.H., Colvin, B.M., Stuart, B.P. et al. (1983). Paralysis in swine due to focal symmetrical poliomalacia: possible selenium toxicosis. *Veterinary Pathology* 20 (3): 265–273.

Hawrylak-Nowak, B. (2013). Comparative effects of selenite and selenate on growth and selenium accumulation in lettuce plants under hydroponic conditions. *Plant Growth Regulation* 70 (2): 149–157.

He, Y., Xiang, Y., Zhou, Y. et al. (2018). Selenium contamination, consequences and remediation techniques in water and soils: a review. *Environmental Research* 164: 288–301.

Hefnawy, A.E.G. and Tórtora-Pérez, J.L. (2010). The importance of selenium and the effects of its deficiency in animal health. *Small Ruminant Research* 89 (2–3): 185–192.

Hill, C.H. (1979). The effect of dietary protein levels on mineral toxicity in chicks. *The Journal of Nutrition* 109 (3): 501–507.

Höghberg, J. and Alexander, J. (2007). Selenium. In: *Handbook on the Toxicology of Metals*, 3e (eds. G. Nordberg, B. Fowler, M. Nordberg and L. Friberg), 482–520. Academic Press Inc.

Hosnedlova, B., Kepinska, M., Skalickova, S. et al. (2017). A summary of new findings on the biological effects of selenium in selected animal species – a critical review. *International Journal of Molecular Sciences* 18 (10): 2209.

Jenkins, K.J. and Hidiroglou, M. (1986). Tolerance of the preruminant calf for selenium in milk replacer. *Journal of Dairy Science* 69 (7): 1865–1870.

Jensen, L.S. (1986). Effect of Monensin on response of chicks to deficient and toxic levels of dietary selenium. *Poultry Science* 65 (6): 1163–1166.

Jensen, L.S. and Chang, C.H. (1976). Fractionation studies on a factor in linseed meal protecting against selenosis in chicks. *Poultry Science* 55 (2): 594–599.

Kessler, J. (1993). Carence en sélénium chez les ruminants: Mesures prophylactiques. *Revue Suisse d'Agriculture* 25: 21–26.

Khan, M.Y. and Gilani, S.H. (1980). Selenium poisoning and embryogenesis: light and electron microscopic studies of the heart. *Environmental Research* 23 (1): 98–109.

Khanal, D.R. and Knight, A.P. (2010). Selenium: its role in livestock health and productivity. *Journal of Agriculture and Environment* 11: 101–106.

Kim, Y.Y. and Mahan, D.C. (2001a). Comparative effects of high dietary levels of organic and inorganic selenium on selenium toxicity of growing-finishing pigs. *Journal of Animal Science* 79 (4): 942–948.

Kim, Y.Y. and Mahan, D.C. (2001b). Prolonged feeding of high dietary levels of organic and inorganic selenium to gilts from 25 kg body weight through one parity. *Journal of Animal Science* 79 (4): 956–966.

Kinder, L.L., Angel, C.R., and Anthony, N.B. (1995). Apparent selenium toxicity in emus (*Dromaius novaehollandiae*). *Avian Diseases* 39 (3): 652–657.

Knott, S.G., McCray, C.W.R., and Hall, W.T.K. (1958). Selenium poisoning in horses in North Queensland. *Queensland Journal of Agricultural Science* 15: 43–58.

Kumar, D., Gautam, A.K., and Sinha, M.K. (2018). Histopathological alterations of selenium toxicity induced in broiler (birds). *Indian Journal of Animal Research* 52 (4): 599–604.

Kury, G., Rev-Kury, L.H., and Crosby, R.J. (1967). The effect of selenous acid on the hematopoietic system of chicken embryos. *Toxicology and Applied Pharmacology* 11 (3): 449–458.

Lakin, H.W. (1972). Selenium accumulation in soils and its absorption by plants and animals. *Geological Society of America Bulletin* 83 (1): 181–190.

Landauer, W. (1940). Studies on the creeper fowl. 13. The effect of selenium and the asymmetry of selenium-induced malformations. *Journal of Experimental Zoology* 83: 431–443.

Latshaw, J.D. (1975). Natural and selenite selenium in the hen and egg. *The Journal of Nutrition* 105 (1): 32–37.

Leibovitz, B., Hu, M.L., and Tappel, A.L. (1990). Dietary supplements of vitamin E, β-carotene, coenzyme Q10 and selenium protect tissues against lipid peroxidation in rat tissue slices. *The Journal of Nutrition* 120 (1): 97–104.

Lenz, M. and Lens, P.N. (2009). The essential toxin: the changing perception of selenium in environmental sciences. *Science of the Total Environment* 407 (12): 3620–3633.

Levander, O.A. (1972). Metabolic interrelationships and adaptations in selenium toxicity. *Annals of the New York; Academy of Sciences (United States)* 192: 181–192.

Lu, J. and Holmgren, A. (2009). Selenoproteins. *Journal of Biological Chemistry* 284 (2): 723–727.

Maag, D.D., Orsborn, J.S., and Clopton, J.R. (1960). The effect of sodium selenite oncattle. *American Journal of Veterinary Research* 21: 1049–1053.

Mahan, D.C. and Moxon, A.L. (1980). Toxic dietary selenium levels for weanling swine. *Journal of Animal Science* 51 (Suppl. 1): 1216–1221.

Mahan, D.C. and Moxon, A.L. (1984). Effect of inorganic selenium supplementation on selenosis in postweaning swine. *Journal of Animal Science* 58 (5): 1216–1221.

Maiyo, F. and Singh, M. (2017). Selenium nanoparticles: potential in cancer gene and drug delivery. *Nanomedicine* 12 (9): 1075–1089.

McDonald, J.W. (1975). Selenium-responsive unthriftiness of young merino sheep in Central Victoria. *Australian Veterinary Journal* 51 (9): 433–435.

McDonald, P., Edwards, R.A., Greenhalgh, J.F.D. et al. (2011). The components of food. In: *Animal Nutrition*, 7e (eds. P. McDonald, R.A. Edwards, J.F.D. Greenhalgh, et al.), 81–85. England: Ashford Colour Press.

McDowell, M. (1992). Selenium. In: *Minerals in Animal and Human Nutrition. Animal Feeding and Nutrition*, 265–275. London, NY: AP Inc.

McQuinn, S.D., Sleper, D.A., Mayland, H.F., and Krause, G.F. (1991). Genetic variation for selenium content in tall fescue. *Crop Science* 31 (3): 617–620.

Mertz, W. and Underwood, E.J. (1986). *Trace Elements in Human and Animal Nutrition*, 5e. Orlando, FL: Academic Press.

Mertz, W., Morris, E.R., Smith, J.C. et al. (1989). Trace elements in the elderly. In: *Nutrition, Aging, and the Elderly* (eds. D. Danford and H.N. Munro), 195–244. Boston, MA: Springer.

Miller, W.T. and Williams, K.T. (1940). Minimum lethal dose of selenium as sodium selenite for horses, mules, cattle and swine. *Journal of Agricultural Research* 60 (3): 163–174.

Moxon, A.L. (1937). Alkali disease or selenium poisoning. *S. Dakota Agric Exp Sta Bull* 311: 22–87.

Moxon, A.L. and Wilson, W.O. (1944). Selenium-arsenic antagonism in poultry. *Poultry Science* 23 (2): 149–151.

National Research Council (NRC). (1983). *Selenium in Nutrition*. 2nd rev. ed. Washington, DC: National Academy of Sciences.

National Research Council (NRC). (1984). *Nutrient Requirements of Beef Cattle*, 6e. Washington, DC, USA: National Academy Press.

National Research Council (NRC). (1985). *Nutrient Requirements of Sheep*, 6e. Washington, DC, USA: National Academy Press.

National Research Council (NRC). (1996). *Nutrient Requirements of Beef Cattle*, 7th rev. ed. (Update 2000). Washington, DC, USA: National Academy Press.

National Research Council (NRC). (2001). *Nutrient Requirements of Dairy Cattle*. 7th rev. ed. Washington, DC, USA: National Academy Press.

National Research Council (NRC). (2007). Minerals. In: *Nutrient Requirements of Horses*, 6e (ed. National Research Council), 94–97. Washington, DC, USA: National Academies Press.

Ohlendorf, H.M. (1989). Bioaccumulation and effects of selenium in wildlife. *Selenium in Agriculture and the Environment* 23: 133–177.

Oldfield, J.E. (2002). A brief history of selenium research: From alkali disease to prostate cancer (from poison to prevention). *Journal of Animal Science* [Online] http://www. asas. org/Bios/Oldfieldhist. pdf. [Accessed May. 15, 2020].

Olson, O.E. and Embry, L B. (1973). Chronic selenite toxicity in cattle. *Nature* 188: 1117.

Ort, J.F. and Latshaw, J.D. (1978). The toxic level of sodium selenite in the diet of laying chickens. *The Journal of Nutrition* 108 (7): 1114–1120.

O'Toole, D. and Raisbeck, M.F. (1995). Pathology of experimentally induced chronic selenosis (alkali disease) in yearling cattle. *Journal of Veterinary Diagnostic Investigation* 7 (3): 364–373.

Pagan, J.D., Karnezos, P., Kennedy, M.A.P. et al. (1999). Effect of selenium source on selenium digestibility and retention in exercised thoroughbreds. *Proceedings of Equine Nutrition and Physiology Society* 76 (4): 135–140.

Palmer, I.S., Thiex, N., and Olson, O.E. (1983). Dietary selenium and arsenic effects in rats. *Nutrition Reports International* 27: 249.

Papazafeiriou, A.Z., Lakis, C., Stefanou, S. et al. (2016). Trace elements content of plant material growing on alkaline organix soils and its suitabilty for small ruminant extensive farming. *Bulgarian Journal of Agricultural Science* 22: 733–739.

Pascual, A. and Aranda, A. (2013). Thyroid hormone receptors, cell growth and differentiation. *Biochimica et Biophysica Acta (BBA) – General Subjects* 1830 (7): 3908–3916.

Peng, F., Guo, X., Li, Z. et al. (2016). Antimutagenic effects of selenium-enriched polysaccharides from pyracantha fortuneana through suppression of cytochrome P450 1A subfamily in the mouse liver. *Molecules* 21 (12): 1731.

Pilon-Smits, E.A. and LeDuc, D.L. (2009). Phytoremediation of selenium using transgenic plants. *Current Opinion in Biotechnology* 20 (2): 207–212.

Poley, W.E. and Moxon, A.L. (1938). Tolerance levels of seleniferous grains in laying rations. *Poultry Science* 17 (1): 72–76.

Poley, W.E., Moxon, A.L., and Franke, K.W. (1937). Further studies of the effects of selenium poisoning on hatchability. *Poultry Science* 16 (4): 219–225.

Poley, W.E., Wilson, W.O., Moxon, A.L., and Taylor, J.B. (1941). The effect of selenized grains on the rate of growth in chicks. *Poultry Science* 20 (2): 171–179.

Poulsen, H.D., Danielsen, V., Nielsen, T.K., and Wolstrup, C. (1989). Excessive dietary selenium to primiparous sows and their offspring. I. Influence on reproduction and growth. *Acta Veterinaria Scandinavica* 30 (4): 371–378.

Prakash, N.T., Sharma, N., Prakash, R., and Acharya, R. (2010). Removal of selenium from Se enriched natural soils by a consortium of bacillus isolates. *Bulletin of Environmental Contamination and Toxicology* 85 (2): 214–218.

Rahim, A.A. (2005). The relationship between whole blood selenium (Se) concentration and the activity of the seleno-enzyme, glutathione peroxidase (GSH-Px, ECI 11.1. 9) in camel (*Camelus* dromedarius). *Journal of Arid Environments* 62 (2): 359–362.

Raisbeck, M.F., O'Toole, D., Sanchez, D.A. et al. (1995). Re-evaluation of selenium toxicity in grazing mammals. In: *Decades Later: A Time for Reassessment* (ed. American Society for Surface Mining and Reclamation), 372–383). Princeton, WV: American Society for Surface Mining and Reclamation.

Rayman, M.P. (2000). The importance of selenium to human health. *The Lancet* 356 (9225): 233–241.

Ridgway, L.P. and Karnofsky, D.A. (1952). The effects of metals on the chick embryo: toxicity and production of abnormalities in development. *Annals of the New York; Academy of Sciences* 55 (2): 203–215.

Rogers, P.A.M., Arora, S.P., Fleming, G.A. et al. (1990). Selenium toxicity in farm animals: treatment and prevention. *Irish Veterinary Journal* 43 (6): 151–153.

Ropstad, E., Osterås, O., Overnes, G., and Frøslie, A. (1988). Seasonal variation of selenium status of Norwegian dairy cows and effects of selenium supplementation. *Acta Veterinaria Scandinavica* 29 (2): 159–164.

Rosenfeld, I. and Beath, O.A. (1947). Congenital malformations of eyes of sheep. *Journal of Agricultural Research* 75 (93–103): 184.

Rosenfeld, I. and Beath, O.A. (1964). *Selenium – Geobotany, Biochemistry, Toxicity and Nutrition*. New York;: Academic Press.

Rosenfeld, I. and Beath, O.A. (2013). *Selenium: Geobotany, Biochemistry, Toxicity, and Nutrition*, 411. Academic Press.

Schiavon, M. and Pilon-Smits, E.A. (2017). The fascinating facets of plant selenium accumulation–biochemistry, physiology, evolution and ecology. *New Phytologist* 213 (4): 1582–1596.

Schutz, D.F. and Turekian, K.K. (1965). The investigation of the geographical and vertical distribution of several trace elements in sea water using neutron activation analysis. *Geochimica et Cosmochimica Acta* 29 (4): 259–313.

Seboussi, R., Al-Hadrami, G., Askar, M., and Faye, B. (2008). Effect of excess selenium on dromedary camel in the United Arab Emirates. In: *Impact of Pollution on Animal Products* (eds. B. Faye and Y. Sinyavskiy), 143–146. Dordrecht: Springer.

Seboussi, R., Faye, B., and Alhadrami, G. (2012). Selenium toxicity in the dromedary camels clinical symptoms and lesions. ISOCARD.

Shortridge, E.H., O'Hara, P.J., and Marshall, P.M. (1971). Acute selenium poisoning in cattle. *New Zealand Veterinary Journal* 19 (3): 47–50.

de Souza, M.P., Chu, D., Zhao, M. et al. (1999). Rhizosphere bacteria enhance selenium accumulation and volatilization by Indian mustard. *Plant Physiology* 119 (2): 565–574.

Spears, J.W. and Weiss, W.P. (2008). Role of antioxidants and trace elements in health and immunity of transition dairy cows. *The Veterinary Journal* 176 (1): 70–76.

Stephen, R.C., Saville, D.J., and Watkinson, J.H. (1989). The effects of sodium selenate applications on growth and selenium concentration in wheat. *New Zealand Journal of Crop and Horticultural Science* 17 (3): 229–237.

Stolzoff, M. and Webster, T.J. (2016). Reducing bone cancer cell functions using selenium nanocomposites. *Journal of Biomedical Materials Research Part A* 104 (2): 476–482.

Sukra, Y., Sastrohadinoto, S., Budiarso, I.T., and Bird, H.R. (1976). Effect of selenium and mercury on gross morphology and histopathology of chick embryos. *Poultry Science* 55 (6): 2424–2433.

Surai, P.F. and Fisinin, V.I. (2015). Selenium in pig nutrition and reproduction: boars and semen quality – a review. *Asian-Australasian Journal of Animal Sciences* 28 (5): 730.

Suttle, N.F. (2010). *Mineral Nutrition of Livestock*. CABI.

Thapar, N.T., Guenthner, E., Carlson, C.W., and Olson, O.E. (1969). Dietary selenium and arsenic additions to diets for chickens over a life cycle. *Poultry Science* 48 (6): 1988–1993.

Tinggi, U. (1999). Determination of selenium in meat products by hydride generation atomic absorption spectrophotometry. *Journal of AOAC International* 82 (2): 364–367.

Tinggi, U., Patterson, C., and Reilly, U. (2001). Selenium levels in cow's milk from different regions of Australia. *International Journal of Food Sciences and Nutrition* 52 (1): 43–51.

Tinggi, U., Reilly, C., and Patterson, C.M. (1992). Determination of selenium in foodstuffs using spectrofluorometry and hydride generation atomic absorption spectrometry. *Journal of Food Composition and Analysis* 5 (4): 269–280.

Tiwary, A.K., Stegelmeier, B.L., Panter, K.E. et al. (2006). Comparative toxicosis of sodium selenite and selenomethionine in lambs. *Journal of Veterinary Diagnostic Investigation* 18 (1): 61–70.

Todorović, M., Jovanović, M., Jokić, Ž. et al. (2004). Alterations in liver and kidneys of chickens fed with high levels of sodium selenite or selenized yeast. *Acta Veterinaria* 54 (2–3): 191–200.

Trelease, S.F. and Beath, O.A. (1949). *Selenium: Its Geological Occurrence and Its Biological Effects in Relation to Botany, Chemistry, Agriculture, Nutrition, and Medicine*. Burlington, VT: The Champlain Printers.

Underwood, E.J. (1979). Trace elements and health: an overview. *Philosophical Transactions of the Royal Society of London B, Biological Sciences* 288 (1026): 5–14.

Wadgaonkar, S.L., Ferraro, A., Race, M. et al. (2018). Optimization of soil washing to reduce the selenium levels of Seleniferous soil from Punjab, Northwestern India. *Journal of Environmental Quality* 47 (6): 1530–1537.

Wahlstrom, R.C. and Olson, O.E. (1959). The relation of pre-natal and pre-weaning treatment to the effect of arsanilic acid on selenium poisoning in weanling pigs. *Journal of Animal Science* 18 (2): 578–582.

Wahlstrom, R.C., Kamstra, L.D., and Olson, O.E. (1956). Preventing selenium poisoning in growing and fattening pigs. Research Bulletins of the South Dakota Agricultural Experiment Station (1887–2011) 456.

Wahlstrom, R.C., Goehring, T.B., Johnson, D.D. et al. (1984). The relationship of hair color to selenium content of hair and selenosis in swine. *Nutrition Reports International* 29 (1): 143–147.

Whelan, B.R., Peter, D.W., and Barrow, N.J. (1994). Selenium fertilizers for pastures grazed by sheep. 1. Selenium concentrations in whole blood and plasma. *Australian Journal of Agricultural Research* 45 (4): 863–875.

White, P.J. (2016). Selenium accumulation by plants. *Annals of Botany* 117 (2): 217–235.

WHO. (1987). *Environmental Health Criteria 58, IPLS International Program on Chemical Safety*. Geneva, Switzerland: World Health Organization.

Wilson, T.M., Scholz, R., and Drake, T. (1982). Porcine focal symmetrical poliomyelomalacia and selenium toxicity: description of a field outbreak and preliminary observations on experimental reproduction [Feeder pigs]. In *Proceedings of the Annual Meeting of American Association of Veterinary Laboratory Diagnosticians*.

Wood, M. (2000). Kenaf and canola – selenium slurpers. *Agricultural Research* 48 (6): 10.

Wu, L., Huang, Z.Z., and Burau, R.G. (1988). Selenium accumulation and selenium-salt cotolerance in five grass species. *Crop Science* 28 (3): 517–522.

Yaeger, M.J., Neiger, R.D., Holler, L. et al. (1998). The effect of subclinical selenium toxicosis on pregnant beef cattle. *Journal of Veterinary Diagnostic Investigation* 10 (3): 268–273.

Young, J.D., Crowley, C., and Tucker, E.M. (1981). Haemolysis of normal and glutathione-deficient sheep erythrocytes by selenite and tellurite. *Biochemical Pharmacology* 30 (18): 2527–2530.

Zong-Ping, L., Zhuo, M., and You-Jia, Z. (1994). Studies on the relationship between sway disease of Bactrian camels and copper status in Gansu province. *Veterinary Research Communications* 18 (4): 251–260.

5

Positive and Negative Impacts of Selenium on Human Health and Phytotoxicity

Gyanendra Tripathi[1], Dhirendra Kumar Srivastava[2], and Vishal Mishra[3]

[1] *Department of Bioengineering, Integral University, Lucknow, Uttar Pradesh, India*
[2] *Council of Science and Technology, Lucknow, Uttar Pradesh, India*
[3] *School of Biochemical Engineering, Indian Institute of Technology (BHU), Varanasi, Uttar Pradesh, India*

5.1 Introduction

Selenium (Se) is a basic supplement having a moderately small range between the intake amounts that bring about the toxicity (>400 μg/day) and the deficiency (<40 μg/day) in living beings (WHO 1996). It is important to deliberately control consumption by people and different animals. It is also important to understand the significance of the connections between natural exposure and health. Topography applies control over the Se level in soil which is used in agriculture and animal that structure the food chain for humans. The selenium concentration status in animals, the human population, and agriculture changes notably around the globe because of various topographical conditions. As in humans, for Se contamination the feed is a significant source, and the knowledge of biogeochemical control on the circulation and versatility of natural selenium is a crucial step in tracing Se-related health toxicity. A high concentration of selenium is related to some phosphatic rocks, coals, sulfide mineralization, and organic-rich black shales, though most other rocks contain low concentration and selenium-inadequate environments are much more common than seleniferous ones. The health results are not just reliant on the overall Se concentration of soils and rock but depend also on Se concentration absorbed by the plants and taken by animals – the bioavailable selenium.

A leap forward in selenium (Se) investigation appeared when a report indicated that expansion of selenium in feed prevented liver cirrhosis and muscular dystrophy in rats (Rayman 2000). The discovery of the glutathione peroxidase enzyme was another defining turn in Se research (Behne and Kyriakopoulos 2001; Rotruck et al. 1973). From that point, the vitality of selenium for humans and animals came into the spotlight and it is now known to be a fundamental nutrient in human food (Hartikainen 2005). Even though Se acts as a major role in many cell cycles, its anticancerous and antioxidant characteristics are of major concern for humankind (Hatfield et al. 2014; Reid et al. 2008).

Selenium Contamination in Water, First Edition. Edited by Pooja Devi, Pardeep Singh, Arindam Malakar, and Daniel Snow.

Selenocysteine (SeCys), selenomethionine (SeMet), and seleno-aminoacids are liable for most reported biomedical impacts of Se in diet (Dumont et al. 2006). For some selenoproteins, Se serves as a catalytic center, for example, iodothyronine-deiodinases, thioredoxin reductase, and glutathione peroxidase (GSHPx), thus it is significant in the strengthening of the immune system, protection against oxidative stress, scavenging of free radicals, and so forth (Kaur et al. 2014). A deficiency of Se in the human diet has caused abnormalities in thyroid function, growth retardation, and metabolism of impaired bone (Reeves and Hoffman 2009). Selenium deficiency in animals is linked to muscular dystrophy, muscular weakness, poor reproductivity, embryonic deformities, poor growth, and poor appetite. These disorders are in general known as white muscle disease (WMD). As a result of these discoveries, deficiency of selenium in livestock and crops is reported worldwide, including in the UK, Finland, the United States, Sri Lanka, Denmark, New Zealand, India, Africa, Australia, Canada, China, and Thailand. Thus, the supplementation of Se in agriculture has turned out to be a common practice by farmers (Levander 1986; WHO 1987).

The first recorded report for selenium toxicity was by Marco Polo when a report by him first told about a disease known as "hoof rot" found in the horses in Turkestan near Tien Shan and Nan Shan mountains, where the selenium concentration was high in soils (Rayman 2000). Digestion of forage and grain containing Se concentration of 5–50 ppm has resulted in a disease called Alkali disease, which was the primary appearance of selenium chronic toxicity or selenosis in horses and cattle (Dickerson and Smith 1994). In 1860 a US army surgeon first reported a document about selenium poisoning in livestock. Fatal disease in horses that gazed around Fort Randall in South Dakota was reported (Martin 1973). The element selenium gained worldwide concern when in 1980 selenium toxicity came into the spotlight after the incidence of Kesterson Wildlife Reservoir and Sanctuary in the San Joaquin Valley of California (Winkel et al. 2012)

Selenium is found in multiple oxidation states or valence states (−2, 0, +4, and +6). These forms of selenium present are the result of its valence states. The selenides (−2) are elemental selenium, a heavy metal unsolvable in water. The selenites and selenates (+4 and +6) are solvable in water, this property makes it more bioavailable (Weast 1988). In different environmental conditions (temperatures, pH, etc.), these oxidation states may vary. The elemental Se is most stable in the soil. A low concentration of it is found in water as a result of its property of coprecipitation with sediments. Selenates that are soluble are absorbed by the plants living in the aquatic ecosystem. These absorbed selenate in plants are changed to carbon-based compounds (organic), for example dimethyl selenide, selenomethionine, selenocysteine, and dimethyl diselenide. As a result, bioaccumulation and biomagnification of selenium in an aquatic organism are observed (Weast 1988).

A report by Superfund Amendments and Reauthorization Act (SARA) Section 313, Toxic Release Inventory (TRI 2002), around 264 to 267 pounds off elemental selenium was released into air, land, and water from processing and manufacturing units in 2000 in the United States. Also, selenium compounds totaling around 7 870 609 pounds were released into the land, water, and air in 2000. All amenities that import, process, and manufacture Se and compounds related to it, are included in this data.

5.2 Exposure of Selenium in the Environment

According to a report by HazDat 2003 in the United States out of 1623 hazardous waste sites, 508 sites have been identified to have selenium as a pollutant, this forced its inclusion in Environmental Protection Agency (EPA) National Priorities List (NPL).

Selenium is omnipresent on earth and is discharged from anthropogenic and natural sources. Major exposure of selenium in the environment is the result of interference by humans that results from the ignition of coal (Figure 5.1). Laborers working in metals industries, health administrations, painters, and mechanics are prone to a more significant concentration of selenium exposure than everyone others in a population. For everyone, the primary exposure to selenium toxicity is water, air, and soil.

5.2.1 Water

The wet deposition by subsurface and surface drainage, and dry deposition by the atmosphere results in the accumulation of selenium in water. In seawater the Se is present at a concentration of 0.04–0.12 μg Se/l which is generally low, while in surface and groundwater the concentration differs widely at 0.06–400 μg Se/l (Chau and Riley 1965; Höghberg and Alexander 1986). In water supplies, the concentration exceeds 10 μgSe/l (NAS 1977). A portion of Se is exposed to water from agricultural runoff, industrial wastewater, and sewage effluent area having high selenium concentration. Selenites and selenates, being water-soluble, can be present in water sources. In the environment, sodium selenite is one of the most movable selenium compounds present.

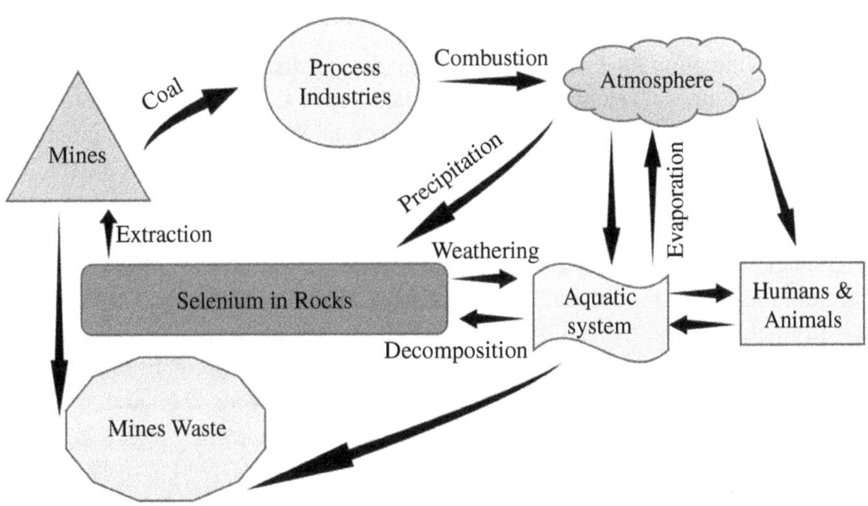

Figure 5.1 Schematic representation of selenium cycle in the environment.

5.2.2 Soil

The main source for exposure of Se to the soil is weathering and leaching of the solid rocks, although wet and dry deposition also give their part to soil Se concentration (Figure 5.1). Se in a form of selenides and selenites with sulfide minerals are present in bedrocks. On water surface and water contained in soil, selenious acid and salts of selenic are found. Weathering of Se-containing rocks to soils, selenium volatilization by bacteria and plant, and volcanic activity are the natural causes of Se exposure to the environment (Fishbone 1983; Lewis et al. 1971). In the soil Se is present as basic ferric selenite and calcium selenate. The decay of animal and plant material also results in the deposition of Se in soil. Operations such as oxidation of Se-bearing pyrite in coal mining also release Se into the soil (Figure 5.1). A level of 0.1–2 mg Se/kg has been recorded in most soils (Dreher and Finkelman 1992). However, this concentration <1200 mg/kg in high-Se areas (Fishbein 1983).

5.2.3 Air

Dimethyl selenide, selenium dioxide, and methyl selenide are the most common forms of Se present in the air (Committee on Medical and Biological Effects, Environmental Pollution Subcommittee 1976). The main reason for Se exposure in the air is the combustion of fossil fuel and coal (Figure 5.1). The volcanic eruption is the main source of Se exposure and biomethylation of Se also releases Se in the air (ASTDR 1997). The plants that run by the energy produced by coal-fired steam are the source of Se discharge via fly ash, and this mostly pollutes water bodies rather than air and lands. Normally, the Se concentration in the air is recorded between 0.1 and 10 ng Se/m^3 and in areas having a high level are recorded up to 500 ng Se/m (Committee on Medical and Biological Effects, Environmental Pollution Subcommittee 1976).

5.2.4 Direct Exposure of Selenium to Humans by Feed

For the general population food material like cereal, grains, fish, meat, and poultry products are the major source of selenium exposure (Mahaffey et al. 1975). According to a survey conducted on a typical US diet, around half of the total Se intake in the human diet is by the consumption of white bread, chicken, beef, eggs, and pork. Animals organs like liver, kidney, and fish contain the highest amount of selenium concentration followed by eggs and meat (Schubert et al. 1987). The bioavailability of selenium is higher in wheat, yeast, and plant materials as compared to animal sources, and this makes them the finest source of Se in the diet (Fairweather-Tait 1997). A low concentration level (<0.01 mg Se/kg) is found in vegetables and fruits. Cereals have an intermediate level of Se. In general, selenate is present in drinking water, but this also depends upon the presence of other metals, pH, E_h of the water. In water, the Se bioavailability is less when compared to the Se bioavailability in food material.

5.3 Effect of Selenium on Human Health

In humans, Se plays a vital role in many biological functions (Figure 5.2).

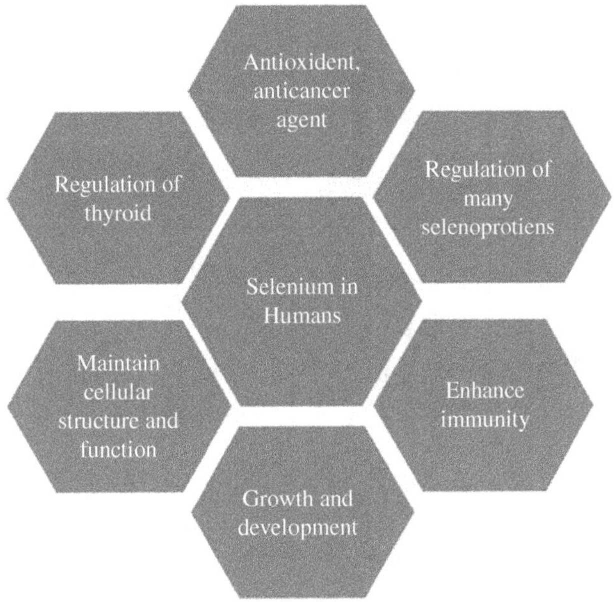

Figure 5.2 Beneficial effect of selenium in humans.

Around 25 selenocysteine containing essential selenoproteins are recognized. Various selenoproteins work as a component of the catalytic cycle and play an important role in redox reaction (Rayman 2012; World Health Organization 1996) (Figure 5.2). Some of the selenoproteins with known health activity are listed in Table 5.1.

The activity of Se in the enzyme is credited to iodothyronine-deiodinase, thioredoxin reductase, glutathione peroxidase (GPx), and selenophosphate synthetase groups (Figure 5.2). Being an important part of GPx, the complex formation of Se with polyunsaturated fatty acid and Vitamin E plays an important role; it catalyzes the lipid organic hydroperoxides or hydrogen peroxide reduction. Selenium also seems to enhance the natural killer cell activity and the production of T cells (Rayman 2012) (Figure 5.2). Se is linked to immune system function, enzyme activation, pancreatic function, detoxifying xenobiotic agent, and DNA repair. In humans and animals, Se has also affected the enzyme P-450 cytochrome system. However, the establishment of an exact biochemical role of selenium is still in progress. Under selenium-deficient conditions, no decrease in P450 enzyme activity has been observed, though inhibition of P-450 enzymes has been suggested in a few studies. In laboratory animals, Se has been reported to prevent cancers; this can be explained as enzymes like P450 activate substances having carcinogenic properties such as phenobarbital and polynuclear aromatic hydrocarbons (Shimada et al. 2011). Arthur and Beckett (1994) suggested that selenium is valuable in the metabolism of thyroid hormone (Figure 5.2). Three iodothyronine-deiodinase (Dio) selenoenzymes are present. Types I and II play roles in the formation of active III, 3- and 5-triiodothyronine (T_3) hormones, while the thyroxine (T_4) conversion to inactive $T_3(rT_3)$ is catalyzed by type III Dio. These hormones work as the main influencer on growth, cellular development, and differentiation, particularly in the child and fetus.

Table 5.1 List of health-associated selenoproteins with known functions.

S.No	Selenoproteins	The function of health result	Selenoprotein-assisted health Effects	References
1.	Selenoprotein P (SEPP1)	10 selenocysteine residues are present; a gauge of selenium status and is a main contributor to plasma Se, transports Se from the liver by mean of the plasma: special receptors are present on the kidney, testis, and brain; has some anticancerous function; needed by brain; deficiency of Se causes spontaneous seizures and irregular movements, in mice; crucial for the fertility of the male; deficiency resulted in infertility with hypomotile and spermatozoa (wrinkled) in mice; related to fasting plasma glucose; it assists as a heavy metal chelator.	Prostate cancer; alter Se level (plasma selenium) and other selenoproteins regulation; adenoma, colorectal cancer	Olson et al. (2005), Schweizer et al. (2004), Yang et al. (2011), Reeves and Hoffman (2009)
2.	Selenoprotein S (SEPS1)	Performs anti-inflammatory function, found in endoplasmic reticulum; ER stress-induced apoptosis effect prevention in cell; linked to the metabolism of glucose and insulin sensitivity.	–	Gao et al. (2003), Reeves and Hoffman (2009)
3.	Selenoprotein N (SelN)	Situated in the endoplasmic reticulum; controls calcium movement essential for early development of muscle; mutations cause multimonitor disease.	–	Reeves and Hoffman (2009)
4.	Thioredoxin reductases (TrxR)	Have a wide range of substrates; redox-active; for DNA synthesis thioredoxin is required.	–	Reeves and Hoffman (2009)
	TrxR1 (cytoplasmic/ nuclear)	Transcription factors activity control, cell apoptosis, proliferation; slower tumor-cell growth is a result of the reduction in expression.	Progressive colorectal adenoma; ancestral amyotrophic lateral sclerosis	Reeves and Hoffman (2009)
	TrxR2 (mitochondrial)	Crucial for the viability of cardiomyocyte.	Gastric cancer: Interaction of TrxR2Assisted to SNP with GPx1 (Leu/ Pro) to result in the risk	Reeves and Hoffman (2009)
	TrxR3 (testis-specific)	–	–	–

5.	Glutathione peroxidases (GPxs)	Antioxidant enzymes: eliminate lipid hydroperoxides, phospholipid cholesterol hydroperoxides, and hydrogen peroxide	–	Reeves and Hoffman (2009)
	GPx1 (cytosolic)	Prevent viral mutations that decrease retroviral virulence; deficiency leads to cardiomyopathy	Cardiometabolic effects: CAD, CVD, metabolic syndrome, coronary artery calcium score, blood pressure, restenosis, intimamedia thickness, intracerebral hemorrhage. Thoracic aortic aneurysm, peripheral vascular disease. Cancer: prostate, liver, lung, bladder; Keshan disease, low blood selenium of GPx1-198Leu carriers and low activity of GPx1; causes Kashin–Beck disease, GPx1-198Leu carriers GPx lower activity; autism	Lei et al. (2009). Beck et al. (2003)
	GPx2 (gastrointestinal)	In the colon, crypts play antiapoptotic function; intestinal mucosal integrity maintenance	–	Florian et al. (2010)
	GPx3 (plasma)	In extracellular fluids as an antioxidant; in plasma, the kidney is the source of GPx3; protection of thyroid in follicular lumen and thyrocytes from hydrogen peroxide	Ischemic stroke; differentiated thyroid cancer	Schiavon et al. (1994), Schmutzler et al. (2007), Reeves and Hoffman (2009)
	GPx4 (phospholipid)	Associated with membrane; in testis for motility and viability of sperm high concentration is present	Colorectal adenocarcinomas, adenomatous polyps; breast cancer; colorectal cancer survival	Foresta et al. (2002), Ursini et al. (1999)
6.	Selenoprotein (SEP15) 15 kDa	Present in the ER; folding of glycoprotein is affected	Lung, rectal, Prostate cancer mortality	Reeves and Hoffman (2009)

(*Continued*)

Table 5.1 (Continued)

S.No	Selenoproteins	The function of health result	Selenoprotein-assisted health Effects	References
7.	Iodothyronine-deiodinases	T_3, and reverse T_3 (rT3) production of an active thyroid.	–	Schomburg and Köhrle (2008)
	Dio1 (kidney, thyroid, liver, etc.)	T_3 Production in the peripheral tissues and thyroid	Muscle strength, lean body mass, Free IGF-1 concentrations	Schomburg and Köhrle (2008)
	Dio2 (pituitary, brain, muscle, heart, ear, etc.)	T_3 production in peripheral tissue	Insulin resistance and type-2 diabetes; bone-mineral density and osteoarthritis; mental retardation	Schomburg and Köhrle (2008)
	Dio3 (placenta, skin, pregnant uterus, cerebral cortex)	rT3 production; prevents the fetus from overexposure of T_3	Osteoarthritis	Schomburg and Köhrle (2008)

5.3.1 Toxicity of Selenium

Three types of exposure history were generally responsible for selenium poisoning. Accumulation of selenium from naturally seleniferous soils in plants is the first type of exposure. The second is from an overdose of Se in feed or by injection. The third is from contaminated waters or plant accumulation, which ultimately results in environmental contamination. The exposure rate of these types of Se poisonings by daily exposure results in chronic, acute, or subacute selenium poisoning. However, it is important to note that the age of animals or humans has a major part in exposure to Se poisoning, as adults are more tolerant than young (Raisbeck 2000).

Acute Se toxicity begins at 8–10 hours of ingestion but delayed is reported of up to 36 hours in clinical manifestation (NRC 1983; Raisbeck 2000; Tiwary et al. 2006). Dimethylselenide has a garlicky smell that can be recognized on the breath in early clinical syndrome. Other clinical signs are also observed, such as lethargy, respiratory distress, head down, anorexia, droopy ears, salivation, fever, tachycardia, sweating, watery diarrhea, gaunt appearance, tetanic spasms, teeth grinding, and stilted gait. After the first observation, the progression of clinical signs develops quickly: histologic lesions show pulmonary edema, systemic congestion, petechial hemorrhages, myocardial necrosis, and skeletal muscle necrosis. In the 1960s, in Enshi District, a human selenosis outbreak was reported. This outbreak was related to a high intake of Se crops cultivated on soils having > 300 mg/kg Se concentration (Table 5.2). Around 1961 to 1964 the rate of illness reached half in the most affected community, which was situated in remote regions among the populations of surviving farmers. The primary symptoms of the disease are nail and hair loss, and some secondary symptoms were also reported, such as disorders of the skin, garlic breath, paralysis, poor dental health, and nervous system symptoms. Though no health examination was done at that time, subsequent investigation were carried out in these regions in the 1970s that uncovered high dietary consumption of 3.2–6.8 mg with 1.3–7.5 mg/l level of Se in the blood and 4.1–100 mg/kg level in the hair (Tan 1989; Yang et al. 1983) (Table 5.2). Clinical observation of chronic selenosis is weakness, emaciation, reproductive failure, weight loss, hoof abnormalities, depression, hair loss, anemia, diarrhea, anorexia, and lameness, (Davis et al. 2014; Rosenfield and Beath 1964). Se-related toxicity to supplementation of mineral has been reported in the US – 12 cases of selenosis as a result of consuming 0.15–0.17 mg labeled selenium tablets for 77 days which actually contained 27–31 mg Se were described in 1984. Clinical observation of patients was nausea, nail damage, fatigue, irritability, hair loss, watery diarrhea, abdominal cramps, vomiting, garlic breath, and skin irritation. Se levels of 0.528 mg/l in the blood serum were reported (WHO 1987). Hawkes and Turek in 2001 reported that in healthy men, a high intake of Se has resulted in a decrease in sperm motility. High Se intake is also reported to enhance the chances with many other types of cancer, such as skin and pancreatic cancer. In some cancers, trials reporting Se supplementation have also reported an amplified risk of type-II diabetes. Possibly high selenium affects the insulin signaling pathways (Rayman 2012). The upper limit of mineral supplementation of selenium according to US-EPA recommendation is 0.1 mg/kg.

Many scientists have done case studies during different periods to study selenium toxicity on human health. For this, they have considered physical losses such as hair loss, nail loss, tooth decay, blackening of teeth, the concentration of Se in urine, blood, a shift in the

Table 5.2 Case histories of selenium toxicity.

S.No	Population study	Health effect	Laboratory indication	References
1.	Tap water with increased levels of Se (50–125 μg/l) was supplied to 120 households.	No noteworthy difference was observed between the unexposed and exposed groups of people; however, women experienced miscarriage.	A high Se content was detected in the urine of exposed group. Much higher Se levels in urine were seen in women with a miscarriage history.	Tsongas and Ferguson (1977)
2.	In a small community 33 residents used drinking water from wells with a higher level (26-1800 μg/l) of Se.	Not assessed.	Increased exposure to Se through potable drinking water can be linked with lower activity of GPx in blood.	Valentine et al. (1980)
3.	In 1961, some communities of Enshi country experienced endemic disease. During 1961–1964, the illness calculated was around 50%, taking 248 residents of the villages that were most affected into account. The composition of Se was 4.99 mg in their diet.	Loss of nails, the decay of the tooth, hair loss, skin infections, and abnormalities in nervous system such as acroparesthesia, pain experienced in extremities, and peripheral anesthesia.	Se level of hair was reported to be 32.2 μg/g in area of selenosis having high Se content. Similarly, Se level in blood was reported to be 3.2 μg/ml.	Yang et al. (1983)
4.	Examination of 65 mothers inhabiting three regions with varying Se dietary intake levels. The Se dietary intake range in Venezuelan seleniferous the area was found to be 250–980 μg/day μg/day.	High dietary Se supply for a long duration on the FT_3, decrease in selenoenzyme iodothyronine-deiodinase (5 DI) activity was seen if found at high levels in the serum, as it catalyzes T_3 and T_4 production.	Mean of serum thyroxin, TSH, and tri-iodothyronine was found to be at a normal level.	Bratter and Negretti de Bratter (1996)
5.	Se is the main component of carbonaceous shale (stone coal) in the region.	Little population of this area experienced hair and nail loss from the early 1930s to 1961. In 1963, symptoms of selenium poisoning were diagnosed in 19 of 23 local inhabitants.	In 1999, the Se level in soil was 4.755 ± 7.43 mg/kg. In inline water, Se content was 58.4 ± 16.8 μg/l.	Zhu et al. (2008)
6.	A comparative study was done in between 80 subjects that were living in the seleniferous area and 80 controls living in the non-seleniferous area. Intake of Se by men was found to be 632 ± 31.2 μg/day and by women 475 ± 52.8 μg/day in that area.	Hair loss, decaying of tooth or black tooth was observed in 17.5% of men and 15% of women.	When calculated, the Se concentration in the sample of hair, urine, and nails in the group was found to be higher than the control group.	Hira et al. (2004)

7.	An outbreak of human Se poisoning occurred in the1960s. High Se level in soils, water, crops, came from the Se-rich carbonaceous rock, thus leading to Se poisoning in humans. Se daily intake was 2144 μg/day; residents' consumption of cereal was considered as the main route for intake of Se, followed by drinking water, vegetables, and meats.	The population experienced unusual symptoms and signs poisoning of Se; 18 of 22 rural communities were affected by selenosis (severe) and showed neurological signs in addition to acroparesthesia and dysesthesia, convulsions, hyperreflexia, motor weakness, hemiplegia, and polyneuritis.	Se level in blood was 3248 μg/l.	Qin et al. (2013)
8.	Around 2065 municipal residents consumed high Se level in drinking water that was reaching 10 μg/l (European standard) in selenate between the years 1974 and 1985.	As a result of inorganic Se, increase mortality rate due to site-specific cancers (multiple myeloma, melanoma, kidney, colorectal, lymphoid neoplasms, cancer) and neurodegenerative diseases like Parkinson's disease.	Not assessed.	Vinceti et al. (2013)
9.	650 human subjects inhabiting the seleniferous area were related to 50 controls (healthy) coming from a non-seleniferous area.	The human population living in the area with Se chronic exposure coming from soil–plant water was found to be at risk of impaired organ function development.	Hair Se content, when calculated for the exposed group, came around 50.9 ± 58.0 μg/g and 22.5 ± 10.7 μg/g in the control population.	Chawla et al. (2015)

hormonal content in the blood, etc. Table 5.2 describes some of the major Se toxicity outbreaks throughout the world. It gives a piece of brief information about the population study exposed to Se, its health effect on an individual or a population, and laboratory indication of the sample.

5.4 Selenium Phytotoxicity

5.4.1 Se Uptake and Accumulation

The roles of Se, positive or negative, in plants are interconnected and together contribute to the overall development and growth of the plant under a non-stressed and stressed condition. Uptake of Se, translocation, and then its distribution depends upon the phases of plant development, species, Se concentration, physiological conditions (pH of soil), membrane transporters activity, and the plant's translocation mechanism (Renkema et al. 2012; Zhao et al. 2005). In agricultural soils, selenate (SeO_4^{2-}) is the most abundantly present and is found to be more water-soluble in comparison to selenite (Missana et al. 2009; Sors et al. 2005). In acidic soils, Se exists mainly as selenite, whereas in alkaline soils it exists as selenate. These Se forms differ only in terms of their absorption and mobility inside the

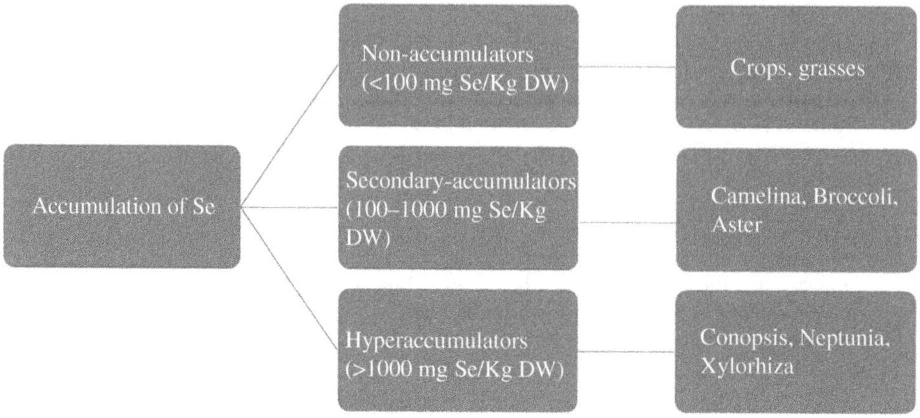

Figure 5.3 Classification based on Se accumulation.

plant and form seven compounds after metabolization. Uptake of selenate in plants is facilitated by sulfur channels (Zhang et al. 2003) and transporters by the mechanism of phosphate transport (Li et al. 2008). In general, during seedling growth, the accumulation of Se is recorded higher in new leaves than in older leaves (Harris et al. 2014). Mostly, Se is accumulated in plant cells in their vacuoles (Gigolashvili and Kopriva 2014). Plants are classified depending upon their capacity to accumulate Se inside them as secondary accumulators, non-accumulators, and hyperaccumulators (Bodnar et al. 2012) (Figure 5.3). In Se-rich regions the accumulation of Se by hyperaccumulators is recorded to be more than 1000 mgSe/Kg DW. The Se tolerance of these plants is conferred because of the presence of methylated forms of SeMet and SeCys, which can be further vaporized as dimethyldiselenide (DMDSe).Species like *Conopsis*, *Stanleya*, *Xylorhiza*, *Neptunia*, *Astragalus species*, etc. are included in hyperaccumulators. Se accumulation by the secondary accumulators has shown no sign of toxicity up to 100–1000 mgSe/Kg of DW. Species like *Brassica napus*, *Brassica juncea*, *Helianthus*, *Broccoli*, *Camelina*, *Aster*, etc. are included in secondary accumulators. Plants that accumulate Se < 100 mgSe/Kg of DW are non-accumulators, like crops and grasses (Bodnar et al. 2012; Galeas et al. 2007) (Figure 5.3).

5.4.2 Effect of Se on Plants

A low dose of selenium protects the plants from many abiotic stresses like desiccation (Pukacka et al. 2011), drought condition (Hasanuzzaman and Fujita 2011), cold (Chu et al. 2010), and metal stress (Kumar et al. 2012; Pandey and Gupta 2015), (Figure 5.4). Under stress conditions, reactive oxygen species (ROS) are formed which disrupt cell membranes, protein, etc. After the application of 50 μM Se a reduction in ROS accumulation in Cd stressed marine algae and 2 μM Se overcome Cd stressed in *Brassica* seedlings (Filek et al. 2008) (Figure 5.4).

A high dose of Se acts as a pro-oxidant and results in oxidative stress in plants, e.g. in *Viciafaba* roots under Pb stress, 1.5 μM Se decreased ROS but at 6 μM an increase in ROS and decrease in cell viability was observed (Mroczek-Zdyrska and Wójcik 2012). Selenium-mediated improved cell integrity and reduction in electrolytic leakage is also observed under many stress conditions (Malik et al. 2012). A low dose of Se has shown a positive

Figure 5.4 Beneficial effect of Se in plants.

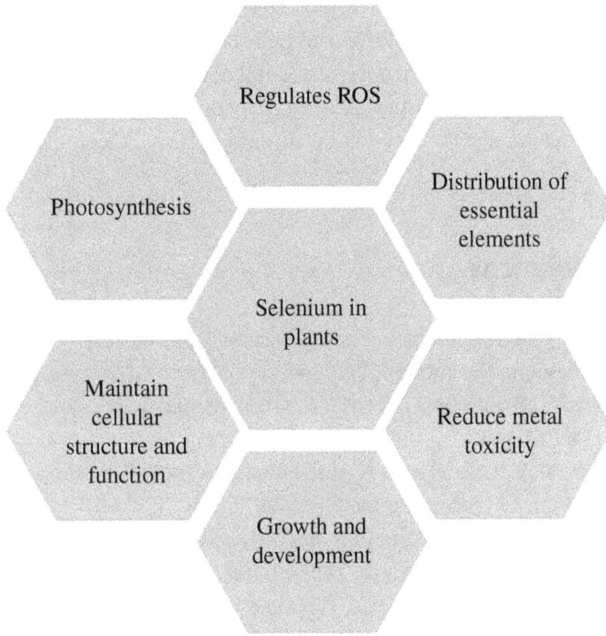

effect on photosynthesis in *Sorghum*, which could be due to increased antioxidant activity and decreased ROS levels upon Se application (Djanaguiraman et al. 2010) (Figure 5.4). In rice seedlings, increased photosynthesis at a low dose, and decreased photosynthesis at a high Se dose are reported (Wang et al. 2012). Se has been also reported to protect the plant from the toxicity of heavy metals (Feng et al. 2011; ZemZembala et al. 2010) (Figure 5.4). Apart from the above-mentioned role, Se has also been reported to increase the nutritive value of potato (Turakainen et al. 2006), increase yield but delay senescence in *Cucurbita pepo* (Xue et al. 2001), protect the plant from the herbivores, pathogens, and insects (Freeman et al. 2006), and increase respiratory potential in *Chicory* (Germ et al. 2007a), *Eruca sativa* (Germ and Osvald 2005), and *Pisum sativum* (Smrkolj et al. 2006).

5.5 Conclusion

Se has multiple and complex effects on human and plant health, which makes it more important to be considered in terms of research to optimize the benefits and reduce the risks related to it. The main factor that needs to be emphasized is the complicated connection with selenium level, and that additional selenium intake (i.e. by supplements or food fortification) may help people with Se deficiency. On the other hand, people with adequate or high Se status could be affected adversely and must not take Se from any other external source (supplementation).

This chapter has confirmed that Se exposure to humans mainly relies on the dietary intake in water and food. For plants, the Se uptake and accumulation is from the agricultural soil. This is significantly controlled by various geological and environmental conditions of the Earth's surface. This guides researchers to do more work to understand not just the total concentration of Se present but also its bioavailability and cycling in the environment. The concentration of Se in the population of humans and animals around the world also contributes to its total environmental concentration. Collaboration between environmental and medical scientists is needed to evaluate the environmental health impact of selenium in diseases like cardiovascular disease, AIDS, cancer, diabetes, etc.

References

Agency for Toxic Substances and Disease Control. (1997). *ATSDR's Toxicological Profiles: Selenium*. Boca Raton, FL: Lewis Publishers, CRC Press, Inc.

Arthur, J.R. and Beckett, G.T. (1994). New metabolic roles for selenium. *Proc. Nutr. Soc.* 53: 615–624.

Beck, M.A., Levander, O.A., and Handy, J. (2003). Selenium deficiency, and viral infection. *J. Nutr.* 133 (suppl 1): 1463S–1467S.

Behne, D. and Kyriakopoulos, A. (2001). Mammalian selenium-containing proteins. *Annu. Rev. Nutr.* 21: 453–473. https://doi.org/10.1146/annurev.nutr.21.1.453.

Bodnar, M., Konieczka, P., and Namiesnik, J. (2012). The properties, functions, and use of selenium compounds in living organisms. *J. Environ. Sci. Health Part C* 30: 225–252. https://doi.org/10.1080/10590501.2012.705164.

Bratter, P. and Negretti de Bratter, V.E. (1996). Influence of high dietary selenium intake on the thyroid hormone level in human serum. *J. Trace Elem. Med. Biol.* 10: 163–166.

Chau, Y.K. and Riley, J.P. (1965). The detcnation of selenium in seawater, silicates, and marine organisms. *Anal. Chim. Acta* 33: 36–49.

Chawla, R., Loomba, R., Chaudhary, R.J. et al. (2015). Impact of high selenium exposure on organ function and biochemical profile of the rural population living in seleniferous soils in Punjab, India. In: *Global Advance in Selenium Research from Theory to Application* (eds. G.S. Banuelos, Z.-Q. Lin, M.F. Moraes, et al.), 93–94. São Paulo: CRC Press.

Chu, J., Yao, X., and Zhang, Z. (2010). Responses of wheat seedlings to exogenous selenium supply under cold stress. *Biol. Trace Elem. Res.* 136: 355–363. https://doi.org/10.1007/s12011-009-8542-3.

Committee on Medical and Biological Effects, Environmental Pollution Subcommittee. (1976). *Selenium*. Washington, DC: National Academy Science.

Davis, T.Z., Stegelmeier, B.L., and Hall, J.O. (2014). Analysis of horsehair as a means of evaluating selenium toxicosis and long-term exposure. *J. Agric. Food Chem.* 62: 7393–7397.

Dickerson, O.B. and Smith, T.H. (1994). Selenium, tellurium, and osmium. In: *Occupational Medicine*, 3e (eds. C. Zenz, O.B. Dickerson and E.P. Horvath Jr.), 577–579. St. Louis, MO: Mosby.

Djanaguiraman, M., Prasad, P.V., and Seppänen, M. (2010). Selenium protects sorghum leaves from oxidative damage under high temperature stress by enhancing antioxidant defense system. *Plant Physiol. Biochem.* 48: 999–1007. https://doi.org/10.1016/j.plaphy.2010.09.009.

Dreher, G.B. and Finkelman, R.B. (1992). Selenium mobilization in a surface coal mine, Powder River Basin. *Environ. Geol. Water Sci.* 19: 155–169.

Dumont, E., Vanhaecke, F., and Cornelis, R. (2006). Selenium speciation from food source to metabolites: a critical review. *Anal. Bioanal. Chem.* 385: 1304–1323. https://doi.org/10.1007/s00216-006-0529-8.

Fairweather-Tait, S.J. (1997). Bioavailability of selenium. *Eur. J. Clin. Nutr.* 51: S20–S23.

Feng, R.W., Wei, C.Y., Tu, S.X. et al. (2011). Detoxification of antimony by selenium and their interaction in paddy rice under hydroponic conditions. *Microchem. J.* 97: 57–61. https://doi.org/10.1016/j.microc.2010.06.003.

Filek, M., Keskinen, R., Hartikainen, H. et al. (2008). The protective role of selenium in rape seedlings subjected to cadmium stress. *J. Plant Physiol.* 165: 833–844. https://doi.org/10.1016/j.jplph.2007.06.006.

Fishbein, L. (1983). Environmental selenium and its significance. *Fundam. Appl. Toxicol.* 3: 411–419.

Florian, S., Krehl, S., Loewinger, M. et al. (2010). Loss of GPx2 increases apoptosis, mitosis, and GPx1 expression in the intestine of mice. *Free Radic. Biol. Med.* 49: 1694–1702.

Foresta, C., Flohé, L., Garolla, A. et al. (2002). Male fertility is linked to the selenoprotein phospholipid hydroperoxide glutathione peroxidase. *Biol. Reprod.* 67: 967–971.

Freeman, J.L., Quinn, C.F., Marcus, M.A. et al. (2006). Selenium-tolerant diamondback moth disarms hyperaccumulator plant defense. *Curr. Biol.* 16: 2181–2192. https://doi.org/10.1016/j.cub.2006.09.015.

Galeas, M.L., Zhang, L.H., Freeman, J.L. et al. (2007). Seasonal fluctuations of selenium and sulfur accumulation in seleniumhyperaccumulators and related non-accumulators. *New Phytol.* 173: 517–525. https://doi.org/10.1111/j.1469-8137.2006.01943.x.

Gao, Y., Walder, K., Sunderland, T. et al. (2003). Elevation in Tanis expression alters glucose metabolism and insulin sensitivity in H4IIE cells. *Diabetes* 52: 929–934.

Germ, M. and Osvald, J. (2005). Selenium treatment affected respiratory potential in Eruca sativa. *Acta Agric. Slov.* 85: 329–335.

Germ, M., Stibilj, V., Osvald, J., and Kreft, I. (2007a). Effect of selenium foliar application on chicory (Cichoriumintybus L.). *J. Agric. Food Chem.* 55: 795–798. https://doi.org/10.1021/jf0629888.

Gigolashvili, T. and Kopriva, S. (2014). Transporters in plant sulphur metabolism. *Front. Plant Sci.* 5: 422. https://doi.org/10.3389/fpls.2014.00442.

Harris, J., Schneberg, K.A., and Pilon-Smits, E.A. (2014). Sulfur–selenium– molybdenum interactions distinguish selenium hyperaccumulator Stanleyapinnata from non-hyperaccumulator *Brassica juncea* (Brassicaceae). *Planta* 239: 479–491. https://doi.org/10.1007/s00425-013-1996-8.

Hartikainen, H. (2005). Biogeochemistry of selenium and its impact on food chain quality and human health. *J. Trace Elem. Med. Biol.* 18: 309–318. https://doi.org/10.1016/j.jtemb.2005.02.009.

Hasanuzzaman, M. and Fujita, M. (2011). Selenium pretreatment upregulates the antioxidant defense and methylglyoxal detoxification system and confers enhanced tolerance to drought stress in rapeseed seedlings. *Biol. Trace Elem. Res.* 143: 1758–1776. https://doi.org/10.1007/s12011-011-8998-9.

Hatfield, D.L., Tsuji, P.A., Carlson, B.A., and Gladyshev, V.N. (2014). Selenium and selenocysteine: roles in cancer, health, and development. *Trends Biochem. Sci.* 39: 112–120. https://doi.org/10.1016/j.tibs.2013.12.007.

Hawkes, W.C. and Turek, P. (2001). Effect of dietary selenium on sperm motility in healthy men. *J. Androl.* 22 (5): 764–772.

HazDat. (2003). Agency for Toxic Substances and Disease Registry (ATSDR). Atlanta, GA: HazDat.

Hira, C.K., Partal, K., and Dhillon, K.S. (2004). Dietary selenium intake by men and women in high and low selenium areas of Punjab. *Public Health Nutr.* 7: 39–43.

Höghberg, J. and Alexander, J. (1986). Selenium. In: *Handbook on the Toxicology of Metals*, 2e (eds. L. Friberg, G.F. Nordberg and V.B. Vouk), 482–520. Amsterdam: Elsevier.

Kaur, N., Sharma, S., and Kaur, S. (2014). Selenium in agriculture: a nutrient or contaminant for crops? *Arch. Agron. Soil Sci.* 60: 1593–1624. https://doi.org/10.1080/03650340.2014.918258.

Kumar, M., Bijo, A.J., Baghel, R.S. et al. (2012). Selenium and spermin alleviates cadmium induced toxicity in the red seaweed Gracilaria dura by regulating antioxidant system and DNA methylation. *Plant Physiol. Biochem.* 51: 129–138. https://doi.org/10.1016/j.plaphy.2011.10.016.

Lei, C., Niu, X., Wei, J. et al. (2009). Interaction of glutathione peroxidase-1 and selenium in endemic dilated cardiomyopathy. *Clin. Chim. Acta* 399: 102–108.

Levander, O.A. (1986). Selenium. In: *Trace Elements in Human and Animal Nutrition* (ed. W. Mertz), 139–197. London: Academic Press.

Lewis, B.G., Johnson, C.M., and Broyer, T.C. (1971). Cleavage of Se-methyl selenomethionine selenonium salt by cabbage leaf enzyme fraction. *Biochim. Biophys. Acta* 237: 603–605.

Li, H.F., McGrath, S.P., and Zhao, F.J. (2008). Selenium uptake, translocation and speciation in wheat supplied with selenate or selenite. *New Phytol.* 178: 92–102. https://doi.org/10.1111/j.1469-8137.2007.02343.x.

Mahaffey, K.R., Corneliussen, P.E., Jelinek, C.F., and Fiorino, J.A. (1975). Heavy metal exposure from foods. *Environ. Health Perspect.* 12: 63–69.

Malik, J.A., Goel, S., Kaur, N. et al. (2012). Selenium antagonizes the toxic effects of arsenic on mungbean (*Phaseolus aureus* Roxb.) plants by restricting its uptake and enhancing the antioxidative and detoxification mechanisms. *Environ. Exp. Bot.* 77: 242–248. https://doi.org/10.1016/j.envexpbot.2011.12.001.

Martin, J.L. (1973). Selenium compounds in nature and medicine. In: *Organic SeleniumCompounds: Their Chemistry and Biology* (eds. D.L. Klayman and W.H.H. Gunther), 663–691. New York;: Wiley.

Missana, T., Alonso, U., and García-Gutiérrez, M. (2009). Experimental study and modeling of selenite sorption onto illite and smectite clays. *J. Colloid Interface Sci.* 334: 132–138. https://doi.org/10.1016/j.jcis.2009.02.059.

Mroczek-Zdyrska, M. and Wójcik, M. (2012). The influence of selenium on root growth and oxidative stress induced by lead in Viciafaba L. minor plants. *Biol. Trace Elem. Res.* 147: 320–328. https://doi.org/10.1007/s12011-011-9292-6.

NAS (1977). *Drinking Water and Health*. Washington, DC: National Academy of Science.

Olson, G.E., Winfrey, V.P., Nagdas, S.K. et al. (2005). Selenoprotein P is required for mouse sperm development. *Biol. Reprod.* 73: 201–211.

Pandey, C. and Gupta, M. (2015). Selenium and auxin mitigate arsenic stress in rice (*Oryza sativa* L.) by combining the role of stress indicators, modulators and genotoxicity assays. *J. Hazard. Mater.* 287: 384–391. https://doi.org/10.1016/j.jhazmat.2015.01.044.

Pukacka, S., Ratajczak, E., and Kalemba, E. (2011). The protective role of selenium in recalcitrant *Acer saccharium* L. seeds subjected to desiccation. *J. Plant Physiol.* 168: 220–225. https://doi.org/10.1016/j.jplph.2010.07.021.

Qin, H.B., Zhu, J.M., Liang, L. et al. (2013). The bioavailability of selenium and risk assessment for human selenium poisoning in high-Se areas, China. *Environ. Int.* 52: 66–74.

Raisbeck, M.F. (2000). Selenosis. *Vet. Clin. North Am. Food Anim. Pract.* 16 (3): 465–480.

Rayman, M.P. (2000). The importance of selenium to human health. *Lancet* 356: 233–241. https://doi.org/10.1016/S0140-6736(00)02490-9.

Rayman, M. (2012). Selenium and human health. *Lancet* 379: 1256–1268.

Reeves, M.A. and Hoffman, P.R. (2009). The human selenoproteome: recent insights into functions and regulation. *Cell. Mol. Life Sci.* 66: 2457–2478. https://doi.org/10.1007/s00018-009-0032-4.

Reid, M.E., Duffield-Lillico, A.J., Slate, E. et al. (2008). The nutritional prevention of cancer: 400 mcg per day Se treatment. *Nutr. Cancer* 60: 155–163. https://doi.org/10.1080/01635580701684856.

Renkema, H., Koopmans, A., Kersbergen, L. et al. (2012). The effect of transpiration on selenium uptake and mobility in durum wheat and spring canola. *Plant Soil* 354: 239–250. https://doi.org/10.1007/s11104-011-1069-3.

Rosenfield, I. and Beath, O.A. (1964). *Selenium, Geobotany, Biochemistry, Toxicity and Nutrition*. New York;: Academic Press.

Rotruck, J.T., Pope, A.L., Ganther, H.E. et al. (1973). Selenium: biochemical role as a component of glutathione peroxidase. *Science* 179: 588–590. https://doi.org/10.1126/science.179.4073.588.

Schiavon, R., Guidi, G.C., Biasioli, S. et al. (1994). Plasma glutathione peroxidase activity as an index of renal function. *Eur. J. Clin. Chem. Clin. Biochem.* 32: 759–765.

Schmutzler, C., Mentrup, B., Schomburg, L. et al. (2007). Selenoproteins of the thyroid gland: expression, localization, and possible function of glutathione peroxidase 3. *Biol. Chem.* 388: 1053–1059.

Schomburg, L. and Köhrle, J. (2008). On the importance of selenium and iodine metabolism for thyroid hormone biosynthesis and human health. *Mol. Nutr. Food Res.* 52: 1235–1246.

Schubert, A., Holden, M.M., and Wolf, W.R. (1987). Selenium content of a core group of foods based on a critical evaluation of published analytical data. *J. Am. Diet. Assoc.* 87: 285–299.

Schweizer, U., Bräuer, A.U., Köhrle, J. et al. (2004). Selenium and brain function: a poorly recognized liaison. *Brain Res. Brain Res. Rev.* 45: 164–178.

Shimada, T., Murayama, N., Tanaka, K. et al. (2011). Spectral modification and catalytic inhibition of human cytochromes P450 1A1, 1A2, 1B1, 2A6, and 2A13 by four chemopreventive organoselenium compounds. *Chem. Res. Toxicol.* 24 (8): 1327–1337.

Smrkolj, P., Germ, M., Kreft, I., and Stibilj, V. (2006). Respiratory potential and Se compounds in pea (*Pisum sativum* L.) plants grown from Se-enriched seeds. *J. Exp. Bot.* 57: 3595–3600. https://doi.org/10.1093/jxb/erl109.

Sors, T.G., Ellis, D.R., Na, G.N. et al. (2005). Analysis of sulfur and selenium assimilation in Astragalus plants with varying capacities to accumulate selenium. *Plant J.* 42: 785–797. https://doi.org/10.1111/j.1365-313X.2005.02413.x.

Tan, J. (ed.) (1989). *The Atlas of Endemic Diseases and their Environments in the People's Republic of China*. Beijing: Science Press.

Tiwary, A.K., Stegelmei, B.L., Panter, K.E. et al. (2006). Comparative toxicosis of sodium selenite and selenomethionine in lambs. *J. Vet. Diagn. Invest.* 18: 60–69.

Tsongas, T.A. and Ferguson, S.W. (1977). Human health effects of selenium in a rural Colorado drinking water supply. In: *Proceedings of Trace Substances in Environmental*

Health – XI. A Symposium (ed. D.D. Hemphill), 30–35. Columbia, MO: University of Missouri.

Turakainen, M., Hartikainen, H., Ekholm, P., and Seppänen, M.M. (2006). Distribution of selenium in different biochemical fractions and raw darkening degree of potato (*Solanum tuberosum* L.) tubers supplemented with selenate. *J. Agric. Food Chem.* 54: 8617–8622. https://doi.org/10.1021/jf0613987.

Ursini, F., Heim, S., Kiess, M. et al. (1999). Dual function of the selenoprotein PHGPx during sperm maturation. *Science* 285: 1393–1396.

Valentine, J.L., Kang, H.K., Dang, P.M., and Schluchter, M. (1980). Selenium concentrations and glutathione peroxidase activities in a population exposed to selenium via drinking water. *J. Toxicol. Environ. Health* 6: 731–736.

Vinceti, M., Crespi, C.M., Bonvicini, F. et al. (2013). The need for a reassessment of the safe upper limit of selenium in drinking water. *Sci. Total Environ.* 443: 633–642.

Wang, Y.D., Wang, X., and Wong, Y.S. (2012). Proteomics analysis reveals multiple regulatory mechanisms in response to selenium in rice. *J. Proteomics* 75: 184–1866. https://doi.org/10.1016/j.jprot.2011.12.030.

Weast, R.C. (ed.) (1988). *CRC Handbook of Chemistry and Physics*, 69e. Boca Raton, FL: CRC Press Incorporated, B-124, B-126, B-132.

Winkel, L.H., Johnson, C.A., Lenz, M. et al. (2012). Environmental selenium research: from microscopic processes to global understanding. *Environ. Sci. Technol.* 46: 571–579. https://doi.org/10.1021/es203434d.

World Health Organization (1987). *Environmental Health Criterion 58 – Selenium*. Geneva: World Health Organization.

World Health Organization (1996). *Trace Elements in Human Nutrition and Health*. Geneva: World Health Organization.

Xue, T.L., Hartikainen, H., and Piironen, V. (2001). Antioxidative and growth-promoting effect of selenium on senescing lettuce. *Plant Soil* 237: 55–61. https://doi.org/10.1023/A:1013369804867.

Yang, G.Q., Wang, S.Z., Zhou, R.H., and Sun, S.Z. (1983). Endemic selenium intoxication of humans in China. *Am. J. Clin. Nutr.* 37: 872–881.

Yang, S.J., Hwang, S.Y., Choi, H.Y. et al. (2011). Serum selenoprotein P levels in patients with type 2 diabetes and prediabetes: implications for insulin resistance, inflammation, and atherosclerosis. *J. Clin. Endocrinol. Metab.* 96: E1325–E1329.

ZemZembala, M., Filek, M., Walas, S. et al. (2010). Effect of selenium on macro and microelement distribution and physiological parameters of rape and wheat seedlings exposed to cadmium stress. *Plant Soil* 329: 457–468. https://doi.org/10.1007/s11104-0090171-2.

Zhang, Y., Pan, G., Chen, J., and Hu, Q. (2003). Uptake and transport of selenite and selenate by soybean seedlings of two genotypes. *Plant Soil* 253: 437–443. https://doi.org/10.1023/A:1024874529957.

Zhao, C., Ren, J., Xue, C., and Lin, E. (2005). Study on the relationship between soil selenium and plant selenium uptake. *Plant Soil* 277: 197–206. https://doi.org/10.1007/s11104-005-7011-9.

Zhu, J., Wang, N., Li, S. et al. (2008). Distribution and transport of selenium in Yutangba, China: impact of human activities. *Sci. Total Environ.* 392: 252–261.

6

Various Analytical Techniques for Se Determination in Different Matrices

Deepak Yadav[1], Rupali Jha[1], Pradeep Kumar[2], and Pardeep Singh[3]

[1] *Department of Chemical Engineering, HBTU, Kanpur, Uttar Pradesh, India*
[2] *Department of Chemical Engineering & Technology, IIT (BHU), Varanasi, Uttar Pradesh, India*
[3] *Department of Environment Sciences, University of Delhi, New Delhi, India*

6.1 Introduction

Selenium ($Se^{79}{}_{34}$) is the 30th most abundant element, and is naturally present as a metalloid form abundantly distributed in the Earth's crust. Selenium is placed under sulfur in the periodic table and is a vital element for living organisms. Selenium oxides are selenate (Se^{6+}) and selenite (Se^{4+}) and are chemically comparable to sulfate. The amount of Se is estimated as 0.05–0.5 mg Se/kg in the Earth's crust and 0.4–1200 mg Se/kg in soil (Lemly 2004), originating primarily from limestone, quartzite, and sandstone (Wang and Gao 2001; Fordyce 2005). In 1817, Swedish doctor Jöns Jakob Berzelius discovered selenium. Since then, selenium's position has developed securely in research and development applications in medicine and material science.

Selenium originates in nature, particularly in carbonate rocks and sedimentary soils in the Earth's crust. The major sources of Se emissions from various agricultural practices, industrial activities like mining, and petrochemical operations cause atmospheric and aquatic pollution. In recent years, studies of various constraints like pH, contact time, and temperature, and improvements in water quality monitoring have revealed that Se is a pollutant of latent ecological anxiety. Selenium originates in metal sulfide ores, which in trace quantities are essential for cellular function in several organisms. This has real-world consequences for business in meeting the United States Environmental Protection Agency's (EPA) stringent Se monitoring expulsion limit of 5 μgSe/l for Se-containing wastewaters.

The main causes of Se pollution are farming, industrial procedures, for example mining, and petrochemical engineering, etc. Zhang et al. 2009 worked on titanium dioxide (TiO_2) nanoparticles for the sorption of selenate and selenite from aqueous solution.

Modern critical techniques now find selenium at concentrations lower than 0.01 mg total Se/kg. Inductively coupled plasma mass spectrometry (ICP-MS) is the best commonly used practice to detect selenium at trace concentrations, as it permits concurrent quantification

Selenium Contamination in Water, First Edition. Edited by Pooja Devi, Pardeep Singh, Arindam Malakar, and Daniel Snow.

of many elements. Only advanced analytical composition provides a suitably high sensitivity to confirm its transferability, chiefly in selenium-deficient schemes.

Se-rich investigations derive from waste rock directly or indirectly. Subsequently, Se can percolate into the ground by rainwater, particularly for soil, surface water, and plant contamination (Izquierdo and Querol 2012; Khamkhash et al. 2017). Former discoveries have taught us that diverse Se absorptions in soil, water, and plant environs are exaggerated by different mining actions. Therefore, the existence of diverse selenium levels in mining discharges found by research in the past (Bujdoš et al. 2005; Khamkhash et al. 2017) is presented in Figure 6.1.

In a recent research review conducted by Etteieb et al. (2020), shown in Figure 6.1, the main points were as follows:

- Discharge of Se at high concentrations is the main cause of ecological distress.
- The motion, bioavailability, and destiny of Se depends on many factors.
- Monitoring of Se pollution in mine engineering is critically essential.
- Endorsements are prepared to diminish Se contamination in the atmosphere.

The vital nature and toxicity of Se in the atmosphere is liable to the effort and the organic systems wherein it is modern. Se and its spinoffs may be widely isolated because of anthropogenic ideas viz. fuel burning and agricultural, industrial, and natural routes. With the intention of assessing the properties of altered Se types on active entities, subtle analytical systems are obligatory to gauge their strength in composite ecological matrices. An outline is stated for the status of selenium speciation and the passage, growth, and amendment of the species (types) in the environment.

There are several rare cases where you must choose between chemical and biological handling, for instance when eliminating selenium (see Figure 6.1). Generally, it has been

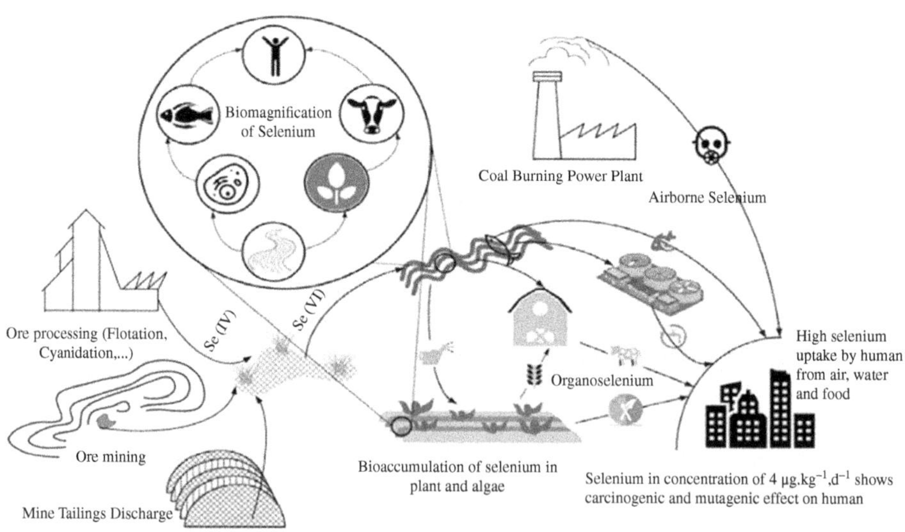

Figure 6.1 Major selenium source discharges, biomagnification, and environmental effects (Etteieb et al. 2020).

detached chemically such as through reducing negotiators and adsorbents, but it can also be detached biologically with anoxic nitrate-reducing bacteria.

Mining and mineral treating industries are the key foundations for releasing selenium into the environment. Extreme levels of selenium might result in toxicity in humans in the form of selenosis, in feeding creatures as alkali syndrome, and in water species as larval and growing abnormalities and as death.

Selenium may be introduced into the atmosphere from natural sources, including volcanic discharges, phosphorites, petroleum source rocks, sedimentary rocks, shales, metallic sulfides, and coal ore deposits (Ryan and Dittrick 2000). On the other side, the key manmade actions include metal, mining, fossil fuel alteration, and irrigation of selenium-rich soils (USEPA 2016). In the past, selenium has been broadly used mostly in the chemical industry, paint, electronics, metallurgy, pharmaceutics, ceramics, glass manufacturing, and in agriculture (Mehdi et al. 2013). Throughout the mining undertakings, selenium can be organized after rock overload excavation and normal withstanding producing important increase in mining discharges. Therefore, selenium may escape into downstream surface water, soil, and plants following the oxidation of waste rocks and raised effluent levels (USEPA 2016). Henceforth, effluent from mining was the early route for introducing selenium into the surface water and soil system through water discharge (Wellen et al. 2015). Consequently, it is critical to evaluate the biogeochemical performance, flexible movement, and bioavailability of selenium in various respective environments based on its fractionation and speciation (Peng et al. 2017). The movement, outcome, and behavior of selenium in soil are exaggerated by its geochemical processes and different speciation, chiefly precipitation, biomethylation, sorption, reduction–oxidation, and complexation.

Figure 6.2 shows a schematic diagram of the advanced analytical practices available to study the performance of Se (Winkel et al. 2012). The US EPA's interim maximum contaminant level (MCL) for selenium pollution (0.05 mg/l) comes over essentially soils and food (EPA US 2006). It is widely used in chemicals, electronics, drugs, fungicide, glass manufacturing, photocopy operations, and feed preservative. Over a long period of time exposure to high-level selenium has caused several adverse health effects, including loss of control over arms and legs and a loss of sensitivity (Fan and Kizer 1990).

The data display a good to strong connection between dissolved Se and total dissolved solids (TDS) in surface water and ground water, as shown in Figure 6.3. To avert the future strength hazard both are linked to selenium excess and it is vital to understand the features. In this feature we will describe enhanced information of the appliances switch the geochemical performance of Se in the exterior situation. Novel practices and approaches permit precise and accurate Se capacities at lesser applications. The adsorption process is established as effective for boundary layer and intraparticle diffusion. Many diverse analytical techniques comprising chromatography, atomic fluorescence spectroscopy (AFS), spectrofluorimetry, UV–visible spectrophotometry, inductively coupled plasma with mass spectrometry (ICP-MS), or ICP with atomic emission (ICP-AES), flameless atomic absorption spectroscopy (FAAS), and electrochemistry offered Se determination in different matrices. Upcoming research on on-site selenium analysis points to an economical and field-deployable monitoring device.

Therefore, monitoring the emerging picture of selenium contamination is essential in order to execute treatment methods successfully to ensure that green salvation ensues from resource

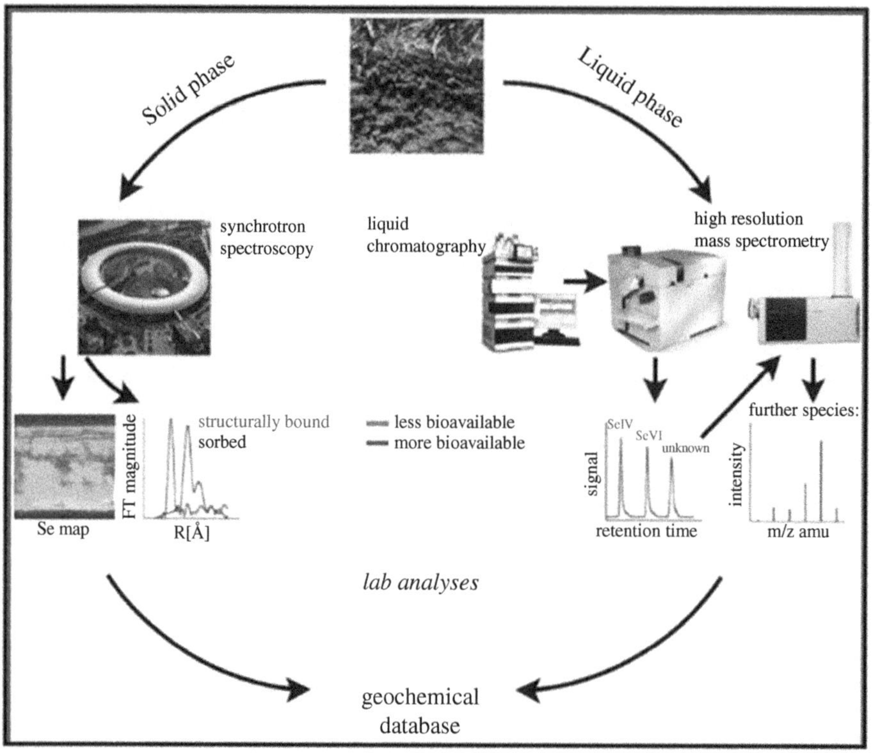

Figure 6.2 Schematic diagram of advanced analytical practices available to study the performance of Se (Winkel et al. 2012).

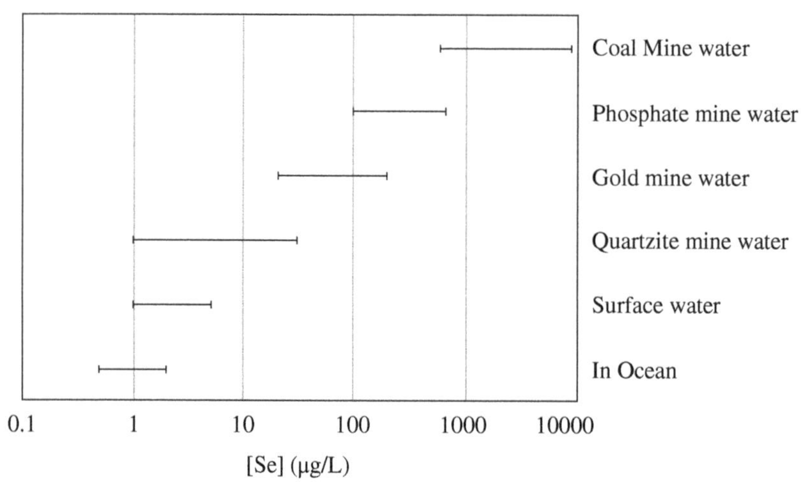

Figure 6.3 Series of selenium concentrations in different aquatic mining effluents (Etteieb et al. 2020).

investigation and fiscal growth. However, many trials have come across the proficiency of Se deduction as this is governed by the cost-effective deliberation, the complexity of wastewater composition mixtures, the intricacy of Se chemistry, and the stringent guidelines to keep to the permitted levels. With the intention of overcoming these experiments in Se elimination, a blend of biological action with physical or chemical methods is projected to elevate Se removal and to certify environmental safety and human health protection.

Considering the source, transportation, and destiny of selenium in oil refinery wastewater, as well as the necessary actions, is a requirement for US oil refineries with a selenium limit in their National Pollutant Discharge Elimination System (NPDES) wastewater permit. Desalter wash water coming from the sour water stripper accounts for most of the selenium entering refinery wastewater. It has been posited that handling low-flow, high-concentration insulated watercourses in a refinery is the most lucrative methodology, both for capital and for O&M costs.

Removal of selenium from wastewater is mainly done by both biological and chemical/physical procedures. This chapter will describe those tools, numerous technology workers, and other causes of material. As is known to all, selenium (Se) is an vital micronutrient for the individual and it is good for our health at low concentration. However, unnecessary Se in the environment can cause serious pollution in atmosphere, soil, and water, and bring clear risk of harm to human health (Sakamoto et al. 2012). Industrial and agricultural undertakings are the leading anthropogenic causes of Se pollution. Selenium absence and toxicity lead to pollution in drinking water and in wastewater, which is a dormant hazard for water systems (Santos et al. 2015). Hence, in recent years, selenium contamination has attracted tremendous interests and has been taken more and more seriously round the world (Santos et al. 2015). The main source of selenium in the environment (e.g. biosphere, hydrosphere, pedosphere, and atmosphere) has both natural and anthropogenic bases. The natural source covers marls, spray, volcanic eruptions, gypsum, sea recycling via biotic cycling, volatilization, the weathering of selenium-containing rocks and soils, etc. (Fairweather-Tait et al. 2011). The anthropogenic bases comprise mining, coal combustion, photocells, insecticide production, agriculture, oil refining, and glass manufacture (Santos et al. 2015).

Generally, potable water is not considered as source of nutrition for the different elements (Vinceti et al. 2013a). Table 6.1 shows some excessive selenium concentrations in the environment.

Different mining activities and mining processes and waste from mining sites including leachates contain Se, which has contributed to spreading Se in downstream environments, reaching surface water, which can contaminate the river (Sharma and Vance 2007).

Se contamination in the Elk River has been exceeded up to 5 µg/l, as reported by Swain (2007), but Hu revealed that Se contamination did not pass beyond 10.38–11.6 µg/l in the Prairie rivers water (|Hu 2008). These studies showed the deep interaction between Se concentration and the effects of mining activities on surface water as well as bodies of water. Canadian Council of Ministers of Environment, has set the standard for Se content in freshwater and aquatic marine life at max to 1 µg/l. The exceeded limit of selenium concentration in surface water beyond 1 µg/l, controlled under the guideline set to protect the eco system.

Table 6.1 Excessive selenium concentrations in the environment (He et al. 2018).

Location (countries)	Selenium sources	Selenium concentrations (average)	References
USA	Seleniferous soils	up to 28 mg/kg	Bajaj et al. (2011).
Ireland	Seleniferous soils	up to 1200 mg/kg	Bajaj et al. (2011)
China	Seleniferous soils	up to 59 mg/kg	Bajaj et al. (2011)
California's San Joaquin Valley	Drainage water of soils	75–1400 μg/l	Zelmanov and Semiat (2013)
Japan	Industrial wastewater	≤ 0.1 mg/l	Soda et al. (2011)
Amman Zarqa Basin (AZB), Jordan	Lower aquifer	1.6–30 μg/l (13 μg/l)	Alkuisi and Abdelfattah (2010)
Amman Zarqa Basin (AZB), Jordan	Middle aquifer	1.5–742 μg/l (35 μg/l)	Alkuisi and Abdelfattah (2010)
Amman Zarqa Basin (AZB), Jordan	Upper aquifer	0.4–246 μg/l (9 μg/l)	Alkuisi and Abdelfattah (2010)
Barwa, Jainpur	Groundwater	341 μg/l	Bajaj et al. (2011).
Hisar	Groundwater	3.2 μg/l	Bajaj et al. (2011)
Chandigarh	Groundwater	0.9 μg/l	Bajaj et al. (2011)
Chennai, India	Groundwater	Se(IV) 0.15–0.43 μg/l	Kumar and Riyazuddin (2011)
Chennai, India	Groundwater	SSe(VI) 0.16–4.73 μg/l	Kumar and Riyazuddin (2011)
Punjab, India	Groundwater	669.5 g/l	Bajaj et al. (2011)
Punjab, India	Soil	6.5 mg/kg	Bajaj et al. (2011)
Punjab, India	Plants	3–670 mg/kg	Bajaj et al. (2011)

Leaching of Se from mine effluent is also dispersed in soil. Even in soil, Se concentration can diverge in the range 0.1–435 mg/kg (Xing et al. 2015; Zhang et al. 2014; Favorito et al. 2017). The variation in Se is strictly connected with biogeochemical processes which take place in the soil, and consequently fractionation and specification also govern the selenium mobility in soil.

On the other hand, mineral deficiency, including of selenium, is a widespread issue. To resolve this issue, researchers recommend boosting the selenium content in the palatable yields, over sprigs or addition of fertilizers in soil (Feng and Wei 2012). Plant roots can easily absorb and incorporate Se, by combining with sulfur (S) and applying biochemical route and volatile when coupled with methylated Se as well.

Selenium Se is a compound that is employed to distribute nutritive Se in the case of mineral shortage but also needs to be eliminated from waste to reduce the impact of pollution from seleniferous sites (Pilonsmits and Quinn 2010).

Apart from all other methods, enrichment of soil with Se is most efficient way to increase the concentration of Se in plants, so Se can easily be incorporated and redeployed to the roots rapidly. However, it is highly influenced by ecological conditions and species of plant. Ma et al. (2018) have revealed that up to 0.1–0.5 ppm Se concentration in forages is acceptable but it is noticeable that there is a high risk of poisoning when concentration exceeds 5 ppm. It has been reported by Wadgaonkar et al. that around 70% of selenite and 85% of selenate are exposed in aerial tissue (Wadgaonkar et al. 2018).

The crucial cause of selenium release in air dispersion is the combustion of coal and oil. Generally, Se content is low in surrounding air. The specific ecological system always occupies important position in the bio-geochemical selenium cycle, since it affect the transport and alteration of Se (Huang et al. 2012). The existence of selenium is linked with anthropological actions, for example fossil fuel combustion and incineration/burning of waste, e.g. tires and paper (Marcocci et al. 2011). It is also related to geographical (topographical) events such as volcanism, soil erosion, and forest fires.

Mehdi et al. (2013) has observed that dimethyl selenide is stable in air, though selenide and selenium dioxide are extremely uneven in exposed air. Hydrogen is the only compound that reacts and oxidizes in its presence to selenium and H_2O. In humid situations, selenium dioxide reacts with water and convert into selenium acid.

As regards diet, cereals, eggs, bread, meat, fish, and milk/dairy are the primary food sources of selenium. Some Brazil nuts are highly enriched sources of Se, with approximately 0.03–512 mg/kg in other weight reported by Silva Junior et al. (2017).

Usually, the Se content in different agricultural products, e.g. vegetables and grains, is determined by the Se content of homologous soils. In some vegetables like beans, peas, potatoes, turnips, beets, and tomatoes, a content of about 6 mg/g of selenium is an extreme, although when crops developed in biological organism Sharma et al. (2017) revealed that due to hostile or synergistic effects, around there were two or three times the level of selenium content in inorganic species.

Fairweather-Tait et al. (2011) found that selenium deficiency was also responsible for some diseases and disorders in human and animals. Deficiency of Se in the human body is directly related to increased health risks of mortality, weak resistance of immune system, and cognitive decline. Consequently, selenium deficiency also trigger some other major diseases such as peripheral myopathy, necrotizing cardiomyopathy, conduction disturbances and muscle fatigue, anemia, thin hair problems, and opacification of the nails. It has been observed by Fairweather-Tait et al. (2011) that low Se concentration in blood plasma (<115 ng/ml) is responsible for a high risk of cancer in individuals (Fairweather-Tait et al. 2011). Figure 6.4 shows the different methods of Se contamination removal.

6.1.1 Health Hazards: Impact of Se on Human Health

It is a well-known fact that minerals show a significant amount as antioxidants, and Se is one of them. Considering the importance of Se for humans, dietitians have proposed that daily ingestion for Se should be 55 μg/d for strong adults in US, 30 μg/d in Europe, and in China Se intake should be around 50–250 μg/d for adults, as determined by the CNS (Chinese Nutrition Society).

Belgium's CSS (Council of Health) recommended selenium daily intake in adults (from 14 years) ranges from 60 μg/day for females to 70 μg/day for males. Pregnant women are

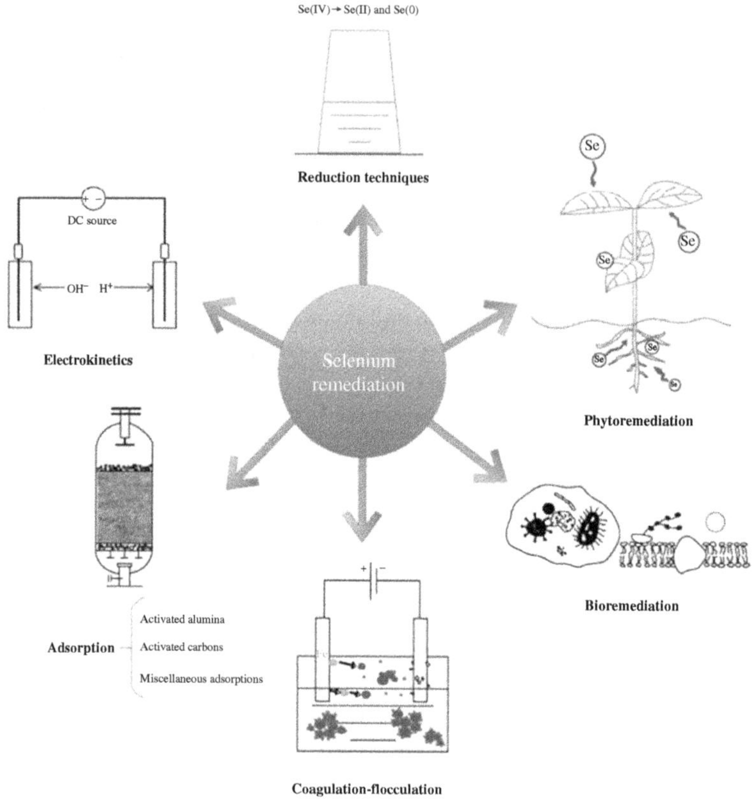

Figure 6.4 Different methods for elimination of selenium from contaminated water and soils (He et al. 2018).

recommended to take 65 μg/d and during lactation intake should increase to 75 μg/d. EFSA (European Food Safety Authority) 2006 puts the guidelines for Acceptable Higher Intake Level at 300 μg/day for adults, 60 μg/day for children aged 1–3 years, and 250 μg/day for children aged 15–17 years (Mehdi et al. 2013; Sun et al. 2014).

It has a broad range of applications and plays a vital role in the human body, including antioxidant defense, human immunity amelioration, cell protection, and carcinogenesis inhibition. On the other end, Zhang et al. (2014) and Lopes et al. (2017) revealed that, on the contrary, the effects of Se on social and animal well-being can be negative when it enters in the food chain. A WHO (2011) report revealed that selenium was also responsible for severe toxicity by producing breathing, hypertension, cardiac complications, and gastrointestinal or prolonged toxicity leading to mental disorder, hair and nail damage, the risk of tooth decay, and chronic alkalosis (WHO 2011; Brodowska et al. 2016; Plak and Bartmiński 2017).

Adverse effects in flora and fauna, e.g. bioaccumulation, imitation of egg-laying vertebrates and teratogenic effects, have also been observed as a result of surplus Se in polluted water bodies. In order to restrict the Se contamination in freshwater, drinking water, agricultural overspill, and soil, and to prevent major environmental impact, every country has ensured limits and issued guidelines to regulate the Se release into the environment, as shown in Table 6.2.

Table 6.2 Selenium monitoring parameters for drinking water and effluent discharge (Tan et al. 2016).

Target	Limit	Agency/country	References
Drinking water	10 μgSe/l	Europe	Santos et al. (2015)
	40 μgSe/l	World Health Organization	WHO (2011)
	1 μgSe/l (as)	Russian Federation	Bassil et al. (2014)
	50 μgSe/l	US EPA	US EPA (2014)
	No regulation	Europe	Luoma and Presser (2009)
	5 μgSe/l (chronic aquatic life toxicity)	US EPA	US EPA (2014)
	20 μgSe/l (acute aquatic life toxicity)	US EPA	
	2 μgSe/l	Canada and British Columbia	Lussier et al. (2003)
	100 μgSe/l	Japan	Fujita et al. (2002)
Effluent water (freshwater discharge)	1 μgSe/l	Korea	Lee et al. (2015)

Selenium is one of the prominent essential minerals for human healthiness and there is an extended history of study relating to it in clinical medicine and ecology of place (Rayman 2012). Selenium has been renowned as an important mineral in our diet, as it is a important element in numerous key metabolic paths like antioxidant defense systems, thyroid hormone balance, and immunity improvement, and Brown and Arthur (2001) agreed that protection of the cell membrane from damage initiated by peroxidation of lipids makes Se a necessary inorganic fraction of food in humans and animals.

As far as countries' policies are concerned, various permissible guidelines for safe and sound uptake of selenium for potable water and contaminated discharge limits of engineering runoffs to freshwater bodies have been accepted, as shown in Table 6.2. Gusek et al. (2008) and Santos et al. (2015) have reported in their research an extreme limit of 50 μg/l for drinking water and 5 μg/l for selenium effluent.

A wide diversity in policies has been detected at the global level and incompatible toxicity consequences among diverse living organisms are clear to see, as a result of the differing restrictions, specification and basis of Se, variable ecological circumstances and hydraulics (river flow rates and habitats) (Swift 2002; Muscatello and Janz 2009; Vinceti et al. 2013b).

6.1.2 Industrial Techniques Used for Selenium Removal

In the last century, Se was not considered a significant element as compared to similar mineral or contamination substances or other heavy metal waste water or organic pollutants. Due to uneven dispersion of Se in different geo-ecological spaces, the European

Union has not considered Se a part of critical raw materials (Nancharaiah et al. 2016), nor was it a worldwide concern (Wen and Carignan 2007). However, in the recent scenario selenium is tied to big concerns in numerous ways. Numerous biological, physical, and chemical treatment processes have been established to reduce the selenium composite content in waste water shown in Figure 6.5. Different physical treatment methods, e.g. ion exchange and membrane filtration, chemical treatments, e.g. adsorption, precipitation, and reduction, and biological treatments, e.g. in-situ microbial reduction, algal volatilization, and wetlands, fixed-bed reactor bioreactors have been evaluated as regards the different ecological conditions for their benefits against restrictions (Holmes and Gu 2016). To remove the Se from contaminated water it is mainly chemical and biological treatment processes that are used in water treatment column, with modest operational success along with cost-effectiveness.

6.1.3 Se Monitoring Recommendations

For an exact Se monitoring and quantification, many analytical instruments are used, and quality controller events must be applied, not only during analysis but also in the pre-analysis phase throughout all steps including sample collection, sample storage, sample preparation for processes, and pretreatment. Initially, in the process of Se analysis in water, according to the US EPA-600/4-82-029 and ISO 5667-3 guideline for storage of liquid,

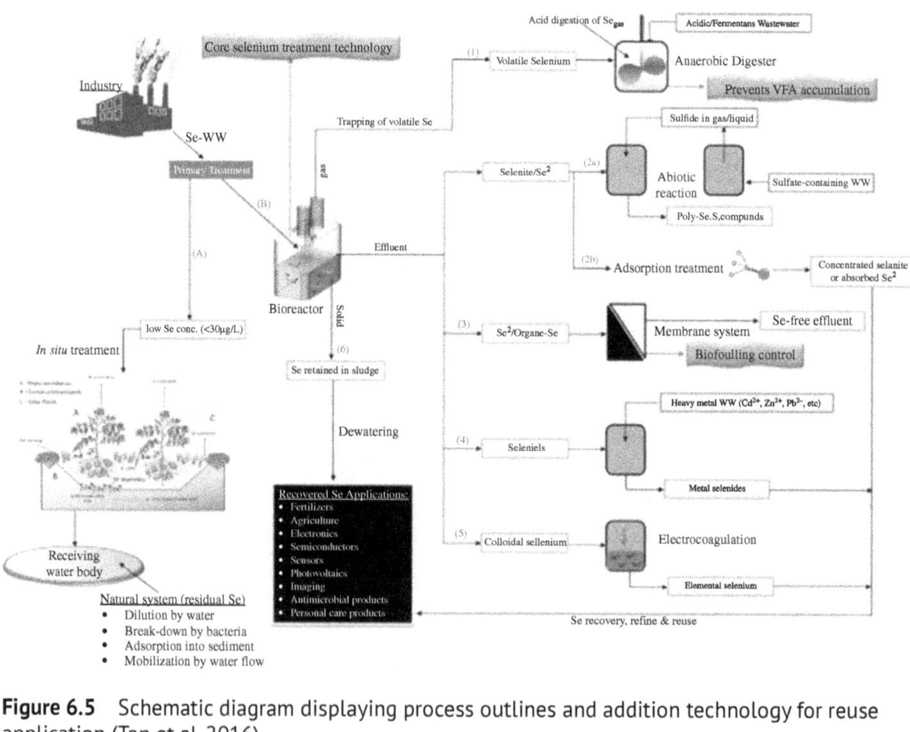

Figure 6.5 Schematic diagram displaying process outlines and addition technology for reuse application (Tan et al. 2016).

samples should be taken in a clean container made of non-reactive material, e.g. borosilicate glass, quartz, high-density polyethylene, propylene, polycarbonate, or fluoroethylene polymer, that should wash clean with detergent, rinsed with nitric acid and hydrochloric acid and soaking in or packed with nitric/hydrochloric acid for around 24 hours before being rinsed with distilled water (Desborough et al. 1999).

Before operating the analytical instrument for Se analysis many quality control measures take place to ensure accurate data. Also, expending analytical appliances for Se analysis must take into account various quality controller processes. Exposure limits definitely vary with type of detector, pre-concentration measures, and injection volume for sample insertion, which should be determined by analyzing up to three blanks (Kumkrong et al. 2018). The systematic data reproducibility must be confirmed by examining models to reproduce exact the same data with the same sample and to avoid the problem of matrix interventions. In order to ensure the precision of analytical technique and to confirm the understanding, the calibration should be done with standard or certified reference ingredients, so that Se concentration and tasters spike in repetitive analysis should be concluded (Haug et al. 2007). Data validation done on the basis of calibration curves should be done on the and theoretical concentrations (Ralston et al. 2008). Selenium in all type of complex must be computed by two discrete Se isotopes with acceptable consistent data as a significant focus. Also, the empathy of Se class centered on retention time of process should similarly be complete by expending matrix spikes and molecular mass spectrometry to escape the shift of retention times of identified Se species because of matrix properties (Ralston et al. 2008). Also, many different kinds of analytical methods or processes have been established to study Se class in water and soil. Inductively Coupled Plasma Mass Spectrometry (ICP-MS) is a high-sensitivity, sophisticated instrument that has multi-element on-line detection ability and speed to reveal selenium (Pettine et al. 2015). The scrutiny of total Se in waste water or fresh water can done with ICP-MS that have endorsed by US EPA procedures 2008 and 6020.

Mass-Spectroscopy (MS) is most sophisticated instrument, and its application allows not only the obtaining of information on the quantitative and qualitative analysis of sample contents, but also on the structure of molecules and molar masses of analyzed chemical composites (Yu et al. 2019). However, application of ICP-MS for Se investigation needs high analytical skills in the credit and modification of spectral, physical and chemical interventions.

Moreover, despite the advancement in analytical speciation methods for quantitative assessment of a number of decreased content species such as polyselenides and Se(0), which is still the key task for Se detection, mainly in reducing environments, due to their ability to create colloidal groups, it is harder to find expert standards. As an alternative, X-ray-based practices, including X-ray Absorption and X-Ray Fluorescence Spectroscopy (XRF), have been suggested for detection of Se (0), which was described by Pettine et al. (2015), accounting for about 30–60% of total Se content sediments in aquatic systems. X-ray absorption spectroscopic techniques, e.g. X-Ray Absorption Near-Edge Spectroscopy (XANES) and Stretched X-Ray Absorption Fine Structure (EXAFS), permit good detection limits to be reached, but that requires the use of synchrotron-based X-ray radiation sources, and is therefore less convenient and not very easy to be inferred.

On the whole, the preference among analytical approaches is highly case-sensitive and depends on interference factors for the quantification of selenium. On the other hand, ICP-MS along with CRC expertise and the hyphenated methods of HG coupled with MS or HPLC and ICP-MS were highly acclaimed for total Se content enquiry in sample and Se speciation individually.

The natural availability of Se for plant uptake, translocation, and growth and its metabolic functioning in plants have been extensively explained in the literature. Different investigative approaches for quantification of Se speciation have also been discussed by many authors. Thus, as such no Se speciation devices have been formally suggested by any government interventions, a team conducting an analysis should select appropriate an analytical method and repeatedly optimize it to reach better sensitivity, especially whenever interfering matrix components are present in the sample. Principally, the hyphenated procedures with HPLC or HG tied to MS and ICP-MS have ben highly commended instruments for Se speciation, given their efficiency in attainment up to very low detection perimeters, their high-precision execution, the reproducibility of their data, and dealing with the difficulties of matrix intrusions. Consequently, observing up-and-coming Se contamination is highly necessary for a consequential execution of effective treatment methodologies to ensure ecological protection, survey reserves, and provide cost-effective growth.

Conversely, in the process of encountering the effectiveness of Se removal, numerous challenges relate to the complexity of Se chemistry, the complex intra-molecular composition of wastewater, economical considerations, and the stringent protocols required to keep under the permissible limit. In order to overcome these trials and remove Se from contaminated water, a well-designed arrangement of biological behavior is need, in addition to chemical or physical approaches applied to enhance the Se elimination at large scale, in order to ensure ecological integrity and human health security.

To better understand the chemical activity of Se in the environment, especially where the Se concentration is inadequate, it is more important to develop a convenient technique for sensible and species-specific analysis. The danger of selenium deficiency in the human body requires urgent addition of Se substances to guarantee the acceptability of Se in human foods. That Se shortages can occur in humans enhances resolve to confirm the competence of Se in human diets and adds weight to the importance of Se supply plans. At this present time, there are a great number of investigative methods available which permit the trace and ultratrace scrutiny of different essential configurations.

To analyze trace elements in waste water in the parts per million (PPM) range, Atomic Absorption Spectrometry (AAS) with flame atomization is the most widely used technique. However, for ultratrace concentrations in the parts per billion (PPB) range and smaller, a number of good analysis techniques for elemental sensitivity are available.

The determination of trace Se element is normally held with voltammetry, potentiometry, atomic spectrometry (AAS), X-ray analysis, and other nuclear methods. The different electrochemical techniques are used to quantify either in the form of free ions in solution (potentiometry) or free ions including ions destined in labile complexes (voltammetry), and these techniques can also offer the ability to analyze the oxidation state of certain of the elements. Atomic spectrometric techniques are extremely sensitive and can be used to quantify the total element content within a sample; in some way, the accurateness of these analytic methods can be dependent on the matrix of the model. X-ray techniques provide a very low limit for detections and are recycled for analysis of the outcomes, because their values are primarily dissimilar from other analytical techniques. Consequently, they are less expected to be predisposed to the identical methodical biases. Advantages and disadvantages of each analytical method should distress the number of analysis techniques are possible to amount with intrusions and problems, detection limits, amount of samples and operating expense.

Some persistent chemicals like perfluorocarboxylic acids (PFCAs) are extensively found in the environment. The accurate analysis of these persistent chemicals in ecological media, mainly soil or sludge, found with low concentration is a most important analytical challenge. The commercialization of electrospray boundaries for liquid chromatography-mass spectrometric examination enabled the analysis of PFCAs, etc. which present in samples at low levels. However, it is so difficult with measurable examination, due to the matrix conquest or improvement still persevering in the sample. The approaches designated in this revision are so modest and speedy taster ablution trial to eliminate the matrix constituents adequately, to facilitate any negligible error found which happens to be due to the presence of colluding matrix peaks in samples and recoveries of PFCAs are consistently possible.

In summary, it should be more exact to escape most false optimistic consequences and create a assenting process to remove practically all false positives data. The methods conveyed in this study have been established to fulfill the basic criteria for a full homologous cycle of PFCAs. Additionally, to remove the medium effects using dispersive solid-phase abstraction is a quick and undemanding cleansing practice that signifies a robust, reliable technique for the assessable analysis of PFCAs, which present in solid environmental conditions.

6.1.4 Development and Challenges during Se Treatment Practices

Due to the bioaccumulative nature of selenium in different tissues, organisms, and the food chain, inducing hazardous effects and toxicity in human and animals, Se treatment is an burning issue for environmental safeguarding. However, Se removal from wastewater is a major challenge for researchers, due to its dissimilar oxidation states, solubility, and its competitive nature with other constituents. Therefore, classifying the Se threats in water bodies and tolerable removal expertise should be necessary for in-depth considerate speciation. In point of fact, each selenium species shows another degree of toxicity. Holmes and Gu (2016) reported in their literature that selenite is moderately uncomplicated to eliminate among conservative water treatment instead of high oxidation state selenate complex matrix removal. Ample elements present in different multi-layered conditions at low levels of content and at the same point no analysis techniques were available for many of these elements due to there presence only of trace elements. After a long time, some analytical methods have been developed as a way to detect the trace elements. In the last half of the twentieth century, with the strong upsurge of element detection skills found in investigative procedures, another novel term called as "ultratrace elements" has surfaces. Even though a similar term has previously occurs and been used, there is no rigid definition given before that. Whenever an element is present at a mass fraction below 1 ppm, that concerns "ultratrace." Knowledge of their presence or appearance in a sample can found with the help of IUPAC, which relates to any element that has a standard concentration >100 ppm atoms or below 100 mg/kg. In the fields of science, technology or industry, before starting the analysis of any element it is very important to know about trace and ultratrace elements.

Ultralow attentiveness of any elements might be an indispensable term for gauging the hazardous dose for organisms, as a number of traces can intensely change their qualities with respect to various purposes, and for that reason the requirement of accurate analysis even at lower concentrations is highly significant. The common use of exceptionally responsive instrumentation necessitates the tolerable device of impurity and their certification for accurate determination within a sample. It is observed that during the analysis, the advance of analytical sensitivity proliferates infection in addition to other complications. Consequently, truthful safeguarding measures should be undertaken by technicians to control trace elements at PPB concentration level and lower. The possibility of inaccuracy during analysis of trace and ultratrace elements can be initiated through unsuitable selection, sample research, storage, and during analysis itself. Thus the accurateness of an analytical determination must depend on continual good practice.

6.2 Spectroscopic Techniques

6.2.1 Atomic Absorption Spectrometry (AAS)

AAS is an assessable technique to analyze metal content and is a suitable instrument for the identification of around 70 different elements. This analysis method determines the presence of an element by passing light of a precise wavelength, which is emitted by a radiation source of a specific element, through a cloud of atoms in a sample. Atoms will engage the light passing through an energy source known as a hollow cathode lamp (HCL).

6.2.2 Flame Atomic Absorption Spectrometer (FAAS)

Flame Atomic Absorption Spectrometer (FAAS) is another appropriate method for determining a metal's concentration at PPM levels with good accuracy for several elements. FAAS uses an air-acetylene and/or nitrous oxide flame atomizer. Samples are introduced to the atomizer in the form of aerosol with the help of a nebulizer. The FAAS technique allows a fast investigation of 10–15 minutes per sample, with high quality accuracy (repeatability), counting moderate intrusions that can be simply altered, and is moderately cost-effective. FAASs have been effective in the analysis of heavy metals within several conditions, but they have limitations when analyzing some elements such as zirconium (Zr) and arsenic (As) which have higher atomization temperature and therefore undergo incomplete atomization.

FAAS is primarily a conservative practice for the analysis of trace metal ions, due to its comparative ease and the cost-effectiveness of the apparatus. Here a sample is put into the flame, where it is dissociated into essential atoms. Electromagnetic radiation present in the UV–Vis light is part of spectrum used in FAAS technique to direct analysis of trace elements present in sample materials. Though in numerous cases the accessible analytical equipment does not contain adequate sensitivity to analyze the usual samples and stuck due to not to break matrix interferences. A number of techniques have been developed for pre-concentration and separation of trace metals as an earlier step for performing successful instrumental analysis at lower detection restrictions, which can improve the precision of analytical outcomes and also convey the analyte concentration keen on the active series of the detector. Different pre-concentration approaches, e.g. ion exchange, adsorption, solvent extraction, and co-precipitation, were often used for trace analysis of copper, gold, palladium, lead, cobalt, cadmium, nickel, tin, chromium, iron, and zinc in diverse study materials.

6.2.3 Inductively Coupled Plasma-Atomic Emission Spectrometry (ICP-AES)

Inductively Coupled Plasma-Atomic Emission Spectrometry (ICP-AES), also referred to as Inductively Coupled Plasma-Optical Emission Spectrometry (ICP-OES), offers an exceptional scope for analysis of trace essentials with high sensitivity. This is because of its high temperatures – up to 8000 K and rising to plasma – that is important for atomization of analyte present in sample. The ICP is formed by means of argon (Ar) gas, which is ionized in a strong electromagnetic field that flows in a precise rotationally symmetric pattern in the direction of a magnetic field produced through a radiofrequency coil. Due to the strong electromagnetic field, steady plasma is generated as an effect of collisions among the neutral Ar atoms and the charged particles. At the same time as the sample is introduced to the plasma, it directly collides with electrons and crashes into charged ions. Numerous molecules break up into their individual atoms through repeated losing and recombining of electrons in the plasma. During the spontaneous excitation of the atom, electromagnetic radiation is emitted that has specific characteristic wavelengths for a particular element. The concentration of this electromagnetic radiation emission indicates the concentration of the element in the sample and is simultaneously detected by an attached semiconductor detector.

6.2.4 Inductively Coupled Plasma Mass Spectrometry (ICP-MS)

Inductively Coupled Plasma Mass Spectrometry (ICP-MS) is a multi-element quantitative analysis instrument that compromises extensive exposure range of elements analysis at trace and ultratrace concentration present in liquid samples with diverse matrices. ICP-MS uses a similar argon plasma basis to dissociate the model into its basic atoms/ions. The ions are unconfined from the plasma and offered into the mass spectrometer, where ions get insulated respective to their atomic mass-to-charge ratio through a quadrupole or magnetic sector analyzer.

Within the analyzer, the metal ions are detected through the faint light that they emit. Samples are presented within a plasma torch which is in either gaseous or aerosol form, and for that reason liquid samples require nebulization. The liquid sample is impelled inside by vial using a peristaltic pump and the high number of ions are shaped and essentially pooled with very low backgrounds, which offers good detection ranges for a good number of elements, normally in the parts per trillion (ppt) range. Due to its exceptional sensitivity and the low limit for detection capability in a small sample, volume ICP-MS, there is the possibility for isotopic analysis. ICP-MS comprises a dynamic range of detection that can widely applied in various fields: clinical and biological, food and environmental, geological, industrial analysis like raw material and effluent analysis, etc., and still a large number of elemental analyses remain to be analyzed. Most of the elemental analysis done by ICP-MS is illustrated in the literature. In the ICP-MS method, the sample is ionized in a similar type of argon plasma as in the ICP-AES technique.

6.2.5 Inductively Coupled Plasma-Optical Emission Spectrometry (ICP-OES)

Inductively Coupled Plasma-Optical Emission Spectrometry (ICP/OES) is another spectroscopic procedure, which is found to be most suitable for trace elements study. The fundamental principle is of unprompted photon emission from atoms and ions that have been excited within the radiofrequency (RF) range. Through the analysis, solid samples entail acid digestion prior to injection, while liquid and gas samples are directly inserted into the instrument. After injecting into the sampler, the sample solution is transformed into an aerosol and then directed into the center of the plasma, which preserves a high atomization temperature of about 8000 K.

After the data analysis it was found that technique of ICP-MS is characterized by a parallel profile to AAS, which is possible because of the generation of hydrides. The intrusion instigated by transition metals should eliminated via removal of metals on ion-exchange resin (cationic) and on-line. Accuracy in the removal of transition metals and applied hydride generation means that the ICP-AES method has a certified discovery limit up to 0.4 ng/ml. By smearing organic compounds like simple alcohols, polyalcohols, and organic acids as transformers, the detection limit of some traces and ultratraces can go down to as little as 0.2 ng/ml. In analysis of selenium through any method with hydride creation it is an important step to decrease selenium to Se(IV). In resolves via ICP-AES by using HCl as reducer, the detection limit achieved was up to 1.5 ng/ml, for the equal as ICP-MS method was recycled. The term known as generation of hydrides pursue with their detention in graphite furnace, the detection limits were found around 0.25, 0.5 ng/ml. It is a most remarkable step when sample injection used with hydride generation and mass spectrometry; the limits of detection of Se(IV) and Se(VI) went down up to 0.012 and 0.016 ng/ml.

6.3 Chromatographic Methods

Gas chromatography (GC) is liquid sample analytical method that provides additional detailed analysis over and above the methods described above. GC analysis provides the total fractions at different degrees of oxidation of metal present in the sample or chromatographic delivery consent to identification of noticeable compounds, both organic and inorganic metal traces. Different sensors are used in chromatographic methods, e.g. Electron Capture Detector (ECD), Spectrophotometers (UV–Vis, IR, UV), Flame Ionization Detector (FID), Photoionisation Detector (PID), Thermal Conductivity Detector (TCD), or Fluorimeters (RF), but these are limited to low-concentration detection. As per the improvement of research in the detection of different analytical pathways, there are many improved and more specific detectors with different sources of excitation: Capacity-Microwave Coupled Plasma (CMP), Microwave-Induced Plasma (MIP), Capacity-Coupled Plasma (CCP), Directly-Coupled Plasma (DCP), Inductively Coupled Plasma (ICP), or atomic fluorescence, are available as required.

6.4 Electroanalytical Methods

Differential Pulse Cathode Stripping Volt-amperometry (DPCSV) for analysis of the total content of selenium with a detection limit of 121 ng/ml has described by Ferri et al. (1989). Another method, Differential Pulse Polarography (DPP), was also used for analysis of selenium Se (IV) in an environment of nitric acid, getting high intensity of the diffusion current. When smearing a strongly basic ($pH > 9$) electrolyte, the limit of determination was 0.08 ng/ml and the effects of the determinations were comparable to those attained by other means.

6.4.1 Other Analytical Methods

The approaches of determination of arsenic, antimony, and selenium defined and discussed above are not all the analytical approaches used for this purpose but have been chosen as the most frequently useful. The other methods revealed in literature comprise molecular absorption spectrometry, direct mass spectrometry (dual MS–MS), capillary electrophoresis as a method for separation of mixtures past to detection by ICP-MS, or ion-exchange chromatography with ICP-AES or MS exposure.

6.4.2 Atomic Fluorescence and Emission Spectroscopy (AFS and AES)

In Atomic Fluorescence and Emission Spectroscopy (AFS and AES) detection techniques, high energy encourages the atoms into spontaneous aggravated electronic states, and they consequently emit light when they get back to the ground electronic state. All elements release particular light at a characteristic wavelength, which is isolated and detected by a spectrometer. The wavelength of the atomic spectral line is always identical for each element, though the intensity of the emitted light is directly relative to the atomic mass of the element. AFS has combined features of both atomic emission and atomic absorption. The ground state atoms, produced by applying flame, are highly excited by a beam of light and transformed into the atomic vapor. The resulting emission from the decay of the atoms is controlled by adjusting the source light as reported by Ramachandran and Kumar (1996) (see also Ramachandran et al. 1993). Mester and Fodor (1997) used the AFS and AES for the detection of different minerals, e.g. calcium (Ca), potassium (K), magnesium (Mg), and sodium (Na), in 23 samples of edible wild mushroom types which appear all over Poland. Arsenic and mercury content have originated in traditional Chinese medicine, so that by using AFS with limits of detection ranging from 0.061–0.19 ppm for mercury and arsenic they are separately observed (Mester and Fodor 1997). On the other hand, atomic absorption has become a practice widely linked to AES and AFS.

6.5 Electrochemical Methods

Potentiometers with Ion-Selective Electrodes (ISEs) and stripping voltammetry by polarographic techniques are binary prime systems for bit discovery. Potentiometric sensors are inert electrochemical procedures where alterations in the balance voltage are checked under zero recent environments. The equilibrium potential of the scheme is preferably labeled as a Nernstian meaning of the action of the goal ions to be restrained.

ISEs are chemical sensor-based technique to resolve selected ions in present compound conditions and have been in practical use in recent years in quantifiable chemistry for the detection of electrolyte concentrations in blood. Electrodes are accessible for the supremely important clinically significant ions, e.g. Na^+, K^+, Ca^{2+}, Cl^-, and H^+. Best of the potentiometric campaigns had their detection limits in micromolar variety, but frequent novel expansions of sensor ranges, poly-ion sensing, novel reference electrodes, and capillary electrophoresis sensors considerably upgraded inferior exposure ranges, providing enlargement of the range for claim, resembling trace and ultratrace metal investigation in ecological and biological trials.

The detection limit for ISE is distinct as where the Nernstian response activity is extrapolated, for the object ion, captures with the inferred potentiometric reaction from extremely dilute solution. Probable biases prevailing in ISE quantities belong in the super- or sub-Nernstian response of the ISE at low absorption and the existence of intrusive ions in the result.

The non-ideal features of the ISE method can be eradicated by design of superior electrodes, practice of new ion-selective resources, and robust standardization of the scheme. The removal of the effect of intrusive ions is more challengeable for correction, subsequently the configuration of the calibration results and rich results may differ from sample to sample. In that case, an enhanced electrode or newly designed electrode and materials designed according to specific requirements can lead to research results at a better level.

6.6 Other Analytical Methods

The distinct and debated methods of detection of the most hazardous metals, e.g. arsenic, selenium, and antimony, have been discussed above. The methods of detection are always specific for different metals or complex groups. Every analysis is always used for a particular group of sample content. The various other methods also available, customized to the demands of the research, such as Direct Mass Spectrometry (dual MS–MS), molecular absorption spectrometry, and capillary electrophoresis, as a method for separation of compounds prior to recognition by ICP-AES or Ion- Exchange Chromatography with MS.

The authors have dedicated to significant organic and inorganic complexes in water, associated with development of investigative technique for detection of arsenic, selenium, and antimony, methods described above such as absorption, spectrophotometric methods, electroanalytical, combined (chromatography with diverse detection methods) approaches, and emission spectroscopy.

This chapter explains the promises of selected analytical methods in determining the concentration of particular metals including arsenic, selenium, and antimony. The implication of expressive the level of these elements in ecological, pharmaceutical, and technological samples is replicated in the sum of planned devices of their purposes, certifying the possibility of optimization for particular cases. The availability of a multiplicity of contributory and delimited effects makes it possible to choose the best method for anticipated levels of consideration.

6.7 X-Ray Techniques

X-ray techniques include a group of spectroscopic techniques of considerable importance. Due to its capability to detect within very low concentration limits with medium sensitivity, X-ray detection techniques have been broadly recycled for trace study of archeological samples, cements, and concrete, environmental samples like detection of airborne particulate matter, forensic samples analysis, geological materials, steels etc. A frequent practice of sample analysis is to use XRF to evaluate all elements present in sample. The sensitivity of XRF may be determined by the intensity of energy of incident energy, the geometry of the apparatus that is used for analysis, and the accuracy of the detector. The accuracy of XRF space is regulated by the figures of identified photons and the element concentration detection limit is a few g/cm^2 for particulate substantial for a extensive variety of elements.

Total Reflection XRF (TXRF) is another type of technique where the entering radiation is incident on the sample at > the critical angle and is totally limited. The short background levels produce an enhanced detection limit and even go down to as little as 2 pg, and it can detect a number of elements with a detection time of lower than around 1000 seconds.

The instrument combined with Synchrotron Radiation Fonts (SR), TXRF with polychromatic beam (particularly useful for the generation of diffraction gratings) for section excitation permits the reduction of detection parameters down to 0.03 pg/m^3 and the practice of new extreme photon unrests will lower the detection limit by numerous orders of scale.

For these practices, samples should be prepared in the form of a thin film and because of this low ground effects equally to the aforementioned techniques. Quantitative analysis is done by the adding of a single element, which should not existing in the sample itself, setting an interior standard, and drawing a correction curve which will be valid in all environments. The lack of strong standard data in these regulation methods indicates large-ordered errors that are extraordinary and related to the very low sums being examined.

6.7.1 Activation Investigation

Activation investigation refers to the understanding and measurable grit by practice of radionuclides shaped after a mark element. It needs very slight or no sample preparation, is nondestructive in environment and, mostly unpretentious by the section matrix, and permits for multi-elemental purpose. Neutron activation analysis (NAA) is the utmost joint alternative where neutrons are cast off to expose and actuate the example. When the size is approved ready-deprived of prior chemical separation, the method is called instrumental NAA (INAA). Consequently, the nuclear reaction between neutrons and the isotope of the element under examination, radionuclides with distinctive half-lives may be distinguished, producing radiation of variable energies that may be detected by an appropriate sensor and are specific to the element's method of formation. In INAA, the decay of the dangerous model is not obligatory; the technique is also nondestructive, which in the case of unique objects is a valued feature.

6.8 Conclusions

Selenium arose as a pollutant of distress from mining activities, in ground and fresh water. It is upsetting both water and global networks and has subsequently been the object of a major ecological deliberation. This chapter has examined Se pollution found in water, soil, and plants as free from mining undertakings. This assessment also debated the effects of speciation, oxidation–reduction, precipitation, and compellation on Se movement in the soil. The disposal of Se for plant acceptance, soil, and water bodies was widely explained and then analytical methods were also presented. Consequently, as a number of Se speciation procedures were formally endorsed by government organizations, a proper analytical technique needed to be designated and developed to have improved sensitivity, particularly in intrusive conditions. Predominantly, the hyphenated practices of HPLC coupled with MS and ICP-MS were vastly endorsed for Se speciation and have proved effective in enormously low discovery restrictions, with great precision, reproducibility, and determining of tricky interventions. Therefore, observing emergent Se pollution is absolutely essential in order that the resultant effective handling methodologies can be successfully executed to ensure eco-friendly salvation and economic expansion. With the purpose of overwhelmed trials in Se subtraction, a series of biochemical treatment or physical techniques were judged to improve Se elimination and to confirm human health safety and environmental protection.

References

Brodowska, M.S., Kurzyna-Szklarek, M., and Haliniarz, M. (2016). Selenium in the environment. *J Elem* 21 (4): 1173–1185.

Brown, K.M. and Arthur, J.R. (2001). Selenium, selenoproteins and human health: a review. *Public Health Nutrition* 4 (2b): 593–599.

Bujdoš, M., Muľová, A., Kubova, J., and Medveď, J. (2005). Selenium fractionation and speciation in rocks, soils, waters and plants in polluted surface mine environment. *Environmental Geology* 47 (3): 353–360.

Desborough, G.A., DeWitt, E., Jones, J. et al. (1999). Preliminary mineralogical and chemical studies related to the potential mobility of selenium and associated elements in phosphoria formation strata, southeastern Idaho (No. 99-129). US Dept. of the Interior, US Geological Survey.

Etteieb, S., Magdouli, S., Zolfaghari, M., and Brar, S. (2020). Monitoring and analysis of selenium as an emerging contaminant in mining industry: a critical review. *Science of the Total Environment* 698: 134339.

Fairweather-Tait, S.J., Bao, Y., Broadley, M.R. et al. (2011). Selenium in human health and disease. *Antioxidants & Redox Signaling* 14 (7): 1337–1383.

Fan, A.M. and Kizer, K.W. (1990). Selenium. nutritional, toxicologic, and clinical aspects. *Western Journal of Medicine* 153 (2): 160.

Favorito, J.E., Luxton, T.P., Eick, M.J., and Grossl, P.R. (2017). Selenium speciation in phosphate mine soils and evaluation of a sequential extraction procedure using XAFS. *Environmental Pollution* 229: 911–921.

Feng, R.W. and Wei, C.Y. (2012). Antioxidative mechanisms on selenium accumulation in Pteris vittata L., a potential selenium phytoremediation plant. *Plant, Soil and Environment* 58 (3): 105–110.

Ferri, T., Morabito, R., Petronio, B.M., and Pitti, E. (1989). Differential pulse polarographic determination of arsenic, selenium and tellurium at μg levels. *Talanta* 36 (12): 1259–1263.

Fordyce, F. (2005). Selenium deficiency and toxicity in the environment. In: *Essentials of Medical Geology* (eds. O. Selinus, B. Alloway, J.A. Centeno, et al.), 373–415. Amsterdam: Elsevier.

Gusek, J., Conroy, K., and Rutkowski, T. (2008). Past, present and future for treating selenium-impacted water. *Tailings and Mine Waste 2008 – Proceedings of the 12th International Conference.*

Haug, J.C., Passeport, E., and Terry, N. (2012). Development of a constructed wetland water treatment system for selenium removal: use of mesocosms to evaluate design parameters. *Environmental Science & Technology* 46 (21): 12021–12029.

He, Y., Xiang, Y., Zhou, Y. et al. (2018). Selenium contamination, consequences and remediation techniques in water and soils: a review. *Environmental Research* 164: 288–301.

Holmes, A.B. and Gu, F.X. (2016). Emerging nanomaterials for the application of selenium removal for wastewater treatment. *Environmental Science: Nano* 3 (5): 982–996.

Hu, X. (2009). Selenium concentration, speciation and behavior in surface waters of the Canadian prairies. *Sci. Total Environ* 407 (22): 5869–5876.

Hu, X. and Wang, F. (2009). Selenium concentration, speciation and behavior in surface waters of the Canadian prairies. *Science of the Total Environment* 407: 5869–5876.

Huag, A., Graham, R.D., Christophersen, O.A., and Graham, H.L. (2007). How to use the world's scarce selenium resources efficiency to increase the selenium concentration in food. *Microbial Ecology in Health and Disease* 19 (4): 209–228.

Izquierdo, M. and Querol, X. (2012). Leaching behaviour of elements from coal combustion fly ash: An overview. *International Journal of Coal Geology* 94: 54–66.

Khamkhash, A., Srivastava, V., Ghosh, T. et al. (2017). Mining-related selenium contamination in Alaska, and the state of current knowledge. *Minerals* 7 (3): 46.

Kumkrong, P., LeBlanc, K.L., Mercier, P.H., and Mester, Z. (2018). Selenium analysis in waters. Part 1: regulations and standard methods. *Science of the Total Environment* 640: 1611–1634.

Lemly, A.D. (2004). Aquatic selenium pollution is a global environmental safety issue. *Ecotoxicology and Environmental Safety* 59 (1): 44–56.

Lopes, A., Ávila, F.W., and Guilherme, L.G. (2017). Selenium behavior in the soil environment and its implication for human health. *Ciênc agrotec* 41: 605–615.

Ma, B., Fernandez-Martinez, A., Grangeon, S. et al. (2018). Selenite uptake by Ca–Al LDH: a description of intercalated anion coordination geometries. *Environmental Science & Technology* 52 (3): 1624–1632.

Marcocci, C., Kahaly, G.J., Krassas, G.E. et al. (2011). Selenium and the course of mild Graves' orbitopathy. *New England Journal of Medicine* 364 (20): 1920–1931.

Mehdi, Y., Hornick, J.L., Istasse, L., and Dufrasne, I. (2013). Selenium in the environment, metabolism and involvement in body functions. *Molecules* 18 (3): 3292–3311.

Mester, Z. and Fodor, P. (1997). Characteristics of the atomic fluorescence signals of arsenic in speciation studies. *Spectrochimica Acta Part B: Atomic Spectroscopy* 52 (12): 1763–1771.

Muscatello, J.R. and Janz, D.M. (2009). Selenium accumulation in aquatic biota downstream of a uranium mining and milling operation. *Science of the Total Environment* 407 (4): 1318–1325.

Nancharaiah, Y.V., Venkata Mohan, S., and Lens, P.N. (2016). Biological and bioelectrochemical recovery of critical and scarce metals. *Trends in Biotechnology* 34 (2): 127–155.

Peng, Q., Wang, M., Cui, Z. et al. (2017). Assessment of bioavailability of selenium in different plant-soil systems by diffusive gradients in thin-films (DGT). *Environmental Pollution* 225: 637–643.

Pettine, M., McDonald, T.J., Sohn, M. et al. (2015). A critical review of selenium analysis in natural water samples. *Trends in Environmental Analytical Chemistry* 5: 1–7.

Pilon-Smits, E.A.H. and Quinn, C.F. (2010). Selenium metabolism in plants. *Cell Biology of Metals and Nutrients* 17: 225–241.

Plak, A. and Bartmiński, P. (2017). The impact of land use on the organic and inorganic selenium content in soils developed from loess. *Journal of Elementology* 22 (4): 1463–1474.

Ralston, N.V., Unrine, J., and Wallschläger, D. (2008). *Biogeochemistry and Analysis of Selenium and its Species*. Washington, DC: North American Metals Council.

Ramachandran, K.N. and Kumar, G.S. (1996). Modified spectrophotometric method for the determination of selenium in environmental and mineral mixtures using 2, 3-diaminonaphthalene. *Talanta* 43 (10): 1711–1714.

Ramachandran, K.N., Kaveeshwar, R., and Gupta, V.K. (1993). Spectrophotometric determination of selenium with 6-amino-1-naphthol-3-sulphonic acid (J-acid) and its application in trace analysis. *Talanta* 40 (6): 781–784.

Rayman, M.P. (2012). Selenium and human health. *Lancet* 379 (9822): 1256–1268.

Ryan, B. and Dittrick, M. (2000). Selenium in the Mist Mountain formation of southeast British Columbia. *British Columbia Ministry of Energy and Mines, Geological Fieldwork*: 2001–1.

Sakamoto, M., Chan, H.M., Domingo, J.L. et al. (2012). Changes in body burden of mercury, lead, arsenic, cadmium and selenium in infants during early lactation in comparison with placental transfer. *Ecotoxicology and Environmental Safety* 84: 179–184.

Santos, S., Ungureanu, G., Boaventura, R., and Botelho, C. (2015). Selenium contaminated waters: an overview of analytical methods, treatment options and recent advances in sorption methods. *Science of the Total Environment* 521: 246–260.

Sharma, S. and Vance, G.F. (2007). Dissolution chemistry of inorganic selenium in alkaline mine soils. *Developments in Environmental Science* 5: 363–380.

Sharma, V.K., McDonald, T.J., Sohn, M. et al. (2017). Assessment of toxicity of selenium and cadmium selenium quantum dots: a review. *Chemosphere* 188: 403–413.

Silva Junior, E.S., Wadt, L.H.O., Silva, K.E. et al. (2017). Natural variation of selenium in Brazil nuts and soils from the Amazon region. *Chemosphere* 188: 650–658.

Sun, H.J., Rathinasabapathi, B., Wu, B. et al. (2014). Arsenic and selenium toxicity and their interactive effects in humans. *Environment International* 69: 148–158.

Swift, M.C. (2002). Stream ecosystem response to, and recovery from, experimental exposure to selenium. *Journal of Aquatic Ecosystem Stress and Recovery* 9 (3): 159–184.

Tan, L.C., Nancharaiah, Y.V., van Hullebusch, E.D., and Lens, P.N. (2016). Selenium: environmental significance, pollution, and biological treatment technologies. *Biotechnology Advances* 34 (5): 886–907.

US Environmental Protection Agency (US EPA). (2006). Inorganic Contaminant Accumulation in Potable Water Distribution Systems. US EPA.

US Environmental Protection Agency (US EPA). (2016). Aquatic Life Ambient Water Quality Criterion for Selenium–Freshwater. US EPA.

Vinceti, M., Crespi, C.M., Bonvicini, F. et al. (2013a). The need for a reassessment of the safe upper limit of selenium in drinking water. *Science of the Total Environment* 443: 633–642.

Vinceti, M., Solovyev, N., Mandrioli, J. et al. (2013b). Cerebrospinal fluid of newly diagnosed amyotrophic lateral sclerosis patients exhibits abnormal levels of selenium species including elevated selenite. *Neurotoxicology* 38: 25–32.

Wadgaonkar, S.L., Nancharaiah, Y.V., Esposito, G., and Lens, P.N. (2018). Environmental impact and bioremediation of seleniferous soils and sediments. *Critical Reviews in Biotechnology* 38 (6): 941–956.

Wang, Z. and Gao, Y. (2001). Biogeochemical cycling of selenium in Chinese environments. *Applied Geochemistry* 16 (11–12): 1345–1351.

Wellen, C.C., Shatilla, N.J., and Carey, S.K. (2015). Regional scale selenium loading associated with surface coal mining, Elk Valley, British Columbia, Canada. *Science of the Total Environment* 532: 791–802.

Wen, H. and Carignan, J. (2007). Reviews on atmospheric selenium: emissions, speciation and fate. *Atmospheric Environment* 41 (34): 7151–7165.

WHO (2011). *World Health Organization, Guidelines for Drinking-Water Quality*. Geneva, Switzerland: WHO Press.

Winkel, L.H., Johnson, C.A., Lenz, M. et al. (2012). Environmental selenium research: from microscopic processes to global understanding. *Environmental Science & Technology* 46: 571–579.

World Health Organization. (2011). *Selenium in Drinking-Water*. Geneva: World Health Organization.

Xing, K., Zhou, S., Xiaoguo, W. et al. (2015). Concentration and characteristics of selenium in soil samples from Dashan Region, a selenium-enriched area in China. *Journal of Soil Science and Plant Nutrition* 61 (6): 889–897.

Yu, X., Liu, C., Guo, Y., and Deng, T. (2019). Speciation analysis of trace arsenic, mercury, selenium and antimony in environmental and biological samples based on hyphenated techniques. *Molecules* 24 (5): 926.

Zhang, Y., Lin, L., Cao, Y. et al. (2009). Phthalate levels and low birth weight: a nested case-control study of Chinese newborns. *The Journal of Pediatrics* 155 (4): 500–504.

Zhang, J., Zhang, Z., Li, Q. et al. (2014). Selenium encapsulated into interconnected polymer derived porous carbon nano-fiber webs as cathode materials for lithium selenium batteries. *Journal of the Electrochemical Society* 161 (14): A2093–A2098.

7

Voltammetric Sensors and Materials for Selenium Detection in Water

Pinki Rani Agrawal and Rahul Sharma

CSIR- National Physical Laboratory, New Delhi, India; Academy of Scientific and innovative Research (AcSIR), Ghaziabad, India

7.1 Introduction

After the presentation of protective effects of selenium on living organisms by Schwartz and Foiz in 1957, selenium was tagged as a trace essential element whose deficiency in diet can result in catastrophic effects on an individual's health (Gupta and Gupta 2010; Margaret 2012; Riaz and Mehmood 2012; Kieliszek and Błażejak 2013; Kieliszek and Błażejak 2016; Ambroziak et al. 2017). Owing to its indispensability and biological functionality, selenium is a critical micronutrient and hence trace amounts of it are highly necessary for the life of humans, animals, and plants (Kune and Watson 2006; Triggiani et al. 2009; Zeng 2009; Wang et al. 2017). In particular, vital biological processes such as enzymatic functions that involve redox homeostasis, thyroid hormone metabolism, DNA repairing, curing of oxidative damage, immunity and reproduction are primarily dependent on selenium-based selenoproteins (Hoffmann and Berry 2008; Duntas 2010; Wrobel et al. 2016; Zoidis et al. 2018). Importantly, its deficiency is known to cause cancer, heart disease, muscular dystrophy, and many other diseases (Dennert et al. 2012; Benstoem et al. 2015; Smith and Garg 2017). Besides the primary biological activity of selenium, its checked intakes play critical physiological roles in preventing cancer and toxicity of many heavy metals such as mercury and cadmium (Berry and Ralston 2008; Spiller et al. 2017). Considering this, an appropriate intake of selenium is highly recommended by many governments and organizations across the world either through natural diet or supplementation. In spite of this, unfortunately, dietary selenium deficiency is still a major concern and is found to be detrimental in many parts of globe, especially in developing nations (Organization 2004; Stoffaneller and Morse 2015; Wang et al. 2018; De Arruda et al. 2019).

However, as per World Health Organization (WHO 2004) guidelines the intake of selenium should not exceed 70 g/day (Fairweather-Tait et al. 2010; Gupta and Gupta 2017). This is due to the catastrophic effects of an overdose of selenium above a certain limit. Excess selenium intake imposes serious toxicological hazards such as vomiting, nausea,

Selenium Contamination in Water, First Edition. Edited by Pooja Devi, Pardeep Singh, Arindam Malakar, and Daniel Snow.

diarrhea, diabetic risks, gastrointestinal upsets, hair loss, white blotchy nails, garlic breath odor, nerve damage, cardio-metabolic and neurological disorders (Macfarquhar et al. 2010; Min et al. 2010; Dennert et al. 2012; Benstoem et al. 2015; Smith and Garg 2017). Furthermore, many recent reports reveal the other cytotoxic effects of selenium toxicity and prolonged intake of selenium supplements such as hypertension, metabolic disorders, and adverse lipid profile. Moreover, selenium mainly exists into natural forms, namely, organic (selenocysteine and selenomethionine) and inorganic (selenites and selenates) forms. However, these are the inorganic forms which exhibit higher toxicity compared to organic ones (Thompson et al. 1984; Kim and Mahan 2001; Nuttall 2006; Popham and Shelby 2007; Pavlata et al. 2012).

Therefore, unlike other trace elements, selenium unexpectedly has a very narrow margin between beneficial and harmful intake doses that ultimately causes selenium deficiency, physiological status, and toxicity. Surprisingly, selenium becomes a potent toxin at concentrations only twice those needed. Normally, the concentration of selenium is about 0.1–0.3 ppb (parts per billion) in surface waters; however, in slightly larger amounts of about 0.9–5 ppb, it undergoes a series of bioaccumulation events (Lemly 1997; May et al. 2008; Lemly 2009; Luoma and Presser 2009; Devi et al. 2017c). In fact, selenium pollution has become a serious concern worldwide in recent times and has become a global phenomenon, though, for decades, selenium has largely been a unrevealed pollutant (Lemly 1997; May et al. 2008; Lemly 2009; Luoma and Presser 2009; Santos et al. 2015; Devi et al. 2017b, c; Tan et al. 2018; He et al. 2018). However, during the pollution surveillance and monitoring programs negative impacts of high selenium amounts have been recognized and consequently selenium as an emerging environmental pollutant has gained ample attention from scientific community, regulatory, and resource management agencies around the world. Moreover, among various resources, aquatic systems are widely considered as the potential selenium intake sources as selenium is strongly bio-accumulated in aquatic habitats and adversely affects aquatic life, nearby wildlife, and human health through drinking water supplies and food chains (Lemly 1997; Hamilton 2004; Luoma and Presser 2009; Santos et al. 2015; Al-Akeel 2017; Tan et al. 2018).

Considering this, the WHO has set a limit of 10 ppb for selenium concentration in drinking water (Lemly 1997; Hamilton 2004). Selenium is naturally found in the Earth's crust in carbonate rocks and volcanic and sedimentary soils, yet about 40% of its emissions to aquatic environments are caused by various anthropogenic and geogenic activities, and hence selenium poisoning is continuously increasing in aquatic systems (Shamberger 1981; Lemly 2002; Maggard 2004; Etteieb et al. 2020).

The major sources of selenium pollution in aquatic systems range from very basic agricultural practices to apex-level industrial processes (Leblanc and Wallschläger 2016; Mateo-Sagasta et al. 2017; Sharma and Bhattacharya 2017) (see Figure 7.1), owing to which aquatic habitats in urban, suburban, and rural domains are continuously undergoing selenium contamination. The drainage water from agricultural irrigation of seleniferous soils into natural water bodies has poisoned aquatic environments by selenium residues (Huang et al. 2009). Moreover, human activities like coal and gold mining, disposal of fossil fuel wastes in thermoelectric power plants, oil refining, phosphate deposits, the use of rock phosphate as fertilizer, the application of sewage sludge to land, the manufacture of detergents and shampoo processing of metal ores and other related processes are rising over the

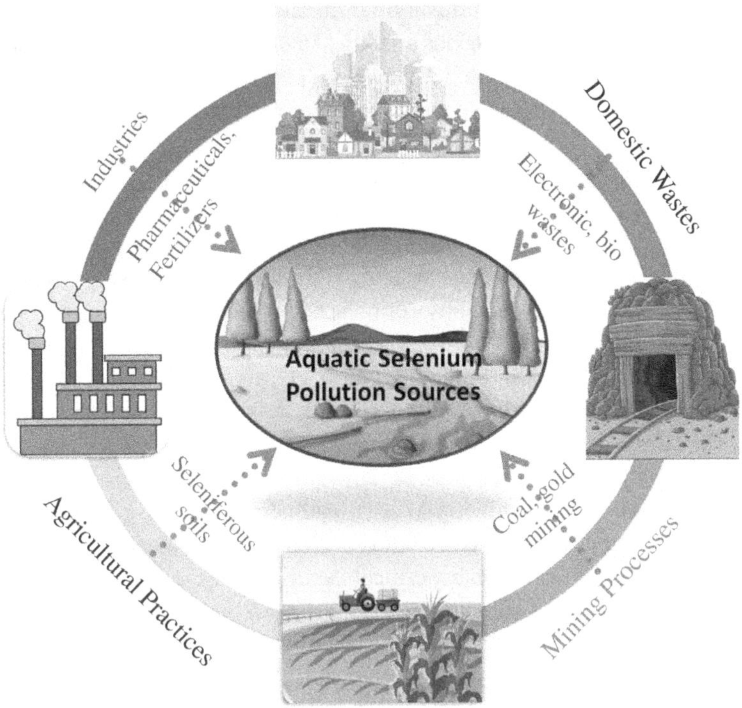

Figure 7.1 Sources of selenium pollution in aquatic systems.

years and quite often lead to the release of naturally present selenium in nearby surface water where a small amount of selenium imposes a significant environmental hazard via bioaccumulation (Keating 2001; Winkel et al. 2012; Ouédraogo and Amyot 2013; El-Ramady et al. 2015; Dorleku et al. 2018; Reta et al. 2018; Etteieb et al. 2020). Furthermore, huge amounts of selenium enter into the environment due to its excessive increase in pharmaceutical, glazing, photocopying, ceramics, paints, and electronics industries (Gutberlet and Baeder 2008; May et al. 2008; Winkel et al. 2012; Needhidasan et al. 2014).

In short, the two crucial aspects of selenium – a required nutrient as an essential element and at the same time a potent toxin – make it particularly critical from the perspective of both human, animal, and plant health. And due to this paradoxical situation, detection of selenium in aquatic systems as well as in drinking water supplies is highly desirable and necessary, both from the selenium nutrition assessment aspect and from the selenium pollution assessment aspect. In particular, the adequate monitoring of selenium aquatic pollution is highly required as its widespread cascading of hazardous impacts to life is more comprehensively increasing than ever before. Therefore, to prevent the cascading of bioaccumulation events of selenium and its toxic effects, highly effective and extremely efficient technology is the need of the hour for detection and quantification of selenium in aquatic, drinking, and wastewater systems. This will lead to sound selenium hazard assessment and water quality goals that ultimately result in lesser biological impacts and therefore adverse impacts to natural resources, public and animal health can be avoided (Bem 1981; May et al. 2008; Pettine et al. 2015).

In the last few decades, due to the strong motivation from the dual character of selenium as an essential trace element and a toxin, various analytical techniques to assess its availability in various environmental matrices have been developed especially for water samples. Inductively coupled plasma (ICP) coupled spectroscopy and electrochemistry-based analytical techniques are the two major methodologies for selenium detection in water (Wang and Rogers 1995; Hymer and Caruso 2006; Sanchez-Rodas et al. 2010; Costas-Mora et al. 2012; Devi et al. 2017a; Fakude et al. 2019; Zam et al. 2019). The various prominent techniques for selenium detection and estimation include high-performance liquid chromatography inductively coupled plasma mass spectrometry (HPLC-ICP-MS) (Pedersen and Larsen 1997), hydride generator atomic absorption spectroscopy (HG-AAS), graphite furnace atomic absorption spectroscopy (GF-AAS) (Meng et al. 1995), electro thermal atomic absorption spectroscopy (ETAAS) (Shaltout et al. 2011), atomic fluorescence spectrometry (AFS) (Wei et al. 1999), gas chromatography (Sentkowska 2019), neutron activation analysis (NAA) (Sakai et al. 1994), capillary electrophoresis (Pyrzyńska 2001), plasmic optical emission spectroscopy (POES) (Webb et al. 2014), and various electroanalytical techniques like anodic stripping voltammetry (ASV), cathodic stripping voltammetry (CSV), and adsorptive stripping voltammetry (AdSV) (Wang and Rogers 1995; Devi et al. 2017a; Fakude et al. 2019). However, many critical factors such as analyte concentration range, result accuracy, number of standardized samples and their volumes, analysis time, physicochemical properties of detection matrices, and the possible interfering species need to be considered during the selection of an analytical technique for selenium detection and quantification (Bolann et al. 2007; Santos et al. 2015; Kruse 2018). Moreover, the validation of analytical methods involves two immensely important parameters: limit of detection (LOD) and limit of quantification (LOQ). The concentration that can be detected, however, is not necessarily quantified while the minimum analytic concentration that can be computed with acceptable accuracy and precision is taken as LOQ. Conventionally, LOD and LOQ are assumed to be 3 and 10 times the noise level respectively; however, several methods of LOD and LOQ determination are reported in various studies. This shows that LOQ is about three times LOD. Moreover, LOD values as a performance parameter in method validation are generally reported in scientific studies instead of LOQ values (Bolann et al. 2007; Shrivastava and Gupta 2011; Bernal 2014; Xiong et al. 2018). In this context, the classical selenium accessing approaches based on colorimetric or spectrophotometric methods like methylene blue and 2,3-diaminonaphthalene are very time-consuming, difficult to apply to a high number of samples, and have low detection capabilities (Magin Jr et al. 1960; West and Ramakrishna 1968; Mathew and Narayana 2006). Furthermore, toxicity and stability of many of the reagents used in these methods are other major concerns. However, these classical methods are still in use because of their simple and facile operation, readily accessibility, non-involvement of expensive and complex equipment (West and Ramakrishna 1968; Mathew and Narayana 2006). Recently, dedicated research in this direction has been going on to improve sensitivity and selectivity of these methods along with reproducibility and better detection limits. On the other hand, instrumental techniques have several advantages over methods discussed above such as shorter analysis time and higher detection capacities.

Principally, selenium or in general any other metal detection in spectrophotometry is carried out through two main modes: the first is flame AAS and the second is

GFAAS. However, for enhancing the sensitivity of the method employed and to reduce interference effects, hydride generation systems are quite often coupled, which convert selenium species into their hydrides. The selenite species Se (IV) can be directly transformed into its hydrides while pre-reduction of selenate form Se (VI) is required for their conversion to hydrides. However, to overcome these issues, multi-commutated flow systems are exploited for enrichment processes, which in turn enhance the performance of these techniques. Specifically, low reagent consumption, flexible and facile operation, and upgradation in sampling frequency and detection limits are the certain advantages of flow systems during selenium detection (West and Ramakrishna 1968; Fernandes et al. 2009; Pistón et al. 2009).

As an alternative to this, in another approach coupling of optical emission spectroscopy (OES) with ICP and selective flow injection solid-phase extraction (SPE) has been practiced recently for speciation of selenium (Gomez-Ariza et al. 1999; Chen et al. 2013). Moreover, detection of selenium to parts per trillion levels (ppt) with high enrichment factor is achieved with this in many environmental water samples. Though these ICP-based techniques exhibit excellent sensitivity and are quite handy in detection of selenium even at very low concentrations, expensive equipment set-ups, long analysis time periods, and the requirement for sophisticated operator personnel limit its commercial viability (Kleckner et al. 2017; Wilschefski and Baxter 2019).

Furthermore, due to generation of harmful and toxic gases in the hydride formation step during the detection of selenium by spectroscopic techniques, these techniques are not advisable in all circumstances (Dedina 1982; Bye 1989; Sanz-Medel et al. 1999; Hineman 2011). As a result, novel chromatographic methods have emerged as efficient analytical tools for speciation and detection of selenium in water samples. However, the performance of chromatography methods is heavily dependent upon many analysis factors such as sample volume and amount, pre-concentration of sample, phase volume of the extractant, and nature and flow rate of eluent. Precisely, coupling of chromatographic methods like HPLC and gas chromatography with selective detectors has been explored in recent times and this approach efficiently allows the detection and estimation of different selenium species even at very low concentration levels (Wake et al. 2004; Bidari et al. 2008; Sentkowska 2019; Yu et al. 2019). Moreover, it is only the volatile selenium species such as organo-selenium compounds that can be quantified using gas chromatography and therefore inorganic selenium needs to be converted into volatile and thermally stable selenium species prior to utilization of gas chromatography (Hunter and Kuykendall 2004). Due to this, HPLC is more commonly used compared to gas chromatography, as most of the selenium species are non-volatile in nature (Sanchez-Rodas et al. 2013). The enhancement of speciation and detection limit of selenium prior to estimation with spectroscopy techniques is presented in various reports by using reversed phase HPLC and dispersive liquid-liquid micro-extraction (DLLME) (Bidari et al. 2008; Chen et al. 2013; Tie et al. 2014). Nevertheless, detection and quantification limits of selenium can be further lowered to ppt level along with selectivity retention by using a pre-concentrating substance in conjunction with the above-mentioned techniques. For instance, a detection limit of 40 ppt is achieved through graphite GFAAS for selenium in water samples by using graphene oxide-TiO_2 nano-composite adsorbent as a pre-concentrator (Zhang et al. 2016). The achieved efficiency of this method is attributed to the outstanding physicochemical properties such as large specific surface area, tunable surface functionalities, and excellent thermal-mechanical-chemical stability of carbon-based adsorbents (Xu et al. 2012; De Lima et al. 2013; Pettine et al. 2015; Zhang et al. 2016). In spite of the excellent sensitivity and selectivity of the above-mentioned conventional techniques, they have many associated limitations and drawbacks such as involvement of expensive and complex instrumentation, expensive and high-grade chemical reagents, trained operators, and tedious and time-consuming sample preparation and system operations. These disadvantages restrict their on-site utilization and real-time monitoring of selenium species in water systems and most of these techniques are limited to the laboratory environment. Moreover, the native characteristics of sample specimen also get dramatically changed during sample collection, transportation, storage, and preparation (Bidari et al. 2008; Xiong et al. 2018; Yu et al. 2019; Etteieb et al. 2020). Compared to this, voltammetric techniques, owing to low cost of instrumentation, minimum and facile sample preparation, less analysis time, and amenability to miniaturization and portability, electrochemistry-based has an edge over the other techniques in selenium detection (Ashournia and Aliakbar 2009; Ashournia and Aliakbar 2010; Sharifian and Aliakbar 2015a, b; Devi et al. 2017a; Porada et al. 2020).

Considering this, electrochemical stripping techniques with extremely high sensitivity and reproducibility have been developed using mercury film electrodes and hanging mercury drop electrodes (HMDE) for selenium detection (Long and Nagaosa 2007; Kolliopoulos et al. 2013; March et al. 2015; Porada et al. 2020). However, it is necessary to design and develop electrodes composed of non-toxic, highly stable, and environmentally friendly materials. Consequently, numerous kinds of voltammetric sensors have been designed in the recent past for selenium detection and quantification using different classes of materials such as metal/metal oxides nanoparticles, carbon materials like graphene, carbon nanotubes, fullerenes, polymers, quantum dots (QDs), and their hybrids and composites. However, due to their large specific surface area, tunable surface functionalities and architecture, extraordinary thermo-mechanical properties, fascinating opto-electronic characteristics, environmental benignity, and biocompatibility carbonaceous nanostructures have exhibited promising potential in terms of sensitivity and selectivity in voltammetric sensor systems for selenium detection (Long and Nagaosa 2007; Kolliopoulos et al. 2013; Pujol et al. 2014; Segura et al. 2015; IDRIS et al. 2017b; Jain et al. 2018).

This chapter presents a comprehensive overview of voltammetric sensors for detection and monitoring of selenium in water systems. Further, it is expected to stretch the readers which are belonging to different areas and domains such as ecologists, environmentalists, biologists, physicists, industrials, people associated with water treatment plants, environmental researchers and students by providing a comprehensive study, state-of-the-art picture of this field. It covers both the theoretical designing principles of voltammetric sensors and the development of active voltammetric electrode materials for selenium detection. Moreover, through a very intensive literature probing, various key physicochemical features of voltammetric electrode materials are discussed, which ultimately contribute toward the selectivity, sensitivity, robustness, fast responsiveness, portability, and stability of the final sensor.

The first section of the chapter begins with an introduction of selenium – its essentiality to life systems and its hazardous effects on human ecosystems along with human and animal health. The second section highlights the basic principles and mechanism of voltammetric techniques and sensors. The third section emphasizes the performance of different kinds of electrode materials such as metal/metal oxide nanoparticles, polymers, carbon-based materials etc. in voltammetric detection of selenium in water systems. The fourth section throws light on specific and general achievements and obstacles associated with the technique and different electrode materials that severally affect the designing and development of effective and efficient voltammetric sensors for selenium detection. Finally, chapter concludes with the comments on the current status, major challenges, and future scope in the direction.

7.2 Voltammetric Method: Basic Principles and Mechanism

The voltammetric method is known as a type of electrochemical method. The Czech chemist Jaroslav Heyrovský first introduced voltammetry, which was called as polarography, in 1992 and he was acknowledged in 1959 by the Nobel Prize in Chemistry for this. The term "voltammetry" derives from "voltamperometry". The voltammetric method is a well known electrochemical technique for the measuring the current as a function of voltage, i.e. electrode potential (Farghaly et al. 2014; Scholz 2015). The highly efficient and commendable voltammetric technique can be utilized for detection of low concentration (nano- to pico- molar concentration) solution, multi-elemental analyses including evolution and in-situ detection of toxic elements (Heyrovský and Shikata 1925; Settle 1997; Barón-Jaimez et al. 2013; Heyrovský and Kůta 2013; Farghaly et al. 2014). The voltammetric technique introduces the three cell processes, which are: working electrode (WE), reference electrode (RE), and auxiliary electrode (AE)/counter electrode (CE). Moreover, the WE is smaller than AE because of the maintaining of the flow of current (Scholz 2015; Devi et al. 2017a). Generally, the WE is grounded because it can be controlled by the potential of the solution. Moreover, these two electrodes give the information about the resulting current. On the other side, the RE is introduced to regulate the potential of WE and measure the voltage between these two electrodes (WE and AE). The RE has also been known as the stable electrode through the current which does not pass because it would lose its own stability (Barón-Jaimez et al. 2013; Farghaly et al. 2014; Scholz 2015). The Figure 7.2 shows a schematic illustration of the working principle of a voltammetric sensor.

The voltammetric instrument was introduced by Heyrovský and Shikata (Heyrovský and Shikata 1925). This instrument helps to monitor the current–voltage curve, which is called a polarograph. Consequently, the current–voltage curve is known as a voltammogram, where the current is shown on the vertical axis and the potential on the horizontal axis. Likewise, the voltammetry is explained by the relationship of current–voltage to time, which is obtained by three electrode cells, such as the voltage that has been given to the WE. Moreover, the subsequent current is passing through an electrochemical cell which would be recorded. Either the applied current can be changed or the subsequent current could be noted down over time periods (t) (Zbinden 1931; Nikelly and Cooke 1957; Demars

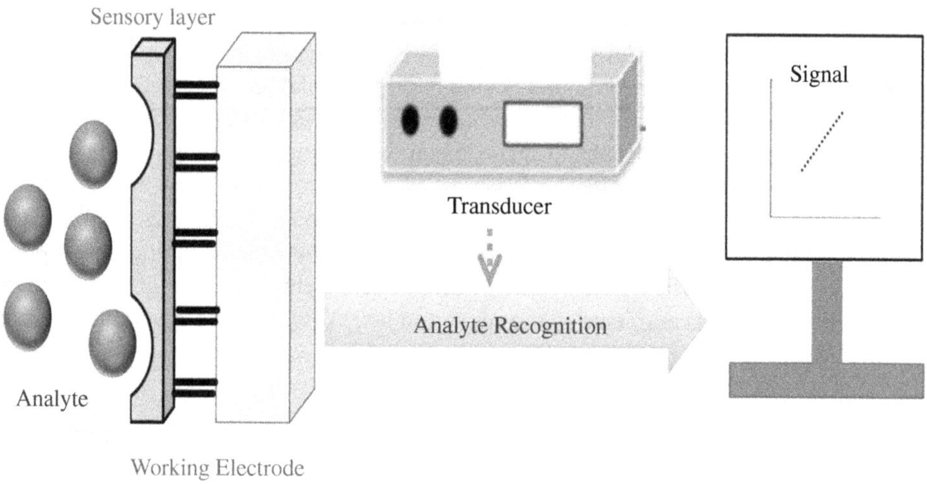

Figure 7.2 Schematic illustration of the working principle of a voltammetric sensor.

and Shain 1957; Hanrahan et al. 2004). We know that the given voltage to the WE works as driving force for the electrochemical reaction which is controlled by the parameter (Eggins 2008; Tripathi et al. 2020). Thereafter, the chemical species present in the solution will be oxidized or reduced at the electrode surface. Platinum wire is used as CE, Ag/AgCl, saturated calomel electrode (SCE) as RE. The widely usable WE is Hg, Au, Pt, and carbon terminals (Harvey 2000). In voltaic cells, the documented current possesses two types of current; for instance, faradaic and non-faradaic. Faradaic current arise from the amendment of oxidation state of the analyte. Moreover, the faradaic current is the section of overall charge which follows Faraday's laws. This current can be examined by interrogating the charge exchange during the redox reaction as portrayed by equation 7.1.

$$O + ne^- \leftrightarrow R \tag{7.1}$$

O and R stands for oxidized and reduced types of redox reaction. As it can be said, the other current which does not follow Faraday's laws is non-faradaic current. Generally, most voltammetric instruments are projected to limit this non-Faradaic current (Brett 2001; Langhus 2002; Ojha 2020; Pooja et al. 2020).

7.3 Type of Voltammetric Methods for Selenium Detection in Water

Generally, the different type of voltammetric techniques can be differentiated by an assortment of chemical reactions, electrochemistry, and physics, for instance rate constant of chemical reaction, high quantitative analysis, electrons engaged with redox reaction, and diffusion constant (Barón-Jaimez et al. 2013). As we know, there are different types of voltammetric sensors, which are reported in the literature, for example, linear sweep voltammetry, staircase voltammetry, square wave voltammetry, cyclic voltammetry (CV), stripping voltammetry, alternating current voltammetry, polarography, rotating electrode voltammetry, and so on (Settle 1997; Zoski 2006; Farghaly et al. 2014). Of these methods, the most utilized voltammetry method to detect the selenium in water is the stripping voltammetry method.

7.3.1 Stripping Voltammetry

The stripping voltammetry method is most useful and utilized technique to examine toxic metals because it gives a wide linear range and direct examination of trace solution of the analyte. Additionally, stripping voltammetry is a worthy technique to examine very low concentrations of analyte (10^{-10}M–10^{-12}M) since it has excellent sensitivity and selectivity. The concept of stripping analysis was first introduced by Zbinden in 1931 (Zbinden 1931). However, stripping voltammetry is based on three steps: pre-concentration, resting, and stripping. The pre-concentration step enhances the sensitivity of the stripping technique more than other reported techniques. Moreover, it can be provided the various modes for pre-concentration of analyte on WE and it can be concluded in two steps such as electrochemically and adsorptive. As reported in literature, ASV, cathode stripping voltammetry, and AdSV are a type of stripping voltammetry (Settle 1997; Kosseva and Kennedy 2003; Hernández 2006; Farghaly et al. 2014).

7.3.1.1 Anodic Stripping Voltammetry

ASV has been demonstrated to examine the various toxic metals and semi metals like Pb, Hg, Cd, As, and Se. The ASV method is a more usable and appropriate technique to detect the toxic metals at various oxidation states including at very low detection limit (LOD), i.e. sub ppb (Figura and Mcduffie 1979). The ASV method consists of three electrode processes. The ASV method is accomplish by the following process steps. The pre-concentration step comes once the selenium has been pre-concentrated on the surface of WE by putting on more negative deposition potential. Thereafter, it is oxidized again in the positive cycle and peak current is obtained and recorded, when stripping/peak current is proportional to the concentration of adsorbed selenium during reduction process (Zoski 2006; Barón-Jaimez et al. 2013; Ivandini and Einaga 2013; Farghaly et al. 2014; Devi et al. 2017a), as described by the reaction in equation 7.2.

$$H_2SeO_3 + 6H^+ + 6e^- \leftrightarrow H_2Se + 3H_2O \tag{7.2}$$

7.3.1.2 Cathodic Stripping Voltammetry (CSV)

In general, the CSV technique is the reverse of the ASV technique. Using the CSV technique, we can detect organic and inorganic salts which are obtained insoluble salts with electrode materials. The sensitivity of CSV is based on three variables: first is the amount that can be plated in given period plate, second is the density of the obtained insoluble salts film, and the last variable is the dissociation rate of insoluble mercury compound during the stripping step (Florence 1979; Farghaly et al. 2014; Devi et al. 2017a). Generally, mercury- and silver-based electrode sensors have been demonstrated as WE for the CSV method.

7.3.2 Square Wave Voltammetry

Square wave voltammetry (SWV) is a similar form of linear potential sweep voltammetry. SWV is a combination of square wave and staircase voltammetry. This technique has been demonstrated in various fields such as medicine and sensing toxic metals (Farghaly et al. 2014). Despite this, it can be used with CSV and ASV technique to improve the working principle of the methods.

7.3.3 Linear Sweep Voltammetry

Linear sweep voltammetry is a well known method to sense heavy metals in a water matrix. In recent years, this technique has been introduced with ASV, CSV, and CV techniques. In this method, the current is measured at the WE. On the other side, the potential is swept linearly between WE and RE over time (Farghaly et al. 2014).

7.3.4 Cyclic Voltammetry

Recently, CV is more widely used and attentive technique in various areas of electroanalytical chemistry because of its low signal to noise ratio, wide range of current over voltage, and fast technique. Additionally, it is also used in the cellular biology field to determine the concentrations in living organisms and organometallic chemistry to investigate redox mechanisms. CV gives much information about the obtained chemical reactions. The cyclic voltammetry technique is introduced as a potentiodynamic electrochemical measurement technique. In CV, The voltammogram obtained from the WE potential slopes linearly versus time. It can be employed in multiple cycle modes. The CV technique rapidly scans the reversed direction of potential of the WE. When the potential is applied in forward and backward direction at the WE, the current obtained is documented. In the CV technique, the backward and forward scan rate is generally similar. In the CV technique, both the anodic and the cathodic reaction take place at the electrode. The measured parameters are anodic and cathodic peak potential (E_{pa} & E_{pc}), anodic and cathodic peak current (I_{pa} & I_{pc}), and the half peak potential ($E_{p/2}$) (Farghaly et al. 2014). Recently, the CV technique has generally been used with the stripping voltammetry technique to get a better LOD of the analyte. Ivandini et al. demonstrated a gold modified boron doped diamond (Au/BDD) electrode to examine the concentration of selenium (Se(IV) and Se(VI)) using the CV technique in acid condition (Ivandini and Einaga 2013). In another approach, Radhi and his team prepared a carbon nanotube modified glassy carbon electrode (GCE) to sense the selenium in a blood matrix using the CV technique (Radhi et al. 2016). In a recent approach, Azizi et al. fabricated a modified pencil graphite electrode using composite film acetophenone (2,4-dinitrophenyl) hydrazone as a WE for the determination of selenium in food and water samples by CV and SWV technique (Azizi and Babakhanian 2018).

7.3.5 Differential Pulse Voltammetry (DPV) Methods

The DPV method is type of pulse voltammetric method which is introduced by Barker et al. It is similar to SWV method (Scott 2016). In the DPV method, a potential pulse is superimposed on changing potential and the current determined at two points, such as before applying the pulse and at the end of pulse (Farghaly et al. 2014).

7.4 Electrodes and Electrode Materials/Modifiers for Voltammetric Detection of Selenium in Water: Designing and Sensing Performance

7.4.1 Electrodes

There are so many characteristic properties of sensing electrodes, for instance, signal/noise ratio, window potential, reproducibility, sensitivity, selectivity, long-term stability, toxicity, and cost, which play a major role in performance to invest the selenium in real water sample by various voltammetric techniques. Over the last decades, the macro electrodes are more attentive materials for investigation of selenium in different matrices. First, HMDE as an electrode is a more attentive and well-used electrode to analyze selenium and other heavy metals in a water matrix. Owing to the toxicity of mercury, several researchers have also demonstrated other electrodes such as a silver, gold, carbon-based electrode (graphite, GCE), and copper as a solid electrode, including ceramic composite. Additionally, the macro electrodes were modified by nanomaterials, polymers, and organic molecules to achieve the best results in analysis of very low concentrations of selenium in a real water matrix and increase the reproducibility and selectivity of the electrodes, including diminishing the toxicity of the mercury electrode. The details of macro electrode are described in section 4.2 and summarized in Tables 7.1 and 7.2.

7.4.2 Electrode Materials

Electrode materials/modifiers are the linchpin of electrochemical sensors as the sensitivity and selectivity of these sensors solely govern the nature and structural features of the active electrode material. From the technological perspective electrode materials having light weight, low density, excellent electrical properties, extraordinary chemical-mechanical-thermal stability, and good environmental benignity are highly desirable. However, the existence of all these properties in a single material is quite rare and optimum performance in the materials is generally introduced through the factual designs of micro/macro and nanostructures. Further, a universal approach that involve synthesis of hybrids and composites from different kinds of novel materials is universally adopted not only for casting the active electrode but also for internalizing the demanded characteristics in the active electrode.

In this section we are throwing light on the current status of various types of electrode materials that have been used in past decades for development of voltammetric sensors for the detection of selenium in water. These materials include metal/metal oxide nanoparticles, polymers, carbon-based materials, and their hybrids and composites.

Mostly, platinum (Pt) wire is used as a CE and Ag/AgCl as RE in reported literature. Researchers are working on modifying the WE to detect the selenium from water. For selenium detection in water, Bertolino et al. demonstrated the square wave voltammetry after

Table 7.1 The summarized different type of electrodes with correspondence their linear range and limit of detection (LOD) in ppb using ASV technique.

Electrode type	Method	Linear range	Limit of detection (ppb)	Matrix	References
BiFE	SWASV	0–100 ppb	0.1	–	Zhang et al. (2010)
MGUA	SWASV	0–100 ppb	0.42	–	Tan and Kounaves (1998)
GCE-AuNPs	SWASV	0.001–50 ppm	0.34	–	Idris et al. (2015)
AuUME	SWASV	0–100 ppb	0.42	Tap water	Tan and Kounaves (1998)
GC/AuNPs/C	SWASV	15–55 ppb	0.175	Sea water	Segura et al. (2015)
GC/AuNPs/E	SWASV	10–50 ppb	0.120	Sea water	Segura et al. (2015)
Graphite SPEs	ASV	10–1000 ppb	4.9	Tap water	Kolliopoulos et al. (2013)
GCE-AuNPs	SWASV	0.1–120 ppb	0.22	Tap water	Idris et al. (2017a)
rGO-GCE	SWASV	10–50 000 ppb	0.85	Tap water	Idris et al. (2017b)
Au modified with poly 3,3′-diaminobenzidine 4HCl-Naflon	DPASV	0.4–158 ppb	0.06	Sea water	Ramadan et al. (2014)
BOD modified with AuNPs	DPASV	10–100 ppb	–	Standards	Fierro et al. (2012)
Poly (3,3′diaminobenzidine) film on Au electrode	DPASV	7.9–79 ppb	0.75	Water quality control	Cai and Khoo (1994)
Renewable silver annular band working electrode	DPASV	1.0–10 ppb	0.15	Surface water	Baś et al. (2014)
Au film/GCE	LSASV	10–60 ppb	0.1	Acidic mixture	Hamilton et al. (1979)
Au film/CD-R	DPASV	0.5–291 ppb	–	Acidic mixture	Pereira et al. (2006)
Au/BDD	CV	2–10 ppm	1.2 ppm	Mineral water, Sea water	Ivandini and Einaga (2013)
Au/BDD	ASV	10–100 ppb	10	Acidic mixture	Fierro et al. (2012)
AuNPs/GCE	SWASV	0.001–50 ppm	0.64	Tap water	Idris et al. (2015)
AuNPs/GCE	SWASV	0.01–12 ppm	0.52	Waste water	Idris et al. (2015)
Methylene blue-Nafion modified/Au	DPASV	0.79–79 ppb	0.394	Acidic mixture	Ramadan et al. (2012)
Vitamin E-Nafion Modified/Au	DPASV	3.94–788 ppb	–	Acidic mixture	Ramadan et al. (2011)
Au/ZnO NCs onto ITO	SWASC	5–100 ppb	2.89	Water	Jain et al. (2018)
GCE-NG	SWASV	1–120 ppb	0.092	Tap water	Fakude et al. (2019)

pre-concentration at activated carbon and the LOD was found to be 0.004 μg/l with linear range 0.01–20 μg/l (Bertolino et al. 2006). Secondly, the flow injection ASV method using gold electrode has been reported to detect selenium (Bryce et al. 1995). In another approach, Zhang et al. introduced the differential pulse AdSV using bismuth film electrode (BiFE) including cetyl trimethyl ammonium bromide and p-amino benzene sulfonic acid as a complexing agent for selenium determination, and the LOD is 0.1 μg/l (Zhang et al. 2010). Moreover, to achieve better results and improve the properties of the voltammetry method, the ASV method is interrogated because of it is more sensitive and selective. Over the last decade, the ASV method has been utilized as a square wave as well as a differential pulse for selenium detection to achieve the LOD. On account of the LOD of selenium, Gold (Au) particles are more a promising material as an electrode to detect the selenium in water owing to its non-toxicity, wide potential window, flow electro kinetics, and reproducibility including favorable redox behavior with the analyte in view of its high stability (Stora et al. 1997; Devi et al. 2017a). Moreover, the geometry of Au has been demonstrated to perform a critical role on current density and thus the detection limit of analyte. To examine selenium in a natural water matrix, the Au electrode has been demonstrated as micro disc, macro disc, and micro band after reviewing the geometry of the Au electrode. First, the prepared Au ultra-micro electrode (AuUME) has been reported by Tan et al. for selenium detection. The AuUME has more sensitivity, fast response time, and a lower detection limit (0.42 ppb) including a fast and reproducible response time (Tan and Kounaves 1998). In another approach, the Au electrode has been modified with carbon materials, organic materials, and nanomaterials for sensing the selenium, which is summarized in Table 7.1 with LOD, linear range, and matrix. In spite of these electrodes, the Au film as an electrode on GCE has also been prepared to achieve better LODs using linear sweep (LS) and differential pulse (DP) ASV. As reported in the literature, Pereira et al. prepared the Au film from a recordable compact disc (CD-R) as electrode to sense the selenium in an acidic mixture using DPASV (Pereira et al. 2006). On the other side, Au film has been deposited on GCE to modify the electrode for selenium detection using the LSASV technique. Au film/GCE has 0.1 ppb as LODs of selenium within the linear range 10–60 ppb (Hamilton et al. 1979). Some researchers are working to use the gold particles with carbon materials and organic materials, which are summarized in Table 7.1. These electrodes have good detection limit of selenium with appropriate linear range as well as good accumulation time. Recently, Au nanoparticles modified GCE as an electrode has been developed electrochemically and chemically, which is demonstrated for selenium detection using SWASV. The LOD of selenium is found to be 0.12 ppb with linear range 10–50 ppb by GCE/AuNPs/E. On the other side, for GCE/AuNPs/C the LOD was observed to be 0.175 ppb with linear range 15–55 ppb (Segura et al. 2015). In another approach, Idris et al. demonstrated the AuNPs modified GCE electrode for selenium and arsenic detection in tap water. This electrode was prepared by electro-deposition through cycling from −400 to 1100 mV and it was utilized as a substrate. The linear range 0.1–120 ppb was observed with LOD 0.22 ppb as well as good reproducibility (Idris et al. 2017a).

The same group also developed the electrochemical sensor based on AuNPs modified GCE and GCE modified with reduced graphene oxide (rGO) (GCE/rGO). The LOD obtained was 0.64 and 0.85 ppb, respectively (IDRIS et al. 2017b). GCE/rGO as electrode was also utilized for real water sample to detect selenium. However, these electrodes

Table 7.2 The summarized different type of electrodes including their correspondence linear range and limit of detection (LOD) in ppb using the CSV technique.

Electrode type	Method	Linear range	Limit of detection (ppb)	Matrix	References
HMDE	DPCSV	0.5–3 ppb	0.06	Rain water	De Gregori et al. (2002), p. 59
HMDE	DPCSV	4–50 ppb	1	Acidic mixture	Vajda (1970), p. 60
HMDE	CSV	0.4–4 ppb 0.23–23.64 ppb 0.04–0.63 ppb	0.4 0.16 0.04	Water sample	Jarzabek and Kublik (1982), p. 62
HMDE	SWCSV	–	0.023	Rain water	Zelić and Branica (1990), p. 64
HMDE	SWCSV	0.79–79 000 ppt	0.63 ppt	Acidic mixture	Sladkov et al. (2002), p. 65
HCADE	DPCSV	0.02–2.8 ppb	0.02	Mineral water, tap water	Piech and Kubiak (2007), p. 68
HMDE	DPCSV	0–80 ppb	0.083	Natural water	Ghoneim et al. (2000), p. 72
HMDE	DPCSV	1–10 ppb, 10–100 ppb	0.64 7.5	Acidic mixture	Mohamed and Leong (2006), p. 75
HMDE	DPCSV	0–30 ppt	0.5 ppt	Sea water	Quentel and Elleouet (1999), p. 76
HMDE	DPCSV	1–1000 ppb	0.07 ppb	Snow, natural water	Papoff et al. (1998), p. 77
MDE	DPCSV	1–150 ppb	0.35	Drinking water, lake water, purified water	De Carvalho et al. (1999), p. 78
HMDE	SWCSV	1–100 ppb	0.030	Sea water, hydrothermal, hemofluids	do Nascimento et al. (2009), p. 80
HMDE	DPCSV	0.6–16 ppb	0.06	Lake water, river water	Grabarczyk and Korolczuk (2010), p. 81
EDTA/HMDE	DPCSV	0–50 ppb	0.02	Acidic mixture	Baltensperger and Hertz (1985), p. 82

HMDE	SWCSV	0.08–3.15 ppb	0.03152	Tap water, river water	Korolczuk and Grabarczyk (2003), p. 84
Hg film/Ag	DPCSV	5–300 ppb	0.1 ppb	Mineral water, natural water, potable water	HamidáYahaya (1992), p. 97
MFE/Ag	DPCSV	0.04–8 ppb	17 ppt	River water	Rubinskaya et al. (2003), p. 98
Hg film/GCE	DPCSV	10–60 ppb	0.37 ppb	Lake water, river water	Piech (2008), p. 99
DAB/HMDE	DPCSV	0–2 ppb	0.1 ppb	River water, sea water	Sharifian and Aliakbar (2015a), p. 105
4-nitro-1,2-phenylendediamine/TMFE	DPCSV	2–20 ppb	0.06 ppb	Sea water	Tanaka et al. (1990), p. 108
Diazonium salts	DPCSV	3–2000 ppb	–	Drinking water, mineral water, tap water	Ashournia and Aliakbar (2010), p. 109
Poly DAB/Au	DPCSV	0.016–16 ppb	0.78 ppb	Water samples	Deryabina et al. (2013), p. 110
DAB-Nafion/MFE/GCE	SWCSV	0.5–50 ppb	0.1 ppb	Sea water	Cai and Khoo (1994), p. 111
DAB-Nafion/GCE	SWASV	1–300 ppb	0.48 ppb	Tap water, rain water	Yang and Sun (2000), p. 112
Ceramic	DPASV	0.1–20 ppb	0.02 ppb	Mineral water, well water	Stozhko et al. (2006), p. 124

suffered from interference from Cu(II) and Cd(II) during interference analysis. Alternatively, Jain et al. demonstrated the Au/ZnO nanocomposites (Au/ZnO NCs) on ITO electrode for sensing the selenium in a water matrix. This electrode has been fabricated by one step electrochemical reduction method (Jain et al. 2018). This electrode has achieved the LOD and quantification was observed to be 2.89 and 8.75 ppb, respectively, including 2.55% ($n = 3$) RSD and 0.19 μA/ppb. Further, this electrode has also utilized for the cations and anions which generally found in water. Consequently, the interference of ions is negligible on selenium stripping current at 10, 50, and 100 ppb concentration of interfering ions. Figure 7.3 demonstrates the effect of various sensing factors concentration: (electrolyte, pH, deposition potential, and deposition time) at 50 ppb Se(IV) on selenite stripping peak current during stripping scan in 0.7–1.0 V (vs Ag/AgCl) potential range. Over the years, owing to porous structure, high surface area, good thermal and electrical properties, low density, and good conductivity, carbon materials (carbon nanotubes, graphene, and activated carbon) have been prominent and versatile materials which are used for various applications such as super-capacitor (Wang et al. 2014), radar-absorbing material (Oh et al. 2004), adsorption of heavy metals (Agrawal et al. 2016, 2018, 2019), EMI shielding (Agarwal et al. 2016), fuel cell (Kaushal et al. 2019), and so on. However, graphene is a good attentive material for electrochemical sensor owing to its apparent properties such as thermal, electrical, optical, mechanical properties, low cost, and high surface area (Benvidi et al. 2015). Despite these properties, graphene has sp^2 bonded structure which is the rationale for chemical and electronic properties which helps to leading to the enlargement of the electron transfer in redox process (Nayak 2016). As seen in the literature, rGO modified GCE has been demonstrated to sense the selenium in real water as discussed above (Idris et al. 2015). To improve the properties of the electrochemical sensor, Fakude et al. demonstrated nitrogen-doped graphene (NG) on modified GCE sensors (GCE-NG) (Fakude et al. 2019). NG was fabricated by thermal treatment of GO in ammonia. Cyclic voltammetry (CV) and electrochemical impedance spectroscopy (EIS) was carried out to analyze the electrochemical properties of unmodified GCE and GCE-NG sensors. A prepared GCE-NG sensor was used to detect selenium in real water samples using SWASV including parameters: 0.1 M $HClO_4$ as supporting electrolyte, −0.8 V deposition potential, and 50 seconds' deposition time. The LOD was obtained 0.092 ppb, which is less than the WHO prescribed limit, in linear concentration range 1–120 ppb. The results are depicted in Figure 7.4: CV, calibration graph with parameter deposition potential and time, and comparison of the performance of unmodified GCE and GCE-NG.

Mercury is a more attentive material as a good sensor owing to good cathodic potential range, reproducibility, highly sensitive, and instant diffusion equilibrium. Some reported mercury-based electrode sensors have been summarized in Table 7.2. In CSV technique, the selenium is found in acidic media, which forms the insoluble mercury (II) selenium complex on mercury electrode surface at a specific voltage rate and is obtained by the following chemical reaction (Devi et al. 2017a):

$$H_2SeO_3 + 4H^+ + Hg + 4e^- \rightarrow Se(Hg) + H_2O$$

$$HgSe + 4H^+ + 2e^- \rightarrow H_2Se + Hg$$

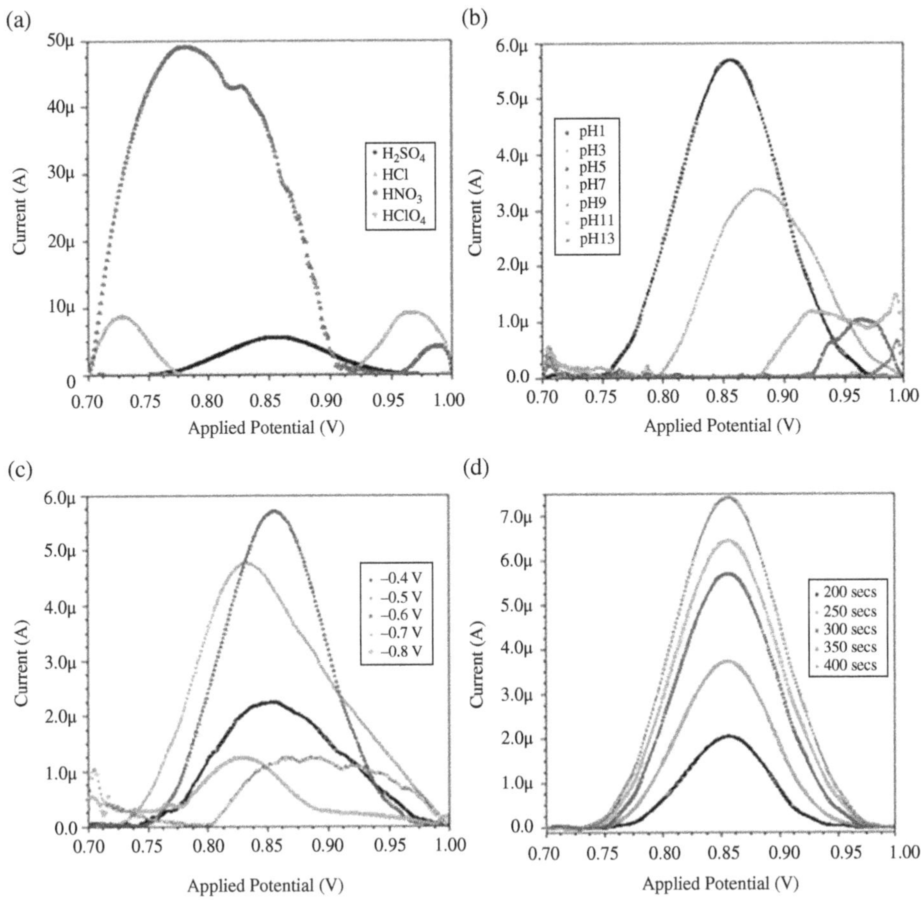

Figure 7.3 Effect of various sensing factors at 50 ppb selenium (IV) concentration: (a) electrolyte; (b) pH; (c) deposition potential; and (d) deposition time on selenite stripping peak current during stripping scan in 0.7e1.0 V (vs Ag/AgCl) potential range (Jain et al. 2018).

Firstly, HMDE was demonstrated for selenite detection up to 1 ppb by Wilson et al. (Dennis et al. 1976). In this approach, the DPCSV technique was employed and during the interference study this method faced interference from oxides and sulfides, which was a major challenge. In another approach, Jarjabek and Kublik prepared the HMDE (Jarzabek and Kublik 1982) as a cathodic electrode to sense the selenium in a different linear range as shown in Table 7.2. As seen in the literature, HMDE is more commonly employed material as a cathodic electrode using SW and DPCSV technique, as summarized in Table 7.2. Several researchers have modified the HMDE with copper metal to achieve the minimum LOD of selenium. Copper has received more attention for selenium detection using the CSV technique owing to it being a good companion for selenium (IV) and it forms the highly soluble selenium copper complex as copper selenide (Cu_2Se, CuSe, Cu_3Se_2) at the electrode including enhancing the sensitivity and reproducibility of a mercury electrode (Zelić and Branica 1990). The reaction is explained below:

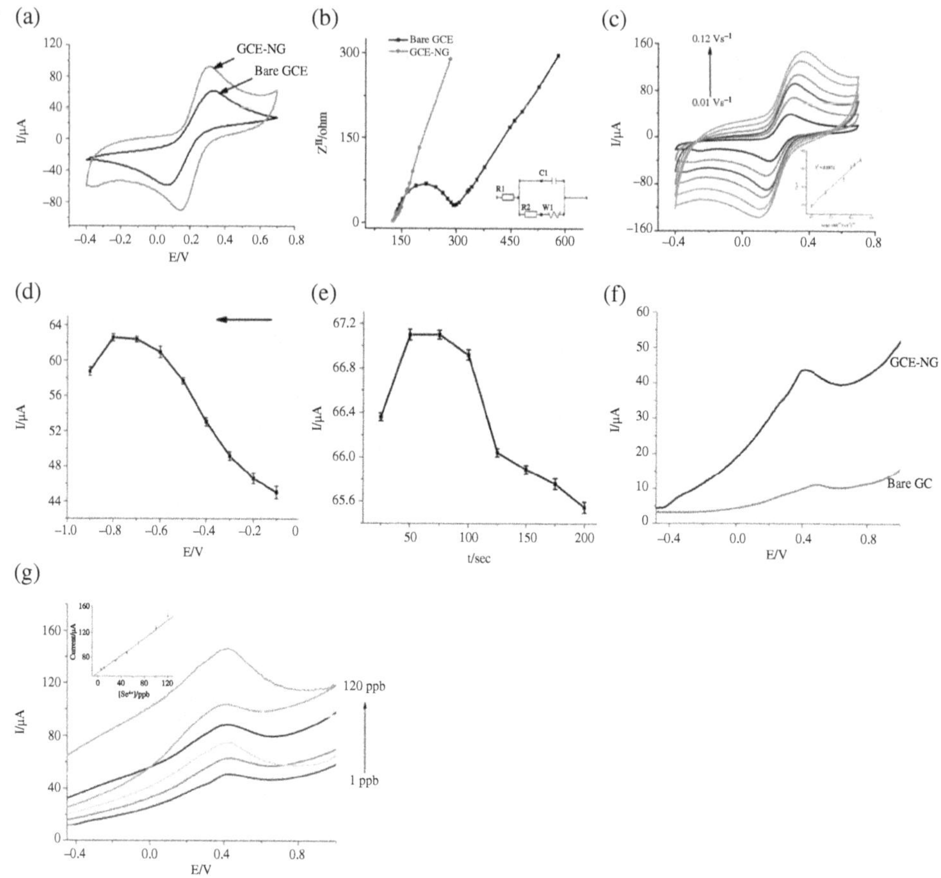

Figure 7.4 Electrochemical response of bare GCE and GCE-in 5 mM $[Fe(CN)_6]^{3-/4-}$ in 0.1 M KCl as supporting electrolyte: (a) CV; (b) Nyquist plot: Inset: equivalent circuit employed when fitting EIS spectra and (c) CV at different scan rates with plot of I_{pa} vs $v_{1/2}$, The effect of (d) deposition potential and the (e) pre-concentration time on peak current during the detection of 30 ppb Se(IV) in 0.1 M $HClO_4$. (f) Comparison of the performance of the unmodified GCE and GCE-NG during the detection of 30 ppb Se(IV). (g) GCE-NG current responses of different Se(IV) concentrations; the insert shows a calibration curve between anodic peak current and Se(IV) concentrations (n = 3) (Fakude et al. 2019).

$$Cu^{2+} + Hg + 4e^- \rightarrow Cu(Hg)$$

$$H_2SeO_3 + 2Cu(Hg) + 8H^+ + 8e^- \rightarrow Cu_2Se(Hg) + 3H_2$$

$$Cu_2Se(Hg) + 4H^+ + 2e^- \rightarrow H_2Se + 2Cu(Hg)$$

To viewing these advantages, the Cu microelectrode (Suz et al. 1997), Cu modified mercury film electrode (Zaitsev et al. 2006), and hanging copper amalgam mercury electrode (Piech and Kubiak 2007) have demonstrated to sense the selenium using cathodic stripping voltammetric analysis. The Cu modified mercury film electrode (Cu/MFE) sensed the LOD of selenium as 40 ng/l in 10 minutes deposition time (Zaitsev et al. 2006). In another approach, the hanging copper amalgam drop electrode (HC) applied to examine the trace concentration of Se(IV) in water matrix. It can be found that the detection limit of selenium is 0.02 μg/l in the linear range 0.02–2.8 ppb at 120 seconds deposition time (Piech and Kubiak 2007). The HCADE has as good a detection limit in minimum deposition time as Cu/MFE. Mattsson et al. demonstrated the digestion method of selenium (Se(VI) to Se(IV)) using an ultraviolet photo-catalytic digestion method. He conducted the selenium detection using the LSCSV method in combination with the UV photo-catalytic method (Mattsson et al. 1995). The LOD is found to be 0.02 ppb in the linear range 0.07–0.9 ppb in the river water matrix. Besides, the other approaches using mercury electrodes to sense the selenium have been summarized in Table 7.2 with LOD and linear range. From the point of view of the environment, the mercury-based WEs have their limitations commercializing the sensing of selenium owing to their toxicity and carcinogenic nature, and its decomposition creates pollution in ground and surface water which affects human life as well as animal life; in addition, its volatility and drop size are dependent on surface tension (Devi et al. 2017a). On account of the toxicity of mercury, several researchers reported a mercury-free electrode such as a silver electrode, and modified by polymer or carbon materials to diminish their toxicity to detect the selenium using the CSV technique. Firstly, Ishiyama et al. reported the silver disk electrode to sense nanograms amounts of selenium using the CSV method. The LOD obtained is 0.20 ng/ml (Ishiyama and Tanaka 1996). Additionally, a silver disk electrode has good repeatability (20 times) and good reproducibility. The Skyllas and Miller (Kazacos and Miller 1980) proposed the mechanism of selenium adsorption at a silver electrode and this reaction occurred at two peaks. The reaction is described below:

$$H_2SeO_3 + 2Ag + 4H^+ + 4e^- \rightarrow Ag_2Se + 3H_2O$$

$$Ag_2Se + 2H^+ + 2e^- \rightarrow H_2Se + 2Ag$$

Likewise, Shan et al. reported a silver electrode to detect selenium in water using a flow-through voltammetric detector (Wang and Shan 2016). The detection limit of selenium is found to be 10 ppb in river and irrigation ditch water. In another side, Baś et al. demonstrated the new electrochemical sensor as renewable silver annular band as a WE which worked to test selenium up to 0.15 ppb within a linear range 1–10 ppb in real water samples (Baś et al. 2014). In another approach to reduce the toxicity of mercury, Rubinskaya et al. developed silver deposited on a Hg thin film electrode and achieved the 0.1 ppb as LOD of selenium (Rubinskaya et al. 2003) and it has no found any kind of interference from the presence of metal ions such as Zn^{2+}, Cd^{2+}, Pb^{2+}, and Cu^{2+} in water. This electrode was successfully applied to testing selenium in mineral water, natural water, and potable water. On the other side, Piech and co-researchers have proposed a new recyclable silver electrode at mercury film in the presence of copper ions to test the selenium in water (Piech 2008). The 17 ppt LOD is obtained in linear range 0.04–8 ppb with 2.7% RSD. The reproducibility of this electrode is comparable to others. Some groups demonstrated the CSV technique in the presence of rhodium to examine the trace amount of selenium in natural water (Wang and Lu 1993; Lange and van den Berg 2000).

Further, to overcome the toxicity of mercury electrode, Bismuth (Bi) is also a more attentive material for testing selenium in a water matrix owing to its non-toxicity. Moreover, Bi is utilized widely in various fields, for instance medicine, cosmetic, electroanalysis of heavy metals, drugs, pesticides, and organic compounds. Owing to these properties, the thin film of bismuth is deposited on a mercury, glassy carbon (GC) electrode. BiFEs are demonstrated to analyze the selenium as various forms in different matrices using adsorptive cathodic stripping voltammetry (AdsCSV), adsorptive differential pulse cathodic stripping voltammetry (AdsDPCSV), and Osteryoung square wave cathodic stripping voltammetry (OSWCSV). As seen in the literature, Long et al. demonstrated the Bi(III) onto an edge plane type of pyrolytic graphite substrate to analyze Se(IV) in a water matrix using OSWCSV (Long and Nagaosa 2007). The detection limit is studied in the initial concentration of Se(IV) in the range of 0.1–1 and 1.0–20 ppb at optimized parameters. The LOD of Se(IV) is observed to be 0.025 ppb in 30 seconds' deposition time. In another work, Sharifian et al. reported a new Bi/Hg electrode to analyze the selenium in vegetables, fruit, and water samples using AdsCSV and evaluated parameters in an open circuit system (Sharifian and Aliakbar 2015b). The result obtained 0.07 ppb as LOD and 0.25 ppb as LOQ. The authors composed the mechanism of adsorption of selenium on the Bi/Hg electrode, which is described as:

$$H_2SeO_{3\left(\text{ads at the Bi/Hg film electrode}\right)} + 2Bi_{(Hg)} + 12H^+ + 12e^- \rightleftharpoons Bi_2Se_{3(Hg)} + 9H_2O$$

$$Bi_2Se_3\left(Hg\right) + 6H^+ + 6e^- \rightleftharpoons 3H_2Se + 2Bi_{(Hg)}$$

Likewise, the same groups reported the in-situ plating of the bismuth electrode to analyze the selenium as a 5-nitropiazselenol complex using AdSV (Sharifian and Aliakbar 2015a). The calibration graph is investigated in the linear range 2–50 ppb under optimized conditions. As a result, the detection limit and quantification limit of Se(IV) is found to be 0.05 and 0.015 ppb, respectively. The detection limit is less than the prescribed limit of selenium by the WHO. The advantage of this electrode is that there is no interference of metal cations and anions in selenium detection found during an interference study of foreign ions. To approach being an alternative for mercury, silver, and bismuth electrodes, the ceramic composite electrode has been investigated to analyze selenium in natural and mineral water (Stozhko et al. 2006). The proposed ceramic composite electrode is highly sensible and highly selective electrode to determine selenium compared to other reported electrodes. The detection limit was 0.02 ppb in the concentration range 0.1–20 ppb at accumulation time 90 seconds.

As discussed, the performance of various electrode materials and modifiers strongly supports the tremendous potential of voltammetric techniques. Consequently, at present, voltammetry-based sensors and sensing devices have become an excellent tool for detection of selenium in water samples and streams. It is their cost-effective instrumentation, robustness, and portability which paved the path of their superiority. However, it is the sensitivity, selectivity, stability, and durability of the sensing electrode which decides the technological feasibility and commercial viability of the voltammetric sensors. It is now well known that the conventional macro electrodes have certain advantages in terms of cost and fabrication but have many serious issues like poor selectivity and sensitivity. In fact, it is the sluggish response of their surfaces toward the selenium species that can be attributed to the kinetically hindered electrode reactions at electrode and analyte (selenium) interface. In this direction, numerous electrode surface modifiers, as reviewed earlier, have emerged as game changers and extremely high sensitivity and selectivity is achieved during selenium detection through the use of these modifiers in electrode designing. Unprecedentedly, the integration of novel nanomaterials in electrode fabrication has resulted in exceptionally ultrasensitive electrode surfaces with long durability and extremely high sensitivity. The reason for their increasing popularity as electrode modifiers lies in their fascinating physicochemical properties, which are far different from their bulk counterparts. Nevertheless, researchers are still paying attention to the synthesis of engineered dimension and morphology-directed nanomaterials, not only to achieve better electrode performance but also highly efficient portable voltammetric sensors. In this context, carbon-based nanomaterials are found to be the star cast members. However, only a few research publications have been reported to date regarding their performance as electrode modifiers. In the series, carbon-based quantum dots (CQDs), which are the latest member of the carbon family, have become marathon materials. Since just after their discovery by Xu et al. in 2014 (Xu et al. 2004), CQDs have garnered ample scientific interest because of their outstanding optoelectronic-electrical-chemical properties. Subsequently, CQDs have been extensively tested and explored in a diverse spectrum of research applications and placed themselves in the category of the most promising and potential candidates, especially in sensing techniques. In general, only very few luminescence sensors, electrochemical sensors, optical biosensors, electrochemical biosensors, and photo-electrochemical biosensors based on CQDs are reported so far. Though CQDs exhibit outstanding properties like large specific surface area, tunable surface functionalities, controllable morphologies, high stability, and better environmental benignity, the role of CQDs-based sensing probes in the development of voltammetric sensors for the detection of selenium in water systems still needs to be explored more. And it is universally expected that CQDs will revolutionize this field in very near future and voltammetry-based inexpensive portable selenium sensors will be as common as present day pH meters (Sharma et al. 2017; Devi et al. 2019).

7.5 Realization of Voltammetric Sensors for Selenium Detection in Water: Concluding Remarks and Future Scope

In this chapter, we have comprehensively discussed the importance and essentiality of selenium as a nutrient for life as well as its toxic and hazardous effects on ecosystems, human, and animal health. Due to a very narrow borderline of concentration between the nutrient and toxicant behavior of selenium, there has recently been an increasingly intense urge as well as need for analysis and monitoring of inorganic selenium in water systems. This has put the research scientists, environmentalists, technologists, industrialists, and governments to work on the design and development of sensing devices. The strong demands and desires of robustness and ultrasensitive portable sensors/sensing devices streamlined the community to focus on voltammetric sensors. These voltammetric approaches are universally considered to be the most promising for field-deployable portable sensors. Consequently, this field continues to exist as a coruscating research area for the coming times for rapid and facile monitoring of selenium in water.

Further, we highlighted the basic principles, mechanisms, and types of voltammetric techniques evolved in recent decades for detection and monitoring of inorganic selenium in water samples. Moreover, the recent achievements in fabrication processes and performance of various electrodes/electrode materials/modifiers such as metal/metal oxide particles, carbon materials (graphite, GCs), and nanomaterial (graphene-based compositions etc.), in selenium analysis are introduced in the present chapter. In particular, the new-generation novel materials of macro/micro and nano-scales (such as metal/metal oxides, carbon, and polymeric materials) along with traditional metal-based electrodes (such as gold and mercury electrodes), have recently provided a new path for electrode modification in order to improve the performance of the voltammetric selenium sensing.

Importantly, the sensitivity of their performance is solely attributed to the pre-concentration of the analyte on the surface of a WE, and the quantification of the accumulated species. Precisely, through significant enhancement in electrical properties, effective electrode surface area for the transfer of the electrochemical signal, these modified materials produce detectable signals suitable for the indirect detection of selenium in water samples.

However, owing to their unique properties, nanomaterials have been emerging as potential candidates in this direction and paving the way for nanotechnology-based development of novel voltammetric sensing platforms. The synergic coupling of the fascinating features of these engineered nanomaterials with conventional voltammetric techniques has resulted in fast response time, extremely ultra-sensitivity and selectivity, data reproducibility, with very LODs in offline and online monitoring of inorganic selenium in water systems.

Thus it can be extrapolated that in the near future these engineered nanostructures will play a vital role in revolutionizing the on-field monitoring quality of selenium detection in water, which will in turn allow collection of reliable data for different sectors for different uses.

Nevertheless, there are few major challenges associated with such compositions in sensing of selenium, such as (i) selectivity in selenium detection in presence of interfering ions, (ii) analysis of selenium in complex forms because in most situations it can easily form complexes, (iii) chemical fouling of the electrode materials, and (iv) reusability and end-of-life utility of electrode/electrode materials. Additionally, toxicity of electrode materials/modifiers, durability, and response to real microenvironments of such nanomaterials are still topics of major concern. Apart from this, some technical issues like continuous inline monitoring of water bodies for selenium detection, commercialization, and bulk-scale electrode material production are still in front of scientists. Besides, there are certain areas which require further investigation and developments such as low-cost sensor manufacturing, installation, maintenance, and data processing. Eventually most of the sensors are quite sensitive to environmental conditions such as humidity, pressure, and temperature, which are known to affect the sensor signals and their quality along with, over time, baseline drifts which cause long-term studies to suffer, and therefore dedicated studies are further needed to meet these challenges.

Nonetheless, another key issue is the speciation process of selenium species in water by these techniques and methods, as they require reagents and resins. This work is also at the initial stage of development and requires considerable attention to be paid so that total selenium determination can be achieved without extra reagents. Lastly, the lab-scale experimentation and testing need to be translated from labs to field applications. The above-mentioned issues and challenges are the major obstacles in the realization of portable and efficient voltammetric sensors for selenium detection in water systems and need to be addressed and demand immediate attention and need to be addressed.

For instance, as an alternative, qualitative and quantitative analysis of inorganic selenium in water through the biosensors can be achieved. Though there are few research reports on that, however, this is an underestimated and unexplored area and could be able to provide a strong platform for real-time selenium monitoring in aqueous environments. The biosensors provide a much better and tremendous potential in terms of environmental benignity, cost-effectiveness, spontaneous commercial viability, and promising end-of-life utility compared to its traditional counterparts. Last but not the least, it can be concluded that voltammetry is a highly cost-effective, very much simpler and more efficient approach to detecting inorganic selenium in water systems than some other available techniques and methods. However, there is still enough to work on and some improvements and development are needed to realize portable sensors in terms of increased sensitivity, selectivity, reproducibility, and reliability. Therefore, future research on voltammetry and voltammetric sensors may build upon these aspects.

References

Agarwal, P.R., Kumar, R., Kumari, S., and Dhakate, S.R. (2016). Three-dimensional and highly ordered porous carbon–MnO_2 composite foam for excellent electromagnetic interference shielding efficiency. *RSC Advances* 6: 100713–100722.

Agrawal, P.R., Kumar, R., Uppal, H. et al. (2016). Novel 3D lightweight carbon foam as an effective adsorbent for arsenic(v) removal from contaminated water. *RSC Advances* 6: 29899–29908.

Agrawal, P.R., Singh, N., Kumari, S., and Dhakate, S.R. (2018). Multiwall carbon nanotube embedded phenolic resin-based carbon foam for the removal of As (V) from contaminated water. *Materials Research Express* 5: 035601.

Agrawal, P.R., Singh, N., Kumari, S., and Dhakate, S.R. (2019). The removal of pentavalent arsenic by graphite intercalation compound functionalized carbon foam from contaminated water. *Journal of Hazardous Materials* 377: 274–283.

Al-Akeel, K. (2017). The pollution of water by trace elements research trends. In: *Advances in Bioremediation and Phytoremediation* (ed. N. Shiomi). Intechopen.

Ambroziak, U., Hybsier, S., Shahnazaryan, U. et al. (2017). Severe selenium deficits in pregnant women irrespective of autoimmune thyroid disease in an area with marginal selenium intake. *Journal of Trace Elements in Medicine and Biology* 44: 186–191.

Ashournia, M. and Aliakbar, A. (2009). Determination of selenium in natural waters by adsorptive differential pulse cathodic stripping voltammetry. *Journal of Hazardous Materials* 168: 542–547.

Ashournia, M. and Aliakbar, A. (2010). Determination of Se (IV) in natural waters by adsorptive stripping voltammetry of 5-nitropiazselenol. *Journal of Hazardous Materials* 174: 788–794.

Azizi, Z. and Babakhanian, A. (2018). Fabricating a new electrochemically modified pencil graphite electrode based on acetophenone (2,4-dinitrophenyl) hydrazone for determining selenium in food and water samples. *Analytical Methods* 10: 5205–5213.

Baltensperger, U. and Hertz, J. (1985). Parameter evaluation for the determination of selenium by cathodic stripping voltammetry at the hanging mercury drop electrode. *Analytica Chimica Acta* 172: 49–56.

Barón-Jaimez, J., Joya, M., and Barba-Ortega, J. (2013). Anodic stripping voltammetry–ASV for determination of heavy metals. Journal of Physics: Conference Series. IOP Publishing, 012023.

Baś, B., Jedlińska, K., and Węgiel, K. (2014). New electrochemical sensor with the renewable silver annular band working electrode: fabrication and application for determination of selenium (IV) by cathodic stripping voltammetry. *Electrochemistry Communications* 49: 79–82.

Bem, E.M. (1981). Determination of selenium in the environment and in biological material. *Environmental Health Perspectives* 37: 183–200.

Benstoem, C., Goetzenich, A., Kraemer, S. et al. (2015). Selenium and its supplementation in cardiovascular disease—what do we know? *Nutrients* 7: 3094–3118.

Benvidi, A., Dehghani-Firouzabadi, A., Mazloum-Ardakani et al. (2015). Electrochemical deposition of gold nanoparticles on reduced graphene oxide modified glassy carbon electrode for simultaneous determination of levodopa, uric acid and folic acid. *Journal of Electroanalytical Chemistry* 736: 22–29.

Bernal, E. (2014). Limit of detection and limit of quantification determination in gas chromatography. *Advances in Gas Chromatography* 3: 57–63.

Berry, M.J. and Ralston, N.V. (2008). Mercury toxicity and the mitigating role of selenium. *EcoHealth* 5: 456–459.

Bertolino, F.A., Torriero, A.A., Salinas, E. et al. (2006). Speciation analysis of selenium in natural water using square-wave voltammetry after preconcentration on activated carbon. *Analytica Chimica Acta* 572: 32–38.

Bidari, A., Hemmatkhah, P., Jafarvand, S. et al. (2008). Selenium analysis in water samples by dispersive liquid-liquid microextraction based on piazselenol formation and GC–ECD. *Microchimica Acta* 163: 243–249.

Bolann, B., Rahil-Khazen, R., Henriksen, H. et al. (2007). Evaluation of methods for trace-element determination with emphasis on their usability in the clinical routine laboratory. *Scandinavian Journal of Clinical and Laboratory Investigation* 67: 353–366.

Brett, C.M. (2001). Electrochemical sensors for environmental monitoring. Strategy and examples. *Pure and Applied Chemistry* 73: 1969–1977.

Bryce, D., Izquierdo, A., and de Castro, M.L. (1995). Flow-injection anodic stripping voltammetry at a gold electrode for selenium (IV) determination. *Analytica Chimica Acta* 308: 96–101.

Bye, R. (1989). Generation of selenium hydride from alkaline solutions: a newf concept of the hydride generation-atomic absorption technique. *Journal of Analytical wMethods in Chemistry* 11: 156–158.

Cai, Q. and Khoo, S.B. (1994). Poly (3, 3′-diaminobenzidine) film on a gold electrode for selective preconcentration and stripping analysis of selenium (IV). *Analytical Chemistry* 66: 4543–4550.

Chen, B., Hu, B., He, M. et al. (2013). Speciation of selenium in cells by HPLC-ICP-MS after (on-chip) magnetic solid phase extraction. *Journal of Analytical Atomic Spectrometry* 28: 334–343.

Costas-Mora, I., Romero, V., Pena-Pereira, F. et al. (2012). Quantum dots confined in an organic drop as luminescent probes for detection of selenium by microfluorospectrometry after hydridation: study of the quenching mechanism and analytical performance. *Analytical Chemistry* 84: 4452–4459.

De Arruda, R.C.T.M., Falcão, C.D.O.L., de Morais, C.M.M. et al. (2019). Processed and ultra-processed foods are associated with high prevalence of inadequate selenium intake and low prevalence of vitamin B1 and zinc inadequacy in adolescents from public schools in an urban area of northeastern Brazil. *PLoS One* 14: e0224984.

De Carvalho, L.M., Schwedt, G., Henze, G., and Sander, S. (1999). Redoxspeciation of selenium in water samples by cathodic stripping voltammetry using an automated flow system. *Analyst* 124: 1803–1809.

De Gregori, I., Lobos, M.A.G., and Pinochet, H. (2002). Selenium and its redox speciation in rainwater from sites of Valparaıso region in Chile, impacted by mining activities of copper ores. *Water Research* 36: 115–122.

De Lima, G.C., do Lago, A.C., Chaves, A.A. et al. (2013). Determination of selenium using atomically imprinted polymer (AIP) and hydride generation atomic absorption spectrometry. *Analytica Chimica Acta* 768: 35–40.

Dedina, J. (1982). Interference of volatile hydride-forming elements in selenium determination by atomic absorption spectrometry with hydride generation. *Analytical Chemistry* 54: 2097–2102.

Demars, R. and Shain, I. (1957). Anodic stripping voltammetry using the hanging mercury drop electrode. *Analytical Chemistry* 29: 1825–1827.

Dennert, G., Zwahlen, M., Brinkman, M. et al. (2012). Selenium for preventing cancer. *Sao Paulo Medical Journal* 130.

Dennis, B., Moyers, J., and Wilson, G. (1976). Determination of selenium as selenide by differential pulse cathodic stripping voltammetry. *Analytical Chemistry* 48: 1611–1616.

Deryabina, V., Slepchenko, G.B., N'ung, F.K., and Lin, H.S. (2013). Voltammetry behavior of iodine and selenium on new organo-modified electrodes: development of a procedure for the determination of iodine and selenium. *Journal of Analytical Chemistry* 68: 896–899.

Devi, P., Jain, R., Thakur, A. et al. (2017a). A systematic review and meta-analysis of voltammetric and optical techniques for inorganic selenium determination in water. *TrAC, Trends in Analytical Chemistry* 95: 69–85.

Devi, P., Kaur, G., Thakur, A. et al. (2017b). Waste derivitized blue luminescent carbon quantum dots for selenite sensing in water. *Talanta* 170: 49–55.

Devi, P., Thakur, A., Chopra, S. et al. (2017c). Ultrasensitive and selective sensing of selenium using nitrogen-rich ligand interfaced carbon quantum dots. *ACS Applied Materials & Interfaces* 9: 13448–13456.

Devi, P., Rajput, P., Thakur, A. et al. (2019). Recent advances in carbon quantum dot-based sensing of heavy metals in water. *TrAC, Trends in Analytical Chemistry* 114: 171–195.

Do Nascimento, P.C., Jost, C.L., de Carvalho, L.M. et al. (2009). Voltammetric determination of Se (IV) and Se (VI) in saline samples—Studies with seawater, hydrothermal and hemodialysis fluids. *Analytica Chimica Acta* 648: 162–166.

Dorleku, M., Nukpezah, D., and Carboo, D. (2018). Effects of small-scale gold mining on heavy metal levels in groundwater in the lower Pra Basin of Ghana. *Applied Water Science* 8: 126.

Duntas, L.H. (2010). Selenium and the thyroid: a close-knit connection. *The Journal of Clinical Endocrinology & Metabolism* 95: 5180–5188.

Eggins, B.R. (2008). *Chemical Sensors and Biosensors*. Wiley.

El-Ramady, H.R., Domokos-Szabolcsy, É., Shalaby, T.A. et al. (2015). Selenium in agriculture: water, air, soil, plants, food, animals and nanoselenium. In: *CO2 Sequestration, Biofuels and Depollution* (eds. E. Lichtfouse, J. Schwartzbauer and D. Robert), 153–232. Springer.

Etteieb, S., Magdouli, S., Zolfaghari, M., and Brar, S. (2020). Monitoring and analysis of selenium as an emerging contaminant in mining industry: a critical review. *Science of the Total Environment* 698: 134339.

Fairweather-Tait, S.J., Collings, R., and Hurst, R. (2010). Selenium bioavailability: current knowledge and future research requirements. *The American Journal of Clinical Nutrition* 91: 1484S–1491S.

Fakude, C.T., Arotiba, O.A., Moutloali, R., and Mabuba, N. (2019). Nitrogen-doped graphene electrochemical sensor for selenium (IV) in water. *International Journal of Electrochemical Science* 14: 9391–9403.

Farghaly, O.A., Hameed, R.A., and Abu-Nawwas, A.-A.H. (2014). Analytical application using modern electrochemical techniques. *International Journal of Electrochemical Science* 9: 3287–3318.

Fernandes, R.N., Reis, B.F., Morales-Rubio, A., and de La Guardia, M. (2009). Multi-pumping mechanised determination of selenium in natural waters by light emitting diode (LED) spectrometry. *Journal of the Brazilian Chemical Society* 20: 1242–1248.

Fierro, S., Watanabe, T., Akai, K. et al. (2012). Anodic Stripping Voltammetry of Se4. *International Journal of Electrochemistry*.

Figura, P. and Mcduffie, B. (1979). Use of Chelex resin for determination of labile trace metal fractions in aqueous ligand media and comparison of the method with anodic stripping voltammetry. *Analytical Chemistry* 51: 120–125.

Florence, T.M. (1979). Cathodic stripping voltammetry: part I. Determination of organic sulfur compounds, flavins and porphyrins at the sub-micromolar level. *Journal of Electroanalytical Chemistry and Interfacial Electrochemistry* 97: 219–236.

Ghoneim, M., Hassanein, A., Hammam, E., and Beltagi, A. (2000). Simultaneous determination of Cd, Pb, Cu, Sb, Bi, Se, Zn, Mn, Ni, Co and Fe in water samples by differential pulse stripping voltammetry at a hanging mercury drop electrode. *Fresenius' Journal of Analytical Chemistry* 367: 378–383.

Gomez-Ariza, J., Giraldez, I., Morales, E., and Pozas, J. (1999). Use of solid phase extraction for speciation of selenium compounds in aqueous environmental samples. *Analyst* 124: 75–78.

Grabarczyk, M. and Korolczuk, M. (2010). Development of a simple and fast voltammetric procedure for determination of trace quantity of Se (IV) in natural lake and river water samples. *Journal of Hazardous Materials* 175: 1007–1013.

Gupta, C.U. and Gupta, C.S. (2010). Selenium deficiency in soils and crops and its impact on animal and human health. *Current Nutrition & Food Science* 6: 268–280.

Gupta, M. and Gupta, S. (2017). An overview of selenium uptake, metabolism, and toxicity in plants. *Frontiers in Plant Science* 7: 2074.

Gutberlet, J. and Baeder, A.M. (2008). Informal recycling and occupational health in Santo André, Brazil. *International Journal of Environmental Health Research* 18: 1–15.

Hamidáyahaya, A. (1992). Determination of selenium (IV) by anodic stripping voltammetry using extraction with pentyl alcohol. *Analyst* 117: 43–45.

Hamilton, S.J. (2004). Review of selenium toxicity in the aquatic food chain. *Science of the Total Environment* 326: 1–31.

Hamilton, T., Ellis, J., and Florence, T.M. (1979). Determination of selenium and tellurium in electrolytic copper by anodic stripping voltammetry at a gold film electrode. *Analytica Chimica Acta* 110: 87–94.

Hanrahan, G., Patil, D.G., and Wang, J. (2004). Electrochemical sensors for environmental monitoring: design, development and applications. *Journal of Environmental Monitoring* 6: 657–664.

Harvey, D. (2000). *Modern Analytical Chemistry*. Boston, MA: McGraw-Hill Companies, Inc.

He, Y., Xiang, Y., Zhou, Y. et al. (2018). Selenium contamination, consequences and remediation techniques in water and soils: a review. *Environmental Research* 164: 288–301.

Hernández, O.L.E. (2006). Furoiltioureas: naturaleza de sus complejos con $CdCl_2$ y $HgCl_2$ y su utilización en sensores electroquímicos. *Universidad de Cádiz*.

Heyrovský, J. and Kůta, J. (2013). *Principles of Polarography*. Elsevier.

Heyrovský, J. and Shikata, M. (1925). Researches with the dropping mercury cathode: part II. The Polarograph. *Recueil des Travaux Chimiques des Pays-Bas* 44: 496–498.

Hineman, A. (2011). *Determination of as, se and hg in Waters by Hydride Generation/Cold Vapor Atomic Absorption Spectroscopy*. Canada: PerkinElmer.

Hoffmann, P.R. and Berry, M.J. (2008). The influence of selenium on immune responses. *Molecular Nutrition & Food Research* 52: 1273–1280.

Huang, S., Ming, H., Jinshun, F. et al. (2009). Assessment of selenium pollution in agricultural soils in the Xuzhou District, Northwest Jiangsu, China. *Journal of Environmental Sciences* 21: 481–487.

Hunter, W.J. and Kuykendall, L.D. (2004). Determination of dimethylselenide and dimethyldiselenide by gas chromatography–photoionization detection. *Journal of Chromatography A* 1038: 295–297.

Hymer, C. and Caruso, J. (2006). Selenium speciation analysis using ICP-MS. *Journal of Chromatography A* 1144: 1.

Idris, A., Mabuba, N., and Arotiba, O. (2015). Electroanalysis of selenium in water on an electrodeposited gold-nanoparticle modified glassy carbon electrode. *Journal of Electroanalytical Chemistry* 758: 7–11.

Idris, A., Mabuba, N., and Arotiba, O. (2017a). Electrochemical co-detection of arsenic and selenium on a glassy carbon electrode modified with gold nanoparticles. *International Journal of Electrochemical Science* 12: 10–21.

Idris, A.O., Mabuba, N., Nkosi, D. et al. (2017b). Electrochemical detection of selenium using glassy carbon electrode modified with reduced graphene oxide. *International Journal of Environmental Analytical Chemistry* 97: 534–547.

Ishiyama, T. and Tanaka, T. (1996). Cathodic stripping voltammetry of selenium (IV) at a silver disk electrode. *Analytical Chemistry* 68: 3789–3792.

Ivandini, T.A. and Einaga, Y. (2013). Electrochemical detection of selenium (IV) and (VI) at gold-modified diamond electrodes. *Electrocatalysis* 4: 367–374.

Jain, R., Thakur, A., Kumar, P., and Pooja, D. (2018). Au/ZnO nanocomposites decorated ITO electrodes for voltammetric sensing of selenium in water. *Electrochimica Acta* 290: 291–302.

Jarzabek, G. and Kublik, Z. (1982). Determination of traces of selenium (IV) by cathodic stripping voltammetry at the hanging mercury drop electrode. *Analytica Chimica Acta* 143: 121–130.

Kaushal, S., Sahu, A., Rani, M., and Dhakate, S. (2019). Multiwall carbon nanotubes tailored porous carbon fiber paper-based gas diffusion layer performance in polymer electrolyte membrane fuel cell. *Renewable Energy* 142: 604–611.

Kazacos, M.S. and Miller, B. (1980). Studies in selenious acid reduction and CdSe film deposition. *Journal of The Electrochemical Society* 127: 869.

Keating, M. (2001). *Cradle to Grave: The Environmental Impacts from Coal*. Clean Air Task Force.

Kieliszek, M. and Błażejak, S. (2013). Selenium: significance, and outlook for supplementation. *Nutrition* 29: 713–718.

Kieliszek, M. and Błażejak, S. (2016). Current knowledge on the importance of selenium in food for living organisms: a review. *Molecules* 21: 609.

Kim, Y.Y. and Mahan, D. (2001). Comparative effects of high dietary levels of organic and inorganic selenium on selenium toxicity of growing-finishing pigs. *Journal of Animal Science* 79: 942–948.

Kleckner, A.E., Kakouros, E., and Robin Stewart, A. (2017). A practical method for the determination of total selenium in environmental samples using isotope dilution-hydride generation-inductively coupled plasma-mass spectrometry. *Limnology and Oceanography: Methods* 15: 363–371.

Kolliopoulos, A.V., Metters, J.P., and Banks, C.E. (2013). Electroanalytical sensing of selenium (IV) utilising screen printed graphite macro electrodes. *Analytical Methods* 5: 851–856.

Korolczuk, M. and Grabarczyk, M. (2003). Determination of Se (IV) in on-line system by Cathodic stripping voltammetry. *Electroanalysis: An International Journal Devoted to Fundamental and Practical Aspects of Electroanalysis* 15: 821–826.

Kosseva, M.R. and Kennedy, J.F. (2003). Handbook of analytical techniques-Vol. I; H. Gunzler, A. Williams (Eds.); WILEY-VCH verlag GmbH, Weinheim, Germanyl, 2001, 559; Vol. II; H. Gunzler, A. Williams (Eds.); WILEY-VCH verlag GmbH, Weinheim, Germany, 2001, 1182. Carbohydrate Polymers 4: 532.

Kruse, P. (2018). Review on water quality sensors. *Journal of Physics D: Applied Physics* 51: 203002.

Kune, G. and Watson, L. (2006). Colorectal cancer protective effects and the dietary micronutrients folate, methionine, vitamins B6, B12, C, E, selenium, and lycopene. *Nutrition and Cancer* 56: 11–21.

Lange, B. and van den Berg, C.M. (2000). Determination of selenium by catalytic cathodic stripping voltammetry. *Analytica Chimica Acta* 418: 33–42.

Langhus, D.L. (2002). Fundamentals of Electroanalytical Chemistry (Monk, Paul MS). *J. Chem. Educ* 79: 1207.

Leblanc, K.L. and Wallschläger, D. (2016). Production and release of selenomethionine and related organic selenium species by microorganisms in natural and industrial waters. *Environmental Science & Technology* 50: 6164–6171.

Lemly, A.D. (1997). Environmental implications of excessive selenium: a review. *Biomedical and Environmenta Sciences* 10: 415–435.

Lemly, A.D. (2002). Selenium pollution around the world. In: *Selenium Assessment in Aquatic Ecosystems* (ed. A.D. Lemly), 3–17. New York;, NY: Springer.

Lemly, D.A. (2009). Aquatic hazard of selenium pollution from coal mining. In: *Coal Mining: Research, Technology and Safety* (ed. G.B. Fosdyke), 167–183. Nova Science.

Long, J. and Nagaosa, Y. (2007). Determination of selenium (IV) by catalytic stripping voltammetry with an in situ plated bismuth-film electrode. *Analytical Sciences* 23: 1343–1346.

Luoma, S.N. and Presser, T.S. (2009). *Emerging Opportunities in Management of Selenium Contamination*. ACS Publications.

Macfarquhar, J.K., Broussard, D.L., Melstrom, P. et al. (2010). Acute selenium toxicity associated with a dietary supplement. *Archives of Internal Medicine* 170: 256–261.

Maggard, R. (2004). The occurrence and impacts of selenium in aquatic systems downstream of a mountaintop mining operation in central Appalachia. *Proceedings America Society of Mining and Reclamation*: 1207–1217.

Magin, G.B. Jr., Thatcher, L.L., Rettig, S., and Levine, H. (1960). Suggested modified method for colorimetric determination of selenium in natural water. *Journal-American Water Works Association* 52: 1199–1205.

March, G., Nguyen, T.D., and Piro, B. (2015). Modified electrodes used for electrochemical detection of metal ions in environmental analysis. *Biosensors* 5: 241–275.

Margaret, P. (2012). Selenium and human health. *The Lancet* 379: 1256–1268.

Mateo-Sagasta, J., Zadeh, S., Turral, H., and Burke, J. (2017). *Water Pollution from Agriculture: A Global Review*. Food and Agricultural Organization.

Mathew, M. and Narayana, B. (2006). An easy spectrophotometric determination of selenium using azure B as a chromogenic reagent. *Indian Journal of Chemical Technology* 13: 455–458.

Mattsson, G., Nyholm, L., Olin, Å., and Örnemark, U. (1995). Determination of selenium in freshwaters by cathodic stripping voltammetry after UV irradiation. *Talanta* 42: 817–825.

May, T.W., Fairchild, J.F., Petty, J.D. et al. (2008). An evaluation of selenium concentrations in water, sediment, invertebrates, and fish from the Solomon River Basin. *Environmental Monitoring and Assessment* 137: 213–232.

Meng, L., Nanto, V., Li, S. et al. (1995). Determination of selenium in natural water by AAS after solvent extraction preconcentration. *Microchemical Journal* 52: 223–230.

Min, J.A., Park, H.J., Cho, B.K., and Lee, J.Y. (2010). Acute generalized exanthematous pustulosis induced by Rhus (lacquer). *Journal of the American Academy of Dermatology* 63: 166–168.

Mohamed, R. and Leong, W.L. (2006). Analysis of selenium species using cathodic stripping voltammetry. *Jurnal Teknologi* 44: 55–66.

Nayak, P.K. (2016). *Recent Advances in Graphene Research*. BoD–Books on Demand.

Needhidasan, S., Samuel, M., and Chidambaram, R. (2014). Electronic waste–an emerging threat to the environment of urban India. *Journal of Environmental Health Science and Engineering* 12: 36.

Nikelly, J.G. and Cooke, W.D. (1957). Anodic stripping polarography. *Analytical Chemistry* 29: 933–939.

Nuttall, K.L. (2006). Evaluating selenium poisoning. *Annals of Clinical and Laboratory Science* 36: 409–420.

Oh, J.-H., Oh, K.-S., Kim, C.-G., and Hong, C.-S. (2004). Design of radar absorbing structures using glass/epoxy composite containing carbon black in X-band frequency ranges. *Composites Part B: Engineering* 35: 49–56.

Ojha, A. (2020). Materials in electrochemical detection of water pollutants. In: *Sensors in Water Pollutants Monitoring: Role of Material* (eds. D. Pooja, P. Kumar, P. Singh and S. Patil), 125–145. Springer.

Ouédraogo, O. and Amyot, M. (2013). Mercury, arsenic and selenium concentrations in water and fish from sub-Saharan semi-arid freshwater reservoirs (Burkina Faso). *Science of the Total Environment* 444: 243–254.

Papoff, P., Bocci, F., and Lanza, F. (1998). Speciation of selenium in natural waters and snow by DPCSV at the hanging mercury drop electrode. *Microchemical Journal* 59: 50–76.

Pavlata, L., Mišurová, L., Pechová, A., and Dvořák, R. (2012). Comparison of organic and inorganic forms of selenium in the mother and kid relationship in goats. *Czech Journal of Animal Science* 57: 361–369.

Pedersen, G.A. and Larsen, E.H. (1997). Speciation of four selenium compounds using high performance liquid chromatography with on-line detection by inductively coupled plasma mass spectrometry or flame atomic absorption spectrometry. *Fresenius' Journal of Analytical Chemistry* 358: 591–598.

Pereira, C.F., Gonzaga, F.B., Guaritá-Santos, A.M., and Souzade, J.R. (2006). Determination of Se (IV) by anodic stripping voltammetry using gold electrodes made from recordable CDs. *Talanta* 69: 877–881.

Pettine, M., Mcdonald, T.J., Sohn, M. et al. (2015). A critical review of selenium analysis in natural water samples. *Trends in Environmental Analytical Chemistry* 5: 1–7.

Piech, R. (2008). Determination of selenium traces on cyclic renewable mercury film silver electrode in presence of copper ions using cathodic stripping voltammetry. *Electroanalysis: An International Journal Devoted to Fundamental and Practical Aspects of Electroanalysis* 20: 2475–2481.

Piech, R. and Kubiak, W.W. (2007). Determination of trace selenium on hanging copper amalgam drop electrode. *Electrochimica Acta* 53: 584–589.

Pistón, M., Silva, J., Pérez-Zambra, R., and Knochen, M. (2009). Determination of total selenium by multicommutated-flow hydride generation atomic absorption spectrometry. Application to cow's milk and infant formulae. *Analytical Methods* 1: 139–143.

Pooja, D., Kumar, P., Singh, P., and Patil, S. (2020). *Sensors in Water Pollutants Monitoring: Role of Material*. Springer.

Popham, H.J. and Shelby, K.S. (2007). Effect of inorganic and organic forms of selenium supplementation on development of larval Heliothis virescens. *Entomologia Experimentalis et Applicata* 125: 171–178.

Porada, R., Jedlińska, K., Lipińska, J., and Baś, B. (2020). Voltammetric sensors with laterally placed working electrodes: a review. *Journal of The Electrochemical Society* 167: 037536.

Pujol, L., Evrard, D., Groenen-Serrano, K. et al. (2014). Electrochemical sensors and devices for heavy metals assay in water: the French groups' contribution. *Frontiers in Chemistry* 2: 19.

Pyrzyńska, K. (2001). Analysis of selenium species by capillary electrophoresis. *Talanta* 55: 657–667.

Quentel, F. and Elleouet, C. (1999). Speciation analysis of selenium in seawater by cathodic stripping voltammetry. *Electroanalysis* 11: 47–51.

Radhi, M.M., Al-Dulimy, W.A.G., and Khalaf, M.S. (2016). Electrochemical study of selenium (IV) mediated by carbon nanotubes modified glassy carbon electrode in blood medium. *Építöanyag* 68: 90–93.

Ramadan, A.A., Mandil, H., and Ozoun, A. (2011). Differential pulse anodic stripping voltammetric determination of selenium (iv) with a vitamin e-nafion modified gold electrode. *Asian Journal of Chemistry* 23: 843.

Ramadan, A.A., Mandil, H., and Ozoun, A. (2012). Differential pulse anodic stripping voltammetric determination of selenium (IV) with a methylene blue-nafion modified gold electrode. *Asian Journal of Chemistry* 24: 391.

Ramadan, A.A., Mandil, H., and Shikh-Debes, A. (2014). Differential pulse anodic stripping voltammetric determination of selenium (IV) at a gold electrode modified with 3, 3′-diaminobenzidine· 4HCl-nafion. *International Journal of Pharmacy and Pharmaceutical Sciences* 6: 148–153.

Reta, G., Dong, X., Li, Z. et al. (2018). Environmental impact of phosphate mining and beneficiation: review. *International Journal of Hydrology* 2: 424–432.

Riaz, M. and Mehmood, K.T. (2012). Selenium in human health and disease: a review. *JPMI: Journal of Postgraduate Medical Institute* 26: 120–133.

Rubinskaya, T., Kovaleva, S., Kulagin, E., and Gladyshev, V. (2003). Determination of selenium (IV) by stripping voltammetry at a mercury-film electrode. *Journal of Analytical Chemistry* 58: 165–170.

Sakai, Y., Tomura, K., and Ohshita, K. (1994). Determination of selenium (IV) and selenium (VI) in natural water samples by neutron activation analysis after chemical pre-collection. *Journal of Radioanalytical and Nuclear Chemistry* 187: 441–450.

Sanchez-Rodas, D., Corns, W., Chen, B., and Stockwell, P. (2010). Atomic fluorescence spectrometry: a suitable detection technique in speciation studies for arsenic, selenium, antimony and mercury. *Journal of Analytical Atomic Spectrometry* 25: 933–946.

Sanchez-Rodas, D., Mellano, F., Morales, E., and Giraldez, I. (2013). A simplified method for inorganic selenium and selenoaminoacids speciation based on HPLC–TR–HG–AFS. *Talanta* 106: 298–304.

Santos, S., Ungureanu, G., Boaventura, R., and Botelho, C. (2015). Selenium contaminated waters: an overview of analytical methods, treatment options and recent advances in sorption methods. *Science of the Total Environment* 521: 246–260.

Sanz-Medel, A., Uría, J.S., and García, A.M. (1999). Continuous aspiration and flow injection in atomic emission plasma spectrometry. *Analytical Spectroscopy Library* 9: 274–308.

Scholz, F. (2015). Voltammetric techniques of analysis: the essentials. *ChemTexts* 1: 17.

Scott, K. (2016). Electrochemical principles and characterization of bioelectrochemical systems. In: *Microbial Electrochemical and Fuel Cells* (eds. K. Scott and E.H. Yu), 26–99. Elsevier.

Segura, R., Pizarro, J., Díaz, K. et al. (2015). Development of electrochemical sensors for the determination of selenium using gold nanoparticles modified electrodes. *Sensors and Actuators B: Chemical* 220: 263–269.

Sentkowska, A. (2019). Chromatographic analysis of selenium species. In: *Importance of Selenium in the Environment and Human Health* (eds. M.M. Rahman et al.). IntechOpen.

Settle, F.A. (1997). *Handbook of Instrumental Techniques for Analytical Chemistry*. Prentice Hall PTR.

Shaltout, A.A., Castilho, I.N., Welz, B. et al. (2011). Method development and optimization for the determination of selenium in bean and soil samples using hydride generation electrothermal atomic absorption spectrometry. *Talanta* 85: 1350–1356.

Shamberger, R.J. (1981). Selenium in the environment. *Science of the Total Environment* 17: 59–74.

Sharifian, P. and Aliakbar, A. (2015a). Determination of Se (IV) as a 5-nitropiazselenol complex by adsorptive stripping voltammetry at an in situ plated bismuth film electrode. *Analytical Methods* 7: 4321–4327.

Sharifian, P. and Aliakbar, A. (2015b). Determination of Se (IV) by adsorptive cathodic stripping voltammetry at a Bi/Hg film electrode. *Analytical Methods* 7: 2121–2128.

Sharma, S. and Bhattacharya, A. (2017). Drinking water contamination and treatment techniques. *Applied Water Science* 7: 1043–1067.

Sharma, V., Tiwari, P., and Mobin, S.M. (2017). Sustainable carbon-dots: recent advances in green carbon dots for sensing and bioimaging. *Journal of Materials Chemistry B* 5: 8904–8924.

Shrivastava, A. and Gupta, V.B. (2011). Methods for the determination of limit of detection and limit of quantitation of the analytical methods. *Chronicles of Young Scientists* 2: 21.

Sladkov, V., Bolyos, A., and David, F. (2002). Modified square-wave cathodic stripping voltammetry for determination of Se (IV) at trace and ultratrace levels. *Electroanalysis: An International Journal Devoted to Fundamental and Practical Aspects of Electroanalysis* 14: 128–134.

Smith, L. and Garg, U. (2017). Disorders of trace metals. In: *Biomarkers in Inborn Errors of Metabolism* (eds. L. Smith and U. Garg), 399–426. San Diego, CA: Elsevier.

Spiller, H.A., Hays, H.L., Burns, G., and Casavant, M.J. (2017). Severe elemental mercury poisoning managed with selenium and N-acetylcysteine administration. *Toxicology Communications* 1: 24–28.

Stoffaneller, R. and Morse, N.L. (2015). A review of dietary selenium intake and selenium status in Europe and the Middle East. *Nutrients* 7: 1494–1537.

Stora, T., Hovius, R., Dienes, Z. et al. (1997). Metal ion trace detection by a chelator-modified gold electrode: a comparison of surface to bulk affinity. *Langmuir* 13: 5211–5214.

Stozhko, N.Y., Morosanova, E., Kolyadina, L., and Fomina, S. (2006). Ceramic composite electrode for the determination of selenium (IV) by stripping voltammetry. *Journal of Analytical Chemistry* 61: 158–165.

Suz, D., Blagojević, S., Vidić, J., and Erceg, M. (1997). Determination of selenium (IV) by cathodic stripping voltammetry using a copper microelectrode. *Microchemical Journal* 57: 255–260.

Tan, S.H. and Kounaves, S.P. (1998). Determination of selenium (IV) at a microfabricated gold ultramicroelectrode array using square wave anodic stripping voltammetry. *Electroanalysis: An International Journal Devoted to Fundamental and Practical Aspects of Electroanalysis* 10: 364–368.

Tan, L.C., Nancharaiah, Y.V., van Hullebusch, E.D., and Lens, P.N. (2018). Selenium: environmental significance, pollution, and biological treatment technologies. In: *Anaerobic Treatment of Mine Wastewater for the Removal of Selenate and its Co-Contaminants* (ed. L.C. Tan), 9–56. CRC Press.

Tanaka, S., Sugawara, K., and Taga, M. (1990). Voltammetry of selenium (IV) based on an adsorptive accumulation of selenium-2, 3-diaminonaphthalene complex. *Analytical Sciences* 6: 475–478.

Thompson, H.J., Meeker, L.D., and Kokoska, S. (1984). Effect of an inorganic and organic form of dietary selenium on the promotional stage of mammary carcinogenesis in the rat. *Cancer Research* 44: 2803–2806.

Tie, M., Li, B., Liu, Y. et al. (2014). HPLC–ICP-MS analysis of selenium speciation in selenium-enriched Cordyceps militaris. *RSC Advances* 4: 62071–62075.

Triggiani, V., Tafaro, E., Giagulli, V.A. et al. (2009). Role of iodine, selenium and other micronutrients in thyroid function and disorders. *Endocrine, Metabolic & Immune Disorders-Drug Targets (Formerly Current Drug Targets-Immune, Endocrine & Metabolic Disorders)* 9: 277–294.

Tripathi, G., Yadav, V.K., Singh, J., and Mishra, V. (2020). Analytical methods of water pollutants detection. In: *Sensors in Water Pollutants Monitoring: Role of Material* (eds. D. Pooja, P. Kumar, P. Singh and S. Patil), 63–78. Springer.

Vajda, F. (1970). Stripping voltammetry of Se (IV) compounds with the hanging drop electrode. *Acta Chirurgica Academiae Scientiarum Hungaricae* 63: 257–265.

Wake, B.D., Bowie, A.R., Butler, E.C., and Haddad, P.R. (2004). Modern preconcentration methods for the determination of selenium species in environmental water samples. *TrAC, Trends in Analytical Chemistry* 23: 491–500.

Wang, J. and Lu, J. (1993). Ultratrace measurements of selenium by cathodic stripping voltammetry in the presence of rhodium. *Analytica Chimica Acta* 274: 219–224.

Wang, J. and Rogers, K. (1995). *Electrochemical Sensors for Environmental Monitoring: A Review of Recent Technology*, US Environmental Protection Agency, Office of Research and Development.

Wang, L.-H. and Shan, X.-E. (2016). Electrochemical reduction of selenium on a silver electrode and its determination in river water. *Journal of Analytical Chemistry* 71: 917–925.

Wang, Q., Yan, J., Wang, Y. et al. (2014). Three-dimensional flower-like and hierarchical porous carbon materials as high-rate performance electrodes for supercapacitors. *Carbon* 67: 119–127.

Wang, N., Tan, H.-Y., Li, S. et al. (2017). Supplementation of micronutrient selenium in metabolic diseases: its role as an antioxidant. *Oxidative Medicine and Cellular Longevity* 2017.

Wang, J., Yang, L., Li, H. et al. (2018). Dietary selenium intake based on the Chinese food pagoda: the influence of dietary patterns on selenium intake. *Nutrition Journal* 17: 50.

Webb, S., Bartos, J., Boles, R., and Hasty, E. (2014). Simultaneous determination of arsenic, cadmium, calcium, chromium, cobalt, copper, iron, lead, magnesium, manganese, molybdenum, nickel, selenium, and zinc in fertilizers by microwave acid digestion and inductively coupled plasma-optical emission spectrometry detection: single-laboratory validation of a modification and extension of AOAC 2006.03. *Journal of AOAC International* https://doi.org/10.1016/j.cej.2013.02.130.

Wei, L., Gupta, P., Hernández, R., and Farhat, F. (1999). Determination of ultratrace selenium and arsenic at parts-per-trillion levels in environmental and biological samples by atomic fluorescence spectrometry with flow injection hydride generation technique. *Microchemical Journal* 62: 83–98.

West, P.W. and Ramakrishna, T.V. (1968). Catalytic method for determining traces of selenium. *Analytical Chemistry* 40: 966–968.

WHO. (2004). *Vitamin and Mineral Requirements in Human Nutrition*. World Health Organization.

Wilschefski, S.C. and Baxter, M.R. (2019). Inductively coupled plasma mass spectrometry: introduction to analytical aspects. *The Clinical Biochemist Reviews* 40: 115.

Winkel, L.H., Johnson, C.A., Lenz, M. et al. (2012). *Environmental selenium research: from microscopic processes to global understanding*. ACS Publications.

Wrobel, J.K., Power, R., and Toborek, M. (2016). Biological activity of selenium: revisited. *IUBMB Life* 68: 97–105.

Xiong, Y., Li, F., Wang, J. et al. (2018). Simple multimodal detection of selenium in water and vegetable samples by a catalytic chromogenic method. *Analytical Methods* 10: 2102–2107.

Xu, X., Ray, R., Gu, Y. et al. (2004). Electrophoretic analysis and purification of fluorescent single-walled carbon nanotube fragments. *Journal of the American Chemical Society* 126: 12736–12737.

Xu, S., Zheng, M., Zhang, X. et al. (2012). Nano TiO_2-based preconcentration for the speciation analysis of inorganic selenium by using ion chromatography with conductivity detection. *Microchemical Journal* 101: 70–74.

Yang, H.-Y. and Sun, I.-W. (2000). Determination of selenium (IV) by a photooxidized 3,3'-diaminobenzidine/perfluorinated polymer mercury film electrode. *Analytical Chemistry* 72: 3476–3479.

Yu, X., Liu, C., Guo, Y., and Deng, T. (2019). Speciation analysis of trace arsenic, mercury, selenium and antimony in environmental and biological samples based on hyphenated techniques. *Molecules* 24: 926.

Zaitsev, N., Osipova, E., Fedulov, D. et al. (2006). Determination of selenium (IV) by cathodic stripping voltammetry using a copper-modified mercury-film electrode modified with copper. *Journal of Analytical Chemistry* 61: 77–83.

Zam, W., Alshahneh, M., and Hasan, A. (2019). Methods of spectroscopy for selenium determination: a review. *Research Journal of Pharmacy and Technology* 12: 6149–6152.

Zbinden, C. (1931). Nouvelle méthode de microdétermination de l'ion cuivre. *Bulletin de la Société de Chimie Biologique* 13: 35–38.

Zelić, M. and Branica, M. (1990). Study of electrochemical reduction of selenium (+4) using square wave voltammetry. *Electroanalysis* 2: 455–462.

Zeng, H. (2009). Selenium as an essential micronutrient: roles in cell cycle and apoptosis. *Molecules* 14: 1263–1278.

Zhang, Q., Li, X., Shi, H., and Yuan, Z. (2010). Determination of trace selenium by differential pulse adsorptive stripping voltammetry at a bismuth film electrode. *Electrochimica Acta* 55: 4717–4721.

Zhang, Y., Chen, B., Wu, S. et al. (2016). Graphene oxide–TiO_2 composite solid phase extraction combined with graphite furnace atomic absorption spectrometry for the speciation of inorganic selenium in water samples. *Talanta* 154: 474–480.

Zoidis, E., Seremelis, I., Kontopoulos, N., and Danezis, G.P. (2018). Selenium-dependent antioxidant enzymes: actions and properties of selenoproteins. *Antioxidants* 7: 66.

Zoski, C.G. (2006). *Handbook of Electrochemistry*. Elsevier.

8

Optical Sensors and Materials for Selenium Determination in Water

Rishabh Jain[1,2], Anupma Thakur[1,2], Praveen Kumar[3], Neerja Mittal[1,2], and Pooja Devi[1,2]

[1] *Academy of Scientific and Innovative Research (AcSIR), Ghaziabad, India*
[2] *Central Scientific Instruments Organisation, Chandigarh, India*
[3] *School of Materials Sciences, Indian Association for the Cultivation of Science, Kolkata, India*

8.1 Introduction

Selenium (Se) was first revealed by J.J. Berzelius in 1817 and is commonly present in the Earth's crust with sulfur-rich mineral deposits. The typical concentrations for selenium are 50–90 μg/kg in soil, but higher concentrations can be due to volcanic and sedimentary rocks (Winkel et al. 2012). The element selenium is recognized as a geologically occurring multifaceted metalloid in nature, with atomic number 34, atomic mass 78.96u, and electronic configuration as $[Ar]3d^{10}4s^{2}4p^{4}$. It belongs to group 16 in the periodic table, which is the group of chalcogens, and thus has similar chemical properties to sulfur. Due to these similar properties, the occurrence and distribution of selenium in the Earth's crust is often linked with the presence of natural sulfides such as sphalerites and pyrites (García-Hernández et al. 2001). However, these footsteps are tiny, and selenium is present as a minor constituent in sulfur-rich minerals deposits/ores. Selenium not only plays a significant role as an essential micronutrient for humans and animals in growth and metabolic activities (Shenkin 2009) but also protects against some forms of cancers (Burk et al. 2006). The selenium-dependent proteins such as glutathione peroxidases protect against oxidative stress, iodothyronine deiodinases regulate thyroid hormone metabolism, and thioredoxin reductases help in the reduction of intramolecular disulfide bonds (Monsen 2000). The health disorders of selenium deficiency in human beings result in Keshan disease and Kashin–Beck disease (Fairweather-Tait et al. 2010). The toxicity of selenium is also well known when consumed above bio-permissible limits. These higher intakes of selenium can lead to toxic health effects, including induced oxidative stress (Fairweather-Tait et al. 2011) and other toxicity effect discussed in next section. Selenium exists in both organic and inorganic forms, as tabulated in Table 8.1. Among these organic and inorganic forms, selenate (Se(VI)) and selenite (Se(IV)) known for their toxicity. Here, Se(IV) is

Selenium Contamination in Water, First Edition. Edited by Pooja Devi, Pardeep Singh, Arindam Malakar, and Daniel Snow.

Table 8.1 Selenium species and their occurrence in the environment.

Species	Molecular formula	Occurrence
Selenate (Se^{4+})	$H_2SeO_4{}^0$, $HSeO_4$, $SeO_4{}^{2-}$	Soil, sediment, and water
Selenite (Se^{6+})	$H_2SeO_3{}^0$, $HSeO_3{}^-$, $SeO_3{}^{2-}$, SeO^{2-}	Acidic and oxidizing environments (such as wastewaters), gases evolved from combustion and volcanic eruptions
Elemental selenium	Se^0	Precipitates of microbial reduction and chemical processes
Selenide (Se^{2-})	H_2Se	Volatile product left after microbial processes
Dimethyl selenide	CH_3-Se- CH_3	Product of bacterial methylation processes
Dimethyl diselenide	CH_3-Se-Se- CH_3	Product of bacterial methylation processes
Dimethylselenium sulfide	CH_3-Se-S- CH_3	Precipitates of microbial methylation method
Dimethylseleniumdisulfide	CH_3-Se-S-S- CH_3	Precipitates of microbial methylation method
Selenodiglutathione	G-S-Se-S-G	Products formed while performing in-vitro study experiments, mimicking synthetic biochemical processes
Selenocysteine	H-Se-(CH-NH_2)-COOH	Organic tissues
Selenomethionine	CH_3-Se-CH_2-CH_2-(CH-NH_2)-COOH	Plants
Trimethylselenonium	$(CH_3)_3$-Se^+	Metabolites in urine
Selenocyanate	Se-CN^-	Wastewater effluents/run-off from industries, especially petroleum

20 times more harmful than Se(VI) as it inhibits the normal functions of the body organs and causes associated with toxic health effects (Sanz Alaejos and Diaz Romero 1995; López de Arroyabe Loyo et al. 2008).

Geologically, selenium is distributed haphazardly on the surface of the Earth in both organic and inorganic forms, as mentioned earlier. This random distribution of selenium and pollution associated with it is known to be caused by geogenic and anthropogenic sources (Risher 2003; Vriens et al. 2014). The risks of elevated concentration of selenium in water resources and toxic health effects connected with it reported in many countries, including North America, Australia, New Zealand, India, and likely unrecognized in other parts of the globe (Figure 8.1) (Chapman et al. 2010; Bhalla et al. 2011). In India, the risks of selenium toxicity in water and soil are mainly from the regions of northern India, including the districts of the Punjab, Haryana, Rajasthan, and Himachal Pradesh (Yadav et al. 2005). The elevated level of selenium in groundwater and soil constitutes a prime resource of its entrance in the food chain, which leads to its bioaccumulation (Elrashidi et al. 1989). Selenate is mainly found in soils, whereas selenite is mainly found in water and is highly toxic to plants grown in solution culture (Gissel-Nielsen and Bisbjerg 1970).

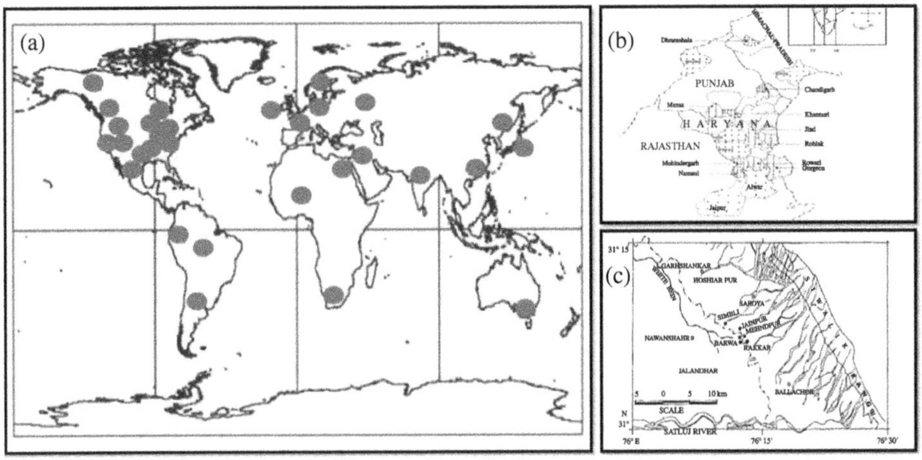

Figure 8.1 (a) International scenario of selenium toxicity (Lemly 2004) and (b and c) Selenium toxicity spread in North India (Yadav et al. 2005; Dhillon and Dhillon 2009) (*Source:* Adapted with permission from the reference provided).

In light of the above facts, it is important to monitor and control the limit of bioavailable selenium in the drinking water. The World Health Organization (WHO) and US Environmental Protection Agency (USEPA) recognize the vulnerabilities associated with elevated concentrations of selenium in water and have thus mandated a maximum acceptable limit of 10 and 50 ppb, respectively, in potable water (World Health Organization 2004). Even though conventional analytical practices are based upon sophisticated high-end equipment such as ICP-MS, AAS, and IC, there is still a need to look for alternate techniques for the primary screening of water resources for the presence of selenium. Optical techniques offer the advantages of high sensitivity and simple hardware design, thus offering the translational potential of this technique into field kits. But the development of optical detection strategies specifically for selenium is still at a nascent stage. Many developed portable methods are not able to achieve these permissible limits with reproducibility, and hence the world at present relies on the laboratory-based analysis of selenium ions in water.

8.2 Health Effects and Sources of Selenium Toxicity

The toxicity of selenium is also well known when consumed above bio-permissible limits. These higher intakes of selenium can lead to toxic health effects, including induced oxidative stress (Fairweather-Tait et al. 2011). This toxicity prominently appears clinically as structural defects, brittle hair, thickened and stratified nails, vomiting, pulmonary edema, and multiple organ disorder (called selenosis), as shown in Figure 8.2. Chronic selenosis is associated with long-term ingestion of selenite-contaminated potable water. At times, elevated concentrations of selenium can take a life (Reid et al. 2004). Thus, it has always gained keen interest among the general public owing to the very narrow range between deficiency (<40 μg/day) and toxicity (>400 μg/day) (Sanz Alaejos and Diaz Romero 1995; Monsen 2000; López de Arroyabe Loyo et al. 2008).

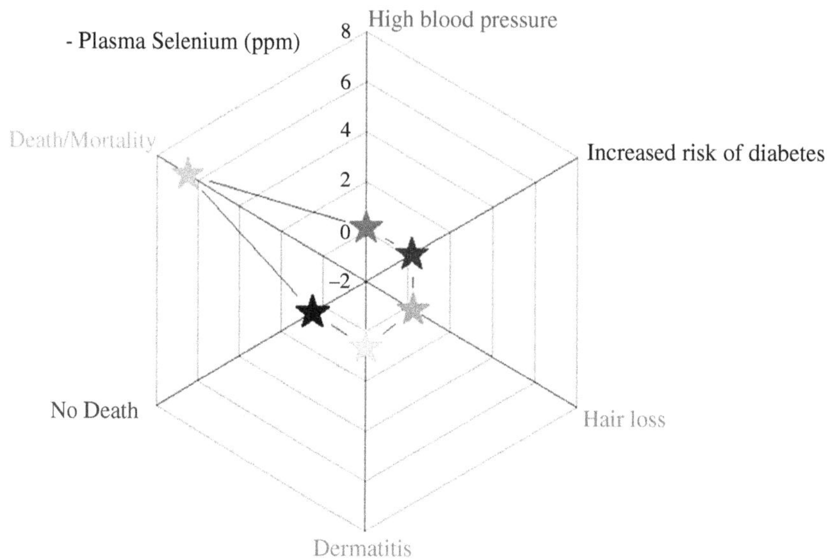

Figure 8.2 Health effects of selenium toxicity based on concentration.

Selenite is generally found under oxidizing and alkaline environmental conditions and is the most dominant species of selenium in adults. The concentration level of total selenium in water varies with the source, including the type of wastewater. Wastewaters resulting from gold mining activities contain the highest amount of selenium, in the range of 0.2–33 ppm. Other mining activities like phosphate and sulfur mining operations contribute selenium in the range of 3 ppb–12 ppm. Flue Gas Desulfurization Treatment processes add up selenium in water in the range of 1– 10 ppm. Further, there are various other industrial wastewater sources which have a significant contribution to selenium contamination. These sources include scrubber waters due to lead smelting (3–7 ppm), discharge from a uranium mine (1600 ppb selenium), drainage water (140–1400 ppb) due to agricultural activities, pond water due to coal mining (8.8–389 ppb), and petrochemical industries wastewater (7.5–55.9 ppb) (Santos et al. 2015).

8.3 Sensing Principles and Design of Optical Sensory Probes

The optical sensors have emerged as a possible solution to meet the global demand for environmental monitoring, medical diagnostics, and toxicological analysis. In general, optical sensors consist of a receptor which changes its optical properties upon interaction with the analyte and is able to convert this change into a human-understandable form. Here, in the case of selenium detection, the receptors are required to selectively and sensitively bind selenium for a recognizable change. The keen interest in the dual role of selenium as an essential trace element and a toxic metalloid has motivated the development of various analytical and field-deployable detection techniques. For choosing an appropriate analytical technique for the determination of selenium, it is critical to consider factors such as high accuracy of the method, analyte concentration detection range, the volume of samples, physiochemical properties of environmental detection matrices, and the possible interfering species. Herein, optical techniques offer merits in terms of the fast response time, better detection limits, reasonable linear dynamic ranges, affordability, easy hardware design, and field deployability. However, the analytical performance of these optical detection techniques is extremely dependent upon the sensory material characteristics. In this context, several organic and inorganic materials have been developed and reported for inorganic selenium analysis, particularly in a water matrix. These sensing materials employs absorbance-based platforms, including colorimetric change (Danzuka and Ueno 1958a; Hawkes 1986; Arikan et al. 1996; Liu et al. 1998; Ensafi and Lemraski 2004), absorbance change (Manish et al. 1994; Pathare and Sawant 1995; Ramachandran and Kumar 1996; Melwanki and Seetharamappa 2000; Krishnaiah et al. 2003; Nakano et al. 2004; Agrawal et al. 2009; Warkocki et al. 2015; Awual et al. 2015b, 2015d; Kuchekar et al. 2015b), and fluorescence change (Costas-Mora et al. 2012; Liang et al. 2013; Huang et al. 2015; Devi et al. 2017a,b) as a measurement approach, as shown in Table 8.2. The measurement of these signals generally requires instrumentation support, except in colorimetry, wherein change is visible by the naked eye.

8.4 Advances in Optical Sensory Probes: A Meta-Analysis on Optical Materials

This section focuses on the advancement of optical sensors probes for the determination of selenium based upon absorbance, color, and fluorescence properties of the sensor platform. In general, these sensing platforms comprise suitable material whose characteristic property changes upon the interaction with the selenium. Both qualitative and quantitative analysis of selenium employs the changes in these characteristic properties of sensing probes. In the latter case, a calibration model was first developed for different concentrations of selenium. Some of the reported probes and their associated mechanisms for selenium detection utilizing the mechanisms discussed are also presented and tabulated in Table 8.2.

8.4.1 Absorbance-Based Platforms

The absorbance-based sensing platforms are generally utilized when the absorbance of materials changes upon the interaction with selenium ions in the water matrix. In general, absorbance (A) refers to the capacity of a material to absorb light at a specific wavelength, and sometimes also refers to the optical density of the material. According to the Beer–Lambert law, the equation shown below is used to calculate the absorbance of any material.

$$A = \log {}^{I_0}\!/_{I}$$

Here, I_0 refers to the reference intensity of light, and I indicates the intensity of the light when interacted with the sample at a specific wavelength. The absorbance characteristics,

Table 8.2 Selenium species and their occurrence in the environment.

Ligand/Dye/Fluorophore	Species detected	Analytical details	Detection matrix	Ref.
Absorbance-based probes (colorimetric)				
3,3′-Diaminobenzidine	Se^0	LOD: 100 ppb	Acidic mixtures	Danzuka and Ueno (1958b)
Tetranitro blue tetrazolium (2,2′,5,5′-tetra-(p-nitrophenyl)-3,3′-(3,3′-diethoxy-4,4′-diphenylene) ditetrazolium chloride)	Se^{4+}	LR: 0.001–0.5 ppb LOD: 0.2 ppb	Acidic mixtures	Hawkes (1986)
Maxilon blue-sulfonazo	Se^{4+}	LR: 0.5–180 ppb LOD: 0.3 ppb	Polluted water, natural water	Ensafi and Lemraski (2004)
Methylene blue	Se^{4+}	LOD: 0.2 ppb	Acidic mixtures	Arikan et al. (1996)
Triphenylmethane	Se^{4+}	LOD: 0.094 ppb	Lake water, river water, springwater, tap water	Liu et al. (1998)
Absorbance-based probes (spectrophotometric)				
J acid, 1-amino-2-naphthol-6-sulphonic acid (6-ANSA)	Se^{4+}	LR: 30–300 ppb LOD: 30 ppb	Polluted water	Manish et al. (1994)
2,3-Diamino-naphthalene	Se^{4+}	LR: 500–12 000 ppb LOD: 12 ppb	Natural water	Ramachandran and Kumar (1996)
Cetylpyridinium chloride	Se^{4+}	LR: 50–1000 ppb LOD: 10 ppb	Wastewater	Agrawal et al. (2009)
Sodium salt of heamethyleneimine carbodithioate	Se^{4+}	LR: 0.5–4 ppm LOD: 1.45 ppb	Polluted water	Pathare and Sawant (1995)
Methdilazine hydrochloride	Se^{4+}	LR: 0.5–2.3 ppm LOD: 0.7 ppm	Acidic mixtures	Melwanki and Seetharamappa (2000)
p-hydrazino-benzenesulfonic acid with N-(1-naphthyl) ethylenediamine	Se^{4+}	LR: 0.2–6 ppb LOD: 3 ppb	Lake water, river water, seawater	Nakano et al. (2004)

2,4-DNPH-N-(1-Naphthalene −1-yl)ethane-1,2-diamine dihydrochloride	Se^{4+}	LOD: 3.5 ppb	Polluted water, natural water	Krishnaiah et al. (2003)
O-methoxyphenyl-thiourea	Se^{4+}	LR: 20–50 ppm LOD: 0.61 ppb	River water	Kuchekar et al. (2015a)
N,N-di(3-carboxy-salicylidene)-3,4-diamino-5-hydroxypyrazole	Se^{4+}	LOD: 1.14 ppb	Mineral water, tap water, simulated water	Awual et al. (2015c)
Mesoporous TiO_2–SiO_2 nanotube	Se^{4+}	LOD: 4.9 ppb	Drinking water	Warkocki et al. (2015)
(3-(3-[methoxycarbonyl] benzylidene) hydrazine) benzoic Acid	Se^{4+}	LOD: 2.02 ppb	Mineral water, tap water, simulated water	Awual et al. (2015a)
Photoluminescence-based probes				
CdSe QD with HAD	Se^{4+}	LR: 5–65 ppb LOD: 0.08 ppb	Lake water, seawater	Costas-Mora et al. (2012)
CdTe QD	Se^{4+}	LR: 5–100 ppb LOD: 0.1 ppb	Urine and hair sample	(Huang et al. 2016)
Silica NPs with 3,3′-diaminobenzidine	Se^{4+}	LR: 1–10 ppm LOD: 0.19 ppm	Water sample	Liang et al. (2013)
Nitrogen-rich ligand/CQDs	Se^{4+}	LOD: 0.1 ppb	Groundwater, lake water, well water	Devi et al. (2017b)
Whey derived CQDs	Se^{4+}	LOD: 0.1 ppb	Groundwater, lake water, well water	Devi et al. (2017a)
N-CQDs m-aminophenol, dinitrobenzenesulfonylchloride (DNS-Cl), p-toluene-sulfonyl chloride	Se^{4+}	LOD: 0.23 ppb LR: 0.2–200 ppb	Aqueous solutions and living cells	Wang et al. (2017)

in general, are measured with UV/Vis spectrophotometer. This instrument sometimes also referred to as an absorption spectrophotometer. This technique measures the change in intensity of light in ultraviolet to the visible region, typically in the wavelength range of 190–1100 nm. The basic principle for absorbance-based platforms/sensor materials for selenium based on its change in UV/Vis spectra, either in terms of intensity at a specific wavelength or shift in its maximum absorption wavelength. The analytical characteristics of the sensor further depend upon pH, conjugation, indicator formation, and aggregation. The two major categories for absorbance-based platforms are spectrophotometer-based and colorimetry-based. Colorimetry chemosensors generally suffer from lower accuracy, reproducibility, and responsivity in comparison to spectrophotometer-based platforms. However, the disadvantages associated with colorimetry can now be taken care of by employing image processing algorithms and color digitalization technologies.

Over the years, several chromogenic reagents reported for the selenium determination. They include organic ligands (3,3-diaminobenzidine (Marczenko 1986), 4-methyl-o-phenylenediamine (Kartal et al. 2010)), dyes (Rhodamine B (Liu et al. 1990), cresyl blue (Ensafi et al. 1997), methylene blue (Nayanova et al. 2016), crystal violet, azure B (Mathew and Narayana 2006)), and plasmonic nanomaterials (silver) (Amjadi et al. 2018). In the early 1990s, the existing methods for the determination of selenium require expensive instruments like inductively coupled plasma mass spectrometry and atomic absorption spectroscopy. The cost-effective methods of detection generally lack sensitivity, reproducibility, and selectivity. Only a few of the reported spectrophotometric methods have shown adequate selectivity and sensitivity for ppb-level quantification of selenium in the water matrix. The popular reagent, 6-amino-1-naphthol-o-sulfonic acid, forms a yellow complex with selenite ions and has an absorption wavelength of 392 nm. This approach could detect selenium to a level of 80 ppb with no interference from iron and copper under acidic conditions (Ramachandran et al. 1993). In acidic medium, selenite ions react with 2,3-diamine-naphthalene to form complexes extracted by cyclohexane or toluene. This method gives the detection limit of 10 ppm approximately at a maximum absorption wavelength of 378.5 nm (Ramachandran and Kumar 1996). However, here color generation is slow and may take up to two hours, which can be enhanced by the addition of bromine ions (Ramachandran and Kumar 1996).

3,3-diaminobenzidine (DAB) has been a widely used organic ligand for selenium detection in water. But it not only lacks sensitivity and selectivity but also is toxic to the environment. These drawbacks of organic ligands led to the development of another simple method referred as iodometry for the detection of selenium. In this method, first iodide is converted to iodine by selenium oxidation followed by the formation of the ion-association complex using basic dyes. Shaopu et al. opted for this practice in the 1990s, for the determination of selenium using Rhodamine B dye with poly(vinyl alcohol) with a low detection limit (Liu et al. 1990). Here, PVA acts as a solubilizer by preventing ion association, which keeps the solution stable for one hour. Here the interference from common interfering ions eliminated by pretreatment of water samples with thiol cotton, which separates the selenite ions from the matrix of ions. This approach successfully applied for screening of selenium in hot spring water, tap water, and river waters (Liu et al. 1990).

Similarly, Matthew and Narayana employed a violet color azure B dye as a chromogenic reagent in place of Rhodamine B dye for selenium detection with iodometry method in water, soil, plant, and human hair samples (Mathew and Narayana 2006). The liberated iodine by selenium (IV) oxidation bleached out azure B dye violet color, and the intensity of the same found to be dependent upon selenium concentration in water samples. This approach offered higher sensitivity and also removed the need for the pre-concentration step reported in the Rhodamine B approach. Moreover, the colorimetric change was observed for a concentration as low as 23 ppb, which was comparatively better when compared with the Rhodamine B dye-based method (Mathew and Narayana 2006). Fascinated by the excellent response of dyes for selenium detection, Nayanova et al. in 2016 developed a whole new approach of using dyes for the detection of selenium without using iodometry. They used the phenomenon of the destructive oxidation of methylene blue dye using selenium (Nayanova et al. 2016). The rate of destructive oxidation of dye depends upon the amount of selenium in the water sample and results in the disappearance of the characteristics color of dye. In other words, the shift in maximum absorption peak wavelength shift of dye from 670 nm to the UV region (190–200 nm) and a maximum absorption peak shift depends on the concentration of products. The authors have also studied the effect of common interfering ions and tested this method for the analysis of selenium in bottled drinking water. This method was able to achieve a detection limit of 1 ppb, satisfying the WHO permissible limit (Nayanova et al. 2016).

Similarly, catalytic-kinetic spectrophotometric methods are an attractive substitute for the determination of trace amounts of selenium. Here, selenite ions catalyze the reduction of maxilon blue-SG or sulfonazo by sulfide (Ensafi and Lemraski 2004). This step is followed by the reduction of methylene blue by seleno-sulfides (Arikan et al. 1996), which contribute to a decrease in absorbance and depend on the selenite ions concentration. These selenite ions also possess catalytic response producing an increase in absorbance, such as the photooxidative pairing of p-hydrazino benzene sulfonic acid with N-(1-naphthyl) ethylenediamine to form an azo dye absorbing at 538 nm (Nakano et al. 2004). In another study, Matamoros and Benning utilized the reaction of Se(IV) ions with hydroxylamine hydrochloride, which diazotizes p-nitroaniline, which in turn combines with N-1-naphthyl-ethylenediamine dihydrochloride (NEDA), to form a red-colored azo compound. This method exhibited a detection limit of 10 ppb (Matamoros and Benning 2008). In another study, researchers designed a selective extraction–spectrophotometric method specifically for selenium(IV) determination (Kuchekar et al. 2015a). They reported O-methoxyphenyl thiourea (OMePT) as a chelating agent for selective selenium, capturing to 0.61 ppb detection level via Se-SH chemistry. This method suffers from little interference from the cations, however, this was overcome by using masking agents (Kuchekar et al. 2015a). Similarly, selenite ions as used as a catalyst for the reduction of bromate by hydrazine dihydrochloride in acidic media (Chand and Prasad 2009). Thus, the liberated bromine method is similar to iodometry discussed earlier (Chand and Prasad 2009). Later, an organic–inorganic nano-hybrid system was also explored, which utilized (3-(3-[methoxycarbonyl] benzylidene)hydrazinyl) benzoic acid as an organic ligand. This ligand indirectly immobilized onto the mesoporous silica as a substrate. This organic–inorganic nano-hybrid system was then successfully used for selenium (IV) detection to a lower 2.02 ppb level (Awual et al. 2015a).

Apart from the basic organic dyes, piazselenol complex derivatives such as 4-methyl-o-phenylenediamine (MOPDA) employed for selenium detection. (Kartal et al. 2010). MODA acts as a chromogenic reagent, which upon interaction with selenium (IV) in acidic condition forms 5-methyl-piazselenol complex having absorbance at 332 nm. Under optimized analytical conditions such as pH, ionic strength, sample volume, and MOPDA concentration, this approach is capable of detecting selenium up to 0.95 ppb. The interference from Cu(II), Fe(III), and Zn(II) ions is taken care of by the addition of EDTA and KSCN in samples before analysis. This method then applied for selenium(IV) analysis in tap water, seawater, wastewater, and soil samples (Kartal et al. 2010).

Besides organic optical probes, plasmonic nanomaterials also studied for selenium detection using the absorbance principle. In a recent report, silver nanoprisms have shown great potential for selenium detection using surface plasmon resonance (SPR) property of nanomaterials (Amjadi et al. 2018). The shift in SPR wavelengths on the addition of selenium is taken as a measurement approach, also visible through the naked eye, as a color change from light blue to violet (Figure 8.3). This SPR effect is associated with the change in morphology (shape deformation to nanodisk) of silver nanoprisms in the presence of selenium species. This method able to achieve a detection range and detection limit of 2.5–100 ppb and 1.2 ppb, respectively, and applied to selenium analysis in food and water samples. The authors also observed Hg(II) as a significant interfering species while analyzing selenium using this method. However, due to its low concentration of Hg(II) in the water and food matrix in comparison to selenium, this interference is negligible and can be ignored (Amjadi et al. 2018).

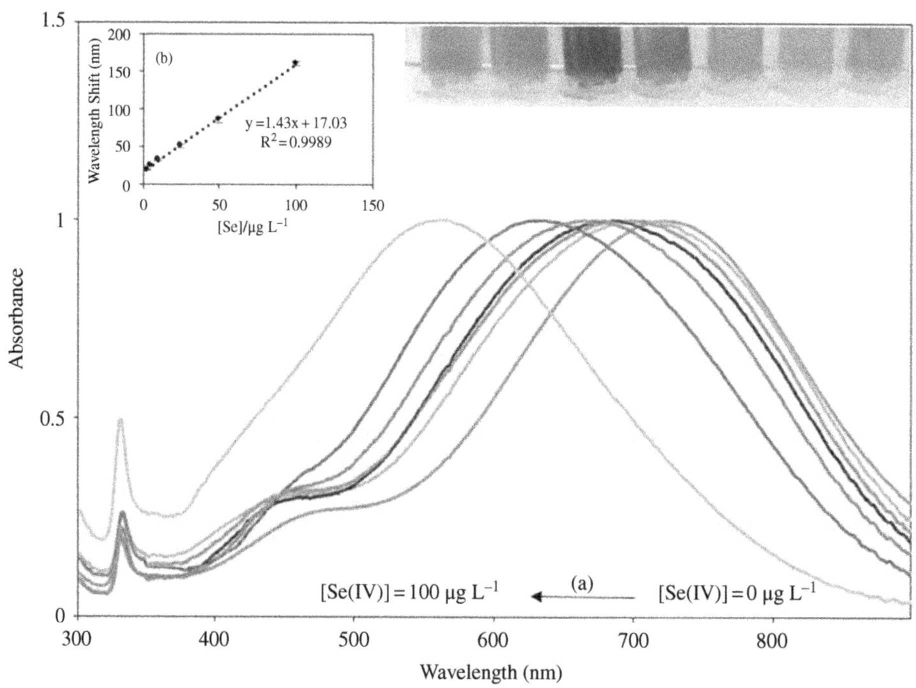

Figure 8.3 UV/Vis Spectral and color change due to change in morphology of silver nanoprisms in the presence of selenium (IV) (*Source:* Adapted with permission from Amjadi et al. (2018).)

8.4.2 Photoluminescence-Based Platforms

In this category, the luminescent or fluorescent property of the sensors utilized to fulfill the interest of the development of fluorescent sensors for the detection of selenium species. The photoluminescence (PL)-based platforms offer merits such as sensitivity, selectivity, accuracy, and cost-effectiveness. These merits make them superior to the absorbance-based probes for the real-time application. PL-based platforms generally have two essential components, i.e. fluorophore, which acts as a transducer to convert the change in fluorescent properties into a human-understandable form and receptor which binds the desired analyte selectively to modulate the photophysical properties of the fluorophore. For selenium, the design of the receptor is such that it should selectively and sensitively bind it using the principles of chemistry.

The design of the receptor, which includes its morphology, surface groups, and structure, depends on the selenium species and analyte matrices such as its ionic strength, pH, and polarity. The ideal fluorophore must have photostability, chemical stability, thermal stability, and a massive Stokes shift. Similarly, the ideal receptor must be reversible to allow continuous monitoring of samples. Usually, fluorophore and receptor linked to each other using covalent linkages. The basic principle for the PL-based platform is the change in emission, classified into two categories, i.e. fluorescence and phosphorescence. The fluorescence presents in molecules with singlet excited state, while latter in molecules with triplet excited state. These states formed when the molecules get excited from the ground state to an excited state. Due to the formation of these states, there exist various sensing mechanisms for fluorescence-based sensing, i.e. fluorescence quenching, lifetime measurements, metal (selenium ions) to ligand charge transfer, and internal charge transfer. Based on these strategies, organic complexes, ligands, and quantum dots (QDs) are the significant candidates to design an efficient fluorescent sensor for selenium.

In 1994, intending to obtain better detection limits for low-level detection of selenium, Wang et al. reported the use of organic ligand 2,3-Diaminonaphthalene (DAN) (Wang et al. 1994). In this method, separation of selenium species done using anion-exchange columns before detection to avoid interference followed by measurement of selenium with respect to the change in fluorescence of DAN at its different concentrations at excitation wavelength 370 nm and emission wavelength 518 nm. Although the method achieved a detection limit of 1.8 ng/l, the complex process of detection employing anion-exchange columns limits its comprehensive utility (Wang et al. 1994). Two years later, DAN employed in the development of a turn-on fluorescent probe where the fluorescence of the DAN-selenite complex evaluated without employing the use of anion-exchange columns at excitation wavelength 370 nm (Rodriguez et al. 1996). The fluorescence intensity of the complex depends on the concentration of selenite ions. This complex was extracted using cyclohexane, which not only improves the sensitivity of selenite detection but also eliminates interference from the common interfering ions. The result of this method was also verified with the existing conventional technique, i.e. hydride generation atomic absorption spectrometry (HG-AAS). Further authors indicate that the enhancement of selectivity or sensitivity of this method by using organic solvents, particularly, 1,4-Dioxane when compared with other solvents (Rodriguez et al. 1996).

Similar to the above strategy, 3,3′-diaminobenzidine (DAB) modified silica nanoparticles employed as a "turn-on" probe for the detection of selenium (Liang et al. 2013). This probe was based on the fact that DAB molecules themselves do not have much luminescence intensity due to the absence of planar π bonds. But on binding with selenite ions, it forms a fluorescent complex, i.e. 3′,4′-diaminophenyl-piazselenol, that can easily be detected using a fluorescence spectrophotometer. Additionally, the high surface area of silica nanoparticles ensured high loading of DAB molecules and thus high sensitivity of the sensor for selenium. For surface modification of silica nanoparticles, the carboxyl groups present on the complex surface activated using 1-ethyl-3-[3-dimethylaminopropyl] carbodiimide hydrochloride (EDC) N-hydroxy-succinimide (NHS) chemistry. The activated carboxyl groups then bind to DAB molecules through amide linkages. For the detection of selenite ions, the fluorescent intensity of the 3′,4′-diaminophenyl-piazselenol complex was investigated using a spectrofluorometer. This method does not require any separation step before the detection of selenite ions in real samples but is time-consuming in comparison to other existing fluorescent sensing systems. Moreover, the effect of copper interference in this method minimized by using EDTA as a chelating agent during the analysis (Liang et al. 2013).

Using the strong chelating interaction of DAB toward selenite, Chen et al. (2016) developed another new probe based upon the DAB modified silica nanosphere embedded with red CdTe QDs. This QDs based probe used for ratiometric fluorescence sensing of selenite (Chen et al. 2016), although in this case, the formation of the CdTe/silica nanoparticle core-shell structure carried out by using the reverse microemulsion method before the activation of carboxyl groups. The developed ratiometric probe exhibited a single emission peak at 635 nm, while on the addition of selenite, a new peak observed at 530 nm, in addition to sensor peak. All the sensing studies carried out at a single excitation wavelength of 415 nm. These studies indicate a dual-color emission, which is explained by the presence of DAB molecules, which on interacting with selenite enhance the emission at 530 nm or yellow-green fluorescence. The change in fluorescence color from red to green-yellow was also observed under UV light on the addition of selenite. This phenomenon results in the appearance of a new peak at 530 nm. Besides, in this method, the authors suggested the use of EDTA to overcome interference from the common ions. Thus the method is also applied successfully for selenite detection in river water, tap water, rice, lettuce, and radish samples. Their results were in agreement with the lab-based ICP-AES technique, supporting its use in-field analysis of selenium (Chen et al. 2016).

Fascinated with the properties of QDs such as massive Stokes shift in QDs helps in creating sulfur vacancies on the surface. Their board emission spectra provide the uniform size distribution and required surface defects. These properties pave their way for sensing applications such as selenium detection, and several other QDs such as CdS, ZnS, and Carbon QDs (CQDs), were reported for luminescence-based detection of selenium in water (Amouzegar et al. 2017). The selectivity of CdS and ZnS QDs toward selenite ions introduced by modifying their surface with sulfur-rich ligands, glutathione, and 3-mercaptopropionic acid, respectively. The performance of glutathione modified CdS QDs was demonstrated against selenite ions detection in the tris-buffer solution (Huang et al. 2008). The sensing mechanism based upon fluorescence quenching of modified CdS QDs due to elimination is sulfur vacancies caused by hydrogen selenide complex formed on the addition of selenium/selenite (Huang et al. 2008). Meanwhile, in the case of MPA modified ZnO QDs, selenite-induced aggregation of QDs is the basis for selenium sensing. MPA acts as a stabilizer for synthesized ZnO QDs. However, binding of selenite ions to –SH groups of MPA causes it to release from the ZnO surface, leading to their aggregation. This aggregation is optically observed by fluorescence quenching of ZnO QDs (emission at 430 nm) in the presence of selenium (IV) ions (Tris HCl buffer, pH 8). These MPA functionalized QDs have a massive Stokes shift, which is a recommended condition for the fluorescent sensor (Amouzegar et al. 2017). More recently, CdTe QDs decorated on paper for the visual detection of selenium reported having a detection limit up to 0.1 ppb (Huang et al. 2015). Despite high sensitivity observed by these semiconductor QDs based PL probes, their inherent toxicity due to their synthetic composition of heavy metals ecologically limits their real-world applications (Derfus et al. 2004). Meanwhile, zero-dimensional carbon quantum dots (CQDs) are an alternative potential optical probe because they surpass the main limitations of semiconductor QDs, namely their biocompatibility, high stability against photobleaching, and environment-friendliness (Sun et al. 2006; Sahu et al. 2012; Shen et al. 2012). Later motivated by this, the authors synthesized a nitrogen-rich ligand and selectively immobilized it onto CQDs surface using Schiff-base chemistry (Devi et al. 2017b). These CQDs interfaced with a nitrogen-rich ligand shown to detect selenium with high sensitivity and selectivity (Figure 8.4). High selectivity of the developed probe assigned to the high bond dissociation energy between N–Se complexes (381 kJ/mol).

The binding of selenium to modified CQDs observed as a periodical change in its fluorescence intensity with respect to selenium concentration. This binding is further explained in terms of non-radiative recombination of charge carriers and investigated by ultrafast spectroscopy. Besides, the inner filter effect and electron transfer were the possible facilitating processes behind observed fluorescence quenching phenomenon in the modified CQDs (Cayuela et al. 2015; Zhang et al. 2015; Li et al. 2016). The association constant of these

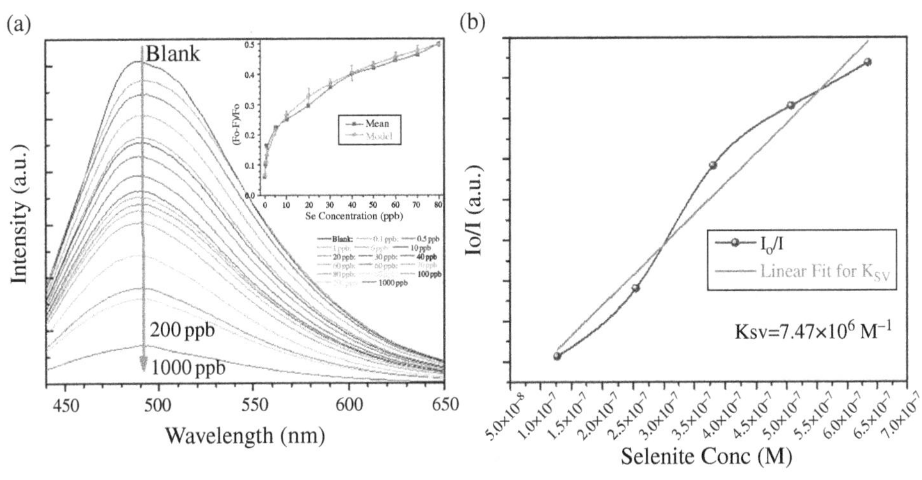

Figure 8.4 (a) Change in PL spectra of CQDs with respect to increasing selenium concentration; (b) Calibration plot for the detection of selenite ions using functionalized carbon quantum dots. (*Source:* Adapted with permission from ref. Devi et al. (2017b).)

modified CQDs-selenite complex was calculated using the Benesi–Hildebrand method and found to be 7.6×10^{6} M^{-1}(Du et al. 2013). This method was proposed for environmental analysis of selenium in real water samples, and results were validated with a reference technique (Devi et al. 2017b).

8.5 Commercial Optical Sensors for Selenium Analysis

In the present scenario, the analysis of selenium in the water matrix is done by laboratory-based methods such as gas chromatography (Najafi et al. 2012), high-performance liquid chromatography coupled with ICP-MS (Niedzielski 2005), neutron activation analysis (Olivas et al. 1994), atomic absorption spectroscopy (Bidari et al. 2007), atomic fluorescence spectroscopy (Li et al. 2012), and electrothermal atomic absorption spectroscopy (Ghasemi et al. 2010). With the advancement in technology, laboratory-based methods have unchallengeable accuracy, precision, and detection limits. However, this process of analysis requires sample transportation to the laboratory, sample storage, a skilled workforce, a dedicated environment, expensive instrumentation, maintenance, and is time-consuming, which results in high analysis cost of the sample. There are several private companies (such as Simplex Health Pvt. Ltd., UK) and research institutions (such as CSIR-CSIO, Chandigarh) that offer a lab-based testing facility for the selenium analysis. However, Hach Instruments developed a commercial method based upon diaminobenzidine for the detection of selenium in the 0.01–1 ppm using an 8194 UV/Vis spectrophotometer (Hach Company 2020). Their two products, i.e. DR2800 and DR1900, are designed for selenium detection, employing diaminobenzidine as a chelating and chromogenic agent. DR2800 has tungsten as a source lamp and works in the transmittance and absorbance modes in the wavelength range of 340–900 nm (Hach Company 2020). This product can detect selenium in the range of 0.01–1 ppm and is applicable for various real-time applications in drinking water, wastewater, and beverages. Likewise, DR1900 has – xenon lamp as a source and also works in the transmittance and absorbance modes in the wavelength range 340–800 nm. This product can detect selenium in the range of 0.01–1 ppm and is applicable for selenium detection in drinking water and industrial wastewater (Hach Company 2020). This process is cost-effective in comparison to conventional techniques and has comparable accuracy, but it is still a tedious, laboratory-based method, and requires skilled human resources. For the determination of selenium in the wastewater, there is a requirement of pretreatment and distillation. The distillation procedure involves the use of a strong acid and oxidizer at high temperatures. Then, an EDTA as a masking agent is added to the sample to remove interfering ions such as iron and copper before the test. Then the addition of a sulfate buffer adjusts the sample to the optimum ~1 pH. Under these conditions, diaminobenzidine reacts with all selenium present as selenite to give a yellow-colored piazselenol complex when extracted and the color intensity measured.

Another product in the market is the Water Selenium Content Detector, model no. ACM-166-SC by ACMAS Technologies Pvt. Ltd. (https://catalogindustry.com/de/catalogs/water-selenium-content-detector). It is an economical single element selenium concentration meter, used for quantitative detection of selenium content in the liquid samples only. It works with the diethyl-p-phenylenediamine (DPD) reagent to form colored compounds on selective absorption of electromagnetic waves having recognized colorimetric assay in the presence of selenium. The equipment comprises of silicon light source, lens, color bottles, and photocell. The signal processor consists of a direct readout LCD meter, which displays the levels of selenium. This product can detect selenium in the range of 0.5–1 ppm with a response time of 0.3 second and is applicable for selenium detection in drinking water, seawater, surface water, and industrial wastewater. Another product from Tintometer GmbH, Germany, is a single-beam portable spectrophotometer known as SpectroDirect (wavelength scan in the range of 330–900 nm) is shown for selenium detection in the 50 ppb–1.6 ppm concentration range. Here, 3,3′-diaminobenzidine in toluene has been utilized as a sensing probe for the detection of selenium (Tintometer GmbH 2020). This method, however, is not able to meet the WHO permissible limits and still requires a skilled workforce for the analysis. As of now, there is no commercial portable selenium sensor satisfying the WHO ASSURED criteria available in the market.

8.6 Summary and Future Outlook

In this chapter, optical sensing platforms were classified into two broad categories: absorbance-based and PL-based platforms. These two classes were then discussed based on their merits and pitfalls concerning their application in the detection of selenium. The absorbance-based platforms have advantages such as ease of use and cost-effective analysis. However, they are often not able to overcome the sensitivity and selectivity of the PL-based platforms. The analytical performance from both the platforms is profoundly affected by the type of solvent, change in pH, and presence of interfering ions. In light of the existing research available, there is still scope to work upon the design of a simple, easy to use, environment-friendly, selective, and sensitive optical chemosensor for selenium. Though it is challenging to design a selective and sensitive chromophores/fluorophore for selective selenium sensing, yet it can be realized using strong chemistry concepts. Despite all these disadvantages, optical strategies have the potential for cost-effective and portable sensing devices design for the real monitoring of selenium. These strategies, in turn, could avoid the high-cost instrumentation and complicated procedures with laboratory-based analysis. One must need to understand the pitfalls of each strategy and look for a potential scientific solution to obtain the characteristics of an ideal selenium sensor.

References

Agrawal, K., Patel, K.S., and Shrivas, K. (2009). Development of surfactant assisted spectrophotometric method for determination of selenium in waste water samples. *Journal of Hazardous Materials* 161: 1245–1249.

Amjadi, M., Hallaj, T., and Salari, R. (2018). A highly sensitive plasmonic sensor for detection of selenium based on the shape transformation of silver nanoprisms. *Sensors and Actuators B-Chemical* 273: 1307–1312.

Amouzegar, Z., Afkhami, A., and Madrakian, T. (2017). Photoluminescence investigation of MPA-ZnS QDs interaction with selenite ion. *Journal of the Iranian Chemical Society* 14: 2475–2483.

Arikan, B., Tunçay, M., and Apak, R. (1996). Sensitivity enhancement of the methylene blue catalytic – spectrophotometric method of selenium (IV) determination by CTAB. *Analytica Chimica Acta* 335: 155–167.

Awual, M.R., Hasan, M.M., and Khaleque, M.A. (2015a). Efficient selenium (IV) detection and removal from water by tailor-made novel conjugate adsorbent. *Sensors and Actuators B: Chemical* 209: 194–202.

Awual, M.R., Hasan, M.M., and Khaleque, M.A. (2015b). Efficient selenium(IV) detection and removal from water by tailor-made novel conjugate adsorbent. *Sensors and Actuators B: Chemical* 209: 194–202.

Awual, M.R., Yaita, T., Suzuki, S., and Shiwaku, H. (2015c). Ultimate selenium (IV) monitoring and removal from water using a new class of organic ligand based composite adsorbent. *Journal of Hazardous Materials* 291: 111–119.

Awual, M.R., Yaita, T., Suzuki, S., and Shiwaku, H. (2015d). Ultimate selenium(IV) monitoring and removal from water using a new class of organic ligand based composite adsorbent. *Journal of Hazardous Materials* 291: 111–119.

Bhalla, A., Singh, G., Kumar, S. et al. (2011). Elemental analysis of groundwater from different regions of Punjab state (India) using EDXRF technique and the sources of water contamination. Proc. of international conference on environmental and computer. *Science* 156–164.

Bidari, A., Jahromi, E.Z., Assadi, Y., and Hosseini, M.R.M. (2007). Monitoring of selenium in water samples using dispersive liquid–liquid microextraction followed by iridium-modified tube graphite furnace atomic absorption spectrometry. *Microchemical Journal* 87: 6–12.

Burk, R.F., Norsworthy, B.K., Hill, K.E. et al. (2006). Effects of chemical form of selenium on plasma biomarkers in a high-dose human supplementation trial. *Cancer Epidemiology Biomarkers & Prevention* 15: 804–810.

Cayuela, A., Soriano, M.L., and Valcárcel, M. (2015). Reusable sensor based on functionalized carbon dots for the detection of silver nanoparticles in cosmetics via inner filter effect. *Analytica Chimica Acta* 872: 70–76.

Chand, V. and Prasad, S. (2009). Trace determination and chemical speciation of selenium in environmental water samples using catalytic kinetic spectrophotometric method. *Journal of Hazardous Materials* 165: 780–788.

Chapman, P.M., Adams, W.J., Brooks, M. et al. (2010). Ecological Assessment of Selenium in the Aquatic Environment. CRC Press.

Chen, L.F., Tian, X.K., Zhao, Y. et al. (2016). A ratiometric fluorescence nanosensor for highly selective and sensitive detection of selenite. *Analyst* 141: 4685–4693.

Costas-Mora, I., Romero, V., Pena-Pereira, F. et al. (2012). Quantum dots confined in an organic drop as luminescent probes for detection of selenium by microfluorospectrometry after hydridation: study of the quenching mechanism and analytical performance. *Analytical Chemistry* 84: 4452–4459.

Danzuka, T. and Ueno, K. (1958a). Determination of trace amounts of selenium in sulfuric acid. Colorimetric method using 3, 3′-Diaminobenzidine. *Analytical Chemistry* 30: 1370–1371.

Danzuka, T. and Ueno, K. (1958b). Determination of trace amounts of selenium in sulfuric acid. Colorimetric method using 3, 3′-Diaminobenzidine. *Analytical Chemistry* 30: 1370–1371.

Derfus, A.M., Chan, W.C., and Bhatia, S.N. (2004). Probing the cytotoxicity of semiconductor quantum dots. *Nano Letters* 4: 11–18.

Devi, P., Kaur, G., Thakur, A. et al. (2017a). Waste derivitized blue luminescent carbon quantum dots for selenite sensing in water. *Talanta* 170: 49–55.

Devi, P., Thakur, A., Chopra, S. et al. (2017b). Ultrasensitive and selective sensing of selenium using nitrogen rich ligand interfaced carbon quantum dots. *ACS Applied Materials & Interfaces* 9: 25847–25856.

Dhillon, K.S. and Dhillon, S.K. (2009). Selenium concentrations of common weeds and agricultural crops grown in the seleniferous soils of northwestern India. *Science of the Total Environment* 407: 6150–6156.

Du, F., Zeng, F., Ming, Y., and Wu, S. (2013). Carbon dots-based fluorescent probes for sensitive and selective detection of iodide. *Microchimica Acta* 180: 453–460.

Elrashidi, M., Adriano, D., and Lindsay, W. (1989). Solubility, speciation, and transformations of selenium in soils. *Selenium in Agriculture and the Environment* 180: 51–63.

Ensafi, A.A. and Lemraski, M.S. (2004). Highly sensitive spectrophotometric reaction rate method for the determination of selenium based on the catalytic reduction of sulfonazo by sulfide. *Analytical Letters* 37: 2469–2483.

Ensafi, A.A., Mansour, H.R., and Soltanie, M.K. (1997). Highly sensitive spectrophotometric determination of ultra trace amounts of selenium. *Analytical Letters* 30: 973–984.

Fairweather-Tait, S.J., Bao, Y., Broadley, M.R. et al. (2010). Selenium in human health and disease. *Antioxidants & Redox Signaling* 14: 1337–1383.

Fairweather-Tait, S.J., Bao, Y., Broadley, M.R. et al. (2011). Selenium in human health and disease. *Antioxidants & Redox Signaling* 14: 1337–1383.

García-Hernández, J., King, K.A., Velasco, A.L. et al. (2001). Selenium, selected inorganic elements, and organochlorine pesticides in bottom material and biota from the Colorado River delta. *Journal of Arid Environments* 49: 65–89.

Ghasemi, E., Najafi, N.M., Raofie, F., and Ghassempour, A. (2010). Simultaneous speciation and preconcentration of ultra traces of inorganic tellurium and selenium in environmental samples by hollow fiber liquid phase microextraction prior to electrothermal atomic absorption spectroscopy determination. *Journal of Hazardous Materials* 181: 491–496.

Gissel-Nielsen, G. and Bisbjerg, B. (1970). The uptake of applied selenium by agricultural plants. *Plant and Soil* 32: 382–396.

Hach Company (2020). *Chemistries, Reagents and Standards* [Online]. United States: Hach Company. Available: https://www.hach.com/chemistries-reagents-and-standards/selenium/family-chemistry.result?pageSize=20&sortBy=sequence&productType=reagents&parameterId=7639975252&productCategory=22219205147&pimContext=USen&focusResults=true [Accessed 2020].

Hawkes, W.C. (1986). Spectrophotometric determinations of traces of selenium by catalytic reduction of tetranitro blue tetrazolium. *Analytica Chimica Acta* 183: 197–205.

Huang, C.B., Wu, C.L., Lai, J.P. et al. (2008). CdS quantum dots as fluorescence probes for the detection of selenite. *Analytical Letters* 41: 2117–2132.

Huang, K., Xu, K., Zhu, W. et al. (2015). Hydride generation for headspace solid-phase extraction with CdTe quantum dots immobilized on paper for sensitive visual detection of selenium. *Analytical Chemistry* 88: 789–795.

Huang, K., Xu, K., Zhu, W. et al. (2016). Hydride generation for headspace solid-phase extraction with CdTe quantum dots immobilized on paper for sensitive visual detection of selenium. *Analytical Chemistry* 88: 789–795.

Kartal, S., Oymak, T., and Tokalioglu, S. (2010). Spectrophotometric determination of selenium(IV) with 4-methyl-o-phenylenediamine based on piazselenol formation. *Journal of Analytical Chemistry* 65: 1221–1227.

Krishnaiah, L., Kumar, K.S., Suvardhan, K., and Chiranjeevi, P. (2003). Simple spectrophotometric determination of traces of selenium in environmental samples. Proceedings of the Third International Conference on Environment and Health, Chennai, India. 15–17 December.

Kuchekar, S.R., Naval, R.M., and Han, S.H. (2015a). Selective determination of selenium (IV) from environmental samples by UV-visible spectrophotometry using O-methoxyphenyl thiourea as a chelating ligand. *International Journal of Environmental Analytical Chemistry* 95: 618–634.

Kuchekar, S.R., Naval, R.M., and Han, S.H. (2015b). Selective determination of selenium (IV) from environmental samples by UV-visible spectrophotometry using O-methoxyphenyl thiourea as a chelating ligand. *International Journal of Environmental Analytical Chemistry*: 1–17.

Lemly, A.D. (2004). Aquatic selenium pollution is a global environmental safety issue. *Ecotoxicology and Environmental Safety* 59: 44–56.

Li, H., Luo, Y., Li, Z. et al. (2012). Nanosemiconductor-based photocatalytic vapor generation systems for subsequent selenium determination and speciation with atomic fluorescence spectrometry and inductively coupled plasma mass spectrometry. *Analytical Chemistry* 84: 2974–2981.

Li, G., Fu, H., Chen, X. et al. (2016). Facile and sensitive fluorescence sensing of alkaline phosphatase activity with Photoluminescent carbon dots based on inner filter effect. *Analytical Chemistry* 88: 2720–2726.

Liang, S., Chen, J., Pierce, D.T., and Zhao, J.X. (2013). A turn-on fluorescent nanoprobe for selective determination of selenium (IV). *ACS Applied Materials & Interfaces* 5: 5165–5173.

Liu, S.P., Zhou, G.M., and Huang, Z.G. (1990). A highly sensitive color-reaction for Se(IV) with the iodide Rhodamine B-PVA system – spectrophotometric determination of trace amounts of selenium in water. *Talanta* 37: 749–752.

Liu, S.P., Liu, Z.F., and Huang, C.Z. (1998). Resonance Rayleigh scattering for an indirect determination of trace amounts of selenium (IV) with iodide-basic Triphenylmethane dye systems. *Analytical Sciences* 14: 799–802.

López de Arroyabe Loyo, R., Nikitenko, S.I., Scheinost, A.C., and Simonoff, M. (2008). Immobilization of selenite on Fe3O4 and Fe/Fe3C ultrasmall particles. *Environmental Science & Technology* 42: 2451–2456.

Manish, R., Ramachandran, K., and Gupta, V. (1994). Extraction spectrophotometric determination of selenium (IV) with J acid in environmental samples. *Talanta* 41: 1623–1626.

Marczenko, Z. (1986). Separation and Spectrophotometric Determination of Elements. New York, NY: Wiley.

Matamoros, A. and Benning, L. (2008). Spectrophotometric determination of low-level concentrations of se in aqueous solutions. *Mineralogical Magazine* 72: 451–454.

Mathew, M. and Narayana, B. (2006). An easy spectrophotometric determination of selenium using azure B as a chromogenic reagent. *Indian Journal of Chemical Technology* 13: 455–458.

Melwanki, M.B. and Seetharamappa, J. (2000). Spectrophotometric determination of selenium (IV) using methdilazine hydrochloride. *Turkish Journal of Chemistry* 24: 287–290.

Monsen, E.R. (2000). Dietary reference intakes for the antioxidant nutrients: vitamin C, vitamin E, selenium, and carotenoids. *Journal of the American Dietetic Association* 100: 637–640.

Najafi, N.M., Tavakoli, H., Abdollahzadeh, Y., and Alizadeh, R. (2012). Comparison of ultrasound-assisted emulsification and dispersive liquid–liquid microextraction methods for the speciation of inorganic selenium in environmental water samples using low density extraction solvents. *Analytica Chimica Acta* 714: 82–88.

Nakano, S., Yoshii, M., and Kawashima, T. (2004). Flow-injection simultaneous determination of selenium (IV) and selenium (IV+ VI) using photooxidative coupling of p-hydrazinobensenesulfonic acid with N-(1-naphthyl) ethylenediamine. *Talanta* 64: 1266–1272.

Nayanova, E.V., Sergeev, G.M., and Elipasheva, E.V. (2016). Selective photometric determination of low conentrations of selenium(IV) and selenium(VI) in bottled drinking water. *Journal of Analytical Chemistry* 71: 379–385.

Niedzielski, P. (2005). The new concept of hyphenated analytical system: simultaneous determination of inorganic arsenic (III), arsenic (V), selenium (IV) and selenium (VI) by high performance liquid chromatography–hydride generation–(fast sequential) atomic absorption spectrometry during single analysis. *Analytica Chimica Acta* 551: 199–206.

Olivas, R.M., Donard, O., Camara, C., and Quevauviller, P. (1994). Analytical techniques applied to the speciation of selenium in environmental matrices. *Analytica Chimica Acta* 286: 357–370.

Pathare, M. and Sawant, A. (1995). Extractive spectrophotometric determination of selenium (IV) using sodium salt of hexamethyleneiminecarbodithioate. *Analytical Letters* 28: 317–334.

Ramachandran, K. and Kumar, G. (1996). Modified spectrophotometric method for the determination of selenium in environmental and mineral mixtures using 2, 3-diaminonaphthalene. *Talanta* 43: 1711–1714.

Ramachandran, K., Kaveeshwar, R., and Gupta, V. (1993). Spectrophotometric determination of selenium with 6-amino-1-naphthol-3-sulphonic acid (J-acid) and its application in trace analysis. *Talanta* 40: 781–784.

Reid, M.E., Stratton, M.S., Lillico, A.J. et al. (2004). A report of high-dose selenium supplementation: response and toxicities. *Journal of Trace Elements in Medicine and Biology* 18: 69–74.

Risher, J. (2003). *Toxicological profile for selenium*, Agency for Toxic Substances and Disease Registry.

Rodriguez, E.M.R., Alaejos, M.S., and Romero, C.D. (1996). Enhancement of the fluorescence intensity of se-2,3-diaminonaphthalene complex in aqueous solution by adding organic solvents. *Analytica Chimica Acta* 334: 161–166.

Sahu, S., Behera, B., Maiti, T.K., and Mohapatra, S. (2012). Simple one-step synthesis of highly luminescent carbon dots from orange juice: application as excellent bio-imaging agents. *Chemical Communications* 48: 8835–8837.

Santos, S., Ungureanu, G., Boaventura, R., and Botelho, C. (2015). Selenium contaminated waters: an overview of analytical methods, treatment options and recent advances in sorption methods. *Science of The Total Environment* 521–522: 246–260.

Sanz Alaejos, M. and Diaz Romero, C. (1995). Analysis of selenium in body fluids: a review. *Chemical Reviews* 95: 227–257.

Shen, J., Zhu, Y., Yang, X. et al. (2012). One-pot hydrothermal synthesis of graphene quantum dots surface-passivated by polyethylene glycol and their photoelectric conversion under near-infrared light. *New Journal of Chemistry* 36: 97–101.

Shenkin, A. (2009). Selenium in intravenous nutrition. *Gastroenterology* 137: S61–S69.

Sun, Y.-P., Zhou, B., Lin, Y. et al. (2006). Quantum-sized carbon dots for bright and colorful photoluminescence. *Journal of the American Chemical Society* 128: 7756–7757.

Tintometer GmbH. (2020). *Selenium (363)* [Online]. Tintometer GmbH. Available: https://www.lovibond.com/en/PW/Water-Testing/Parameter/Selenium/Selenium-363 [Accessed 2020].

Vriens, B., Lenz, M., Charlet, L. et al. (2014). Natural wetland emissions of methylated trace elements. *Nature Communications* 5: 3035.

Wang, D.C., Alfthan, G., and Aro, A. (1994). Determination of total selenium and dissolved selenium species in natural-waters by fluorometry. *Environmental Science & Technology* 28: 383–387.

Wang, Q., Zhang, S., Zhong, Y. et al. (2017). Preparation of yellow-green-emissive carbon dots and their application in constructing a fluorescent turn-on nanoprobe for imaging of selenol in living cells. *Analytical Chemistry* 89: 1734–1741.

Warkocki, W., EL-Safty, S.A., Shenashen, M.A. et al. (2015). Photo-induced recovery, optical detection, and separation of noxious SeO 3 2− using a mesoporous nanotube hybrid membrane. *Journal of Materials Chemistry A* 3: 17578–17589.

Winkel, L.H.E., Johnson, C.A., Lenz, M. et al. (2012). Environmental selenium research: from microscopic processes to global understanding. *Environmental Science & Technology* 46: 571–579.

World Health Organization (2004). Guidelines for drinking-water quality: recommendations. World Health Organization.

Yadav, S.K., Singh, I., Singh, D., and Han, S.-D. (2005). Selenium status in soils of northern districts of India. *Journal of Environmental Management* 75: 129–132.

Zhang, Q., Zhang, C., Li, Z. et al. (2015). Nitrogen-doped carbon dots as fluorescent probe for detection of curcumin based on the inner filter effect. *RSC Advances* 5: 95054–95060.

9

Biosensors for the Detection of Selenium in Environment

Paramdeep Kaur[#], Deepanshu Monga[#], and Baljinder Singh

Department of Biotechnology, Panjab University, Chandigarh, India
☆ Deepanshu Monga and Paramdeep Kaur made equal contributions

9.1 Introduction

Selenium is an essential micronutrient for animals and humans but if present at high concentrations it also acts as an environmental pollutant. The gap between the essentiality and toxicity of selenium is very narrow. Therefore, the levels of selenium in the environment and their exposure to living organisms are gaining more attention nowadays. Selenium is identified all over the world in almost all environmental matrices including air, soil, water plants, and foods (Bailey 2017; Tan et al. 2018). Sedimentary and volcanic rocks, metallic sulfides, and fossil fuel deposits act as naturally occurring rich sources of selenium (Tabelin et al. 2018). Therefore, natural activities such as volcanic eruptions and weathering of rocks release large amounts of selenium into the surrounding environment (El-Ramady et al. 2015). On the other hand, anthropogenic activities such as metal and coal mining (Khamkhash et al. 2017; Lemly 2018), use of fossil fuels (Drout et al. 2018), and various agricultural and industrial activities (Staicu et al. 2017) are also major sources of selenium pollution. Selenium is a vital nutrient and uptake of adequate amount of selenium is required for several crucial functions in the body. It has antioxidant, anti-carcinogenic activities and is important for hormonal regulation (Preda et al. 2016). However, high exposure to selenium can lead to various respiratory, gastrointestinal, cardiovascular, and mental complications and an increased risk of some serious health conditions such as alopecia, dermatitis, non-melanoma skin cancer, prostate cancer, type-2 diabetes, and hypertension (Rayman 2017). High concentrations of selenium have also shown potential adverse effects on the physiology and development of plants (Kolbert et al. 2016; Silva et al. 2018). Selenium exists in environment in different inorganic and organic forms that vary significantly in their toxicity potential depending upon the chemical form (Mehdi et al. 2013). In its elemental form selenium is considered biologically inert due to low solubility in aqueous media. An inorganic compounds such as selenite (SeO^{2-}) and selenate (SeO_4^{2-}) are the most toxic due to their high solubility, whereas

Selenium Contamination in Water, First Edition. Edited by Pooja Devi, Pardeep Singh, Arindam Malakar, and Daniel Snow.

selenides are comparatively less toxic (Garousi 2015). The toxicity of organic compounds is also relatively low compared to inorganic compounds (Yücel et al. 2020). Toxic effects of these compounds have been reported at levels of intake as low as around 16 μg/day for inorganic selenium and around 260 μg/day for organic selenium (Vinceti et al. 2017). The USEPA has set the maximum contaminant level of 0.05 mg/l for selenium in drinking water. Selenium almost entirely exists in inorganic forms in water (He et al. 2018). Also, these selenium species are quite frequent in soil. The contamination of environmental resources with selenium is a serious environmental as well as health concern. Thus, timely monitoring and regulation of selenium contamination is important to achieve the proposed safety limits (Saidur et al. 2017).

Highly sensitive and reliable methods of detection are required for accurate measurements of selenium (Bora et al. 2013). Various instrumental techniques such as mass spectrophotometry (Afton et al. 2008), ultraviolet–vis (UV) spectroscopy (Ulusoy 2015), gas chromatography (Bidari et al. 2008), fluorescence spectrometry (Fernandez-Perez et al. 1999), inductively coupled plasma mass spectrometry (ICP-MS) (Martinez-Bravo et al. 2001), or atomic absorption spectrometry/emission spectroscopy (Maleki et al. 2005) have been used for selenium detection. These methods are highly precise and specific but they have various drawbacks associated such as high cost and complexity, and long processing times (Mehta et al. 2016; Yu et al. 2017). Also, these methods are laboratory-bound due to need for large equipment and resources. On the other hand, biosensors are exceptionally precise and selective as well as portable, inexpensive, fast processing, and easy to use. Therefore, biosensors appear to be suitable analytical tools for detection of environmental pollutants such at selenium (Baeumner 2003).

9.2 Biosensors and Their Types

Biosensors are devices which are capable of converting biological response into electrical signal. These devices are analytical in nature. "Biosensor" term was coined by Cammann (Cammann 1977). According to the IUPAC (International Union of Pure and Applied Chemistry) definition, biosensor is "a self-contained integrated device that is capable of providing specific quantitative or semi-quantitative analytical information using a biological recognition element (biochemical receptor) which is in direct spatial contact with a transduction element" (Thévenot et al. 2001). A biosensor has three main elements, which are:

1) Recognition element
2) Signal transducing structure
3) Amplification/processing element

9.2.1 Working

The recognition element is biological in nature and it is immobilized on the transducer by covalent binding, non-covalent binding, or membrane entrapment. Environmental contaminant acts as an analyte that binds and interacts with the recognition element. It results in the production of a signal which is taken up by the transducer. The signal may arise due to change in proton concentration, release or uptake of electrons, release or uptake of gases, emission of heat, change in mass, release or uptake of gases, and so on. This signal is converted by the transducer element to an appropriate measurable parameter such as current, voltage, fluorescence, luminescence etc. Finally, the electronic circuit amplifies and processes the signal into digital values (Evtugyn et al. 1998; Karube and Nomura 2000).

9.2.2 Types of Biosensors

The biosensors can be classified into different types on the basis of the type of the biological component or the transducer used in the system. On the basis of recognition or the biological component, biosensors can be categorized into two types. One category of biosensors includes those in which there occurs some catalytic process which results in the primary sensing event (it includes enzymes or whole cells as the recognition elements). The other category of biosensors is generally affinity-based. This includes affinity sensors based on antibodies or nucleic acids as the main recognition elements.

Based upon the transducer used the biosensors can be of different types such as amperometric (based on measuring current at constant potential), potentiometric (based on measuring potential at constant current), piezoelectric (detects changes in mass), thermal (detects changes in temperature), or optical (detect changes in light transmission). An effective biosensor can be prepared by the combination of any appropriate receptor with any relative transducer.

Biosensors based on enzymes, whole cells, and DNA probes coupled with a suitable transducer have been developed for selenium detection in water (Table 9.1).

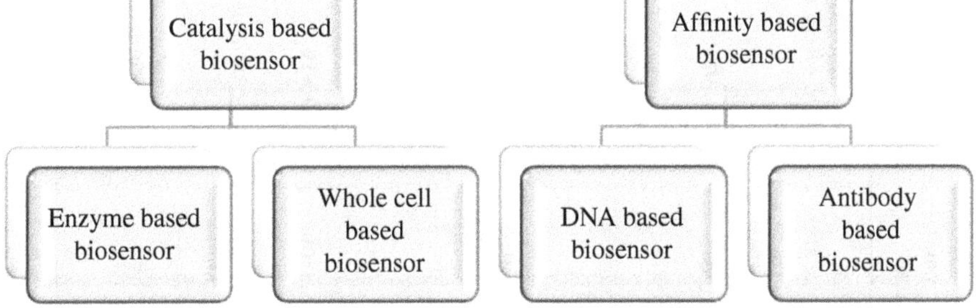

Figure 9.1 Classification of biosensors on the basis of recognition/biological element.

Table 9.1 Different biosensors for selenium detection and their properties.

Selenium compounds detected	Biological component	Detector/detection method	Detection sensitivity/ limit of detection (LOD)	Response time	Application	References
Sodium selenite and selenium dioxide	*Thevetia peruviana* seed carboxyl esterase	Paper chromatography	5 µg/µl	–	Detection of selenium compounds in soil and water samples	Saritha and Kumar (2001)
Selenite	*E. coli*-roGFP2	Spectrophotometric analysis	3.1 ppm	80 to 210 seconds	Detection of various toxicants inducing cellular oxidation	Arias-Barreiro et al. (2010)
Selenite	*Escherichia coli* cells expressing roGFP2 protein	Spectrofluorophotometer	5.8×10^{-6} mg/l	0–9 min	Detection of metalloids inducing oxidative stress	Ooi et al. (2015)
Selenium (Se^{4+})	*Vibrio fisheri* strain DSM7744	Luminometer	1 ppb	2 min	Detection of selenium and other biocides in water	Attar and Afshar (2010)
Selenium	*Acinetobacter baylyi* Tox2	Luminescence dete ctor	–	15–30 min	Toxicity assessment of heavy metal contaminated water	Cui et al. (2018), Rampley et al. (2020)
Selenate and selenite	*Pseudomonas aeruginosa* strain JS-11	Colorimetric determination	2 µM	2–72 hours	Detect and convert selenium oxyanions to Se^0 nanospheres	Dwivedi et al. (2013)
Selenium and sodium selenite	Sulfur-oxidizing bacteria	EC meter	$EC_{50} = 0.32$ mg/l	2 hours	Detection of heavy metals and related toxicity in water	Eom et al. (2019)
Selenite	Fluorescently labeled DNA attached to metal oxides	Fluorescence plate reader	0.6 µM	1 hour	Detection of selenium oxyanions	Yu et al. (2017)

9.3 Biosensors for Selenium Detection

9.3.1 Enzyme-Based Biosensors

High sensitivity and specificity of enzymes make them a preferred biomolecule for development of biosensors (Verma and Singh 2005). The response of enzyme-based biosensors is evaluated by either measuring the co-substrate consumption or the yield of the product and this method of analysis is referred to as direct monitoring of analytes. But an application of these biosensors for environmental monitoring remains limited as very few environmental pollutants act as the substrate for the enzymes, whereas many pollutants and heavy metals act as enzyme inhibitors. Indirect monitoring of analytes using enzyme inhibition-based biosensors is a great success as the enzymes are mostly highly susceptible to very low concentration of inhibitors, owing for their better sensitivity (Turdean 2011; Bachan Upadhyay and Verma 2013). A simple and sensitive enzyme-based biosensor for detection of selenium compounds in soil and water samples has been prepared using the carboxylesterase enzyme from the seeds of *Thevetia peruviana* coupled with paper chromatography for signal detection. This enzyme is sensitive to various soluble selenium compounds such as sodium selenite and selenium dioxide. Various other carboxylesterase enzymes derived from different animal and plant sources were tested for this purpose but only *T. peruviana* carboxylesterase exhibited sensitivity for selenium compounds. The carboxylesterase enzyme catalyzes the substrate 1-naphthyl acetate to 1-naphthol and produces a magenta color in the presence of diazonium salt fast blue B, whereas the inhibition of the enzyme by selenium compounds leads to development of white inhibition spots on the chromatogram. The limit of detection for both sodium selenite and selenium dioxide was determined to be 5 μg/μl (Saritha and Kumar 2001).

Hinderance in reusability is the main limitation of inhibition-based biosensors as the enzymes' structure gets distorted and subsequently there is loss of activity. High purification cost, sensitivity to change in physical parameters such as temperature and pH, and activity decrease due to immobilization are the other limitations of enzyme-based biosensors.

9.3.2 Whole Cell-based Biosensors

Living cells produce a variety of molecules (receptors) which can be applied for detection of diverse range of analytes. Use of whole cells as biosensors rather than biologically derived molecules or enzymes is better as the native cell environment provides optimum conditions for the activity of the receptor molecules and increases the stability of the biosensors. Moreover, it eliminates the need for extraction and purification of the receptor molecules making the process easier and less time consuming (Gupta et al. 2019; Saini et al. 2019). But the culture of these cells requires selective media which may contribute to its higher cost of production. The other components present in the whole cells may interfere with the proper detection of the target analyte. Therefore, selection of appropriate cells and their optimization is an important step in designing DNA-based biosensors. The cells can also be modified desirably using genetic engineering methods to facilitate the activity of the biosensor (Gui et al. 2017). Whole cell-based biosensors often use the cellular response to estimate toxic effects of the pollutant for its detection. A bacterial biosensor has been developed using *Escherichia coli* cells transformed with roGFP gene, that can detect SeO_3^{2-} by their effect of inducing cellular oxidation stress at concentrations as low as of 3.1 ppm within a response time of 80–210 seconds (Arias-Barreiro et al. 2010). Oxidative stress can be induced in a cell due to various factors or environmental stressors (such as selenite) that cause an increase in the reactive oxygen species (ROS) in the cell and thus an imbalance in the cellular redox status (Betteridge 2000; Birben et al. 2012). Green fluorescent protein (GFP) from *Aequorea Victoria* can be converted to redox-sensitive GFP (roGFP) by substitution of surface-exposed residues with cysteines at suitable positions. The intracellular redox changes in the cell induce formation of disulfide bond between two cysteine residues present on the surface of the roGFP2 at the positions 147 and 204 (Meyer and Dick 2010; Müller et al. 2017). The 4-hydroxybenzylidene-imidazolinone chromophore or Y66 chromophore, is the chromophore in roGFP2, that emits fluorescence. Formation of disulfide bond induces changes in the protonation status of the chromophore leading to excitation peak shift from 400 to 490 nm thus allowing for a ratiometric quantification (Dooley et al. 2004; Hanson et al. 2004). On the basis of the same principle another biosensor for selenium has been reported using roGFP2 expressing *E. coli* cells fabricated in an immobilizing matrix called *K*-carrageenan. The use of immobilizing matrix provides the system with greater stability and sensitivity compared to a suspension biosensor. The biosensor has been reported to be stable for 46 days which is far better than previously reported suspension biosensors with lifespan of around 10 hours. The LOD (Limit of Detection) observed for selenite was 5.8×10^{-6} mg/l and response time was between 0 and 9 minutes. Using a multiplate reader, the response time can be as short as 10 seconds (Ooi et al. 2015).

Bioluminescent bacterium *Vibrio fischeri* strain DSM 7744 coupled with a luminometer has also been used for detection of selenium as well as other heavy metals, pesticides, and herbicides in water. The whole cell optical biosensor is highly sensitive, with LOD of 1 ppb (parts per billion) for selenium. It is based on the toxic effect of the pollutant on the bacterial cells such that an increase in the concentration of the pollutant decreases the intensity of bioluminescence (Attar and Afshar 2010). *V. fischeri* as well as other bioluminescent bacteria like *Photobacterium phosphoreum* and *Vibrio harvey* have been widely used for various detection purposes (Nunes-Halldorson and Duran 2003). The reaction responsible for the blue green light emission by these bacteria is as follows:

$$\mathrm{FMNH_2 + O_2 + R-CHO \rightarrow FMN + H_2O + R-COOH + Light\left(\sim 495\,nm\right)}$$

The reaction catalyzed by enzyme luciferase oxidizes coenzyme ($FMNH_2$) and emits light along with the production of FMN and the carboxylic acid. The source of bioluminescence remains constant as the oxidized flavin (FMN) and the carboxylic acid are continuously recycled in corresponding reactions (Erzinger et al. 2018).

Another toxicity biosensor based on detection of bioluminescence has been developed by transformation of *Acinetobacter baylyi* ADP1 by plasmid pWH1274_lux to form *A. baylyi* Tox2 to detect the cytotoxicity of various heavy metals. The plasmid consisted of a luxCDABE gene cassette expressed constitutively under the control of Ptet promoter. The bioluminescence intensity of the biosensor was seen to be directly affected by increase in the concentrations of toxic compounds. This biosensor is also suitable for use in seawater as

A. baylyi Tox2 is tolerant to salinity (Cui et al. 2018). This technology has already been employed in field studies, for water quality assessment and detection of high concentrations of selenium as well as other toxic metals including aluminum, cadmium, chromium, iron, zinc, lithium, and nickel (Rampley et al. 2020).

Pseudomonas aeruginosa strain JS-11 is tolerant to selenite up to the concentration of 100 µM with an EC_{50} value of 140 µM. The bacteria release a metabolite phenazine-1-carboxylic acid (PCA) and redox agents like NADH and NADH-dependent reductases into the culture supernatant, which cause the conversion of soluble and colorless selenite into insoluble red elemental selenium (Se^0) nanoparticles. These nanoparticles can be easily detected and quantified using colorimetric methods of detection. This study is actually focused on the formation of selenium nanoparticles and their application in biosensors for detection of other heavy metals (Dwivedi et al. 2013). Due to their nontoxic nature selenium nanoparticles have been widely used in various therapeutic or biosensing technologies (Wang et al. 2010; Maiyo and Singh 2017). More than 16 species of bacteria and Archaea can reduce colorless selenite/selenite to red elemental selenium. Plants and fungi have also shown their potential in the formation of selenium nanoparticles (Li et al. 2007; Vetchinkina et al. 2013).

Recently a study presented another detection method for the toxicity induced by selenium and some other metals like mercury, hexavalent chromium, and arsenic to sulfur-oxidizing bacteria (SOB). The inhibition of SOB in a fed batch bioreactor due to the effect of contaminants was evaluated by measuring electrical conductivity. The 2 hours half maximum effective concentration (EC_{50}) of SOB for selenium was reported to be 0.33 mg/l (Eom et al. 2019). The detection time as well as the sensitivity for selenium was observed to be better compared to any previous studies (Ibrahim and Spacie 1990; Hsieh et al. 2004). SOB such as *Acidithiobacillus*, *Thiospirillopsis*, and *Thiovulum* are chemoautotrophic bacteria that obtain energy by the oxidation of inorganic sulfur (S°) following this reaction:

$$S + H_2O + 1.5O_2 \rightarrow SO_4^{2-} + 2H^+, \Delta G^{o\prime} = -587.1\,\text{kJ / reaction}.$$

Formation of sulfate (SO_4^{2-}) increases the EC (electrical conductivity) of the medium but in the presence of a toxicant the activity of SOB is reduced, which is reflected by a decrease in the EC value which can be easily measured using a EC meter (Hassan et al. 2010; Oh et al. 2011). The sensitivity of SOB to various environmental pollutants has resulted in their extensive use in biosensor technology (Yang et al. 2017; Hassan et al. 2019).

9.3.3 DNA-based Biosensors

These are the most emerging type of biosensor and are of particular interest in the case of detection of heavy metals. They are highly specific and sensitive in nature. They also have higher shelf life and stability in extreme conditions than enzymes or antibodies (Lan et al. 2010; Barthelmebs et al. 2011). They are less toxic and more biodegradable in nature. The DNA-based biosensors can be modified easily by the use of fluorophores or quenchers to generate strong fluorescent signals. DNA being negatively charged has been widely used for detection of metal cations whereas charge repulsion between DNA and anions limits the direct application of DNA for detection of anions (Bala and Górski 2016; Zhang et al. 2018). Therefore, a different strategy using metal oxides has been employed in designing DNA-based anion detectors. There is only single report of a DNA-based biosensor specific for selenium detection. This biosensor constitutes of fluorescently-labeled DNA molecules adsorbed to metal oxide nanoparticles, resulting in quenched fluorescence. In the presence of anions or molecules with higher affinity for the metal oxides, the DNA is displaced that leads to recovery of the fluorescence signal. This arrangement is highly efficient for selenite detection as it shows high affinity for various metal oxides. Initially 10 different metal oxides (Mn_2O_3, NiO, Cr_2O_3, TiO_2, CeO_2, $CoOFe_2O_3$, Fe_3O_4, In_2O_3, and ZnO) were tested for adsorption of both selenite and selenate adsorption. Selenite showed high absorbance to six out of the 10 metal oxides, with Fe_2O_3 giving the most efficient results in optimum conditions. However, selenate failed to give a satisfactory response in any case due to its lower affinity. In addition, these oxides might also be helpful for removal of selenium from the environment. The biosensor can detect selenite concentrations as low as $0.6\,\mu M$. This biosensor is also sensitive to phosphate and arsenate but they can be distinguished from selenite using the sensor array strategy (Yu et al. 2017). The differences in the affinity of metal oxides for the different analytes helps in pattern-recognition-based sensor array for discrimination of the anions (Liu and Liu 2015).

9.4 Conclusion

Biosensing technology is still evolving and various new biosensors for detection of environmental pollutants are coming up. The use of biosensors for selenium detection has not yet been widely accepted but some reports have shown the practicality of this technology to be efficiently used for the purpose (Rampley et al. 2020). The biosensor systems proposed for selenium detection so far have proved to be highly sensitive and can match detection limits comparative to the safety standards. Portability of biosensors makes them preferred choice of method for on-site examination of an environmental pollutant. Although there are some drawbacks associated with some biosensors such as low shelf life or possible interference of results due to various factors, optimization of biosensors is achievable by further considerations (Liu and Liu 2015; Ooi et al. 2015). Thus, biosensors exhibit great scope and applicability in the future for environmental studies.

References

Afton, S., Kubachka, K., Catron, B., and Caruso, J.A. (2008). Simultaneous characterization of selenium and arsenic analytes via ion-pairing reversed phase chromatography with inductively coupled plasma and electrospray ionization ion trap mass spectrometry for detection: applications to river water, plant extract and urine matrices. *Journal of Chromatography A* 1208: 156–163.

Arias-Barreiro, C.R., Okazaki, K., Koutsaftis, A. et al. (2010). A bacterial biosensor for oxidative stress using the constitutively expressed redox-sensitive protein roGFP2. *Sensors (Basel, Switzerland)* 10: 6290–6306.

Attar, H. and Afshar, S. (2010). Design of sensible biosensor for rapid detection of biocides in potable water. *Asian Journal of Biotechnology* 2: 120–126.

Bachan Upadhyay, L.S. and Verma, N. (2013). Enzyme inhibition based biosensors: a review. *Analytical Letters* 46: 225–241.

Baeumner, A.J. (2003). Biosensors for environmental pollutants and food contaminants. *Analytical and Bioanalytical Chemistry* 377: 434–445.

Bailey, R.T. (2017). Selenium contamination, fate, and reactive transport in groundwater in relation to human health. *Hydrogeology Journal* 25: 1191–1217.

Bala, A. and Górski, Ł. (2016). Determination of mercury cation using electrode modified with phosphorothioate oligonucleotide. *Sensors and Actuators B Chemical* 230: 731–735.

Barthelmebs, L., Hayat, A., Limiadi, A.W. et al. (2011). Electrochemical DNA aptamer-based biosensor for OTA detection, using superparamagnetic nanoparticles. *Sensors and Actuators B: Chemical* 156: 932–937.

Betteridge, D.J. (2000). What is oxidative stress? *Metabolism: Clinical and Experimental* 49: 3–8.

Bidari, A., Hemmatkhah, P., Jafarvand, S. et al. (2008). Selenium analysis in water samples by dispersive liquid-liquid microextraction based on piazselenol formation and GC–ECD. *Microchimica Acta* 163: 243–249.

Birben, E., Sahiner, U.M., Sackesen, C. et al. (2012). Oxidative stress and antioxidant defense. *The World Allergy Organization Journal* 5: 9–19.

Bora, U., Sett, A., and Singh, D. (2013). Nucleic acid based biosensors for clinical applications. *Biosensors Journal* 2: 1–8.

Cammann, K. (1977). Bio-sensors based on ion-selective electrodes. *Fresenius' Zeitschrift für analytische Chemie* 287: 1–9.

Cui, Z., Luan, X., Jiang, H. et al. (2018). Application of a bacterial whole cell biosensor for the rapid detection of cytotoxicity in heavy metal contaminated seawater. *Chemosphere* 200: 322–329.

Dooley, C.T., Dore, T.M., Hanson, G.T. et al. (2004). Imaging dynamic redox changes in mammalian cells with green fluorescent protein indicators. *The Journal of Biological Chemistry* 279: 22284–22293.

Drout, R.J., Howarth, A.J., Otake, K.-I. et al. (2018). Efficient extraction of inorganic selenium from water by a Zr metal–organic framework: investigation of volumetric uptake capacity and binding motifs. *CrystEngComm* 20: 6140–6145.

Dwivedi, S., Alkhedhairy, A.A., Ahamed, M., and Musarrat, J. (2013). Biomimetic synthesis of selenium nanospheres by bacterial strain JS-11 and its role as a biosensor for nanotoxicity assessment: a novel Se-bioassay. *PLoS One* 8: e57404.

El-Ramady, H.R., Domokos-Szabolcsy, É., Shalaby, T.A. et al. (2015). Selenium in agriculture: water, air, soil, plants, food, animals and nanoselenium. In: *CO2 Sequestration, Biofuels and Depollution* (eds. E. Lichtfouse, J. Schwartzbauer and D. Robert). Springer.

Eom, H., Hwang, J.-H., Hassan, S.H. et al. (2019). Rapid detection of heavy metal-induced toxicity in water using a fed-batch sulfur-oxidizing bacteria (SOB) bioreactor. *Journal of Microbiological Methods* 161: 35–42.

Erzinger, G.S., Schmoeller, F., Pinto, L.H. et al. (2018). Bioluminescence systems in environmental biosensors. In: *Bioassays: Advanced Methods and Applications* (eds. D.-P. Häder and G. Erzinger), 241–262. Elsevier.

Evtugyn, G., Budnikov, H., and Nikolskaya, E. (1998). Sensitivity and selectivity of electrochemical enzyme sensors for inhibitor determination. *Talanta* 46: 465–484.

Fernandez-Perez, V., Jiménez-Carmona, M., and de Castro, M.L. (1999). Continuous extraction with acidified subcritical water of arsenic, selenium and mercury from coal prior to on-line derivatisation-atomic fluorescence detection. *Journal of Analytical Atomic Spectrometry* 14: 1761–1766.

Garousi, F. (2015). The toxicity of different selenium forms and compounds–review. *Acta Agraria Debreceniensis* 64: 33–38.

Gui, Q., Lawson, T., Shan, S. et al. (2017). The application of whole cell-based biosensors for use in environmental analysis and in medical diagnostics. *Sensors (Basel, Switzerland)* 17: 1623.

Gupta, N., Renugopalakrishnan, V., Liepmann, D. et al. (2019). Cell-based biosensors: recent trends, challenges and future perspectives. *Biosensors & Bioelectronics* 141: 111435.

Hanson, G.T., Aggeler, R., Oglesbee, D. et al. (2004). Investigating mitochondrial redox potential with redox-sensitive green fluorescent protein indicators. *The Journal of Biological Chemistry* 279: 13044–13053.

Hassan, S.H., van Ginkel, S.W., Kim, S.-M. et al. (2010). Isolation and characterization of Acidithiobacillus caldus from a sulfur-oxidizing bacterial biosensor and its role in detection of toxic chemicals. *Journal of Microbiological Methods* 82: 151–155.

Hassan, S.H., Gurung, A., Kang, W.-C. et al. (2019). Real-time monitoring of water quality of stream water using sulfur-oxidizing bacteria as bio-indicator. *Chemosphere* 223: 58–63.

He, Y., Xiang, Y., Zhou, Y. et al. (2018). Selenium contamination, consequences and remediation techniques in water and soils: a review. *Environmental Research* 164: 288–301.

Hsieh, C.-Y., Tsai, M.-H., Ryan, D.K., and Pancorbo, O.C. (2004). Toxicity of the 13 priority pollutant metals to Vibrio fisheri in the Microtox® chronic toxicity test. *The Science of the Total Environment* 320: 37–50.

Ibrahim, A. and Spacie, A. (1990). Toxicity of inorganic selenium to the green alga Selenastrum capricornutum Printz. *Environmental and Experimental Botany* 30: 265–269.

Karube, I. and Nomura, Y. (2000). Enzyme sensors for environmental analysis. *Journal of Molecular Catalysis B Enzymatic* 10: 177–181.

Khamkhash, A., Srivastava, V., Ghosh, T. et al. (2017). Mining-related selenium contamination in Alaska, and the state of current knowledge. *Minerals* 7: 46.

Kolbert, Z., Lehotai, N., Molnár, Á., and Feigl, G. (2016). "The roots" of selenium toxicity: a new concept. *Plant Signaling & Behavior* 11: e1241935.

Lan, T., Furuya, K., and Lu, Y. (2010). A highly selective lead sensor based on a classic lead DNAzyme. *Chemical Communications (Cambridge, England)* 46: 3896–3898.

Lemly, A.D. (2018). Selenium poisoning of fish by coal ash wastewater in Herrington Lake, Kentucky. *Ecotoxicology and Environmental Safety* 150: 49–53.

Li, S., Shen, Y., Xie, A. et al. (2007). Rapid, room-temperature synthesis of amorphous selenium/protein composites using Capsicum annuum L extract. *Nanotechnology* 18: 405101.

Liu, B. and Liu, J. (2015). Comprehensive screen of metal oxide nanoparticles for DNA adsorption, fluorescence quenching, and anion discrimination. *ACS Applied Materials & Interfaces* 7: 24833–24838.

Maiyo, F. and Singh, M. (2017). Selenium nanoparticles: potential in cancer gene and drug delivery. *Nanomedicine (London, England)* 12: 1075–1089.

Maleki, N., Safavi, A., and Doroodmand, M.M. (2005). Determination of selenium in water and soil by hydride generation atomic absorption spectrometry using solid reagents. *Talanta* 66: 858–862.

Martinez-Bravo, Y., Roig-Navarro, A., Lopez, F., and Hernandez, F. (2001). Multielemental determination of arsenic selenium and chromium (VI) species in water by HPLC ICP MS. *Journal of Chromatography. A* 926: 265–274.

Mehdi, Y., Hornick, J.-L., Istasse, L., and Dufrasne, I. (2013). Selenium in the environment, metabolism and involvement in body functions. *Molecules (Basel, Switzerland)* 18: 3292–3311.

Mehta, J., Bhardwaj, S.K., Bhardwaj, N. et al. (2016). Progress in the biosensing techniques for trace-level heavy metals. *Biotechnology Advances* 34: 47–60.

Meyer, A.J. and Dick, T.P. (2010). Fluorescent protein-based redox probes. *Antioxidants & Redox Signaling* 13: 621–650.

Müller, A., Schneider, J.F., Degrossoli, A. et al. (2017). Systematic in vitro; assessment of responses of roGFP2-based probes to physiologically relevant oxidant species. *Free Radical Biology & Medicine* 106: 329–338.

Nunes-Halldorson, V.D.S. and Duran, N.L. (2003). Bioluminescent bacteria: lux genes as environmental biosensors. *Brazilian Journal of Microbiology* 34: 91–96.

Oh, S.-E., Hassan, S.H., and van Ginkel, S.W. (2011). A novel biosensor for detecting toxicity in water using sulfur-oxidizing bacteria. *Sensors and Actuators B. Chemical* 154: 17–21.

Ooi, L., Heng, L.Y., and Mori, I.C. (2015). A high-throughput oxidative stress biosensor based on Escherichia coli roGFP2 cells immobilized in a k-carrageenan matrix. *Sensors (Basel, Switzerland)* 15: 2354–2368.

Preda, C., Vasiliu, I., Bredetean, O. et al. (2016). Selenium in the environment: essential or toxic to human health? *Environmental Engineering and Management Journal* 15: 913–921.

Rampley, C., Whitehead, P., Softley, L. et al. (2020). River toxicity assessment using molecular biosensors: heavy metal contamination in the Turag-Balu-Buriganga river systems, Dhaka Bangladesh. *The Science of the Total Environment* 703: 134760.

Rayman, M. (2017). Selenium intake and status in health & disease. *Free Radical Biology and Medicine* 112: 5.

Saidur, M., Aziz, A.A., and Basirun, W. (2017). Recent advances in DNA-based electrochemical biosensors for heavy metal ion detection: a review. *Biosensors & Bioelectronics* 90: 125–139.

Saini, R., Hegde, K., Brar, S.K., and Verma, M. (2019). Advances in whole cell-based biosensors in environmental monitoring. In: *Tools, Techniques and Protocols for Monitoring Environmental Contaminants* (eds. S.K. Brar, K. Hegde and V.L. Pachapur), 263–284. Elsevier.

Saritha, K. and Kumar, N.N. (2001). Qualitative detection of selenium in fortified soil and water samples by a paper chromatographic–carboxyl esterase enzyme inhibition technique. *Journal of Chromatography. A* 919: 223–228.

Silva, V.M., Boleta, E.H.M., Lanza, M.G.D.B. et al. (2018). Physiological, biochemical, and ultrastructural characterization of selenium toxicity in cowpea plants. *Environmental and Experimental Botany* 150: 172–182.

Staicu, L.C., Hullebusch, E.D.V., Rittmann, B.E., and Lens, P.N.L. (2017). Industrial selenium pollution: sources and biological treatment technologies. In: *Bioremediation of Selenium Contaminated Wastewater* (ed. E.D.V. Hullebusch), 75–101. Cham: Springer International Publishing.

Tabelin, C.B., Igarashi, T., Villacorte-Tabelin, M. et al. (2018). Arsenic, selenium, boron, lead, cadmium, copper, and zinc in naturally contaminated rocks: a review of their sources, modes of enrichment, mechanisms of release, and mitigation strategies. *The Science of the Total Environment* 645: 1522–1553.

Tan, L.C., Nancharaiah, Y.V., van Hullebusch, E.D., and Lens, P.N. (2018). Selenium: environmental significance, pollution, and biological treatment technologies. In: *Anaerobic Treatment of Mine Wastewater for the Removal of Selenate and its Co-Contaminants* (ed. L.C. Tan), 9–56. CRC Press.

Thévenot, D.R., Toth, K., Durst, R.A., and Wilson, G.S. (2001). Electrochemical biosensors: recommended definitions and classification. *Analytical Letters* 34: 635–659.

Turdean, G.L. (2011). Design and development of biosensors for the detection of heavy metal toxicity. *International Journal of Electrochemistry* 2011: 343125.

Ulusoy, H.İ. (2015). Simple and useful method for determination of inorganic selenium species in real samples based on UV-VIS spectroscopy in a micellar medium. *Analytical Methods* 7: 953–960.

Verma, N. and Singh, M. (2005). Biosensors for heavy metals. *Biometals: An International Journal on the Role of Metal Ions In Biology, Biochemistry, and Medicine* 18: 121–129.

Vetchinkina, E., Loshchinina, E., Kursky, V., and Nikitina, V. (2013). Reduction of organic and inorganic selenium compounds by the edible medicinal basidiomycete Lentinula edodes and the accumulation of elemental selenium nanoparticles in its mycelium. *Journal of Microbiology* 51: 829–835.

Vinceti, M., Filippini, T., Cilloni, S. et al. (2017). Health risk assessment of environmental selenium: emerging evidence and challenges. *Molecular Medicine Reports* 15: 3323–3335.

Wang, T., Yang, L., Zhang, B., and Liu, J. (2010). Extracellular biosynthesis and transformation of selenium nanoparticles and application in H_2O_2 biosensor. *Colloids and Surfaces. B, Biointerfaces* 80: 94–102.

Yang, S.-H., Cheng, K.-C., and Liao, V.H.-C. (2017). A novel approach for rapidly and cost-effectively assessing toxicity of toxic metals in acidic water using an acidophilic iron-oxidizing biosensor. *Chemosphere* 186: 446–452.

Yu, T., Liu, B., and Liu, J. (2017). Adsorption of selenite and selenate by metal oxides studied with fluorescent DNA probes for analytical application. *Journal of Analysis and Testing* 1: 2.

Yücel, U.M., Başbuğan, Y., Uyar, A. et al. (2020). Use of an antiarrhythmic drug against acute selenium toxicity. *Journal of Trace Elements in Medicine and Biology: Organ of the Society for Minerals and Trace Elements (GMS)* 59: 126471.

Zhang, J., Lin, Y., Peng, H. et al. (2018). Dual signal amplification electrochemical biosensor for Lead cation. *Electroanalysis* 30: 955–961.

10

Physical and Chemical Methods for Selenium Removal

S. Kaviya

Department of Applied Chemistry, Cochin University of Science and Technology (CUST), Kerala, India

10.1 Introduction

10.1.1 Selenium and Origin of Contamination

Selenium (Se) is in the chalcogen group and has characteristic physical and chemical properties in line with arsenic and sulfur, respectively. It can appear in organic and inorganic forms in the ecosystem. Inorganic Se compounds are present in ground and surface water such as selenium (Se^0), selenide (Se^{2-}), selenate (SeO_4^{2-}), and selenite (SeO_3^{2-}). Usually, organic Se exists in plants and land as trimethylselenonium cation, selenoamino acid, dimethylselenium sulfide, methylselenides, and methylselenol (Natasha et al. 2018) (Figure 10.1). Se exists naturally in soil and rock as well as in the form of sulfide mineral, burning of Se-rich coal, utilization of insecticide, agriculture, and metal and oil refining processes (Santos et al. 2015; Vesper et al. 2008). Even so, soil is a primary source of contamination and further it is transferred to water bodies.

10.1.2 Why Se Removal is Essential

Selenite and selenate are the most common forms in the environment at normal pH and redox circumstances (Winkel et al. 2012). Selenite likely to adsorb on iron/aluminum-containing minerals (soil minerals). In everyday life, 0.1–0.5 ppm of Se is considered a micronutrient whereas it becomes poisonous at >3 ppm. It is widely used as a fertilizer, antidandruff shampoo, electronic, fungicides, anti-biofilm, etc., (Ramya et al. 2015). The toxicity of Se depends on the chemical state, oxidation, and concentration. Selenite is more toxic when compared to other oxyanions (Martínez and Charlet 2009). Exposure to Se cause extensive fibrosis, neurological disorders, gastrointestinal disorders, and pulmonary congestion (Martin and Griswold 2009) to humans. Disposal of Se into water bodies increases the concentration of Se level, which effects feeding, reproduction, breeding, and survival process of fish, birds and other aquatic and non-aquatic animals (Hopkins

Selenium Contamination in Water, First Edition. Edited by Pooja Devi, Pardeep Singh, Arindam Malakar, and Daniel Snow.

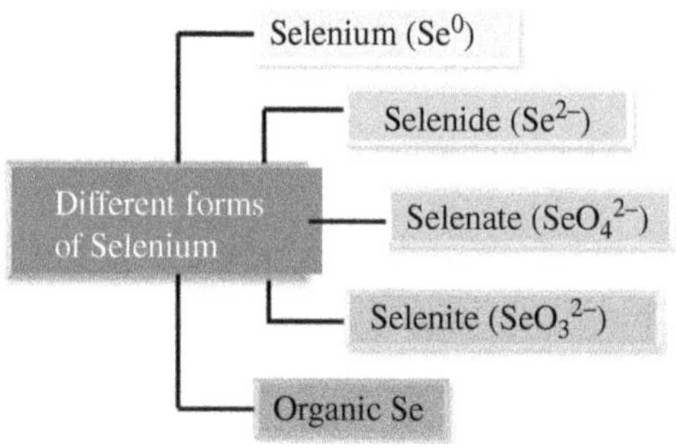

Figure 10.1 Schematic description of different forms of selenium.

et al. 2000). Sulfur-rich vegetables, for example *Allium* species and *Brassica*, have a capacity to accumulate Se and replace sulfur (S) in protein (Temmerman et al. 2014). Accumulation of Se in plants effects its nutrient absorption and metabolism, and subsequently increases the Se consumption by living organisms (Saha et al. 2017). Hence, monitoring and removal of Se from the contaminated area can effectively reduce the risk factor.

10.2 Methods Available for Se Removal

Several technologies have been investigated for the removal of Se. At present, the available options are classified into physical, chemical, and biological treatment (Figure 10.2). All these processes are extensively varied in operating cost, parameters, and competency. In biological treatment selenium oxyanions are converted into less biologically obtainable particulate element Se^0 (Tan et al. 2016), which comprises reduction and volatilization process. These reactions take place in bioreactor or wetlands. In particular, in biological system selenium is removed by alga (Liu et al. 2019), plants (Carvalho and Martin 2001), fungi (Urík et al. 2016), and bacteria (aerobic and anaerobic) (Tana et al. 2018) (Mal et al. 2016) under ecological conditions. Whereas, in biological reactors renewable organic carbon sources act as an electron donor for the removal of Se. It may be added directly to the waste stream or added externally.

Though biological treatment is a cheaper alternative, it has several limitations such as inadequate number of large-scale treatment plants, and biologically delivered Se^0 exhibits colloidal properties which endures in aquatic ecosystems and leads to potential exposure to aquatic receptors (e.g. fish and filter-feeding mollusks) (Santos et al. 2015). Hence, direct ejection of Se oxyanions from contaminated water would consequently be a superior substitute. However, operating cost, utilization of hazardous chemicals, maintenance charge, and sludge formation limit the applications of the direct ejection method (Bleiman and Mishael 2010). Here, we are focusing in detail on various physical and chemical methods involved in the direct removal of Se and its oxyanions. The details of various methods and materials used for their removal is listed in Table 10.1.

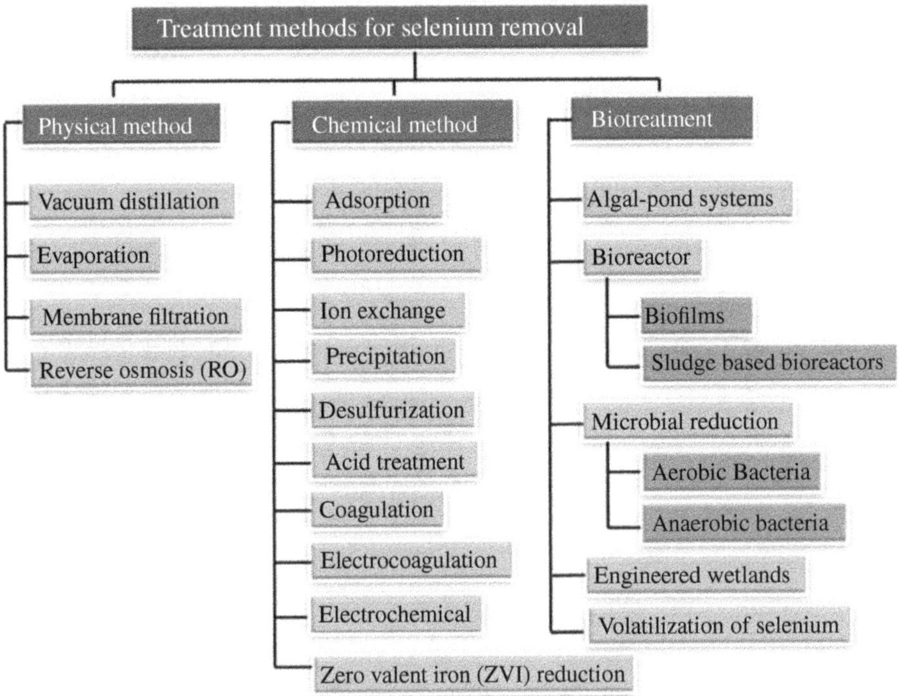

Figure 10.2 Graphical illustration of different techniques available for the removal of Se.

10.3 Physical Methods for Se Removal

In the physical method, evaporation, vacuum distillation, and elusion are mostly used for the purification/separation/extraction of Se from other products (Rakhimov et al. 2013; Shiryaev et al. 2010). Different types of evaporation techniques are existing, such as evaporation ponds, improved evaporation system, and crystallization/evaporation technique. Some of the problems associated with the evaporation techniques are that they demand a large area, are time-consuming, and cause contamination of ground water (Manning and Burau 1995). Membranes are involved for the purification as well as the removal of Se from the contaminated water. There are four physical techniques that work based on the membrane: microfiltration (MF), ultrafiltration (UF), reverse osmosis (RO), and nanofiltration (NF) (Doménech et al. 2020). RO and NF have shown better removal efficacy Se. Hence we are going to discuss these techniques in this part.

10.3.1 Reverse Osmosis

Reverse osmosis (RO) gives a good rate of Se separation without incorporation of chemicals and thermal energy. Ro varies from other membrane methodologies in that it includes a semi-permeable membrane so that filtration efficacy is reliant on influent solute concentration, flux, and pressure. Flux is the rate of water movement across a membrane,

Table 10.1 Various physical and chemical methods used for the removal of Se.

Form of Se removed	Method of removal	Material used for the removal	pH of study	Observation	Ref.
Se (VI)	Sorption	Maghemite (γ-Fe_2O_3)	3.5–8	The selenate oxyanions are sorbed onto the γ-Fe_2O_3 via electrostatic interaction	Jordan et al. (2013)
Selenite	Sorption	Magnetite (Mag-P), commercial magnetite (Mag-C), goethite, activated carbon (AC), and a composite material containing 19% magnetite supported on activated carbon (CM-19)	5.26	The sorptive kinetic conditions are goethite > Mag-P > Mag-C > CM-19	Kwon et al. (2015)
Selenite and selenate	Sorption	Fe_3O_4 nanomaterials	2–6	Capacities of binding were found to be 1923 and 1428 mg Se/kg of non-microwave-assisted Fe_3O_4 and 2380 and 2369 mg Se/kg of microwave-assisted Fe_3O_4 for selenite and selenate, respectively	Gonzalez et al. (2012)
Selenite and selenate	Sorption	Mg-Al layered double hydroxide	–	Maximum sorption capacity ranged from 494 to 563 meq kg^{-1} for selenite and from 609 to 659 meq kg^{-1} for selenate	Constantino et al. (2017)
Selenate	Sorption	Anatase, a polymorph of titanium dioxide	3.5–11	Formation of outer-sphere complexes on the whole pH range	Jordan et al. (2011)
Selenite and selenate	Sorption	Mg-Al-CO_3 LDH nanoparticles and chitosan, a sustainable and bio-degradable biopolymer	4–10	The maximum adsorption capacities are 17 mg/g for Se (IV) and ~ 12 mg/g for Se (VI) with respect to the mass of LDH	Li et al. (2018)
Selenite and selenate	Adsorption	Nanocrystalline aluminum oxide (n-Al_2O_3) impregnated chitosan beads (AICB)	6.5–7	For AICB-selenite system, the embedded n-Al_2O_3 is responsible for adsorption and chitosan acts as a main adsorbent	Yamani et al. (2014)
Se(IV) and Se(VI)	Adsorption	MFe_2O_4 ($CuFe_2O_4$, $CoFe_2O_4$, $MnFe_2O_4$)	2–11	Adsorption capacity of Se (IV) and Se(VI) follows the order of $CuFe_2O_4 > CoFe_2O_4 >> MnFe_2O_4$	Sun et al. (2015)
Selenate and selenite	Adsorption	Zirconium-based metal–organic frameworks (MOFs)	6	NU-1000 is a superior adsorbent for selenate and selenite	Howarth et al. (2015)

Selenite	Adsorption	Graphene oxide modified by iron oxide-based microparticles	2	Adsorption efficiency is 90%	Koudelkova et al. (2019)
Selenite	Adsorption	Hematite-coated magnetic nanoparticles	4–9	Adsorption and desorption efficiencies are 97% and 82%, respectively	Ma et al. (2018)
Selenite and selenate	Adsorption	Iron (Fe^{3+}) oxide/hydroxide nanoparticle	8	95–98% regeneration efficiency of the selenium	Zelmanov and Semiat (2013)
Selenite and selenate	Adsorption	Zirconium metal–organic frameworks (MOFs, UiO-66 and UiO-66-NH2)	7	Se (IV) formed strong inner sphere complex with MOFs and Se (VI) formed outer-sphere complex with MOFs	Wei et al. (2018)
Selenite	Adsorption	3D ZrO_2 microsphere	4	The removal efficacy is 99.48%	Wu et al. (2019)
Selenate and selenite	Adsorption	Graphene oxide polymeric nanocomposite hydrogel beads	3–12	The adsorption capacity is 1.62 mg/g at 25°C	Bandara et al. (2019)
Selenite	Adsorption	6-((2-(2-hydroxy-1 naphthoyl) hydrazono) methyl) benzoic acid onto mesoporous inorganic silica	1–10	Maximum adsorption capacity was 103.73 mg/g	Awuala et al. (2014)
Selenium	Adsorption	Chitosan–montmorillonite composites and oxides (Al-oxide, Fe-oxide)	4.5	Ai-oxide and Fe-oxide removal capacity are 18.4, 17.2 and 8.2 mg/g, respectively	Bleiman and Mishael (2010)
Selenite and selenate	Adsorption	Nanocrystalline aluminum oxide	6.5–7	The association constants for selenate binding on n-Al_2O_3 1.215×10^{-2}	Yamani et al. (2014)
Selenite and selenide	Adsorption	Poly (allylamine)-modified magnetic graphene oxide (PAA-MGO)	5.8	The adsorption capacity of nanocomposite is 120.1 mg/g for Se (IV) and 83.7 mg/g for Se (VI)	Lu et al. (2017)
Selenate	Adsorption	Multiwall carbon nanotube (MWCNT)	7	Excellent selenate adsorbent with an adsorption capacity of up to 1.865 $mg\,g^{-1}$	Kamaraj and Vasudevan (2015)
Selenite and selenate	Adsorption	Poly (amido amine) functionalized graphene oxide (PAMAM-GO).	6	Removal capacity of 60.9 $mg\,g^{-1}$ in Se (IV) and 77.9 $mg\,g^{-1}$ in Se (VI)	Xiao et al. (2016)

(*Continued*)

Table 10.1 (Continued)

Form of Se removed	Method of removal	Material used for the removal	pH of study	Observation	Ref.
Selenate and selenite	Nanomaterials	$MnFe_2O_4$	2–6	The nano-Jacobsite had a selenite and selenate binding capacity of 6573.76 and 769.23 mg Se/kg of NM, respectively	Gonzalez et al. (2010)
Selenite and selenate	Ion exchange	Fe (III) and Mn (III) hydrous oxides	4–7	High phosphate and silicate concentrations may have decreased the removal efficiency of Se (IV)	Szlachta and Chubar (2013)
Selenate	Ion exchange	Mg-Al-CO_3 layered double hydroxide (LDH)	–	LDH is a promising material for the removal of Se at high level of sulfate	Li et al. (2017)
Selenate	Ion exchange	Polyamine-type weakly basic ion exchange resin (Eporasu K-6)	3–12	Saturation capacity of 1.7 mmol/g	Nishimura et al. (2007)
Selenite and selenate	Adsorption and iron exchange	Mg–Al–CO_3 layered double hydroxide (LDH)	5	Mg-Al LDH could purify 16 200 and 4200 bed volumes (BVs) of the selenite/selenate-containing solutions	Chubar and Szlachta (2015)
Selenite	Adsorption and ion exchange	Nano hydrated ferric oxide (HFOs)/polymer hybrid	3–8	Removal of selenite from 2 to 0.01 mg/l was achieved by the hybrid	Pan et al. (2010)
Selenite	Reductive precipitation	Sodium sulfide	7	Se was effectively removed in sulfide solution at both neutral and acidic pH	Jung et al. (2016)
Selenate and selenite	Desulfurization, precipitation and ion exchange	FerrIX™ A33E	5.5	FerrIX™ A33E is effective in removing Se from fuel gas desulfurization wastewater the only after the desulfurization of raw effluent with $BaCl_2$	Staicu et al. (2017)
Selenate	Electrochemical	Reactive iron anodes and copper plate cathode in a bicarbonate medium	–	Selenate removal rates varied from 45.1 to 97.4%, depending on current density and selenate concentration	Baek et al. (2013)

Selenate (Se(VI)) and selenite (Se(IV))	Photocatalytic reduction	TiO_2 with organic hole scavenger (formic acid, methanol and ethanol)	3.5 and 4 (formic acid); 2.2–4 for methanol and ethanol	The fastest rate of Se ions photoreduction was observed in the presence of formic acid than methanol and ethanol	Tan et al. (2003a)
Selenate (Se(VI)) and selenite (Se(IV))	Photocatalytic reduction	Degussa P25 TiO_2, Millennium PC500 and PC50	3.5	Millennium PC500, with the highest surface area demonstrated the fastest photoreduction of Se (IV) and Se(VI), compared to P25 and Millennium PC50	Nguyen et al. (2005a)
Selenate (Se(VI))	Photocatalytic reduction	TiO_2 with formic acid	3.5	The molar adsorption ratio of 3:1 of formate-to-selenate on the TiO_2 surface. The elemental Se was further reduced to hydrogen selenide (H_2Se)	Tan et al. (2003b)
Selenite and selenate	Photocatalytic reduction	TiO_2 with formic acid	3.5	The formation of Se(0) particles was due to a chemical reaction between Se(IV) and Se(2−).	Nguyen et al. (2005b)
Selenite and selenate	Photocatalytic reduction	TiO_2 with EDTA	4 (Selenite) and 6 (selenite)	Selenate was removed from 20 to 100 mg/l at pH 4 within 3 h	Labaran and Vohra (2014)
Selenate	Photocatalytic reduction	Pt on metal oxides (TiO_2, ZrO_2, and Al_2O_3), TiO_2/carbon nanotube with hydrazine monohydrate	Acidic	Pt/TiO_2/CNT catalyst have shown much conversion rate than other catalyst	Zhao et al. (2017)
Selenate	Photocatalytic reduction	TiO_2 supported Pt catalyst with hydrazine	7	The Pt catalysts were deactivated by the deposition of Se metal, but still showed excellent reusability	Zhao et al. (2015)
Selenite, selenate and selenocyanate	Solar-photocatalytic reduction	TiO_2 with EDTA	4	Selenocyanate is removed via a two steps process, i.e. SeCN-complex destruction followed by stepwise oxidation to selenite/selenate and then reduction to Se^0	Labaran and Vohra (2017)
Selenate	Zero-valent iron reduction	zero-valent iron/Fe_3O_4/Fe^{2+}	6.5	The hybrid ZVI/Fe_3O_4/Fe^{2+} system (hZVI) have shown better activity due to synergetic effect	Tang et al. (2016)

(Continued)

Table 10.1 (Continued)

Form of Se removed	Method of removal	Material used for the removal	pH of study	Observation	Ref.
Selenite	Zero-valent iron reduction	Nanoscale zero-valent iron	–	1.3 mM selenite was quickly removed from water within 3 min with 5 g/l nZVI.	Ling et al. (2015)
Selenate	Zero-valent iron reduction	Pillared bentonite (Al-bent) supported nanoscale zero-valent iron	8	Removal efficiency (95.7%) much higher than NZVI reduction (62.1%) and Al-bent adsorption (9.86%)	Li et al. (2015)
Selenocyanate ($SeCN^-$)	Zero-valent iron reduction	Fe(0)	6	$SeCN^-$ and Fe(0) included deselenation of $SeCN^-$ and electrochemical reduction of Se(0) to selenide	Meng et al. (2002)
Selenite	Zero-valent iron reduction	ZVI with ultrasonic irradiation	3–9	Ultrasound promotes the generation of $^{\cdot}OH$ and accelerates the removal of Se with ZVI	Fu et al. (2016)
Selenate	Zero-valent iron reduction	Mn^{2+} and Co^{2+} with ZVI	5.9–7.6	Co^{2+} promoted Fe^{2+} release and the formation of a bimetallic (Co^0–Fe^0) catalytic system by redox reaction	Tang et al. (2014a)
Selenate	Zero-valent iron reduction	As (V) and Mo (VI) with ZVI	6.8–7.9	Addition of As (V) and Mo(VI) significantly reduce the removal of selenate	Zhang et al. (2005)
Selenium	Nanofiltration	Polyamide nanofiltration membrane	2–12	98% of removal is achieved	Malhotra et al. (2020)
Selenate and Selenite	Nanofiltration	Polyhedral oligomeric silsesquioxane (POSS) thin-film nanocomposite	–	Removal efficacy toward Se (IV) and Se(VI) is 93.9 and 96.5%, respectively	He et al. (2016)
Selenium	Nanofiltration	Polyamide core shell biofunctionalized (PA-CSBF) matrix	–	The membrane rejection of 98% toward Se	Zeeshana et al. (2020)
Selenate	Chemical reduction	Nanosized Fe and NiFe	Lower than 8	100% selenate removal was obtained in 5 h	Mondal et al. (2004)
Oxoselenium anions	Physico-chemical method	Iron (II) hydroxide	8.8	Iron precipitates carry the Se and removed from the solution	Zingaro and Dufner (1997)

conveyed in units of gallons per square foot of membrane area per day. RO is a membrane partition method that works at high pressure to force a solution across a membrane that preserves the Se and its oxoanions and other dissolved salts on the refuse side of the membrane and lets the purified water flow to the permeant side (Figure 10.3) (CH2M HILL 2010). The proficiency by which water is detached from the contaminant is usually described as recovery rate, which is generally expressed in percentage (%). The membrane has the smallest pore size with a capability to filter all molecules which are greater than 0.0015 μm in size. It does desalination of seawater, treatment of industrial wastewater prior to release into the environment. The membrane plays a role of ion exchange medium and a hurdle to neutral ions. Density of charged compounds, concentration, feed pressure, and membrane compression are the factors affect the performance of the membrane. Each membrane has a unique rejection rate to specific ions. Using two high-pressure thin-film composite RO membranes was scrutinized for the treatment of multicomponent electrolyte solution (Marifias and Selleck 1992). Reverse osmosis membrane gives an elevated purity effluent. Even so, it has the limitations such as cost, large installation, lower water output, pretreatments, and higher pressure (Yousif et al. 1996). Studies have been carried out to improve the performance of the RO membrane by surface modifications (Xu et al. 2013a).

10.3.2 Nanofiltration

Nanofiltration technique refers the technique transitional between filtration and reverse osmosis which eliminates particles with a size in the order of 1 nm (Doménech et al. 2020). The membrane with the pore size of 0.5–10 nm and the working pressures between 5 and 40 bar (Eriksson 1988). The composite of nanomembrane varies from inorganic, organic, and the combination of organic and inorganic materials. Typically, inorganic membrane is made on ceramic such as aluminum oxide (Al_2O_3), titanium dioxide (TiO_2), aluminum oxide-silicon carbide Al_2O_3-SiC, silicon dioxide (SiO_2), zirconium dioxide (ZrO_2) (Cao et al. 2012; DeFriend et al. 2003; Facciotti et al. 2014), etc., whereas organic membranes are produced from polymeric materials such as polysulfones, polyvinylidene fluoride (Ahmad

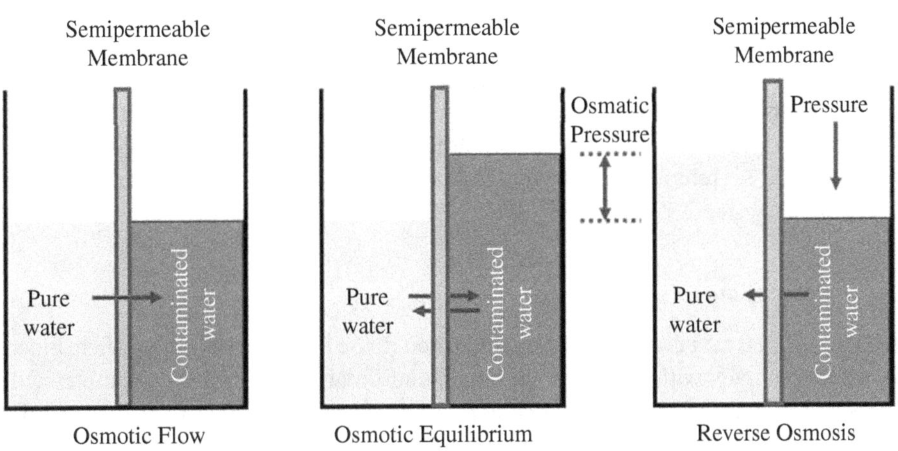

Figure 10.3 Osmotic pressure in a reverse osmosis (RO) system.

et al. 2013; Liu et al. 2011), etc., In the case of hybrid organic–inorganic membrane, this is made by combining polymeric materials with ceramic (Goh et al. 2015; Yin and Deng 2015). Nanomaterials are also used in the nanofiltration process, such as carbon nanotube, graphene oxide materials, etc. (Hu and Mi 2013; Zhao et al. 2016). In this process, the mechanism of separation/removal is controlled by the microhydrodynamic and interfacial process which occurs on the surface and pores on the membrane (Mohammad et al. 2015).

Nanofiltration membrane entails lower hydraulic pressures and fewer pretreatments than a typical reverse osmosis membrane, which possibly reduces the cost for the removal of contaminants. Mostly, nanofiltration membrane comprises a porous inert layer and a hydrophobic rejection layer. Kharaka and his team (Kharaka et al. 1996) worked on the removal of selenium from the agricultural drainage water using nanofiltration method where they have removed more than 95% and 90% of Se and multivalent anions, respectively. In another work, polyamide nanofiltration membrane is used for 98% removal of Se from drinking water (Malhotra et al. 2020). They have achieved the efficiency under pressure at 15 kgf/cm^2 for half an hour and flux of 140 $l/(m^2h)$. Polyhedral oligomeric silsesquioxane (POSS), a thin-film nanocomposite membrane, is developed for the removal of Se and As which has a diameter of 0.65 nm, weight of 302 Da, and permeability of 5.4 LMH/bar and elimination of Se (VI) and Se (IV) is 93.9 and 96.5%, respectively (He et al. 2016). 98% of membrane rejection toward Se was obtained by polyamide core shell biofunctionalized (PA-CSBF) matrix membrane with enhanced hydrophilicity (Zeeshana et al. 2020).

10.4 Chemical Methods for Se Removal

The available chemical methods for Se treatment operations are adsorption, chemical reduction, precipitation, ion exchange, acid treatment, and desulfurization. In this part, we have discussed the most common method used for Se removal.

10.4.1 Precipitation

Se (IV) is adsorbed more effectively than Se (VI) because Se (VI) is affected by the competing anions such as chloride, sulfate, phosphate, etc. The supplement of ferric chloride with sodium hydroxide (NaOH) leads the generation of ferrihydrite, which strongly adsorbed Se (IV) and precipitates out of the solution. It causes the formation of a greater quantity of sludge and the water is to be removed by belt press/frame press/centrifugation (Holmesa and Gu 2016). Se removal by precipitation method is not viable to remove a level below 50 ppm.

10.4.2 Adsorption

Se (IV) and Se (VI) adsorbed to the adsorbent through the formation of inner sphere bidentate surface complex with two covalent bonds and outer sphere hydrated complex with electrostatic interaction, respectively (Rovira et al. 2008). Adsorption of Se was studied with the combination of chitosan–montmorillonite composites, aluminum (Al)-oxide and iron (Fe)-oxide (Bleiman and Mishael 2010). Adsorption of Se on chitosan-clay composite is increased with polymer stacking on the clay, which is higher than the oxide's adsorption. But the adsorption is not affected by pH for composites, whereas the adsorption is decreased in oxides at higher pH. 6-((2-(2-hydroxy-1naphthoyl)hydrazono)methyl) benzoic acid immobilized mesoporous silica is used for the detection and removal of selenite from water (Awuala et al. 2014). They have also investigated the effect of pH, concentration, interfering ions, contact time on the performance. The material has higher surface to volume and pores on the structure which made it as a good adsorbent to capture selenite. The maximum adsorption capacity is 103.73 mg/g and residual selenite concentration is less than 0.01 mg/l. Moreover, it can remove the selenite even in the presence of interfering anions and cations. Bandara and his team (Bandara et al. 2019) used the graphene oxide-based polymeric nanocomposite hydrogel beads for selenite removal. The efficacy of removal is enhanced by the polymer composite (chitosan and polyethylenimine) owing to the plentiful amine and alcoholic functional group. They have conducted the experiment with various effecting factors and found the better removal efficiency is with an acidic pH. The maximum adsorption capacity is 1.62 mg/g, which is based on Langmuir isotherm (Figure 10.4a).

Nanomaterials have been used for the detection, removal, and purification of hazardous materials from water (Kaviya 2018; Kaviya et al. 2017). It is preferred for its higher surface to volume ratio and more activated cites to absorb and remove the contaminants from water (Oliveira et al. 2003). $MnFe_2O_4$ nanomaterial with the size of 27.5 nm is explored for the removal of selenite and selenate (Gonzalez et al. 2010). They have examined the effect of pH (Awuala et al. 2014; Baek et al. 2013; Bandara et al. 2019; Bleiman and Mishael 2010; Blöcher et al. 2003), contesting anions such as chloride ion (Cl^-), sulfate ion (SO_4^{2-}), phosphate ion (PO_4^{3-}), and nitrate ion (NO_3^-), reaction time, and adsorption property of the nanomaterials. The results showed that Cl^- and NO_3^- have no effect on the adsorption of selenite and selenate but SO_4^{2-} and PO_4^{3-} drastically reduce the adsorption of Se (VI) and Se (IV) to 87 and 20%, respectively. This is due to the weak binding attraction of Cl^- and NO_3^- ions to iron oxide compared to the complex formed between the Se oxoanions. On the other hand, SO_4^{2-} and selenate have similar binding affinity on metal oxide surface. Hence,

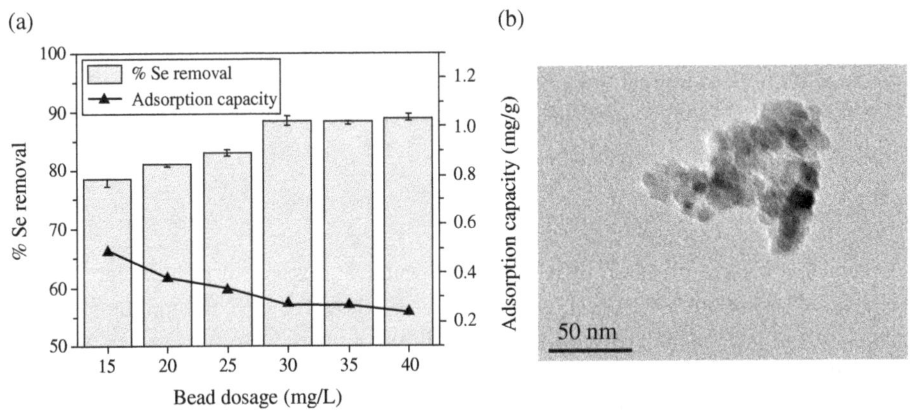

Figure 10.4 (a) Percentage Se removal and adsorption capacity (mg/g) with bead dosage (Bandara et al. 2019), (b) TEM image of magnetic nanoparticle @ hematite (Ma et al. 2018).

higher concentrations of SO_4^{2-} in solution compete for active binding sites on $MnFe_2O_4$ and affect the adsorption of Se oxoanions, whereas PO_4^{3-} ions accumulate or precipitate on the surface of $MnFe_2O_4$, which results in change in the surface potential of the material to negative charge. Further, it prevents and avoids binding of the Se oxoanions on the surface of $MnFe_2O_4$ (Goh and Lim 2004). Three types of MFe_2O_4 are synthesized and investigated for the removal of Se (VI) and Se (IV) (Sun et al. 2015). The results revealed that the adsorption capacity is in the order of $CuFe_2O_4 > CoFe_2O_4 >> MnFe_2O_4$. In another study, the interference of carbonate, nitrate, and silicate on the adsorption site of hematite-coated magnetic nanoparticles for the removal of selenite was examined (Ma et al. 2018) (Figure 10.4b). They also recycled and regenerated the adsorbent. They got a result of 97% and 82% of adsorption and desorption, respectively, in five consecutive cycles. The major limitation of adsorption method is disposal and regular replacement of the adsorbents.

10.4.3 Ion Exchange

Ion exchange is a flexible physicochemical approach for wastewater treatment. During the process, Se (IV) and Se (VI) in aqueous phase are replaced with ions such as PO_4^{3-}, Cl^-, SO_4^{2-}, etc., when the contaminated water flows through ion exchange resins/material filled in a column. Ion exchange resins (carbon-based, minerals, etc.) are strongly affected by variation in the solution pH, which drastically affects the Se removal (Nishimura et al. 2007; Santos et al. 2015). The main drawback of this method is loss of activity in presence of interfering ions and cost.

10.4.4 Photocatalytic Reduction

In photocatalysis, the catalysis reaction takes place in presence of light (UV/visible/sunlight). There are four steps happening when the semiconductor-photocatalyst is used for environmental remediation. Those are: (i) transformation of pollutant (e.g. heavy metal) from water to the surface, (ii) contaminant should stick on the surface of the photocatalyst, (iii) photocatalytic reaction carried out between the pollutant and the catalyst, (iv) pollutant is reduced/decomposed/released from the catalyst (Deng et al. 2019). Photocatalytic reaction is commenced by the charge carriers such as holes (h^+) and electrons (e^-) when the photocatalyst is irradiated with the light (Kaviya and Prasad 2015a) (Eq. 10.1). The photogenerated electron is responsible for the photocatalytic reduction (Eq. 10.2). The rate of reduction depends on the process

$$\text{Photocatalyst} + h\nu \rightarrow e^- h^+ \tag{10.1}$$

and on adsorbate concentration, light irradiation, catalyst dosage, surface property of the catalyst, interfering elements, and pH (Kaviya 2020a,b), though the recombination of charge carriers limits the photocatalytic reaction. The addition of hole/electron scavenger and metals on the surface of photocatalyst would effectively retard the recombination process (Kaviya and Prasad 2016). Photocatalytic reduction was investigated for the removal of inorganic ions such as chromate anions (Ku and Jung 2001), cadmium ions (Cd^{2+}) (Chenthamarakshan et al. 2000), lead ions (Pb^{2+}) (Chen et al. 2015), mercury ions (Hg^{2+})

(Khalil et al. 2002), etc., and organic compounds such as hydrazine (Chatterjee 2000), benzoquinone (Richard 1994), 4-nitrophenol (Kumar et al. 2018), etc. Photocatalyst is also utilized for the reduction/removal of selenide (Se^{2-}), selenate (SeO_4^{2-}), and selenite (SeO_3^{2-}). Nguyen and his research group examined the photocatalytic reduction of selenite and selenate to selenium where TiO_2 is used as a photocatalyst (Nguyen et al. 2005a,b). They have carried out the reaction with various TiO_2 catalysts such as Millennium PC500, PC50, and Degussa P25 TiO_2. The experiment was performed with 1000 ml of contaminated solution which is irradiated with 380 nm UV light using 200 W Hg lamp at pH 3.5. As a result, superior adsorption of selenate than selenite was observed for all the catalyst, which is due to the various chemical structure of two ions. Similarly, selenate is reduced faster than selenite to Se (0) because Se (IV) to Se (0) requires four electrons, whereas Se (VI) to Se (0) needs six electrons to reduce. The results revealed that Millennium PC500 has shown higher activity owing to the larger surface area and causes a greater rate of photoreduction.

In another study, the photoreduction of Se anions was performed on a TiO_2 catalyst with the addition of organic hole scavengers such as ethanol, salicylic acid, methanol, formic acid, sucrose, and acetic acid (Tan et al. 2003a,b). The reaction was performed under UV light. The mechanism was described in terms of various rate of reducing radicals generation when these compounds were responded with hydroxyl radical. For example, the process takes place in format and alcohol is given below (equations 10.2 and 10.3).

$$HCOO^- + OH^{\cdot} \rightarrow {}^{\cdot}COO^- + H_2O \tag{10.2}$$

$$R - CH_2 - OH + OH^{\cdot} \rightarrow R - {}^{\cdot}CH - OH + H_2O \tag{10.3}$$

The solvent properties like polarity and viscosity also play a role in the electron transfer process (Hecht and Fawcett 1995). They also examined the effect of pH on the adsorption of Se anions on TiO_2 by varying the pH from 2.2 to 5. Se anion adsorption is decreased with increasing pH and format adsorption is opposite to this trend. The higher rate of photoreduction and mineralization observed in the presence of formic acid is due to the higher ability to react with the lattice oxygen in TiO_2 (Muggli and Backes 2002). Likewise, ethylenediaminetetraacetic acid (EDTA) is used as a hole scavenger for the photocatalytic removal of Se anions and has achieved the removal at pH 4 from its initial concentration from 20 to 100 mg/l within 3 h (Labaran and Vohra 2014). Platinum (Pt) supported metal oxides such as Al_2O_3, TiO_2, and ZrO_2 were used for the reduction of selenate in the presence of hydrazine monohydrate (Zhao et al. 2017). The better performance was noticed on Pt/TiO_2 catalyst, though deactivation of the catalyst was spotted due to the exposure of Pt by the reduction product. They have coated TiO_2 on carbon nanotube (CNT) to overcome the above-mentioned issue and increase the activity and durability (Figure 10.5a).

10.4.5 Zero-Valent Ion Reduction

Zero-valent iron (ZVI) reduction method is a chemical reduction technique which is considered to be an optimistic methodology owing to its low cost and reduced toxicity (Li et al. 2015). It has been utilized for the removal of organic and inorganic pollutants (Guan et al. 2015; Kaviya and Prasad 2015b). It has been given more consideration for resolving Se in ground water. The reduction potential of Se by ZVI depends on the surface property of

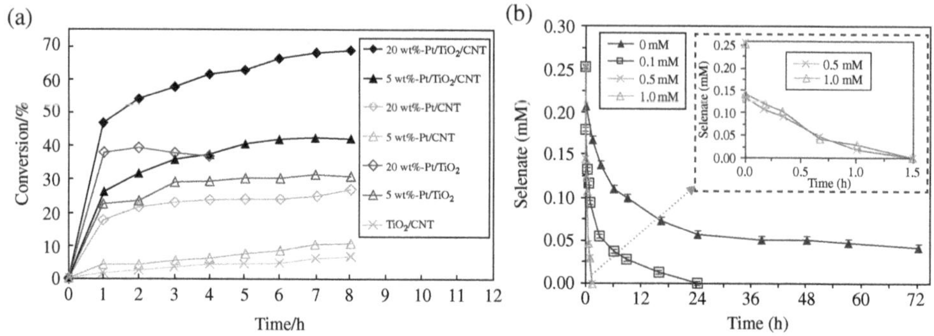

Figure 10.5 (a) Conversion of Se (VI) as a function of time over different types of catalyst (Zhao et al. 2017). (b) Effect of Fe^{2+} and magnetite dosage on selenate removal in the $ZVI/Fe_3O_4/Fe^{2+}$ system (Tang et al. 2016).

Fe and the concentration of dissolved oxygen in ground water. For instance, 60 nm-sized zero-valent ion (5 g/l) with the surface area of 30–35 m^2/g was used for the removal of 1.3 mM selenite from water for three minutes (Ling et al. 2015). The study further disclosed that oxygen and Se can diffuse through the defects of nano ZVI and assists the solid phase diffusion and Fe(0)-stimulated reduction of selenite. Experiments also performed to increase the adsorption of selenium oxyanions on nano ZVI (nZVI), positively charged pillared bentonite (Al-bent) is added as a support (Li et al. 2015). The removal efficiency of nZVI/Al-bent (95.7%) is higher than the individual (nZVI 62.1% and Al-bent 9.86%), which is due to the good adsorption property of Al-bent and accelerating the reduction. The sequestration of Se (IV) and Se (VI) by NZVI under various environmental conditions has also been investigated (Xie et al. 2017). The formation of iron oxide on its surface leads to the loss of its performance. In order to avoid the above issue, some additives were added to ZVI such as ZVI/Fe_3O_4 (Xu et al. 2013b), $ZVI/Fe_2O_3/Fe^{2+}$ (Liu et al. 2014), ZVI/Fe_3O_4/graphene (Lv et al. 2014), ZVI/sand (Btatkeu et al. 2014), etc.

Removal of selenate in a $ZVI/Fe_3O_4/Fe^{2+}$ system was examined by Tang and his team (Tang et al. 2016). They compared the performance of the hybrid system with ZVI, Fe^{2+}, Fe_3O_4, ZVI/Fe^{2+}, Fe_3O_4/Fe^{2+}, and ZVI/Fe_3O_4. The hybrid ZVI system has shown outstanding selenate removal compared to other systems, which is due to the synergic effect of ZVI, Fe (II), and Fe_3O_4. When they increased the quantity of magnetite from 10 to 20 g/l, the time taken for selenate removal was drastically reduced from 1.5 to 0.9 hours (Figure 10.5b). Hence, the activity is increased due to magnetic property, variable composition (Fe^{2+} and Fe^{3+}), electron conductivity, reaction site of Fe_3O_4, and the electron donor ability of ZVI in the presence of Fe^{2+}. The X-ray photoelectron spectroscopy studies exhibited the step-by-step reduction of selenate, i.e. Se(VI) to Se(IV) and to Se(0). The capability of iron corrosion product with ZVI on the removal of selenate was also exploited (Tang et al. 2014b). The addition of iron corrosion product with ZVI removes the selenate in the range of 4–7.5%. It is further increased with the external addition of Fe^{2+} to the system. The result indicates that selenate removal proficiency is increased with increasing Fe (II) concentration and stoichiometry ratio for the reduction is 1:1 (Fe (II)/Se). The effect of other metal ions on the removal of Se oxyanions was explored with ZVI. Tang et al. (2014a) evaluated the effect of addition of molybdenum (Mo) (II) and cobalt (Co) (II) with ZVI/Fe^{2+} on the removal of selenate. Co (II) considerably increases the removal of selenate by accelerating the release of Fe (II) and formation of Co (0)-Fe (0) catalytic arrangement via redox reaction. For selenate reduction, 69% and 30% of electron are offered by Co (II) and Mo (II), respectively. Similarly, the effect of arsenic (As) (IV) and (Mo) (VI) on the removal of Se (VI) with ZVI (Zhang et al. 2005). But, As (IV) and Mo (VI) reduce the efficacy of selenate removal.

10.4.6 Electrochemical Methods

Electrochemical methods have been used to remove the contaminations. The electrochemical reduction of Se (VI) to Se (0) and oxidation of Fe (II) and Fe $(OH)_2$ could happen as follows (Eqs. 10.4–10.7) (Figure 10.6) (Baek et al. 2013):

$$SeO_4^{2-} + 2e^- + 3H^+ \rightarrow HSeO_3^{2-} + H_2O\,E^0 = 1.060\,V \tag{10.4}$$

$$HSeO_3^- + 4e^- + 5H^+ \rightarrow Se^0 + 3H_2OE^0 = 0.903\,V \tag{10.5}$$

$$Fe^{2+} \rightarrow Fe^{3+} + e^-\,E^0 = 0.771\,V \tag{10.6}$$

$$Fe(OH)_2 + OH^{\cdot} \rightarrow Fe(OH)_3 + e^-\,E^0 = 0.54\,V \tag{10.7}$$

In electrochemical method, the removal tendency depends on several parameters such as types of the electrode and electrolyte, electric current, initial concentration, etc. Iron anode, copper plate as a cathode in a bicarbonate medium is used for the removal of selenate (Baek et al. 2013). Ferrous hydroxide is formed at the anode and retards the formation of oxygen, in the removal process of which it plays a major role. Concentration of the selenate

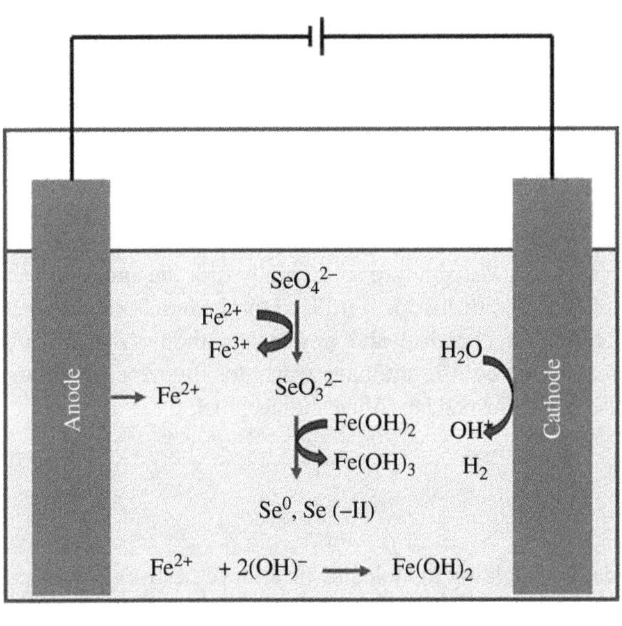

Figure 10.6 (a) Graphical illustration of electrochemical removal of selenate from aqueous solution.

and current density varies the removal rate from 45.1% to 97.4%. They observed that the existence of dissolved oxygen did not cause any effect on the removal of selenate. Electrocoagulation (EC) is simple version of electrochemical methods to treat water and pollutant water where it utilizes electrochemical cell with dc voltage which is applied to Fe or Al electrode and water/contaminated water as an electrolyte. In the EC process, termination of metal from the anode happened and instantaneous generation of hydroxyl ions and hydrogen (H_2) will be produced. The gas formed assists the particle in floating on the surface of the water. Hence, this process contains three fundamentals such as electrochemistry, coagulation, and flotation. Moreno et al. (2009) have studied the relationship between green rust and electrocoagulation (EC). They have used iron electrodes because of its performance and durability over an Al electrode and have measured the pH of the electrode at various zones during the EC process. Effect of background electrolytes such as lithium chloride (LiCl), sodium chloride (NaCl) and potassium chloride (KCl) on the removal of selenate and selenite was studied at different pH and ionic strength (Zhang et al. 2018).

10.5 Combination of Physical and Chemical Methods

In physical and chemical methods, the residual concentration of Se, the combination of multiple heavy metals, the huge amount of sludge, and the expense of the reagents or membrane are all drawbacks. A hybrid process of combination of physical and chemical methods is a chance to overcome the issues and enhance the performance of the process. Examples of hybrid processes are crossflow microfiltration/coagulation, microfiltration/adsorption, and microfiltration/flotation/adsorption (Blöcher et al. 2003; Matis et al. 2005; Mavrov et al. 2004), etc.

Mavrov and his coworkers (Mavrov et al. 2006) studied the combination of electrocoagulation and microfiltration methods to remove Se from wastewater which consists of mixture of heavy metal anions and other metal cations such as As, Cd, lead (Pb), Fe, copper (Cu), and Zinc (Zn). In this study, iron hydroxide which is produced from electrocoagulation method shows good adsorption toward Se removal but low precipitation due to the fines of sludge. Iron hydroxide is completely removed from the sludge by microfiltration where 0.3 μm pore-sized ceramic membrane was used for this process. The added advantage of Se is removed from the cake which is formed on the microfiltration membrane acts as a barrier for adsorption of Se. In this hybrid treatment 98.7% of Se, 99.9% of As, 98% of Cu and lead, 99.9% of zinc and cadmium was removed successfully with the anode current density of 4.8 mA/cm^2 for 20 minutes. Iron hydroxide is utilized for the reduction of oxoselenium anions by the incorporation of chemical and physical methods (Zingaro and Dufner 1997). They have chosen an agriculture drainage water for the experiment and found an excellent result because of the formed landfilled sludge which filters the Se^0.

10.6 Conclusion

The progress of cost-effective and reliable methodologies to remove Se from water and other sources is a priority environmental concern. To accomplish a desirable rate of Se removal from the ecosystem, there is likely a need to progress a timely monitoring steps,

Se management approach, prevention of release strategies from the industries, source control, and pre-treatment, which will also minimize the adverse effects of Se contamination. Various physical, chemical, and biological treatment methods have been shown to remove Se from water. However, hybrid process or the combination of these treatment methods have shown a favorable outcome for removing Se and other pollutants from industrial wastewater. In future, methodologies should be improved for removing Se from its interfering matrix materials; and reusability and recycling of the materials used for the removal should also be a focus for concern.

Acknowledgments

S.K. thanks UGC, Govt. of India, for Dr. D. S. Kothari post-doctoral fellowship.

References

Ahmad, A.L., Abdulkarim, A.A., Ooi, B.S., and Ismail, S. (2013). Recent development in additives modifications of polyethersulfone membrane for flux enhancement. *Chemical Engineering Journal* 223: 246–267. https://doi.org/10.1016/j.cej.2013.02.130.

Awuala, M.R., Hasan, M.M., Ihara, T., and Yaita, T. (2014). Mesoporous silica based novel conjugate adsorbent for efficient selenium (IV) detection and removal from water. *Microporous and Mesoporous Materials* 197: 331–338. https://doi.org/10.1016/j.micromeso.2014.07.005.

Baek, K., Kasem, N., Ciblak, A. et al. (2013). Electrochemical removal of selenate from aqueous solutions. *Chemical Engineering Journal* 215-216: 678–684. https://doi.org/10.1016/j.cej.2012.09.135.

Bandara, P.C., Perez, J.V.D., Nadres, E.T. et al. (2019). Graphene oxide nanocomposite hydrogel beads for removal of selenium in contaminated water. *ACS Applied Polymer Materials* 1: 2668–2679. https://doi.org/10.1021/acsapm.9b00612.

Bleiman, N. and Mishael, Y.G. (2010). Selenium removal from drinking water by adsorption to chitosan-clay composites and oxides: batch and columns tests. *Journal of Hazardous Materials* 183 (1–3): 590–595. https://doi.org/10.1016/j.watres.2013.10.054.

Blöcher, C., Dorda, J., Mavrov, V. et al. (2003). Hybrid flotation-membrane filtration process for the removal of heavy metal ions from wastewater. *Water Research* 37: 4018–4026. https://doi.org/10.1016/S0043-1354(03)00314-2.

Btatkeu, K.B.D., Vargas, H.O., Tchatchueng, J.B. et al. (2014). Determining the optimum Fe^0 ratio for sustainable granular Fe^0/sand water filters. *Chemical Engineering Journal* 247: 265–274. https://doi.org/10.1016/j.cej.2014.03.008.

Cao, X.P., Li, D., Jing, W.H. et al. (2012). Synthesis of visible-light responsive C, N and Ce codoped TiO_2 mesoporous membranes via weak alkaline sol-gel process. *Journal of Materials Chemistry* 22: 15309–15315. https://doi.org/10.1039/c2jm31576a.

Carvalho, K.M. and Martin, D.F. (2001). Removal of aqueous selenium by four aquatic plant. *Journal of Aquatic Plant Management* 39: 33–36. ISSN: 0146-6623.

CH2M HILL (2010). Review of Available Technologies for the Removal of Selenium from Water. North American Metal Council.

Chatterjee, D. (2000). Photocatalytic reduction of hydrazine to ammonia catalysed by $[Ru^{III}(edta)(H_2O)]^-$ complex in a Pt/TiO_2 semiconductor particulate system. *Journal of Molecular Catalysis A: Chemical* 154 (1–2): 1–3. https://doi.org/10.1016/S1381-1169(99)00291-5.

Chen, H., Li, T., Zhang, L. et al. (2015). Pb(II) adsorption on magnetic γ-Fe_2O_3/titanate nanotubes composite. *Journal of Environmental Chemical Engineering* 3 (3): 2022–2033. https://doi.org/10.1016/j.jece.2015.07.010.

Chenthamarakshan, C.R., Yang, H., Ming, Y., and Rajeshwar, K. (2000). Photocatalytic reactivity of zinc and cadmium ions in UV-irradiated titania suspensions. *Journal of Electroanalytical Chemistry* 494: 79–86. https://doi.org/10.1016/S0022-0728(00)00374-0.

Chubar, N. and Szlachta, M. (2015). Static and dynamic adsorptive removal of selenite and selenate by alkoxide-free sol-gel-generated Mg-Al-CO_3 layered double hydroxide: effect of competing ions. *Chemical Engineering Journal* 279: 885–896. https://doi.org/10.1016/j.cej.2015.05.070.

Constantino, L.V., Quirino, J.N., Monteiro, A.M. et al. (2017). Sorption-desorption of selenite and selenate on Mg-Al layered double hydroxide in competition with nitrate, sulfate and phosphate. *Chemosphere* 181: 627–634. https://doi.org/10.1016/j.chemosphere.2017.04.071.

DeFriend, K.A., Wiesner, M.R., and Barron, A.R. (2003). Alumina and aluminate ultrafiltration membranes derived from alumina nanoparticles. *Journal of Membrane Science* 224: 11–28. https://doi.org/10.1016/S0376-7388(03)00344-2.

Deng, Y., Feng, C., Tang, L. et al. (2019). *Nanohybrid Photocatalysts for Heavy Metal Pollutant Control. Nanohybrid and Nanoporous Materials for Aquatic Pollution Control*, 125–153. Elsevier https://doi.org/10.1016/B978-0-12-814154-0.00005-0.

Doménech, N.G., Milton, F.P., and Gun'ko, Y.K. (2020). Recent progress and future prospects in development of advanced materials for nanofiltration. *Materials Today Communications* 23: 100888. https://doi.org/10.1016/j.mtcomm.2019.100888.

Eriksson, P. (1988). Nanofiltration extends the range of membrane filtration. *Environmental Progress* 7: 58–62. https://doi.org/10.1002/ep.3300070116.

Facciotti, M., Boffa, V., Magnacca, G. et al. (2014). Deposition of thin ultrafiltration membranes on commercial SiC microfiltration tubes. *Ceramics International* 40: 3277–3285. https://doi.org/10.1016/j.ceramint.2013.09.107.

Fu, F., Lu, J., Cheng, Z., and Tang, B. (2016). Removal of selenite by zero-valent iron combined with ultrasound: Se (IV) concentration changes, Se(VI) generation, and reaction mechanism. *Ultrasonics Sonochemistry* 29: 328–336. https://doi.org/10.1016/j.ultsonch.2015.10.007.

Goh, K. and Lim, T. (2004). Geochemistry of inorganic arsenic and selenium in a tropical soil: effect of reaction time, pH, and competitive anions on arsenic and selenium adsorption. *Chemosphere* 55: 849–859. https://doi.org/10.1016/j.chemosphere.2003.11.041.

Goh, P.S., Ng, B.C., Lau, W.J., and Ismail, A.F. (2015). Inorganic nanomaterials in polymeric ultrafiltration membranes for water treatment. *Separation and Purification Reviews* 44: 216–249. https://doi.org/10.1080/15422119.2014.926274.

Gonzalez, C.M., Hernandez, J., Parsons, J.G., and Torresdey, J.L.G. (2010). A study of the removal of selenite and selenate from aqueous solutions using a magnetic iron/manganese oxide nanomaterial and ICP-MS. *Microchemical Journal* 96: 324–329. https://doi.org/10.1016/j.microc.2010.05.005.

Gonzalez, C.M., Hernandez, J., Videa, J.R.P. et al. (2012). Sorption kinetic study of selenite and selenate onto a high and low pressure aged iron oxide nanomaterial. *Journal of Hazardous Materials* 211-212: 138–145. https://doi.org/10.1016/j.jhazmat.2011.08.023.

Guan, X., Sun, Y., Qin, H. et al. (2015). The limitations of applying zero-valent iron technology in contaminants sequestration and the corresponding countermeasures: the development in zero-valent iron technology in the last two decades (1994–2014). *Water Research* 75: 224–248. https://doi.org/10.1016/j.watres.2015.02.034.

He, Y., Tang, Y., and Chung, T.S. (2016). Concurrent removal of selenium and arsenic from water using POSS-polyamide thin-film nanocomposite nanofiltration membranes. *Industrial and Engineering Chemistry Research* 55 (50): 12929–12938. https://doi.org/10.1021/acs.iecr.6b04272.

Hecht, M. and Fawcett, W.R. (1995). Solvent effects in the electroreduction of ethylenediaminetetraacetatochromium (III) at a mercury electrode. *The Journal of Physical Chemistry* 99: 1311–1316. https://doi.org/10.1016/0022-0728(95)04060-2.

Holmesa, A.B. and Gu, F.X. (2016). Emerging nanomaterials for the application of selenium removal for wastewater treatment. *Environmental Science: Nano* 3: 982–996. https://doi.org/10.1039/c6en00144k.

Hopkins, W.A., Congdon, J.D., and Ray, J.K. (2000). Incidence and impact of axial malformations in larval bullfrogs (Rana catesbeiana) developing in sites polluted by a coal burning power plant. *Environmental Toxicology and Chemistry* 19: 862–868. https://doi.org/10.1002/etc.5620190412.

Howarth, A.J., Katz, M.J., Wang, T.C. et al. (2015). High efficiency adsorption and removal of selenate and selenite from water using metal-organic frameworks. *Journal of the American Chemical Society* 137 (23): 7488–7494. https://doi.org/10.1021/jacs.5b03904.

Hu, M. and Mi, B. (2013). Enabling graphene oxide nanosheets as water separation membranes. *Environmental Science & Technology* 47: 3715–3723. https://doi.org/10.1021/es400571g.

Jordan, N., Foerstendorf, H., Weiß, S. et al. (2011). Sorption of selenium(VI) onto anatase: macroscopic and microscopic characterization. *Geochimica et Cosmochimica Acta* 75 (1519): 1530. https://doi.org/10.1016/j.gca.2011.01.012.

Jordan, N., Ritter, A., Foerstendorf, H. et al. (2013). Adsorption mechanism of selenium (VI) onto maghemite. *Geochimica et Cosmochimica Acta* 103: 63–75. https://doi.org/10.1016/j.gca.2012.09.048.

Jung, B., Safan, A., Batchelor, B., and Wahab, A.A. (2016). Spectroscopic study of se (IV) removal from water by reductive precipitation using sulfide. *Chemosphere* 163: 351–358. https://doi.org/10.1016/j.chemosphere.2016.08.024.

Kamaraj, R. and Vasudevan, S. (2015). Decontamination of selenate from aqueous solution by oxidized multi-walled carbon nanotubes. *Powder Technology* 274: 268–275. https://doi.org/10.1016/j.powtec.2015.01.043.

Kaviya, S. (2018). Size dependent ratiometric detection of Pb (II) ions in aqueous solution by light emitting biogenic CdS NPs. *Journal of Luminescence* 195: 209–215. https://doi.org/10.1016/j.jlumin.2017.11.031.

Kaviya, S. (2020a). Evolution of ZnO-based Photocatalyst for the degradation of pollutants. In: *Green Photocatalysts for Energy and Environmental Process. Environmental Chemistry for a Sustainable World*, vol. 36, 109–139. Springer https://doi.org/10.1007/978-3-030-17638-9_5.

Kaviya, S. (2020b). Progression in Fenton Process for the wastewater treatment. In: *Green Methods for Wastewater Treatment*, vol. 35 (eds. M. Naushad, S. Rajendran and E. Lichtfouse), 87–120. Springer https://doi.org/10.1007/978-3-030-16427-0_5.

Kaviya, S. and Prasad, E. (2015a). Biogenic synthesis of ZnO-Ag nano custard apples for efficient photocatalytic degradation of methylene blue by sunlight irradiation. *RSC Advances* 5 (22): 17179–17185. https://doi.org/10.1039/C4RA15293J.

Kaviya, S. and Prasad, E. (2015b). Sequential detection of Fe^{3+} and As^{3+} ions by naked eye through aggregation and disaggregation of biogenic gold nanoparticles. *Analytical Methods* 7: 168–174. https://doi.org/10.1039/C4AY02342K.

Kaviya, S. and Prasad, E. (2016). Eco-friendly synthesis of ZnO nanopencils in aqueous medium: a study of photocatalytic degradation of methylene blue under direct sunlight. *RSC Advances* 6 (40): 33821–33827. https://doi.org/10.1039/C6RA04306B.

Kaviya, S., Kabila, S., and Jayasree, K.V. (2017). Room temperature biosynthesis of greatly stable fluorescent ZnO quantum dots for the selective detection of Cr^{3+} ions. *Materials Research Bulletin* 97: 163–168. https://doi.org/10.1016/j.materresbull.2017.07.025.

Khalil, L.B., Ropheal, M.W., and Mourad, W.E. (2002). The removal of the toxic hg(II) salts from water by photocatalysis. *Applied Catalysis B: Environmental* 36: 125–130. https://doi.org/10.1016/S0926-3373(01)00285-5.

Kharaka, Y.K., Ambats, G., and Presser, T.S. (1996). Removal of selenium from contaminated agricultural drainage water by nanofiltration membranes. *Applied Geochemistry* 11 (6): 797–802. https://doi.org/10.1016/S0883-2927(96)00044-3.

Koudelkova, Z., Bytesnikova, Z., Xhaxhiu, K. et al. (2019). Electrochemical evaluation of selenium (IV) removal from its aqueous solutions by unmodified and modified graphene oxide. *Molecules* 24: 1063–1076. https://doi.org/10.3390/molecules24061063.

Ku, Y. and Jung, I.L. (2001). Photocatalytic reduction of Cr (VI) in aqueous solutions by UV irradiation with the presence of titanium dioxide. *Water Research* 35 (1): 135–142. https://doi.org/10.1016/S0043-1354(00)00098-1.

Kumar, S., Pandit, V., Bhattacharyya, K., and Krishnan, V. (2018). Sunlight driven photocatalytic reduction of 4-nitrophenol on Pt decorated ZnO-RGO nanoheterostructures. *Materials Chemistry and Physics* 214: 364–376. https://doi.org/10.1016/j.matchemphys.2018.04.113.

Kwon, J.H., Wilson, L.D., and Sammynaiken, R. (2015). Sorptive uptake of selenium with magnetite and its supported materials onto activated carbon. *Journal of Colloid and Interface Science* 457: 388–397. https://doi.org/10.1016/j.jcis.2015.07.013.

Labaran, B.A. and Vohra, M.S. (2014). Photocatalytic removal of selenite and selenate species: effect of EDTA and other process variables. *Environmental Technology* 35 (9): 1091–1100. https://doi.org/10.1080/09593330.2013.861857.

Labaran, B.A. and Vohra, M.S. (2017). Solar Photocatalytic removal of selenite, Selenate, and Selenocyanate species. *Clean: Soil, Air, Water* 45 (10): 1600437. https://doi.org/10.1002/clen.201600268.

Li, Y., Cheng, W., Sheng, G. et al. (2015). Synergetic effect of a pillared bentonite support on SE(VI) removal by nanoscale zero valent iron. *Applied Catalysis B: Environmental* 174–175: 329–335. https://doi.org/10.1016/j.apcatb.2015.03.025.

Li, M., Farmen, L.M., and Chan, C.K. (2017). Selenium removal from sulfate-containing groundwater using granular layered double hydroxide materials. *Industrial and Engineering Chemistry Research* 56 (9): 2458–2465. https://doi.org/10.1021/acs.iecr.6b04461.

Li, M., Dopilka, A., Kraetz, A.N. et al. (2018). Layered double hydroxide/chitosan nanocomposite beads as sorbents for selenium oxoanions. *Industrial and Engineering Chemistry Research* 57 (14): 4978–4987. https://doi.org/10.1021/acs.iecr.8b00466.

Ling, L., Pan, B., and Zhang, W.X. (2015). Removal of selenium from water with nanoscale zero-valent iron: mechanisms of intraparticle reduction of Se (IV). *Water Research* 71: 274–281. https://doi.org/10.1016/j.watres.2015.01.002.

Liu, F., Hashim, N.A., Liu, Y. et al. (2011). Progress in the production and modification of PVDF membranes. *Journal of Membrane Science* 375: 1–27. https://doi.org/10.1016/j.memsci.2011.03.014.

Liu, W., Ai, Z., Cao, M., and Zhang, L. (2014). Ferrous ions promoted aerobic simazine degradation with Fe@Fe_2O_3 core–shell nanowires. *Applied Catalysis B: Environmental* 150-151: 1–11. https://doi.org/10.1016/j.apcatb.2013.11.034.

Liu, F., Huanga, J.C., Zhoua, C. et al. (2019). Development of an algal treatment system for selenium removal: effects of environmental factors and post-treatment processing of Se-laden algae. *Journal of Hazardous Materials* 365: 546–554. https://doi.org/10.1016/j.jhazmat.2018.11.017.

Lu, Z., Yu, J., Zeng, H., and Liu, Q. (2017). Polyamine-modified magnetic graphene oxide nanocomposite for enhanced selenium removal. *Separation and Purification Technology* 183: 249–257. https://doi.org/10.1016/j.seppur.2017.04.010.

Lv, X., Xue, X., Jiang, G. et al. (2014). Nanoscale zero-Valent iron (nZVI) assembled on magnetic Fe_3O_4/graphene for chromium (VI) removal from aqueous solution. *Journal of Colloid and Interface Science* 417: 51–59. https://doi.org/10.1016/j.jcis.2013.11.044.

Ma, Z., Shan, C., Liang, J., and Tong, M. (2018). Efficient adsorption of selenium (IV) from water by hematite modified magnetic nanoparticles. *Chemosphere* 193: 134–141. https://doi.org/10.1016/j.chemosphere.2017.11.005.

Mal, J., Nancharaiah, Y.V., van Hullebusch, E.D., and Lensad, P.N.L. (2016). Effect of heavy metal co-contaminants on selenite bioreduction by anaerobic granular sludge. *Bioresource Technology* 206: 1–8. https://doi.org/10.1016/j.biortech.2016.01.064.

Malhotra, M., Pal, M., and Pal, P. (2020). A response surface optimized nanofiltration-based system for efficient removal of selenium from drinking water. *Journal of Water Process Engineering* 33: 101007. https://doi.org/10.1016/j.jwpe.2019.101007.

Manning, B.A. and Burau, R.G. (1995). Selenium immobilization in evaporation pond sediments by in situ precipitation of ferric oxyhydroxide. *Environmental Science & Technology* 29: 2639–2646. https://doi.org/10.1021/es00010a028.

Marifias, B.J. and Selleck, R.E. (1992). Reverse osmosis treatment of multicomponent electrolyte solutions. *Journal of Membrane Science* 72: 211–229. https://doi.org/10.1016/0376-7388(92)85050-S.

Martin, S. and Griswold, W. (2009). Human health effects of heavy metals. *Environmental Science and Technology Briefs for Citizens* 15: 1–6. http://www.engg.ksu.edu/CHSR.

Martínez, A.F. and Charlet, L. (2009). Selenium environmental cycling and bioavailability: a structural chemist point of view. *Reviews in Environmental Science and Biotechnology* 8: 81–110. https://doi.org/10.1007/s11157-009-9145-3.

Matis, K.A., Lazaridis, N.K., Zouboulis, A.I. et al. (2005). A hybrid flotation-microfiltration process for metal ions recovery. *Journal of Membrane Science* 247: 29–35. https://doi.org/10.1016/j.memsci.2004.07.031.

Mavrov, V., Erwe, T., and Chmiel, H. (2004). Selective separation of heavy metals from industrial wastewater streams by means of heavy metal bonding agents. *Water, Air, and Soil Pollution: Focus* 4 (4–5): 147–155. https://doi.org/10.1023/B:WAFO.0000044794.93260.c3.

Mavrov, V., Stamenov, S., Todorova, E. et al. (2006). New hybrid electrocoagulation membrane process for removing selenium from industrial wastewater. *Desalination* 201: 290–296. https://doi.org/10.1016/j.desal.2006.06.005.

Meng, X., Bang, S., and Korfiatis, G.P. (2002). Removal of selenocyanate from water using elemental iron. *Water Research* 36: 3867–3873. https://doi.org/10.1016/S0043-1354(02)00086-6.

Mohammad, A.W., Teow, Y.H., Ang, W.L. et al. (2015). Nanofiltration membranes review: recent advances and future prospects. *Desalination* 356: 226–254. https://doi.org/10.1016/j.desal.2014.10.043.

Mondal, K., Jegadeesan, G., and Lalvani, S.B. (2004). Removal of selenate by Fe and NiFe nanosized particles. *Industrial and Engineering Chemistry Research* 43: 4922–4934. https://doi.org/10.1021/ie030715l.

Moreno, H.A.C., Cocke, D.L., Gomes, J.A.G. et al. (2009). Electrochemical reactions for electrocoagulation using iron electrodes. *Industrial and Engineering Chemistry Research* 48: 2275–2282. https://doi.org/10.1021/ie8013007.

Muggli, D.S. and Backes, M.J. (2002). Two active sites for photocatalytic oxidation of formic acid on TiO_2: effects of H_2O and temperature. *Journal of Catalysis* 209: 105–113. https://doi.org/10.1006/jcat.2002.3640.

Natasha, M., Shahid, N.K., Niazi, S. et al. (2018). A critical review of selenium biogeochemical behavior in soil-plant system with an inference to human health. *Environmental Pollution* 234: 915–934. https://doi.org/10.1016/j.envpol.2017.12.019.

Nguyen, V.N.H., Amal, R., and Beydoun, D. (2005a). Photocatalytic reduction of selenium ions using different TiO_2 photocatalysts. *Chemical Engineering Science* 60: 5759–5769. https://doi.org/10.1016/j.ces.2005.04.085.

Nguyen, V.N.H., Beydoun, D., and Amal, R. (2005b). Photocatalytic reduction of selenite and selenate using TiO_2 photocatalyst. *Journal of Photochemistry and Photobiology A: Chemistry* 171: 113–120. https://doi.org/10.1016/j.jphotochem.2004.09.015.

Nishimura, T., Hashimoto, H., and Nakayama, M. (2007). Removal of selenium (VI) from aqueous solution with polyamine-type weakly basic ion exchange resin. *Separation Science and Technology* 42: 3155–3167. https://doi.org/10.1080/01496390701513107.

Oliveira, L.C.A., Rios, V.R.A., Fabris, J.D. et al. (2003). Clay-oxide magnetic composites for the adsorption of contaminants in water. *Applied Clay Science* 22: 169–177. https://doi.org/10.1016/S0169-1317(02)00156-4.

Pan, B., Xiao, L., Nie, G. et al. (2010). Adsorptive selenite removal from water using a nano-hydrated ferric oxides (HFOs)/polymer hybrid adsorbent. *Journal of Environmental Monitoring* 12: 305–310. https://doi.org/10.1039/b913827g.

Rakhimov, A.V., Warot, G., Karaivanov, D.V. et al. (2013). Purification of selenium from thorium, uranium, radium, actinium and potassium impurities for low background measurements. *Radiochimica Acta* 101: 653–659. https://doi.org/10.1524/ract.2013.2060.

Ramya, S., Shanmugasundaram, T., and Balagurunathan, R. (2015). Biomedical potential of actinobacterially synthesized selenium nanoparticles with special reference to antibiofilm, anti-oxidant, wound healing, cytotoxic and anti-viral activities. *Journal of Trace Elements in Medicine and Biology* 32: 30–39. https://doi.org/10.1016/j.jtemb.2015.05.005.

Richard, C. (1994). Photocatalytic reduction of benzoquinone in aqueous ZnO or TiO_2 suspensions. *New Journal of Chemistry* 18: 443–455. https://doi.org/10.1590/S0103-50532012000200007.

Rovira, M., Giménez, J., Martínez, M. et al. (2008). Sorption of selenium (IV) and selenium (VI) onto natural iron oxides: goethite and hematite. *Journal of Hazardous Materials* 150 (2): 279–284. https://doi.org/10.1016/j.jhazmat.2007.04.098.

Saha, U., Fayiga, A., and Sonon, L. (2017). Selenium in the soil-plant environment: a review. *International Journal of Applied Agricultural Sciences* 3: 1–18. https://doi.org/10.11648/j.ijaas.20170301.11.

Santos, S., Ungureanu, G., Boaventura, R., and Botelho, C. (2015). Selenium contaminated waters: An overview of analytical methods, treatment options and recent advances in sorption methods. *The Science of the Total Environment* 521-522: 246–260. https://doi.org/10.1016/j.scitotenv.2015.03.107.

Shiryaev, V.S., Pimenov, V.G., Lipatova, M.M. et al. (2010). Removal of barium impurities from selenium by vacuum distillation. *Inorganic Materials* 46 (3): 314–317. https://doi.org/10.1134/S0020168510030192.

Staicu, L.C., Crini, N.M., and Crini, G. (2017). Desulfurization: critical step towards enhanced selenium removal from industrial effluents. *Chemosphere* 172: 111–119. https://doi.org/10.1016/j.chemosphere.2016.12.132.

Sun, W., Pan, W., Wang, F., and Xu, N. (2015). Removal of Se (IV) and Se (VI) by MFe_2O_4 nanoparticles from aqueous solution. *Chemical Engineering Journal* 273: 353–362. https://doi.org/10.1016/j.cej.2015.03.061.

Szlachta, M. and Chubar, N. (2013). The application of Fe-Mn hydrous oxides based adsorbent for removing selenium species from water. *Chemical Engineering Journal* 217: 159–168. https://doi.org/10.1016/j.cej.2012.11.100.

Tan, T., Beydoun, D., and Amal, R. (2003a). Effects of organic hole scavengers on the photocatalytic reduction of selenium anions. *Journal of Photochemistry and Photobiology A: Chemistry* 159: 273–280. https://doi.org/10.1016/S1010-6030(03)00171-0.

Tan, T.T.Y., Beydoun, D., and Amal, R. (2003b). Photocatalytic reduction of Se(VI) in aqueous solutions in UV/TiO_2 system: importance of optimum ratio of reactants on TiO_2 surface. *Journal of Molecular Catalysis A: Chemical* 202: 73–85. https://doi.org/10.1016/S1381-1169(03)00205-X.

Tan, L.C., Nancharaiah, Y.V., van Hullebusch, E.D., and Lens, P.N.L. (2016). Selenium: environmental significance, pollution, and biological treatment technologies. *Biotechnology Advances* 34 (5): 886–907. https://doi.org/10.1016/j.biotechadv.2016.05.005.

Tana, Y., Wang, Y., Wang, Y. et al. (2018). Novel mechanisms of selenate and selenite reduction in the obligate aerobic bacterium *Comamonas testosteroni* S44. *Journal of Hazardous Materials* 359: 129–138. https://doi.org/10.1016/j.jhazmat.2018.07.014.

Tang, C., Huang, Y.H., Zeng, H., and Zhang, Z. (2014a). Promotion effect of Mn^{2+} and Co^{2+} on selenate reduction by zero-valent iron. *Chemical Engineering Journal* 244: 97–104. https://doi.org/10.1016/j.cej.2014.01.059.

Tang, C., Huang, Y.H., Zeng, H., and Zhang, Z. (2014b). Reductive removal of selenate by zero-valent iron: the roles of aqueous Fe^{2+} and corrosion products, and selenate removal mechanisms. *Water Research* 67: 166–174. https://doi.org/10.1016/j.watres.2014.09.016.

Tang, C., Huang, Y., Zhang, Z. et al. (2016). Rapid removal of selenate in a zero-valent iron/Fe_3O_4/Fe^{2+} synergetic system. *Applied Catalysis B: Environmental* 184: 320–327. https://doi.org/10.1016/j.apcatb.2015.11.045.

Temmerman, L.D., Waegeneers, N., Thiry, C. et al. (2014). Selenium content of Belgian cultivated soils and its uptake by field crops and vegetables. *Sci. Total Environ.* 468 (469): 77–82. https://doi.org/10.1016/j.scitotenv.2013.08.016.

Urík, M., Boriová, K., Bujdoš, M., and Matúš, P. (2016). Fungal selenium (VI) accumulation and biotransformation-filamentous fungi in Selenate contaminated aqueous media remediation. *CLEAN–Soil, Air, Water* 44 (6): 610–614. https://doi.org/10.1002/clen.201500100.

Vesper, D.J., Roy, M., and Rhoads, C.J. (2008). Selenium distribution and mode of occurrence in the Kanawha formation. southern West Virginia U.S. *International Journal of Coal Geology* 73: 237–249. https://doi.org/10.1016/j.coal.2007.06.003.

Wei, J., Zhang, W., Pan, W. et al. (2018). Experimental and theoretical investigations on Se (IV) and Se (VI) adsorption to UiO-66-based metal-organic frameworks. *Environmental Science: Nano* 5: 1441–1453. https://doi.org/10.1039/C8EN00180D.

Winkel, L.H.E., Johson, C.A., Lenz, M. et al. (2012). Environmental selenium research: from microscopic processes to global understanding. *Environmental Science & Technology* 46: 571–579. https://doi.org/10.1021/es203434d.

Wu, X., Guo, X., and Zhang, L. (2019). Fabrication of porous zirconia microspheres as an efficient adsorbent for removal and recovery of trace Se (IV) and Te(IV). *Industrial and Engineering Chemistry Research* 58 (1): 342–349. https://doi.org/10.1021/acs.iecr.8b04288.

Xiao, W., Yan, B., Zeng, H., and Liu, Q. (2016). Dendrimer functionalized graphene oxide for selenium removal. *Carbon* 105: 655–664. https://doi.org/10.1016/j.carbon.2016.04.057.

Xie, Y., Dong, H., Zeng, G. et al. (2017). The comparison of Se(IV) and Se(VI) sequestration by nanoscale zero-valent iron in aqueous solutions: the roles of solution chemistry. *Journal of Hazardous Materials* 338: 306–312. https://doi.org/10.1016/j.jhazmat.2017.05.053.

Xu, G.R., Wang, J.N., and Li, C.J. (2013a). Strategies for improving the performance of the polyamide thin film composite (PA-TFC) reverse osmosis (RO) membranes: surface modifications and nanoparticles incorporations. *Desalination* 328: 83–100. https://doi.org/10.1016/j.desal.2013.08.022.

Xu, J., Tang, J., Baig, S.A. et al. (2013b). Enhanced dechlorination of 2,4-dichlorophenol by Pd/Fe-Fe_3O_4 nanocomposites. *Journal of Hazardous Materials* 244: 628–636. https://doi.org/10.1016/j.jhazmat.2012.10.051.

Yamani, J.S., Lounsbury, A.W., and Zimmerman, J.B. (2014). Adsorption of selenite and selenate by nanocrystalline aluminum oxide, neat and impregnated in chitosan beads. *Water Research* 50: 373–381. https://doi.org/10.1016/j.watres.2013.10.054.

Yin, J. and Deng, B. (2015). Polymer-matrix nanocomposite membranes for water treatment. *Journal of Membrane Science* 479: 256–275. https://doi.org/10.1016/j.memsci.2014.11.019.

Yousif, K.K., Gil, A., Theres, S.P., and Roy, A.D. (1996). Removal of selenium from contaminated agricultural drainage water by nanofiltration membranes. *Applied Geochemistry* 6: 797–802. https://doi.org/10.1016/S0883-2927(96)00044-3.

Zeeshana, M.H., Khana, R.U., Shafiq, M., and Sabir, A. (2020). Polyamide intercalated nanofiltration membrane modified with biofunctionalized core shell composite for efficient removal of arsenic and selenium from wastewater. *Journal of Water Process Engineering* 34: 101175–101183. https://doi.org/10.1016/j.jwpe.2020.101175.

Zelmanov, G. and Semiat, R. (2013). Selenium removal from water and its recovery using iron (Fe^{3+}) oxide/hydroxide-based nanoparticles sol (NanoFe) as an adsorbent. *Separation and Purification Technology* 103: 167–172. https://doi.org/10.1016/j.seppur.2012.10.037.

Zhang, Y., Amrhein, C., and Frankenberger, W.T. Jr. (2005). Effect of arsenate and molybdate on removal of selenate from an aqueous solution by zero-valent iron. *Science of the Total Environment* 350: 1–11. https://doi.org/10.1016/j.scitotenv.2005.02.032.

Zhang, N., Gang, D.D., McDonald, L., and Lin, L.S. (2018). Background electrolytes and pH effects on selenate adsorption using iron-impregnated granular activated carbon and surface binding mechanisms. *Chemosphere* 195: 166–174. https://doi.org/10.1016/j.chemosphere.2017.11.161.

Zhao, J., Matsune, H., Takenaka, S., and Kishida, M. (2015). Reductive reaction of Selenate with hydrazine over TiO_2-supported Pt catalyst in aqueous solution. *Chemistry Letters* 44: 1563–1565. https://doi.org/10.1246/cl.150695.

Zhao, E.Y., Ji, Y.L., Weng, X.D. et al. (2016). High-flux positively charged nanocomposite nanofiltration membranes filled with poly(dopamine) modified multiwall carbon nanotubes. *ACS Applied Materials & Interfaces* 8: 6693–6700. https://doi.org/10.1021/acsami.6b00394.

Zhao, J., Matsune, H., Takenaka, S., and Kishida, M. (2017). Reduction of selenate with hydrazine monohydrate over Pt catalysts in aqueous solution. *Chemical Engineering Journal* 308: 963–973. https://doi.org/10.1016/j.cej.2016.09.143.

Zingaro, R.A. and Dufner, D.C. (1997). Reduction of oxoselenium anions by iron (II) hydroxide. *Environment International* 23 (3): 299–304. https://doi.org/10.1016/S0160-4120(97)00032-9.

11

Chemical Methods for Removal and Treatment of Selenium from Water

Surinder Singh[1], Lavisha Bashambu[1], Pooja Devi[2], and Sushil Kumar Kansal[1]

[1] *Dr. S.S. Bhatnagar University Institute of Chemical Engineering and Technology, Panjab University, Chandigarh, India*
[2] *Academy of Scientific and Innovative Research (AcSIR), Ghaziabad, India; Central Scientific Instruments Organisation, Chandigarh, India*

11.1 Introduction

The element selenium (Se) happens to be a metalloid, discovered by Jöns Jacob Berzelius, a chemist from Sweden in 1918 (Wiberg et al. 2001). The name "selenium" was given to this element after the Greek word for moon (i.e. "Selene") due to its appearance of a crescent moon. Selenium is a vital trace mineral-nutrient required in low amounts by the body (Xiao et al. 2016), and has been found to be distributed in the Earth's crust very unevenly (Dhillon and Dhillon 2003; Wen and Carignan 2007). Selenium is considered to be non-metallic substance bearing atomic mass 78.97 and lying in the 16th group of the periodic table. Its properties are halfway between the elements lying above and below it in the periodic table, i.e. sulfur and tellurium, and some are common with arsenic as well. The physical properties of selenium resemble arsenic and the chemical properties are nearly equivalent to sulfur. It is located along with other elements in natural ores and mostly as methyl derivatives; it seldom exists in its pure form+. It is found in the form of metallic gray to black crystals in its elemental form. There is a very thin margin between selenium's vitality and toxicity. Selenium occurs in Earth's environment in four different forms viz. selenite or Se(IV) [(SeO_3^{2-})], selenate or Se(VI) [(SeO_4^{2-})], selenide (Se^{2-}), and elemental selenium (Se^0), out of which (Se^0) is very uncommon (Winkel et al. 2011). The less commonly found species of selenium are cysteine and methionine and volatile methylated compounds. Some of such compounds are dimethyl selenide, dimethyl selenone, and dimethyl diselenide (Sinharoy et al. 2019). The oxyanions Se(IV), selenite, and Se(VI), selenate, are generally existing in aquatic bodies, and the Se(IV) form is more harmful than Se(VI) (selenate) (Awual et al. 2015a). Se(VI), the oxidized form of element selenium, is dominant in an oxidizing environment and the selenite form of selenium is dominant when a reducing environment is present. The different selenium species (compounds) and their types are shown in Table 11.1.

Selenium Contamination in Water, First Edition. Edited by Pooja Devi, Pardeep Singh, Arindam Malakar, and Daniel Snow.

Table 11.1 Different selenium species existing in the environment.

Selenium species	Description
Inorganic species	
Elemental selenium (Se^0)	Naturally non-soluble with at least 11 different allotropes. Formed as a precipitate after chemical and microbial reduction processes
Selenide (Se^{2-} or H_2Se) Se (-II), (Se^- or HSe) Se (-I)	Stable in strongly reducing conditions as metal selenides and other organic compounds. Volatile compounds formed by microbial processes
Selenate (SeO_4^{2-}), Se (+VI)	Exists as a tetrahedral oxyanion form that shares similarities with sulfate and is mainly predominant in soils, sediments, and waters
Selenite (SeO_3^{2-}), Se (+IV)	Exists as a pyramidal oxyanion that is more reactive, present in mildly oxidizing acidic environments and major mobile species governed by sorption or desorption processes on solid surfaces
Organic species	
Dimethylselenide (Me_2Se)	Volatile compounds that can be formed from biomethylation processes
Dimethyldiselenide (Me_2Se_2)	
Trimethylselenonium cation (Me_3Se^+)	Metabolite from urine
Dimethylselenosulfide (MeSSeMe)	Product of microbial methylation process
Selenocyanate ($SeCN^-$)	Can be found in wastewater from petroleum refineries
Amino acids and low-molecular mass species	
Selenomethionine (SeMet)	Main selenium form found in plants
Selenocysteine (SeCys)	21st amino acid component of selenoproteins. Main selenium form in organic tissues
Selenoglutathione	Can be formed during *in vitro* experiments mimicking biochemical processes
Selenoproteins	Selenium is a component of various forms of enzymes, peptides, and sugars that can be found in a biological system (i.e. GPX, TR, etc.)

Modified from Fernández-Martínez and Charlet (2009) and Kieliszek and Błazejak (2013).
Source: Reproduced with permission from Tan et al. (2016) (License no. 4851171065329).

Bioavailability of selenium is due to environmental and geochemical factors such as soil properties, pH, ferrous hydroxide content, organic carbon, redox conditions, and clay minerals and selenium speciation (Winkel et al. 2011). Selenium is nowadays an emerging water contaminant, through geogenic activities like rock seepage, natural sources like erosion of the sedimentary rocks, volcanic matter, and extreme weathering, along with industrial activities, e.g. copper smelting, refining, minerals mining, power generation, etc. (Donner et al. 2018; Robberecht and Grieken 1982; Santos et al. 2015). The maximum permitted concentration of selenium for the purpose of drinking water has been set to 0.05 parts per million (ppm) by the United States Environment Protection Agency (EPA). The upper limit related to selenium consumption is set at 400 mg/day. The allowed intake of selenium for humans ranges from 40 to 400 μg/day and permissible concentration for drinking water intake is set at 40 μg/l. Hence the effects of selenium are dually harmful as both deficient and excess amounts are detrimental and cause severe health implications (Lopes et al. 2017).

Selenium is mostly found as selenate in soils which can percolate into rivers very easily. Concentration of selenium varies naturally for different regions (Sharma et al. 2018). These can lead to both soil as well as water contamination. Although there are permissible values set by different safety and health administrations and government agencies, selenium contamination is even so prevailing all over the world. Selenium is a very useful nutrient required for both humans and animals when taken in lower (permissible) amounts. Appropriate intake of it can even prevent cancer (Brodowska et al. 2016; Combs and Gray 1998) but when present in higher amounts it itself acts as a toxin. Se is also bioaccumulative and its concentration increases as we move up the trophic levels (Lemly 1999; Santos et al. 2015).

Plants and animals both require selenium as a micronutrient, and multivitamins also contain selenium. It helps in proper functioning of the body by playing a vital role in metabolism and thyroid function. However, higher concentrations can turn out to be toxic and can cause adverse health effects. It can lead to fever, nausea, respiratory, cardiovascular, gastrointestinal, thyroid related and even kidney dysfunctions, and liver- and heart-related problems (Brodowska et al. 2016; Etteieb et al. 2020; Lopes et al. 2017). High concentration of selenium in diet can cause toxic health effects such as selenosis in humans, alkali disease in grazing animals, tissue damage, reproductive failure, and in aquaculture, developmental deformities and severe damage to fish (Etteieb et al. 2020; Gonzalez et al. 2010; Lopes et al. 2017). If a very high dose is consumed it can even lead to death. Se concentration naturally found in the human body ranges from 13–20 mg. The US recommended dietary allowance for consumption of Se has been set at a value of 55 μg/day for teenagers as well as adults (He et al. 2018; Lopes et al. 2017; Zeng 2009). In animals too a minimum concentration is required for survival, although they won't die if it is present in excess. Selenium can cause acute health effects such as hair damage, fingernail changes, tiredness or fatigue, derogatory effect on peripheral nervous system, and irritability (Etteieb et al. 2020; Ma et al. 2018). The long-term effects include kidney and liver failure and damage to nervous system. Due all these health issues and harmful implications in both humans and animals, abatement of selenium from water and water bodies is a matter of emergent concern.

Selenium sources originating from human activities include coal burning, mining, smelting, and some other industrial manufacturing operations (Etteieb et al. 2020; Nancharaiah and Lens 2015). Also, fossil fuels are found to be prominent sources of selenium. Release of Se into water bodies in more than the permissible limit is considered harmful because of its consumption by aquatic life which can further cause biomagnification. Agricultural runoff containing herbicides and insecticides, etc. also contains selenium compounds (Bleiman and Mishael 2010). Therefore, treating wastewater with high concentration of Se before discharge is a necessity. Selenium is usually present as Se(VI) and Se(IV) ions in water. Selenium occurs within the environmental regions due to natural events like eruption of volcanoes, rock sedimentation process, mineral mining, geogenic seepage, agriculture, and anthropogenic sources like burning of selenium rich fossil fuels, use of insecticides and herbicides (Fernández-Martínez and Charlet 2009; Santos et al. 2015; Wen and Carignan 2007). Se is a significant element, having a lot of applications in areas such as genetic, medicine, biochemical, molecular, and material science (Tan et al. 2016). The overall cycle of Se in the environment and its transport is described in Figure 11.1.

Selenium pollution from industries is mainly caused by coal combustion in power plants, petroleum, and fuel oil for energy usage, metals processing, insecticides production, mining related operations, oil refining and glass manufacture (Chapman et al. 2010; Nancharaiah and Lens 2015; Santos et al. 2015). Within the fossil fuels, coal happens to be a major fuel utilized in USA, accounting for 26 827 thousand tons of consumption of coal for US power produced in 2019–2020 up to May 2020 (EIA 2020). Another known source of selenium contamination is the flue gas desulfurization water, which contains high quantities of Se. This process is used in mining industries to wipe out sulfur from exhaust flue gases, but then the process water gets enriched with sulfur as well as other heavy metals including selenium. Varied chemical, membrane-based, physical, and biological techniques have been adopted for the same but they do not remove the ions completely,

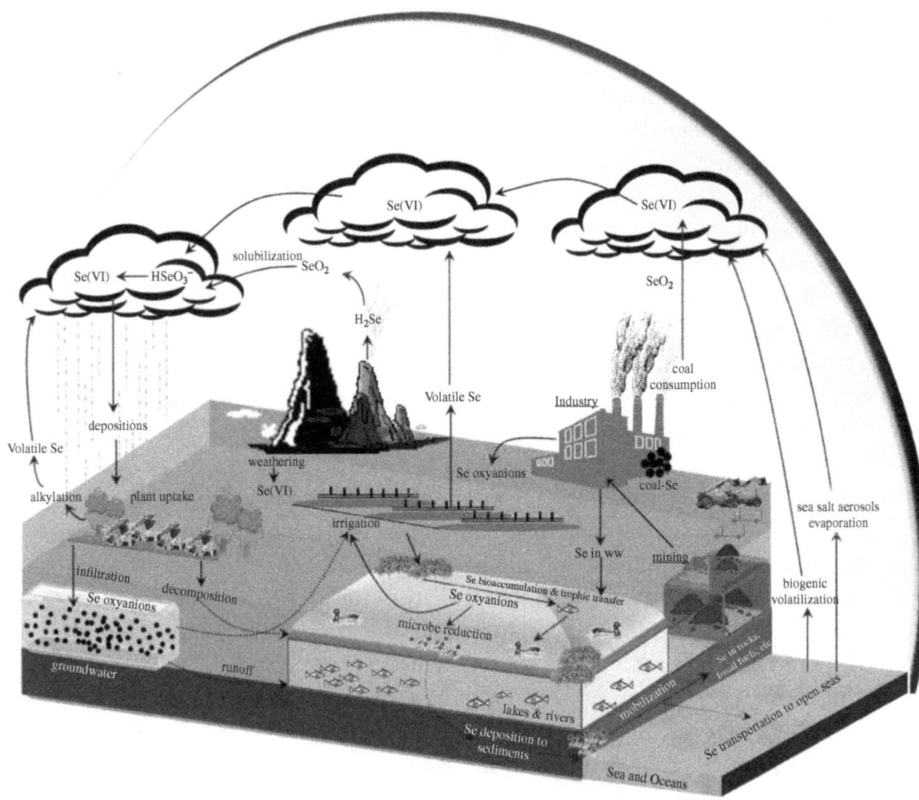

Figure 11.1 Selenium cycle in the environment. *Source:* Reproduced with permission from Tan et al. (2016) (License no. 4851171065329).

instead reducing them into more neutral Se. Ion exchange combined with adsorption can manifest better results (Staicu et al. 2017).

In many cases industrial wastewater is found to have approximately 1–7000 μg/l of Se in it, which is higher than the permissible limit (Kapoor et al. 1995). Remediation methods of Se are of high interest in coal mining industries. Although many methods have been developed and are being researched, major challenges being faced include economical viability and low concentration levels in wastewater causing difficulty in separation for selenium. Also, the huge operating and maintenance expenditure, and the selectivity for Se and chemical toxicity limit the utility of these remediation methods (Dorraji et al. 2016; Gonzalez et al. 2012).

Various physical, biological, and/or chemical treatment methods have been studied for selenium removal. Physical methods include nanofiltration, reverse osmosis, ion exchange, membrane separation etc. Chemical methods include adsorption and reduction of Se ions. Recent techniques utilized for selenium removal include reduction, co-precipitation, coagulation, and flocculation, adsorption, electrocoagulation (EC), electrochemical methods, bioconcentration, electro-kinetics, membrane technology, photocatalysis, chemical precipitation, reduction, bio-mineralization, solvent extraction, biological treatment, and anion exchange filtering; most of these are environment-friendly as well (Adio et al. 2017; Awual et al. 2015a, b; Geoffroy and Demopoulos 2011; Gonzalez et al. 2010; Ma et al. 2018; Markwart et al. 2019; Nancharaiah and Lens 2015; Santos et al. 2015; Wen and Carignan 2007). Adsorption technique has become popular due to ease of operation, separation efficiency, and cost-effectiveness (Xiao et al. 2016). Various adsorbents such graphene oxides and magnetic nanomaterials not only show great affinity for Se adsorption but are also reusable with insignificant loss in their capacity (Ma et al. 2018). Nanomaterials show great adsorption capacity for Se ions and many have succeeded in being selective toward selenium along with the presence of competitive ions. Biological methods such as use of microbes for degradation or even producing adsorbents from biomass are very environment-friendly but come along with limitations in their efficiency and some other constraints. Selenium exists in many chemical forms which are not of equal toxicity. Inorganic Se is much more toxic than its organic form. In water it is mostly found in inorganic form. Another way of treating water having high concentration of selenium is by reducing its ions to a more stable and neutral form. Selenate and selenite ions had been effectively reduced to elemental Se(0) by non-zero-valent iron and even by combining precipitation and reduction reactions (Geoffroy and Demopoulos 2011). Iron oxides and hydroxides show appreciable reduction potential for selenium ions. Most of the reduction methods have been experimented predominantly for Se(IV). This is attributed to the fact that toxicity of the ions in Se(IV) form is generally existing in higher amounts than Se(VI) form (Sharrad et al. 2012). Thus, one of the reasons adsorption methods get an edge over reduction or precipitation is that they are not biased considerably by the type of ion present in water. For example, iron sulfide showed adsorption affinity for both selenate and selenite although being dependent on the initial concentration present in water (Han et al. 2011). This chapter outlines the different chemical methods and techniques for selenium removal, the selection of process parameters, and the advantages and disadvantages of the selenium removal using chemical techniques.

11.2 Selenium Removal Methods

Selenium is a new and emerging pollutant in water. It did not gain much attention in earlier times because of its presence in very low concentrations. With the increase in industrial activities such as energy production, mining activities, agriculture etc., its concentration in water bodies has increased to a significant extent. The toxic effects of selenium are fully known and selenium contamination in water affects people in the form of different health hazards. The Se removal processes are classified into physical, biological, and chemical methods. Physical methods include ion exchange, nanofiltration, reverse osmosis, carbonation, and EC or electrochemical sedimentation (He et al. 2018; Lenz and Lens 2009; Santos et al. 2015). In the electrochemical-cyclone twinning method stainless steel cathodes are used to get electrochemical deposition of Se. About 98% Se can be recaptured using this method. The disadvantage is that only treatment of dilute solutions is possible using this method. Membranes made of nanocomposites are also used as selective filters. The characteristics of the separation membranes are mainly determined by the pore size and number and thickness of layers. Biological methods such as phytoremediation, bacterial reduction, bio-mineralization, bioreactor-based separation, and bioremediation can also be used for Se removal. Chemical treatment methods are also widely utilized for removal of Se.

11.3 Chemical Treatment

Chemical treatments can be broadly classified into adsorption- and reduction-based techniques. Adsorption happens to be widely explored technique nowadays owing to its low cost and adequate efficiency but is greatly affected by presence of oxyanions. Adsorbents such as activated carbon, nanocomposites, and magnetic-coated nanoparticles have been studied for abatement of Se(VI) and Se(IV) since these ions are the most prominent in aqueous systems. Photocatalysts also act as reducing agents indirectly. Almost all the chemical techniques are affected due to pH and also presence of other ions pose a competitive environment. The adsorption losses of Se are dependent upon factors like pH, initial concentration of adsorbate, contact time, material of adsorber, chemical form of Se ions, and suspended ions (Robberecht and Grieken 1982). Different adsorbents and reducing agents show different preferences for pH. Adsorption had been utilized as alternative treatment technology for remediation of Se for the past decade owing to its sustainable usage along with reutilization of the used adsorbents (Dorraji et al. 2016; Lu et al. 2017).

11.4 Adsorption

11.4.1 Magnetic Graphene Oxide Nanocomposite

Various workers have reported abatement of Se(IV) and Se(VI) utilizing different adsorbents such as chitosan, activated carbon, graphene oxide, magnetic nanoparticles (MNPs), different hydroxides, chitosan, clay composites, metal-based oxides, and nanocomposites (Bleiman and Mishael 2010; Dorraji et al. 2016; Fu et al. 2014; Yamani et al. 2014). One of the major complications of using nanoscale particles as adsorbents is separating them after the process of adsorption is over for their reuse. MNPs bring the benefit of easy separation by external magnetic force. Graphene in its oxide form is an excellent sorbent material owing to its surface properties and can be modified using MNPs coating to improve its functionality. For its synthesis graphene is first oxidized and then magnetic iron oxide nanoparticles are immobilized on it. Fu et al. (2014) utilized MNPs-coated graphene oxide (MGO) for abatement of Se in their study. MGO offered high adsorptive removal for both Se(IV) (selenite) and Se(VI) (selenate) owing to its morphological characteristics and varied functional elements/groups at the surface. It was observed that iron oxide helped greatly for adsorption of Se. The obtained magnetic hysteresis loop (magnetization curve) provided the evidence of MGO displaying superparamagnetic behavior. The strong magnetic effect of MGO enabled the effective magnetic separation. The MGO nanocomposites exhibited greater than 99.9% removal for selenite and around 80% adsorptive removal for selenate using 1 g/l dosage and 300 ppb of initial selenium concentration, at pH of 6 within 10 seconds. It was observed that decreasing the pH to the acidic region increased sorption of Se onto MGO, consequently the sorption of Se(VI) rose to 95% when utilizing pH of 2.

11.4.2 Dendrimer Functionalized Graphene Oxide

Graphene as such can be applied for adsorptive removal of organic impurities owing to its large π-electron system, but it is not effective to eradicate metal ions. Hence GO is utilized in Se abatement owing to diverse functional elements available at its surface (Xiao et al. 2016). GO exhibits enhanced surface area and selectivity for Se which labels it as prominent adsorbent. Xiao et al. (2016) utilized dendrimer poly amino amide (PAMAM) functionalized GO for abatement of Se from simulated wastewater. Dendrimers happen to be a new class of polymers which are repetitively branched molecules. Their morphological and chemical characteristics such as enhanced surface area to volume ratios make them appropriate for many industrial applications. Dendrimer PAMAM can be functionalized on GO owing to its property of possessing large amounts of functional groups and thus high reactivity. GO possess different functional groups, e.g. carbonyl and hydroxyl, etc., and these groups combine with the APTES to provide amine-substitution of GO. Various generations of PAMAM have been compared to GO and MGO for their adsorption capability. Various chemical processes had been adopted for synthesis and Se(IV) (selenite) and Se(VI) (selenate) show different amounts of affinity to these PAMAM generations. GO–G4 shows the maximum adsorption capacity among all the generations and MGO. Se(VI) shows enhanced affinity as generations increase, unlike Se(IV) for which affinity starts to drop after GO–G3. The adsorption is mainly controlled by electrostatic interactions. Low pH values are favorable for both keeping amino groups protonated and for adsorption. Removal capacity up to 60.9 mg/g for selenite and 77.9 mg/g for selenate were observed.

11.4.3 Poly (allylamine) (PAA) –Graphene Oxide

Lu et al. (2017) utilized water soluble poly (allylamine) (PAA), grafted on MGO nanocomposite for abatement of selenium from wastewater. PAA was functionalized on GO using EDC/NHS reaction while magnetic iron oxide was immobilized by co-precipitation.

A different dendrimer can also be used other than PAMAM to obtain modified graphene oxide. Poly allylamine (PAA) can be functionalized on GO in a way similar to mentioned above. For synthesis EDC solution and co-precipitation is used. PAA-GO was seen to have better adsorption capacity than PAMAM-GO. It was observed to be 120.1 mg/g in the case of Se(IV) (selenite) and 83.7 mg/g for sorption of Se(VI), i.e. selenate ions. This also shows that studies on different modifications to existing methods can improve the efficiency to a great extent.

11.4.4 Hematite-Coated Magnetic Nanoparticles (MNPs)

Magnetic coating over nanoparticles helps not only in adsorption but also in separation after adsorption. Likewise hematite-modified MNPs can be used as an adsorbent. MNPs such as magnetite show low efficiency and therefore a requirement of modification was felt. Thus, hematite was used for modification. Ma et al. (2018) conducted adsorption experiments utilizing hematite-coated MNPs for investigating the influence of pH and competing anions such as chloride, sulfate, nitrate, etc. The sorption capability was found out to be 25.0 mg/g for Se(IV). The adsorbent could treat 100 μg/l Se(IV) to below 10 μg/l in 10 minutes utilizing by 0.1 g/l of adsorbent. It was observed that adsorbent worked extremely well in pH range of 4–9 but as pH increases the capacity starts decreasing. It was reported that competing ions had negligible influence over the sorption capacity, although increasing concentration of natural organic matter lowers the removal rate of Se. Reusability of the adsorbent also proved to be satisfactory as it reduced only about 10% after the first cycle of regeneration but continued to be the same afterwards.

11.4.5 Organic Ligands

The organic–inorganic-based conjugate materials are favorable sorbents for the reason that they are very selective toward a particular ion. These materials are preferred owing to their better characteristics like enhanced surface area, better pore volume, and high internal diffusion rates. The growing need for easy and efficient methods which can be observed visually (colorimetric) has raised the demand for new optical adsorbents. Awual et al. (2015a) utilized organic ligand immobilized on mesoporous silica for abatement of Se from aqueous solutions. Immobilization of a ligand (3-(3-(methoxycarbonyl)benzylidene)hydrazinyl) benzoic acid (MBHB) on mesoporous silica produces an adsorbent with such qualities. The detection and sorption of Se ions was highly vulnerable to change in pH. The limit of detection obtained was 2.02 ppb at optimal conditions (pH-1.5, time-40 minutes). Due to extreme selectivity the presence of various ions even at high concentrations did not affect adsorption capacity substantially. The obtained sorption capability was 93.56 mg/g. Regeneration of adsorbent also showed impeccable results. The sorption of Se(IV) at 2 mg/l initial concentration in the presence of other cations having 15 mg/l conc. and anions 50 mg/l conc. and regeneration of the conjugate adsorbent over multiple cycles is depicted in Figure 11.2.

Awual et al. (2014) utilized organic ligand 6-((2-(2-hydroxy-1 naphthoyl) hydrazono) methyl) benzoic acid (HNHMB), which was impregnated on mesoporous nanosilica. Factors affecting the sorption process were optimized and it was deduced that the adsorbent had specific affinity for removal of Se(IV). The detection limit was 2.16 ppb and

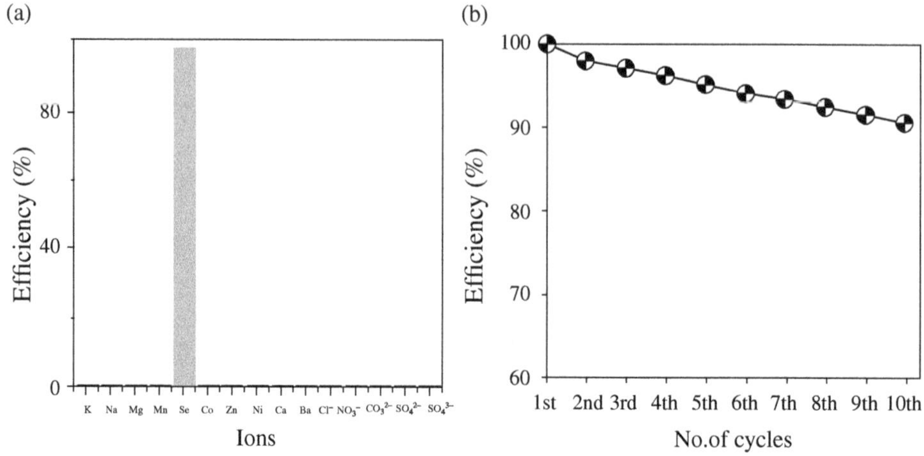

Figure 11.2 (a) Adsorption of Se(IV), selenite ions with co-existing competing ions. (b) Regeneration studies. *Source:* Reproduced with permission from Awual et al. (2015a) (License no. 4890020213884).

competing ions effect was negligible. On comparing results with MBHB based adsorption, however an enhancement in adsorption capability was observed which 103.73 mg/g was. This could be explained by strong surface complexation of the adsorbent.

Similar to the methods above Awual et al. (2015b) utilized another different organic ligand which was also immobilized on mesoporous silica. The ligand used was N, N-di (3-carboxy salicylidene)-3, 4-diamino-5-hydroxypyrazole. The purpose of using an organic ligand was the same, i.e. changes could be observed visually and the adsorbent was economical and eco-friendly. Competing ions did not interfere much with the Se removal rates. The limit of detection came out to be 1.14 ppb and pH value of 2 was observed to serve the most favorable conditions. It showcased enhanced absorption capacity of about 111.12 mg/g and was reusable for up to eight successive cycles. Competing ions in the concentration range of 10–100 ppm had no effect on the selenite adsorption. Appearance of color and signal intensity change in reflectance spectra were detected. Different concentrations of Se were noticed by difference in colors by naked eye without employing any sophisticated instrument. The fully reliable and sensitive detection was the biggest achievement of this composite adsorbent-based Se monitoring system for water/wastewater analysis. Thus, it was concluded by authors that this ligand accomplished the Se abatement better than the previously utilized ligands.

11.4.6 Se Adsorption Using Metal Oxides/Hydroxides

Zelmanov and Semiat (2013) utilized nano Fe oxide/hydroxide for adsorptive abatement of Se. Iron oxide/hydroxide nano-matrices were produced using hydrolysis, precipitation, and the sol–gel techniques. High pH values were observed to be unfavorable for the process and maximum adsorption of Se(IV) occurred at pH of 4. Increase in pH caused repellence between Se and surface of adsorbent as they both acquire negative charge. The adsorption capacity using iron oxide/hydroxide nanoparticles for Se(IV) obtained was 95 mg/g and for that of Se(VI) was 15 mg/g. One duly noted fact was that adsorption capacity obtained proved to be better than many adsorbents such as goethite, magnetite, rice husk, hematite, etc. Recovery of adsorbent was also easy, which made it cost-effective as well. Around 95–98% removal efficiency for Se could be obtained using the adsorbent at higher pH values.

11.4.7 FeOOH

Large particle size and surface properties of FeOOH make it a potential adsorbent. Its low cost of production is also in its favor. Sharrad et al. (2012) utilized FeOOH adsorbent for removal of Se(IV). The surface area obtained for this adsorbent was 361.4 m^2/g. Se(IV) sorption was observed to enhance with decline of pH and optimized (max) sorption occurred at pH of 5. Its pH dependency was attributed to OH^- ions releasing during adsorption of anions like selenite on FeOOH; however, its optimum pH range remains still broad enough (5–9). The competing ions like phosphate and silicate reduced the adsorption capability at high concentrations (50 μg/l); however, vanadium to the contrary increased it. The adsorption capacity obtained using Langmuir isotherm ($R^2 = 0.999$) was 29.4 mg/g for pH 5, temperature 35 °C and FeOOH dose of 0.5 g/l. The influence of co-existing ions on selenite adsorption at 50 μg/l is shown in Figure 11.3.

11.4.8 Natural Goethite

A sorbent similar to above is natural goethite, i.e. α-FeOOH. The advantage of this adsorbent is that it is inexpensive and readily available. Although it has a good adsorption potential, but it is less than the synthetic goethites. Jacobson and Fan (2019) utilized natural goethite (55%) having surface area measuring to 13.13 m^2/g for the sorptive abatement of Se(IV) along with As(V) using 6 l, PVC batch reactor to obtain real-time analysis of the data. The experimental investigations were executed at 4 pH while the adsorbate concentration was confined to 150 μg/L. The maximal sorption capability transpired to be 7.74 mg/g for the pH 4 using the Langmuir model, which was less than that obtained for FeOOH.

11.4.9 Biosorption

Modified biomass also finds its way to the list of adsorbents as they are environment-friendly and promote the utilization of waste into something worthwhile and profitable. Using biomass without modification is not preferred due to its low efficiency even though it has an enormous number of open bonds. Hong et al. (2020) utilized food waste-based biochar impregnated with iron (Fe-FWB) for adsorption of Se(VI) onto it. The prepared food waste after removal of impurities and drying was loaded with Fe using $FeCl_3$ solution and then pyrolyzed at 300, 450, and 600 °C. Use of biomass seems to be a very promising solution, but developments have to be made to increase the capacity for making its practical use economical. The maximum adsorption capacity obtained using 0.425 mm sized Fe-FWB at 25 °C, at pH 7 was 11.72 mg/g. It was much more than the sorption capability exhibited by other adsorbents like hematite, magnetite, nano-jacobsite, and n-TiO_2 for similar particle sizes. Effect of interactions of variables, i.e. contact time, process temperature, and amount of Fe affecting the sorbed Se(VI) concentration were depicted in a 3-D

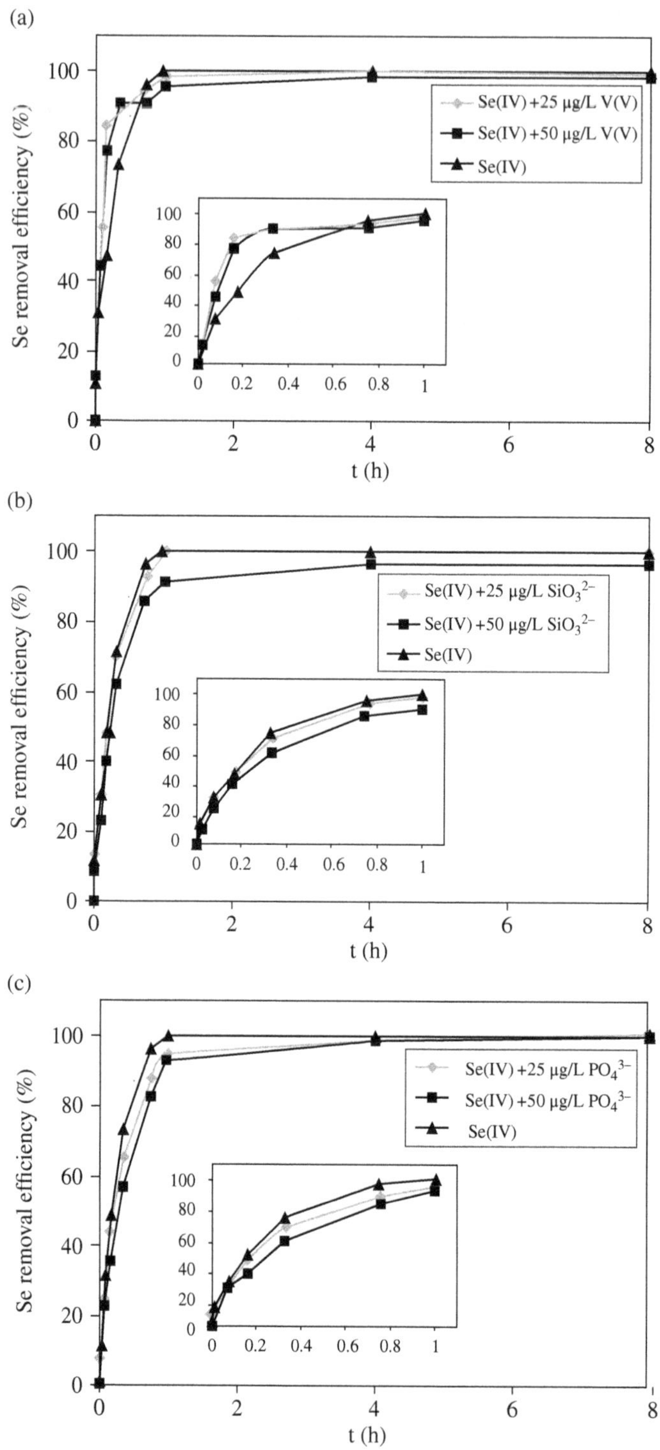

Figure 11.3 Effect of competing ions on Se(IV) adsorption (a) V(V), vanadium (b) SiO_3^{2-}, silicate (c) PO_4^{3-}, phosphate (adsorbent dosage: 0.5 g/l; initial selenium concentration: 50 μg/l; pH = 5.0). *Source:* Reproduced with permission from Sharrad et al. (2012) (License number 4890040448174).

representation using response surface method (RSM) curves vs independent variables, while others factors were kept constant. Se(VI) abatement was depicted between two experimental factors and other factors are constant. The graphical RSM curves are shown in Figure 11.4

11.4.10 Chitosan-Nanocomposite Hollow Fibers

Dorraji et al. (2016) utilized iron impregnated chitosan-based nanocomposite hollow fibers for adsorption of Se(IV). Chitosan is a linear polysaccharide obtained from chitin, found in exoskeletons of crustaceans (crabs, shrimps, etc.). Chitosan hollow fibers (CS-HFs) can be modified by impregnation of Fe_3O_4 nanoparticles to produce an efficient and cost-effective adsorbent. $FeCl_3{\cdot}9H_2O$ solution was used for synthesis of modified CS-HFs. Factors that affected the adsorption experiment most were adsorbent dosage, pH value, and concentration of Se ions. The process parameters were optimized using response surface methodology. Available sulfate ions could not hinder the Se adsorption much. However, presence of ions like chloride and nitrite were observed to be unfavorable. The process produces an eco-friendly adsorbent and the adsorption capability obtained was 1.34 μg/mg.

11.4.11 Modified Cellulose Aerogels

Aerogel is a nanoporous and low-density substance with large surface area. Newly developed cellulose aerogels are of great interest being environment-friendly, easy availability. This promotes sustainable growth. These, however, have lower adsorptive sites and thus need to be modified to meet an efficiency that makes it capable for practical operations. Liu et al. (2018) utilized modified carbon aerogels based on sisal fibers for removal of selenium. In this case cellulose has been extracted from sisal fibers. Carbon aerogels synthesized from these fibers were modified using AlOOH solution. $Mg(NO_3)_2.6H_2O$ was used to derive Mg■Al LDHs carbon aerogels. These exhibited an adsorption capacity of about 73.6 mg of Se per gram. Presence of strong acidic ions such as SO_4^{2-}, Cl^-, and NO_3^- enhances the capacity unlike the case of most adsorbents. However, presence of basic ions such as CO_3^- depletes the rate of removal considerably. Similarly, acidic conditions, i.e. low pH values increased efficiency. These also showed up to the mark recyclability.

11.4.12 Ion-Imprinted Polymeric Adsorbents

Ion-imprinted polymers have been utilized as novel sorbents owing to their exceptional property of very high selectivity for a particular ion. Their specialized pores are the reason behind this property. Se and As ion-imprinted polymers were synthesized chemically using bulk polymerization and studied for their adsorption capacity and various factors that affect it, by Mafu et al. (2016). Both batch and column studies were performed for selective adsorption of Se and As. During batch adsorption pH values beyond 7 and temperature above 25 °C showed a decrease in the sorption capacity. The maximum adsorption of around 447 μg/g was obtained at optimized parameters, i.e. pH-7, temperature of 25 °C and contact time of 280 minutes. Selectivity obtained was remarkable. For the column experiments maximum adsorption capacity obtained was 530 μg/g at same temperature and pH.

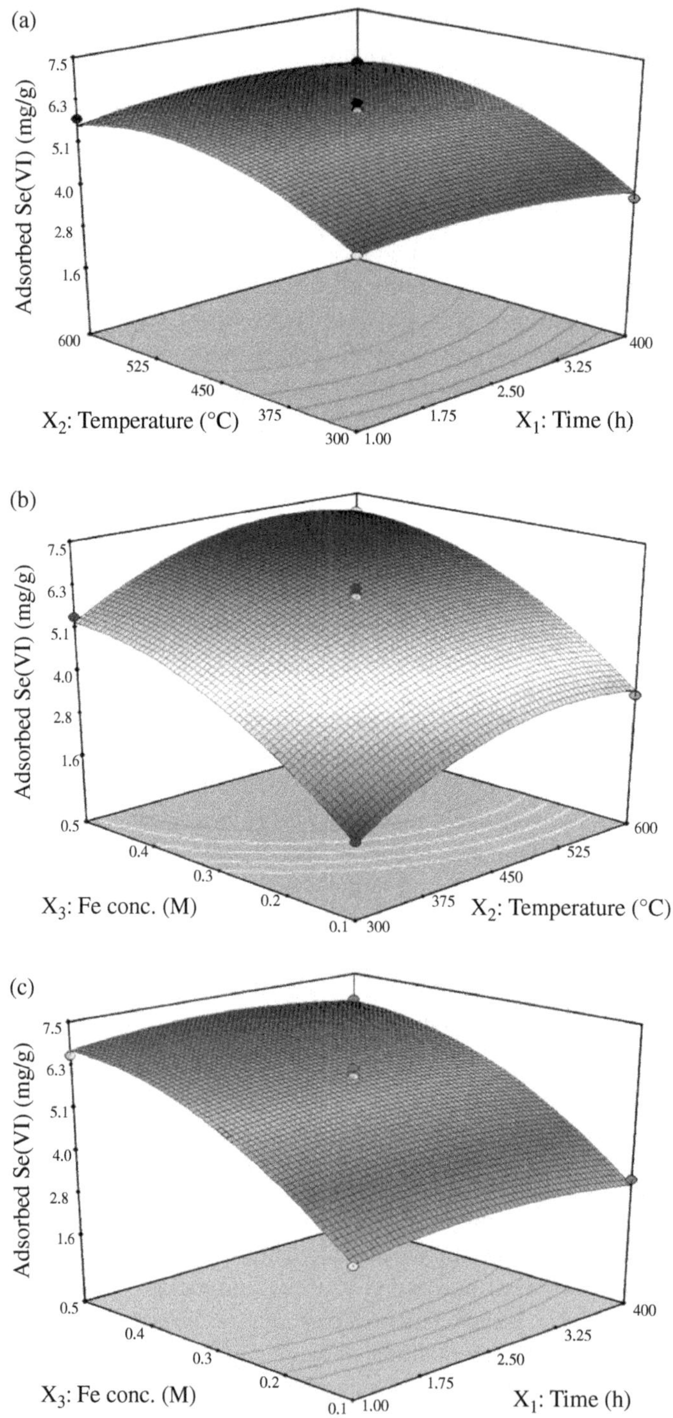

Figure 11.4 Response surface method (RSM) analysis concerning Se(VI) abatement by Fe-FWB: (a) influence of contact time and process temperature; (b) influence of process temperature and ferrous concentration; (c) influence of ferrous concentration and contact time. *Source:* Reproduced with permission from Hong et al. (2020) (License no. 4890010711526).

11.5 Combination of Adsorption and Ion Exchange

11.5.1 Iron Oxide-anion Exchange Resin

Sulfur dioxide is one of the major air pollutants. Many industries generate SO_2 as a result of their processes and thus equipment such as wet scrubbers is used. One of its byproducts is flue gas desulfurization (FGD) wastewater, which is rich in many heavy metals and non-metals including Se. Staicu et al. (2017) utilized Iron oxide functionalized anion exchange resin (FerrIX A33E) which was a commercial ion exchange resin serving the purpose of an adsorbent in their study Adsorption and ion exchange are methods that share many common features. Combination of these two was utilized to achieve better results. Experimental investigations were undertaken using varied concentrations of the resin and pH. It was observed that acidic pH of less than 4 was favorable. Sulfate ions (present in three times more amount in FGD) decreased the removal of Se due to its resemblance in nature and acting as strong competitor for adsorption. Due to this factor FerrIX was not very selective for Se. It was then necessary to treat the FGD with $BaCl_2$ to precipitate sulfur and resultant $BaSO_4$ using ferric chloride. After desulfurization, removal of Se up to 80% was achieved along with elimination of several other toxic metals as well, e.g. Cr, Zn, and Cd (Figure 11.5). The abatement of Se ions along with other existing ions after treatment with $BaCl_2$ and $FeCl_3$ is shown in Figure 11.5.

11.5.2 Ferrous-Manganese Hydrous Oxide

Ion exchange adsorbents consisting of metal hydrous oxides show great affinity and selectivity for selenium. Szlachta and Chubar (2013) utilized ferrous-manganese hydrous oxides based adsorption for selenite and selenate ions. The adsorbent contained Fe and Mn in the ratio 1:1. The process parameters affecting the adsorption and competitive ions were studied. Adsorbing material was highly permeable in nature. The adsorbent had high affinity for Se(IV) and sorption capability amounting to 41.02 mg/g was obtained for pH 4 as compared to a 19.84 mg/g capacity for the Se(VI) ions at same pH. The difference in adsorbed amounts of Se(IV) and Se(VI) were observed due to unlike strength of bonding with the surface of adsorbent. After diffusion into the pores, ion exchange is responsible for controlling the rate of adsorption. At high pH values higher concentration of negative ions of Se causes repulsion between the adsorbent and the ions. Hence the adsorption got decreased (18.45 mg/g) when the pH was enhanced to pH 8 for Se(IV) ions; however, it remained almost same (19.41 mg/g) for the Se(VI). Phosphate and silicate ions caused the removal rate to drop considerably.

11.6 Combination of Adsorption and Reduction

11.6.1 Zero-Valent Iron

More research had been made on removal of selenite in comparison to selenate. Zero-valent iron at the nano size range is commonly used for treatment of groundwater, soil, etc.

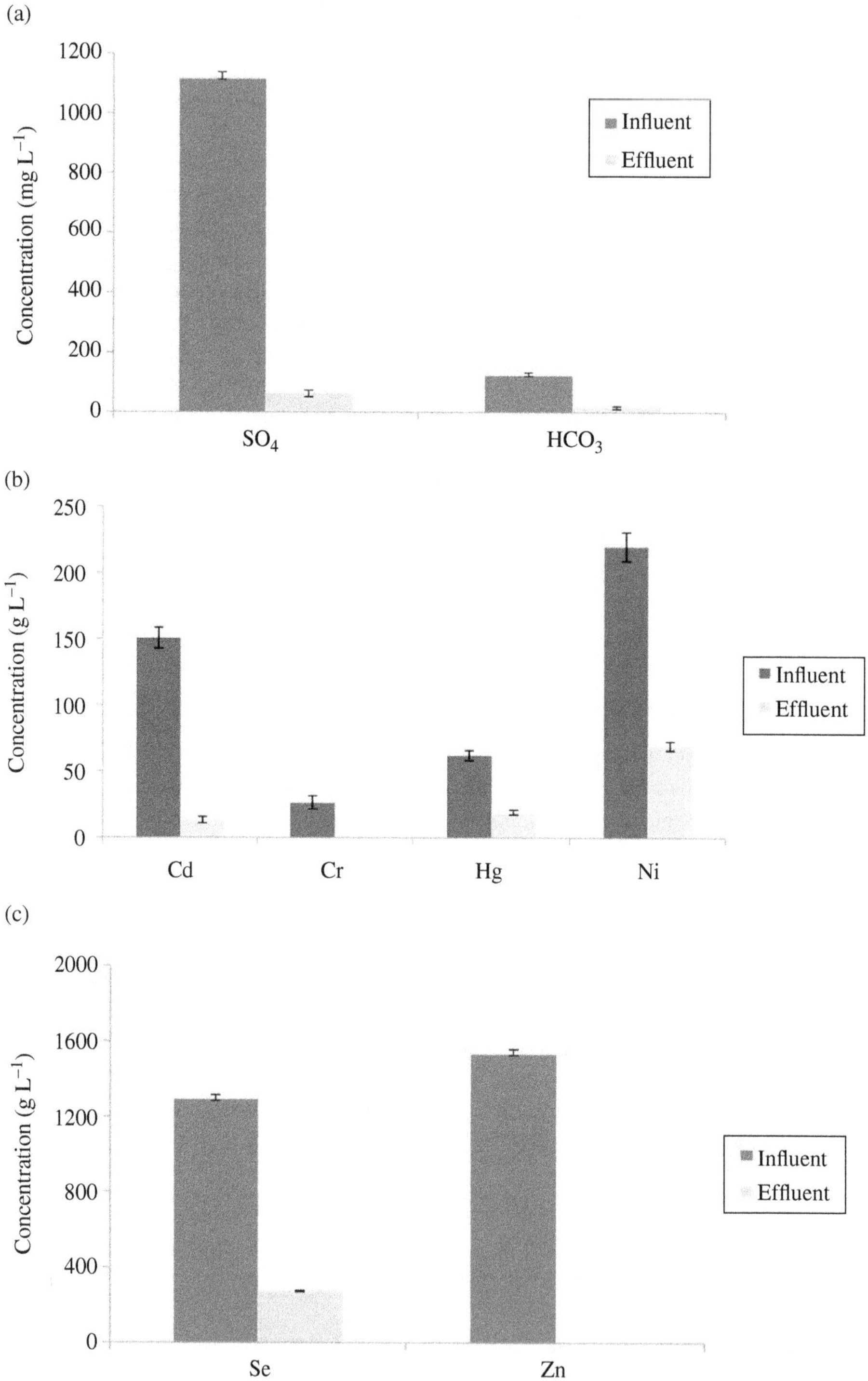

Figure 11.5 Abatement of different contaminants including Se from FGD using pretreatment ($BaCl_2$ and $FeCl_3$): (a) SO_4^{2-} and HCO_3^- (b) Cd, Cr, Hg, and Ni, (c) Se and Zn. (effluent in graphs refers to treated solution). *Source:* Reproduced with permission from Staicu et al. (2017) (License no. 4890031262779).

and detoxification of many harmful pollutants. Ling et al. (2015) utilized nano-sized, zero-valent (n-ZVI) having size 60 nm and surface area amounting to 30–35 m^2/g for adsorption of Se(IV). Reduction of Se(IV) to Se(0) is key mechanism exhibited by n-ZVI based Se adsorption and it showed very high removal rates. Surface properties of zero-valent iron made it a good adsorbent as well and higher diffusion rates were observed; 1.3 mM, Se(IV) was rapidly removed from water, utilizing 5 g/l n-ZVI in time of three minutes. High reactivity and retention capacity were also due to the defects exhibited by surface of n-ZVI. A thin film of selenium emerged over the n-ZVI surface, which was studied using surface analysis using XPS. Experiments showed that depletion of iron was observed as an outcome of iron oxidation. Se was collected between the pure iron core and oxidized iron.

11.6.2 Biosynthesized Nanoscale Zero-Valent Iron

Another method very similar to the one mentioned above is using nanoscale zero-valent iron but producing it via biosynthesis. This makes the process even more environmentally friendly. Adio et al. (2017) utilized *Aloe vera* extracts for the synthesis of nanoscale zero-valent iron. Experiments were carried out for both Se and As removal. Chemicals used in synthesizing zero-valent iron pose high risk as they are very hazardous, which makes biosynthesis a safer technique to follow. The process was observed to be mostly dependent on adsorbent dosage and shaker speed. Low pH values also showed high removal rates. 95% arsenic and 90% Se removal rates were obtained using the above adsorbent.

11.7 Precipitation and Reduction

Sulfur dioxide can reduce selenium ions to its neutral states but the reaction is too sluggish. Sulfur ions, however, are known to be able to precipitate Se with some drawbacks. Geoffroy and Demopoulos (2011) employed sodium sulfide to reduce Se(IV) from aqueous solutions. Experimental investigations were conducted at sulfide to selenium ratio above 1.8 at pH of 7. The reactions involved oxidation of sulfide ions and reduction of Se. They form a solid phase solution together. The precipitate formed was very stable and reaction was favorable below the pH value of 7. Also, the presence of sodium sulfide in excess works conversely for the reaction.

11.8 Bioreactor (Reduction)

Bioreactors use microbes for the reducing of selenium ions into elemental selenium. Soda et al. (2011) utilized a lab-scale bioreactor having anaerobic sludge as medium for reduction of soluble Se into elemental selenium. Genesis of neutral Se was attributed to the appearance of a dark red color in the bioreactor. Carbon sources such as acetate, ethanol, methanol etc. assist in reduction by acting as the electron donors. It took around two weeks to drop the concentration of Se to 0.1 mg/l from 1.5–3.6 mg/l. Using biological ways of reduction can prove to be inexpensive compared to other chemical methods but still many

Table 11.2 Methods/media for treatment/reduction of selenium from water/waste-water.

Selenium removal method	Brief description	Merits	Demerits	Reference
Ion exchange and adsorption	Using polyamine type ion exchange resins in a batch process and continuous flow packed column	Eco-friendly technology, pH range 3–12	Presence of other ions affect capacity significantly, Removal rate sensitive to feed concentration.	Nishimura et al. (2007)
Adsorption using FeS	Using mackinawite (FeS) adsorbent due to surface reaction on the active sites	Easy separation after adsorption experiment; can be reused. Sulfate ions had negligible effect on Se(IV) adsorption.	Selenium (VI) adsorption was low; only 10%	Han et al. (2011)
Adsorption using polymer-clays	Chitosan–montmorillonite clay composites for adsorbing Se	Highly effective and selective for Se; not pH dependent, not affected by competing ions like sulfur	Adsorption by oxides decreased at high pH	Bleiman and Mishael (2010)
Ferrihydrite adsorption	Uses both adsorption and precipitation as a technique for removal	High efficiency	Not effective for very dilute solutions	Higgins et al. (2009)
Photocatalytic reduction	Using binary oxide type photocatalysts under visible light, TiZr-10 and Degussa P-25	mesoporous, high surface area, formic acid and EDTA acted as hole scavenger	Under UV light Degussa P-25 is effective	Aman et al. (2011)
Phytoremediation	Uses phytoextraction and phytovolatilization to remove Se from contaminated water	Environment-friendly	Expensive to set up hydroponic conditions, constant monitoring required	He et al. (2018)
Adsorption using LDH	Ca–Al–Cl layered double hydroxides having plate like structure	Effective for Se(VI) −186.6 mg/g at 50 °C, promising for industrial application, high efficiency	Cost of preparation of LDHs	Xu et al. (2019)
Bioremediation	Using *Chlorella vulgaris* and algal treatment	Effective for Se(IV) and Se(VI), increased Se volatilization, helps in chromium removal also	—	Zou et al. (2020)
Adsorption using Al based water treatment residuals	Water treatment residuals-clays-Al hydroxide utilized as adsorbent to treat Se ions	Cost-effective; easy process, pH range 5–9	—	Ippolito et al. (2009)
Adsorption using commercial cation exchange resin	Fe^{3+} cation exchange resin used Se(IV) removal.	Highly effective for Se(IV) adsorption	Se(VI) is not adsorbed	Kawamoto et al. (2019)

cost factors and other parameters need to be well researched before implementing it on a large scale.

The important chemical treatment methods along with different adsorbents/matrices/composites have been compiled and enlisted in Table 11.2.

11.9 Conclusions

The main chemical treatment technologies for effective selenium ions removal are adsorption, reduction, and a combination of adsorption and ion exchange. The nature of adsorbents, chemical modification, functionalization, or impregnation with suitable agents/compounds, their characteristics, and the right choice of process parameters (optimized) are considered to be the decisive factors in the abatement of selenium. The chemical removal is also affected by the affinity of selenite or selenate ions toward the removing matrix, i.e. nanocomposites, ligands, resins, or a combination of these ions/adsorbents along with the competing ions existing in the wastewater. Adsorption happens to be effective chemical technique providing high removal efficiencies (95–99%) and capacity of more than 100 mg/g for many of the adsorbents.

Natural adsorbents need to be chemically activated or modified to provide efficient adsorption capabilities. Reduction is also an effective technique for chemical removal of Se. The key mechanism for removal using this technique happens to be reducing Se(IV) (selenite) and Se(VI) (selenate) ions into their elemental form, i.e. Se(0). Zero-valent iron and iron oxides/hydroxides have also showcased promising results. Nanomaterials and MNPs are also emerging as potential sorbents. Impregnation of iron nanoparticles by coupling with metals also improves the reduction of selenium ions. Photoreduction and/or photocatalysis is also an effective methodology for treatment of selenium-contaminated waters. The highly active photocatalysts working under both visible and UV irradiation are promising materials for effective reduction of selenium.

Modified-biochars and metal impregnated resins also happen to be great choice for chemical removal of Se ions. Sometimes the wastewaters/feedwaters need to be pre-treated for certain competing ions, e.g. phosphates, sulfates, and carbonates etc. which hinder or reduce the effective removal of Se. Selection of suitable adsorbent material in presence of competing ions is key step toward achieving effective removal efficiencies. The advent of novel and reusable waste adsorbent materials and recent modification at nanoscale have revolutionized the chemical abatement of Se. Hence for the optimum treatment of Se from aqueous mediums the chemical techniques discussed above can be utilized without fail and significant removal rates can be obtained by employing the suitable matrix (removing material) and optimizing the process parameters.

Acknowledgment

The authors are highly thankful to Panjab University, Chandigarh for proving the online databases and resources for preparing the book chapter.

References

Adio, S.O., Omar, M.H., Asif, M., and Saleh, T.A. (2017). Arsenic and selenium removal from water using biosynthesized nanoscale zero-valent iron: a factorial design analysis. *Process Saf. Environ. Prot.* 107: 518–527. https://doi.org/10.1016/j.psep.2017.03.004.

Aman, N., Mishra, T., Hait, J., and Jana, R.K. (2011). Simultaneous photoreductive removal of copper (II) and selenium (IV) under visible light over spherical binary oxide photocatalyst. *J. Hazard. Mater.* 186: 360–366. https://doi.org/10.1016/j.jhazmat.2010.11.001.

Awual, M.R., Hasan, M.M., Ihara, T., and Yaita, T. (2014). Mesoporous silica based novel conjugate adsorbent for efficient selenium(IV) detection and removal from water. *Microporous Mesoporous Mater.* 197: 331–338. https://doi.org/10.1016/j.micromeso.2014.07.005.

Awual, M.R., Hasan, M.M., and Khaleque, M.A. (2015a). Efficient selenium(IV) detection and removal from water by tailor-made novel conjugate adsorbent. *Sens. Actuators B* 209: 194–202. https://doi.org/10.1016/j.snb.2014.11.010.

Awual, M.R., Yaita, T., Suzuki, S., and Shiwaku, H. (2015b). Ultimate selenium(IV) monitoring and removal from water using a new class of organic ligand based composite adsorbent. *J. Hazard. Mater.* 291: 111–119. https://doi.org/10.1016/j.jhazmat.2015.02.066.

Bleiman, N. and Mishael, Y.G. (2010). Selenium removal from drinking water by adsorption to chitosan – clay composites and oxides: batch and columns tests. *J. Hazard. Mater.* 183: 590–595. https://doi.org/10.1016/j.jhazmat.2010.07.065.

Brodowska, M.S., Kurzyna-Szklarek, M., and Haliniarz, M. (2016). Selenium in the environment. *J. Elem.* 21: 1173–1185. https://doi.org/10.5601/jelem.2016.21.2.1148.

Chapman, P.M., Adams, W.J., Brooks, M. et al. (2010). *Ecological Assessment of Selenium in the Aquatic Environments*. Pensacola, FL: SETAC Press.

Combs, G.F. and Gray, W.P. (1998). Chemopreventive agents: selenium. *Pharmacol. Ther.* 79: 179–192.

Dhillon, K.S. and Dhillon, S.K. (2003). Distribution and management of seleniferous soils. *Adv. Agron.* 79: 119–184.

Donner, M.W., Cuss, C.W., Poesch, M. et al. (2018). Selenium in surface waters of the lower Athabasca River watershed: chemical speciation and implications for aquatic life. *Environ. Pollut.* 243: 1343–1351. https://doi.org/10.1016/j.envpol.2018.09.067.

Dorraji, M.S.S., Hanifehpour, Y., Joo, S.W. et al. (2016). Performance of chitosan based nanocomposite hollow fibers in the removal of selenium (IV). *Chem. Eng. Res. Des.* 117: 309–317. https://doi.org/10.1016/j.cherd.2016.10.043.

Energy Information Administration (EIA). (2020). https://www.eia.gov/electricity/?t=epmt_1_01.

Etteieb, S., Magdouli, S., Zolfaghari, M., and Brar, S.K. (2020). Monitoring and analysis of selenium as an emerging contaminant in mining industry: a critical review. *Sci. Total Environ.* 698: 134339. https://doi.org/10.1016/j.scitotenv.2019.134339.

Fernández-Martínez, A. and Charlet, L. (2009). Selenium environmental cycling and bioavailability: a structural chemist point of view. *Rev. Environ. Sci. Biotechnol.* 8: 81–110. https://doi.org/10.1007/s11157-009-9145-3.

Fu, Y., Wang, J., Liu, Q., and Zeng, H. (2014). Water-dispersible magnetic nanoparticle – graphene oxide composites for selenium removal. *Carbon* 77: 710–721. https://doi.org/10.1016/j.carbon.2014.05.076.

Geoffroy, N. and Demopoulos, G.P. (2011). The elimination of selenium (IV) from aqueous solution by precipitation with sodium sulfide. *J. Hazard. Mater.* 185: 148–154. https://doi.org/10.1016/j.jhazmat.2010.09.009.

Gonzalez, C.M., Hernandez, J., Parsons, J.G., and Gardea-Torresdey, J.L. (2010). A study of the removal of selenite and selenate from aqueous solutions using a magnetic iron/manganese oxide nanomaterial and ICP-MS. *Microchem. J.* 96: 324–329. https://doi.org/10.1016/j.microc.2010.05.005.

Gonzalez, C.M., Hernandez, J., Peralta-Videa, J.R. et al. (2012). Sorption kinetic study of selenite and selenate onto a high and low pressure aged iron oxide nanomaterial. *J. Hazard. Mater.* 211–212: 138–145. https://doi.org/10.1016/j.jhazmat.2011.08.023.

Han, S.D., Batchelor, B., and Abdel-Wahab, A. (2011). Sorption of selenium (IV) and selenium (VI) to mackinawite (FeS): effect of contact time, extent of removal, sorption envelopes. *J. Hazard. Mater.* 186: 451–457. https://doi.org/10.1016/j.jhazmat.2010.11.017.

He, Y., Xiang, Y., Zhou, Y. et al. (2018). Selenium contamination, consequences and remediation techniques in water and soils: a review. *Environ. Res.* 164: 288–301. https://doi.org/10.1016/j.envres.2018.02.037.

Higgins, T.E., Sandy, A.T., and Givens, S.W. (2009). Flue gas desulfurization wastewater treatment. *Power* 153: 1–8.

Hong, S., Ndumbe, F., Kang, J. et al. (2020). Chemosphere synthesis of Fe-impregnated biochar from food waste for selenium (VI) removal from aqueous solution through adsorption: process optimization and assessment. *Chemosphere* 252: 126475. https://doi.org/10.1016/j.chemosphere.2020.126475.

Ippolito, J.A., Scheckel, K.G., and Barbarick, K.A. (2009). Selenium adsorption to aluminum-based water treatment residuals. *J. Colloid Interface Sci.* 338: 48–55. https://doi.org/10.1016/j.jcis.2009.06.023.

Jacobson, A.T. and Fan, M. (2019). Evaluation of natural goethite on the removal of arsenate and selenite from water. *J. Environ. Sci. (China)* 76: 133–141. https://doi.org/10.1016/j.jes.2018.04.016.

Kapoor, A., Tanjore, S., and Viraraghavan, T. (1995). Removal of selenium from water and wastewater. *Int. J. Environ. Stud.* 49 (2): 137–147. https://doi.org/10.1080/00207239508711016.

Kawamoto, D., Yamanishi, Y., Ohashi, H. et al. (2019). A new and practical Se(IV) removal method using Fe 3 + type cation exchange resin. *J. Hazard. Mater.* 378: 120593. https://doi.org/10.1016/j.jhazmat.2019.04.076.

Kieliszek, M. and Błazejak, S. (2013). Selenium: significance, and outlook for supplementation. *Nutrition* 29: 713–718. https://doi.org/10.1016/j.nut.2012.11.012.

Lemly, A.D. (1999). Selenium transport and bioaccumulation in aquatic ecosystems: a proposal for water quality criteria based on hydrological units. *Ecotoxicol. Environ. Saf.* 42: 150–156.

Lenz, M. and Lens, P.N.L. (2009). The essential toxin: the changing perception of selenium in environmental sciences. *Sci. Total Environ.* 407: 3620–3633. https://doi.org/10.1016/j.scitotenv.2008.07.056.

Ling, L., Pan, B., and Zhang, W.X. (2015). Removal of selenium from water with nanoscale zero-valent iron: mechanisms of intraparticle reduction of Se(IV). *Water Res.* 71: 274–281. https://doi.org/10.1016/j.watres.2015.01.002.

Liu, Y., Jing, Z., Zhang, T. et al. (2018). Fabrication of functional biomass carbon aerogels derived from sisal fibers for application in selenium extraction. *Food Bioprod. Process.* 111: 93–103. https://doi.org/10.1016/j.fbp.2018.07.004.

Lopes, G., Ávila, F.W., and Guilherme, L.R.G. (2017). Selenium behavior in the soil environment and its implication for human health. *Ciênc. Agrotecnol.* 41: 605–615. https://doi.org/10.1590/1413-70542017416000517.

Lu, Z., Yu, J., Zeng, H., and Liu, Q. (2017). Polyamine-modified magnetic graphene oxide nanocomposite for enhanced selenium removal. *Sep. Purif. Technol.* 183: 249–257. https://doi.org/10.1016/j.seppur.2017.04.010.

Ma, Z., Shan, C., Liang, J., and Tong, M. (2018). Efficient adsorption of selenium(IV) from water by hematite modified magnetic nanoparticles. *Chemosphere* 193: 134–141. https://doi.org/10.1016/j.chemosphere.2017.11.005.

Mafu, L.D., Mamba, B.B., and Msagati, T.A.M. (2016). Synthesis and characterization of ion imprinted polymeric adsorbents for the selective recognition and removal of arsenic and selenium in wastewater samples. *J. Saudi Chem. Soc.* 20: 594–605. https://doi.org/10.1016/j.jscs.2014.12.008.

Markwart, B., Liber, K., Doig, L.E. et al. (2019). Selenium oxyanion bioconcentration in natural freshwater periphyton. *Ecotoxicol. Environ. Saf.* 180: 693–704. https://doi.org/10.1016/j.ecoenv.2019.05.004.

Nancharaiah, Y.V. and Lens, P.N.L. (2015). Selenium biomineralization for biotechnological applications. *Trends Biotechnol.* 33: 323–330. https://doi.org/10.1016/j.tibtech.2015.03.004.

Nishimura, T., Hashimoto, H., and Nakayama, M. (2007). Removal of selenium (VI) from aqueous solution with polyamine – type weakly basic ion exchange resin. *Sep. Sci. Technol.* 42: 3155–3167. https://doi.org/10.1080/01496390701513107.

Robberecht, H. and Grieken, R.V. (1982). Selenium in envirnmental waters: determination, speciation and concentration levels. *Talanta* 29: 823–844.

Santos, S., Ungureanu, G., Boaventura, R., and Botelho, C. (2015). Selenium contaminated waters: an overview of analytical methods, treatment options and recent advances in sorption methods. *Sci. Total Environ.* 521–522: 246–260. https://doi.org/10.1016/j.scitotenv.2015.03.107.

Sharma, V.K., Sohn, M., and McDonald, T.J. (2018). Chapter 8 – Remediation of selenium in water: a review. In: *Advances in Water Purification Techniques. Meeting the Needs of Developed and Developing Countries* (ed. A. Satinder), 203–218. https://doi.org/10.1016/B978-0-12-814790-0.00008-9.

Sharrad, M.O.M., Liu, H., and Fan, M. (2012). Evaluation of FeOOH performance on selenium reduction. *Sep. Purif. Technol.* 84: 29–34. https://doi.org/10.1016/j.seppur.2011.07.011.

Sinharoy, A., Saikia, S., and Pakshirajan, K. (2019). Biological removal of selenite from wastewater and recovery as selenium nanoparticles using inverse fluidized bed bioreactor. *J. Water Process Eng.* 32: 100988. https://doi.org/10.1016/j.jwpe.2019.100988.

Soda, S., Kashiwa, M., Kagami, T. et al. (2011). Laboratory-scale bioreactors for soluble selenium removal from selenium refinery wastewater using anaerobic sludge. *Desalination* 279: 433–438. https://doi.org/10.1016/j.desal.2011.06.031.

Staicu, L.C., Morin-Crini, N., and Crini, G. (2017). Desulfurization: critical step towards enhanced selenium removal from industrial effluents. *Chemosphere* 172: 111–119. https://doi.org/10.1016/j.chemosphere.2016.12.132.

Szlachta, M. and Chubar, N. (2013). The application of Fe–Mn hydrous oxides based adsorbent for removing selenium species from water. *Chem. Eng. J.* 217: 159–168. https://doi.org/10.1016/j.cej.2012.11.100.

Tan, L.C., Nancharaiah, Y.V., van Hullebusch, E.D., and Lens, P.N.L. (2016). Selenium: environmental significance, pollution, and biological treatment technologies. *Biotechnol. Adv.* 34: 886–907. https://doi.org/10.1016/j.biotechadv.2016.05.005.

Wen, H. and Carignan, J. (2007). Reviews on atmospheric selenium: emissions, speciation and fate. *Atmos. Environ.* 41: 7151–7165. https://doi.org/10.1016/j.atmosenv.2007.07.035.

Wiberg, E., Wiberg, N., and Holleman, A.F. (2001). *Inorganic Chemistry*. New York;, NY: Academic Press.

Winkel, L.H.E., Johnson, C.A., Lenz, M. et al. (2011). Environmental selenium research: from microscopic processes to global understanding. *Environ. Sci. Technol.* 46: 571–579. https://doi.org/10.1021/es203434d.

Xiao, W., Yan, B., Zeng, H., and Liu, Q. (2016). Dendrimer functionalized graphene oxide for selenium removal. *Carbon N. Y.* 105: 655–664. https://doi.org/10.1016/j.carbon.2016.04.057.

Xu, S., Li, D., Guo, X. et al. (2019). Selenium (VI) removal from caustic solution by synthetic Ca−Al−Cl layered double hydroxides. *Trans. Nonferrous Met. Soc. Chin.* 29: 1763–1775. https://doi.org/10.1016/S1003-6326(19)65084-3.

Yamani, J.S., Lounsbury, A.W., and Zimmerman, J.B. (2014). Adsorption of selenite and selenate by nanocrystalline aluminum oxide, neat and impregnated in chitosan beads. *Water Res.* 50: 373–381. https://doi.org/10.1016/j.watres.2013.10.054.

Zelmanov, G. and Semiat, R. (2013). Selenium removal from water and its recovery using iron (Fe 3+) oxide/hydroxide-based nanoparticles sol (NanoFe) as an adsorbent. *Sep. Purif. Technol.* 103: 167–172. https://doi.org/10.1016/j.seppur.2012.10.037.

Zeng, H. (2009). Selenium as an essential micronutrient: roles in cell cycle and apoptosis. *Molecules* 14 (3): 1263–1278. https://doi.org/10.3390/molecules14031263.

Zou, H., Huang, J., Zhou, C. et al. (2020). Science of the total environment mutual effects of selenium and chromium on their removal by Chlorella vulgaris and associated toxicity. *Sci. Total Environ.* 724: 138219. https://doi.org/10.1016/j.scitotenv.2020.138219.

12

Biological Treatment Advancements for the Remediation of Selenium from Wastewater

Vishal Singh[1], *Gyanendra Tripathi*[2], *and Vishal Mishra*[1]

[1] *School of Biochemical Engineering, Indian Institute of Technology (BHU), Varanasi, Uttar Pradesh, India*
[2] *Department of Bioengineering, Integral University, Lucknow, Uttar Pradesh, India*

12.1 Introduction

Selenium (Se) finds its wide applications in various industrial sectors such as photoelectric cells, electronics, stainless steel, glass manufacturing, metallurgical additives, and pesticides as well as in the field of genetic, molecular, health, and biochemical areas. This process leads to the generation of Se laden wastewater containing selenate (SeO_4^{2-}) and/or selenite (SeO_3^{2-}) (Chung et al. 2006a, c; Lenz and Lens 2009; Tan et al. 2016). Also, combustion of coals having selenium during the generation of power in plants releases SeO_3^{2-} and/or SeO_4^{2-} in the flue gas (Yudovich and Ketris 2006; Lenz and Lens 2009). Some phenomena occurring in natural environments such as pyrite oxidation release a large amount of SeO_4^{2-} and sulfate in the groundwater and mine drainage (Pettine et al. 2012). Release of wastewater containing SeO_3^{2-} and/or SeO_4^{2-} from agricultural sources was the cause of increase in Se pollution of wetlands in the San Joaquin Valley, CA (Presser 1994). Selenium pollution widely affects the human life, agricultural processes, and ecosystems. Plant species absorb Se present in the soil and water bodies via roots, through which Se enters into the food chains, raising concerns for the long-term impact on the environment (Brown and Arthur 2001; Staicu et al. 2015).

Various authors have reported different treatment technologies based upon physiochemical and biological principles for the removal of Se from wastewater (Gerhardt et al. 1991; Twidwell et al. 2005; Zhang et al. 2008; Hill 2013). Application of iron salts for chemically precipitating Se species from the industrial wastewater have been widely applied. But these types of chemical techniques generate contaminated sludge which is not recyclable as well as requiring extreme care during the handling process (Soda et al. 2011). The biologically mediated treatment has been widely researched treatment technology for the reduction of Se concentration from wastewater. Biological treatment process presents a cheaper alternative to physical and chemical treatment process with an added advantage

Selenium Contamination in Water, First Edition. Edited by Pooja Devi, Pardeep Singh, Arindam Malakar, and Daniel Snow.

of the ability to adapt to the varying characteristics of wastewater and conditions of environment. While the large-scale biological treatment process has been limited to a few places only, multiple pieces of research conducted at the small-scale level have proved that biologically mediated treatment is efficient in achieving the prescribed limit of Se discharge into the water bodies (Gusek et al. 2008; Rutkowski et al. 2013).

Biological treatment technologies work on the principle of reducing and/or volatilizing the Se oxyanions via different configurations of the bioreactor, construction of wetlands or through the process of phytoremediation. These technologies utilize aerobic and anaerobic bacteria, fungi, algae, and plants that can perform the catalyzation of Se under varying environmental conditions (Hill 2013). These technologies are also able to use the organic sources of electron donor that are present itself in the wastewater stream or added from external sources (Hageman et al. 2013). Biological treatment technologies work by reducing the soluble and toxic oxyanions of Se to less toxic and insoluble elemental form of Se (Se^0) and the process is controlled by various parameters such as temperature, pH, and electron donors (Lenz and Lens 2009). Se^0 is then recovered from the treated water via filtration, gravity settling, electrocoagulation, or precipitation by using other metals (metal co-precipitation) (Staicu et al. 2015; Mal et al. 2016).

This chapter reports the various biological treatment technologies that utilizes bacteria, algae, plant, and fungi. It also briefly describes the mechanism of Se uptake by these biological species.

12.2 Bacteria-Mediated Selenium Removal

Several bacterial species capable of reducing and tolerating Se oxyanions have been isolated from both contaminated and natural environment in the recent years, which enables the better comprehension of the process involved in the Se bioremediation process (Nancharaiah and Lens 2015). These Se-reducing bacterial species perform the Se removal process in several ways such reduction of Se oxyanions into elemental Se (Se^0), volatilizing the selenium species (dimethyl selenide or hydrogen selenide) or directly incorporating them into Se amino acids (methyl selenocysteine). The process of Se reduction by bacteria has been explained in the further sections (El-Ramady et al. 2014).

12.2.1 Mechanism

In the case of bacteria (eubacteria and archaea), Se is metabolized via two pathways: (i) dissimilation reaction for the Se oxyanions reduction to Se^0 and further its reduction to selenides; (ii) incorporation of reduced Se directly into amino acids (SeCys) which results in the accumulation of selenoproteins (Fisher and Hollibaugh 2008). In the biogeochemical cycle of selenium, anaerobic microorganism performs the dissimilation reduction reaction, which is the primary reduction reaction performed to inhibit anoxic deposits (Oremland 2004). Several SeO_3^{2-} reducing bacteria has been identified that readily reduce the selenite compound (Losi and Frankenberger 1997), while only a smaller number of SeO_4^{2-} reducing bacteria have been identified (Maiers et al. 1988; Oremland et al. 1994; Fujita et al. 1997). A SeO_3^{2-} reduction mechanism has been discovered in both anaerobic and aerobic bacteria (Viamajala et al. 2006). Conversion of SeO_3^{2-} to Se^0 is performed by the detoxification mechanism mediated by glutathione (Turner et al. 2012), siderophores (Zannoni et al. 2007), glutaredoxin (Nogueira and Rocha 2011), fumarate reduction (Li et al. 2014), or nitrite reductase (Schröder et al. 1997). Selenate-reducing bacteria have shown widely different reduction mechanism, e.g. SES-3 and *Thauera selenatis* performs the dissimilatory anaerobic reduction of SeO_4^{2-}, while *Pseudomonas stutzeri* performs the reduction of SeO_4^{2-} only for detoxification (Oremland et al. 1994). Figure 12.1 represents a simplified representation of the biochemical cycle of selenium and its reduction pathway performed by bacteria.

As shown in Figure 12.1, anthropogenic and geologic sources such as coal and oil mining and combustion facilities, selenium refining industries, milling operations, and semiconductor manufacture are responsible for the Se release as SeO_4^{2-} in the water bodies. SeRB (selenium-respiring bacteria) are present everywhere in environment and perform an essential function in the biochemical cycle of selenium. Both the reduction of Se oxyanions and organic matter degradation are coupled together in the aerobic sediments. Bacteria can reduce both SeO_3^{2-} and SeO_4^{2-} to Se^0 under anaerobic condition. Also, few bacteria perform the reduction SeO_4^{2-} and directly incorporate them into organic compounds such as selenoproteins. Some SeRBs such as *Sulfurospirillum barnesii* and *Bacillus arseniciselenatis* reduce Se species (Se (IV) and Se (VI)) and result in Se^0 formation, which is insoluble in nature. This Se^0 is reduced further by other bacteria (e.g. *Bacillus selenitireducens*) resulting in the production of selenides. SeOB (selenium oxidizing bacteria) such as *Thiothrix*, *Thiobacillus*, etc. perform the oxidization of selenite in the aerobic environment. Volatile species of Se, which includes $(CH_3)_2Se$ and $(CH_3)_2Se_2$, are produced through alkylation reaction. Bacteria also perform the vital dealkylation reactions both in the water and soil (Stolz et al. 2006; Lenz and Lens 2009).

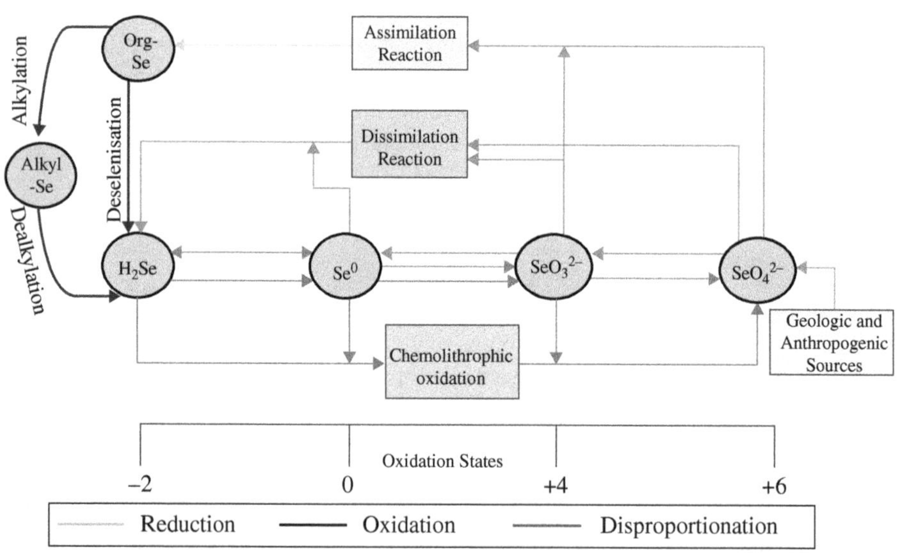

Figure 12.1 Simplified representation of the biochemical cycle of selenium and its reduction pathway performed by bacteria (Nancharaiah and Lens 2015).

12.2.2 Treatment Technologies

Table 12.1 represents various technologies used for the treatment of Se-utilizing bacteria along with technology type, experimental condition, and Se removal efficiency.

12.2.2.1 Biofilm Bioreactor

Bioreactors are capable of treating selenium and related co-oxyanions by the application of microorganism without the use of harmful chemicals, thus generating less chemically contaminated sludge. Moreover, it allows the flexibility for the manipulation of operating parameters, thus optimizing the treatment conditions. Effluents generated from multiple sources including power generation, mining, agricultural run-off, and other industries have been treated using different kinds of bioreactors as discussed in the upcoming section (Tan et al. 2016).

Biofilm-based bioreactors including rotating biological contactors, fluidized bioreactors, and static or moving bed filter have been recently researched for the treatment of metal-laden wastewater. Biofilm is a type of attached growth system in which microbes grow on a solid substrate such as rock, sand, activated carbon and moistened by wastewater. Microorganisms growing on the surface produce EPS (extracellular polymeric substances) that assist them in the attachment with carrier or support material and also helps the microbes in the process of entrapping and complexing metal cations and biomineralizing the Se^0 precipitates. These bioreactors also require solid–liquid separators, e.g. membrane separator, gravity clarifies, and plate separator for the separation of suspended microbes (Hill 2013). Three types of biofilm-based bioreactors: (i) ABMet (Advanced Biological Metals removal); (ii) FBR (Fluidized biofilm reactor); (iii) MBfR (Membrane biofilm reactors).

12.2.2.1.1 ABMet ABMet was developed by General Electric (GE), USA that utilizes the anoxic biological reactor for the removal of Se. It is a type of fixed-film, full-scale, and packed bed biofilm-based bioreactor. It consists of biological and filtration units where porous granular activated carbon is used as the solid substratum for the formation of biofilm and preventing their washout. Se reduction is performed by naturally occurring consortia of bacteria, and molasses is used as an e− donor. This bacterial consortium requires low nutrient supply, therefore reducing the cost of operation and maintenance. Also, the design of the reactor is based upon plug-flow criteria that leads to the development of an ORP (oxidation–reduction potential) gradient along the vertical axis of the system whose range varies from 0 to 200 mV. This gradient provides varied redox potential, thus increasing the efficiency of reduction of oxyanions (Sonstegard et al. 2010).

This bioreactor has been able to decrease the concentration of Se within 2 to 16 hours empty bed contact time (EBCT) via the process of reduction and selenium. Here, EBCT predicts the wastewater residence time in the bioreactor, during which it is assumed that there is no development of biofilm or presence of solids in the bioreactor. Reduced selenium is then removed by a backwash cycle from the bioreactor. It has been reported that this system can remove 99.3% of selenium from the wastewater (Sonstegard et al. 2008). It was also able to eliminate nitrates and other heavy metals. The system is also independent of the temperature, as it has shown consistent reduction of Se throughout the year. The

Table 12.1 Treatment technologies utilizing bacterial species for the treatment of selenium.

Technology type	Technology	Wastewater source	Experimental condition	Se removal efficiency (%)	Reference
Biofilm bioreactor					
	ABMet®	Flue-gas desulfurization	15–40 °C, pH 7.0, 2–16 h HRT, molasses (e^- donor)	~97%	Sonstegard et al. (2008, 2010)
	Envirogen FBR	Mine and selenium refinery plant	0.5–1.0 h HRT, 8–10 °C	~96%	Hill (2013)
	H_2-MBfR	Various sources	pH 7.0–9.0	~90%	Chung et al. (2010), Van Ginkel et al. (2011)
	BSeR	Mining	Molasses (e^- donor)	~97%	Nancharaiah and Lens (2015)
Sludge-based bioreactors					
	UASB reactor	Synthetic wastewater	pH 7.0, 30 °C and HRT 6 h, lactate (e^- donor)	99% (methanogenic condition) and 97% (SO_4^{2-} reducing condition)	Lenz et al. (2008)
	UASB reactor	Selenium refinery plant	24 h HRT, 30 °C, ethanol (e^- donor)	64–70%	Soda et al. (2011)
	SBR	Synthetic sewage	24 h HRT, glucose, fructose, sucrose, etc. (e^- donor)	90%	Kashiwa et al. (2000, 2001)
Chemostat reactor		Synthetic Wastewater	95.2 h HRT, pH 7.8–8.0, and 30 °C	99%	Fujita et al. (2002)
Jar fermenter		Synthetic wastewater	48 h HRT, pH 9.0, 38 °C, 250 rpm	80%	Kagami et al. (2013)

ABMet: Advanced Biological Metals removal; UASB: Upflow anaerobic sludge blanket reactor; MBfR: Membrane biofilm reactor; FBR: fluidized bed reactor; SBR: Sequencing batch reactor.

disadvantage of the system includes the large requirement of land, wastewater having high suspended solids (SS) loads requires pretreatment, and carbon source competition by the other ions present in the wastewater (Sonstegard et al. 2010).

12.2.2.1.2 FBR (Fluidized Bed Reactor) System Fluidized Bed Reactor (FBR) system was developed by where wastewater is passed at high velocities and media suspended on the solid carrier support through recirculation, thus maintaining a plug flow system within a suspended bed column. In this system, a thin film is maintained on the solid support allowing a high rate of mass transfer and high volumetric efficiency, generating a large amount of microbial biomass. Microbes are grown on a solid support made up of sand or activated charcoal which provide them with a large surface area for the creation of an anoxic environment for the Se reduction (Hill 2013). This system can operate at very low temperature of 10 °C and short HRT varying between 0.5 and 1 hour, which depends upon the expanded bed height. A solid–liquid separator is required for the removal of generated biomass and reduced selenium. A bench-scale FBR has shown up to 96% of removal efficiency for the Se (Hill 2013). Due to the complete mixing in the bioreactor, this system offers the advantage of low land requirement and low cost of installation, but have some disadvantages similar to ABMet technology such as loss of biomass and competition for carbon source (Tan et al. 2016).

12.2.2.1.3 MBfR (Membrane Biofilm Reactor) System Use of Hydrogen gas based MBfR system for the removal of Se at high removal rate has been reported by various authors (Chung et al. 2006a, b, c, 2010; Van Ginkel et al. 2011). In this system, H_2 gas is allowed to reach to the crossing point of biofilm and hollow-fiber film (membrane) via gas transfer membranes. These gas transfer membranes are generally bubble-less. Biofilm consists of autohydrogenotrophic bacteria such as *Cupriavidus metallidurans* (Van Ginkel et al. 2011). Following reaction (Eq. 12.1) takes place in which selenium oxyanions are reduced by accepting an electron from H_2 which in turns gets oxidized:

$$SeO_4^{2-} + 3H_2 + 2H^+ \rightarrow Se^0 + 4H_2O \qquad \Delta\mathrm{G} = -71\,\mathrm{KJ/electron} \qquad (12.1)$$

Microorganisms are able to harvest energy through the above coupling redox reaction. Supply of H_2 directly to the interface has shown increased efficiency of Se removal. Main cost component of the system is the membrane which can be made up of both inorganic or organic material such as hollow-fiber bundles or sheet. Hollow-fiber membranes are capable of operating at high pressure of gas without the bubble formation and provides high surface/volume ratio due to small outside diameter (0.1 mm) (Tan et al. 2016). Chung et al. (2006c) performed removal studies of Se using the MBfR system and obtained more than 95% removal of Se. Since H_2 gas is utilized by MBfR system as the e^- donor, it offers an added advantage of almost complete reduction of chemical oxygen demand (COD). However, use of H_2 gas increases the cost of operation and also system may dysfunction at lower pH (Chung et al. 2006b).

12.2.2.1.4 BSeR System In the BSeR system, anaerobic bioreactors having fix-film biofilm developed on the exterior area of granular activated carbon are connected in series and applied for the removal of Se. A pilot-scale BSeR system was set up at Garfield Wetlands-Kessler Springs, UT, having the capacity of 1892 l and used for the treatment of water having composition 49 μg SeO_3^{2-} per l and 1870 μg SeO_4^{2-} per l. This reactor utilizes agriculture-grade based molasses as an electron donor. Biofilm was separated from the effluent of the reactor using a slow sand filter. Final concentration of Se in the effluent was 2 μg/l, which was very low in comparison to the regulatory discharge limit set to 50 μg/l. Almost 97% of Se^0 was obtained from this system (Nancharaiah and Lens 2015).

12.2.2.2 Sludge-Based Bioreactors

Various types of sludge generated from the secondary wastewater treatment process, denitrifying process, sulfate-reduction process, and methanogenic process harbors several Se-reducing bacteria (Soda et al. 2011a). Some of the sludge-based treatment processes are described in the given sections.

12.2.2.2.1 UASB Reactor UASB reactors (upflow anaerobic sludge blanket reactors) are implemented worldwide for the remediation of contaminants generated from industries through the anaerobic process (Nnaji 2014). In these reactors, a thick layer of sludge generates at the bottom in which the biological treatment of effluent takes place (Seghezzo et al. 1998). Under the favorable conditions such as pH and temperature, bacteria naturally aggregate to form a granule having the size in the range of millimeters that settles at the bottom of the reactor (Nnaji 2014). This phenomenon decreases the biomass loss from the system when operated at industrial scale (Bhunia and Ghangrekar 2007). Also, due to the retention of biomass in the reactor, it can be operated even when the organic load of the wastewater is high. The direction of the influent flow in the reactor is toward in the upward direction which causes turbulence during the flow, maintaining good interaction between the wastewater and biomass and also enhances the production of biogas during the anaerobic treatment of wastewater (Seghezzo et al. 1998). Reduced Se may wash out with the treated wastewater or get precipitated into the sludge, which is then recovered via filtration or solid–liquid separator (Staicu et al. 2015).

In order to optimally design the UASB reactor, following factors are taken into consideration: (i) composition of the industrial effluent; (ii) upflow velocity; (iii) loading rates; and (iv) pH. UASB reactor having pilot-scale capacity was utilized for the Se removal from the agricultural wastewater generated from the San Joaquin Valley. This system used methanol as an electron donor, and 58–90% reduction in Se concentration was obtained (Golder Associates Inc 2009). The addition of Se-reducing strains such as *S. barnesii* to the UASB reactor increased the overall process of Se removal and also increased the removal efficiency of nitrates (Lenz and Lens 2009).

One problem that arises during the treatment process is the retention of the biogenic Se in the reactor. A study reported that during the operation of the reactor at a higher temperature (55 °C) increased the retention time of biogenic Se as well as the removal efficiency of the Se (Dessì et al. 2016). Therefore, operating the reactor at higher temperature can offer an advantage for the treatment of effluents having high temperatures such as in a WFGD (wet flue-gas desulfurization) scrubbing waste. Up to 94% removal efficiency of the

Se was obtained while operating the reactor at 55 °C. Also different types of biogenic Se such as nanospheres and nanorods were developed during the operation of the reactor in thermophilic conditions (Dessì et al. 2016).

Some of the advantages of a UASB reactor includes the capacity of treating wastewater having a higher organic load, low land requirement, and less biomass loss, as explained earlier. As in the other reactor configurations, carbon source competition also occurs in this reactor (Tan et al. 2016).

12.2.2.2.2 SBR Reactor A SBR reactor (sequencing batch reactor) is a type of bioreactor in which both the treatment and separation of solids occur in the same reactor. A SBR reactor is operated in batch mode in a sequence of steps which includes: (i) filing of the reactor with the wastewater; (ii) bioremediation of Se; (iii) settling of reduced Se; (iv) decanting; and (v) reactor in idle mode. All of the above steps are operated in cyclic mode, providing complete aeration or operating in an alternate mode of aerobic-anaerobic phase during the bioremediation process (Chan et al. 2009). *Bacillus sp.* (Strain-SF-1) was used for the Se species removal from synthetic sewage cultivated in a lab-scale SBR. Various types of electron donors such as glucose, fructose, sucrose, citrate, acetate, lactate, glycerol, and pyruvate were used, and up to 90% Se removal was obtained in a 24-hour cycle of operation (Kashiwa et al. 2000b, 2001). Rege et al. (1999) isolated denitrifying bacterial consortia from wastewater treatment plant situated in Pullman (Washington, USA) and utilized for the reduction of Se during cultivation in a 1.5 l SBR, and acetate was used as the e^- donor (Rege et al. 1999).

SBR offers various advantages such as high Se removal efficiency even at low operation maintenance and allows flexibility in the influent flow. Also, there is less requirement for land as it allows the integration of both aerobic and anaerobic operations. Complex set-up of operation and difficulty in controlling the aerobic and anaerobic microbial consortia are the main disadvantages of the SBR system (Chan et al. 2009).

12.2.2.3 Selenium Volatilization

In this technique, reduction of Se is done by transforming it into the gaseous phase and later recovering it by gas trapping. But microbes perform volatilization process at a slow rate, due to which this approach cannot be applied to treat large volumes of effluents generated from the industries. Also, volatile selenium compounds generated during the process are toxic in nature and this limits its applicability (Tan et al. 2016).

P. stutzeri (Strain: NT-1) isolated from the wastewater of drainage generated from Se refinery plant was able to reduce Se concentration during cultivation in aerobic conditions (Kuroda et al. 2011). This strain was further cultivated in 5 l fermenter, reducing the Se concentration by 80%. Volatize Se was recovered by the gas trapping technique using 70% HNO_3, and the recovery efficiency was around 82% (Kagami et al. 2013).

12.3 Algae-Mediated Se Removal

Algal species have been thoroughly studied for their ability to treat wastewater (Singh and Mishra 2019). There are various reports which proved that algae could be efficiently utilized for the Se treatment (Gerhardt et al. 1991; Huang et al. 2013). The main advantages for utilizing microalgae for the treatment of Se include low-cost technology, harvesting of the algal biomass for reuse (biofuel, food supplement, fertilizer), and in situ application (Tan et al. 2016). Mechanism of Se uptake and treatment technologies utilizing algae are discussed later in the section.

12.3.1 Mechanism

Schematic diagram of Se uptake by algal cells and its metabolism by the cells is shown in Figure 12.2. As per the research conducted by various authors, specifically no transporter for the Se has been reported in algae, and its uptake is usually through transporter proteins that are not specific for particular ions (Sors et al. 2005; Umysová et al. 2009; Schiavon and Pilon-Smits 2017). Dominant Se species present in the soluble form are SeO_4^{2-} and SeO_3^{2-} (Cutter and Bruland 1984; Plant et al. 2004), whereas the biological form of selenium such as selenoproteins, methyl selenides, and selenoamino acids rarely exist in the aquatic bodies (Fan et al. 2002). Therefore, microalgae can uptake either selenite or selenate. But selenite compounds are less available to microalgae than the selenate compounds as they are

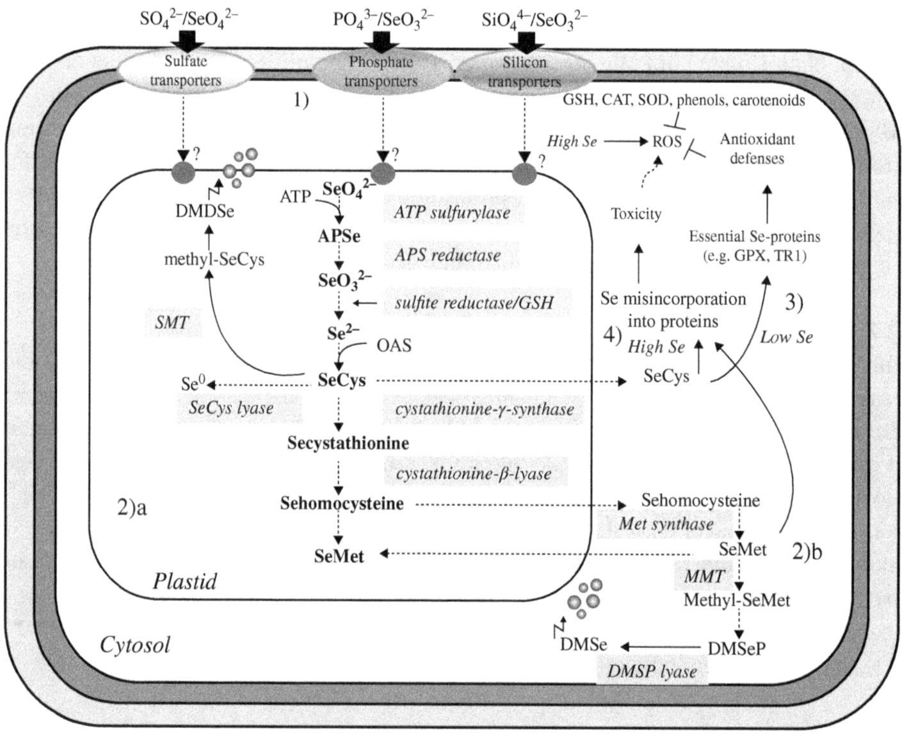

Figure 12.2 Schematic representation of Se uptake by algal cells and their metabolism. (1) Uptake of Se via sulfate, silicate, and phosphate transporters; (2) Se/sulfur assimilation pathway: OAS: O-acetylserine; APSe: adenosine phosphoselenate; GSH: glutathione; (Se)Cys: (seleno)cysteine; (Se) Met: (seleno)methionine; DMSeP: dimethylselenoproprionate; MMT: methylmethionine methyltransferase; SMT: selenocysteine methyltransferase; DM(D)Se: dimethyl(di)selenide (volatile) (Schiavon et al. 2017).

readily absorbed on the rocky surfaces, including both organic and mineral matter (Chapman et al. 2010; Vriens et al. 2016). Kinetic analysis has revealed that various systems operate across the surface of the cell membrane of the microalgal cells for the uptake of Se (Morlon et al. 2006). Two main mechanisms of Se uptake have been identified in the microalgae by selenite [^{75}Se] radiotracer technique: (i) saturated transport system having high affinity and being ATP-dependent which becomes active in a short period after the exposure Se; (ii) passive transport system having low affinity that is proportional to the external concentration of Se and becomes active after the long exposure of Se (Obata et al. 2004; Araie et al. 2011).

Availability of selenium compounds depends upon the other factors such as pH and oxidizing/reducing environment of the wastewater (Geering et al. 1968; Riedel and Sanders 1996; Tuzen and Sari 2010) as SeO_4^{2-} is highly stable under strong oxidizing condition while SeO_3^{2-} is stable under weak oxidizing condition (Geering et al. 1968). Regarding the effect of pH, it determines the level of protonation of the ionic species of the selenium controlling the formation of biologically active species of selenium which promotes the uptake of Se compounds by microalgae. Under the pH range of 5–9, selenic acid completely dissociates, thus increasing the uptake of selenate by algae (Riedel and Sanders 1996), while selenous acid belongs to a weak acid category and dissociates only when pH is lower than 7 and Se is absorbed in the form of selenite by algae (Riedel and Sanders 1996; Morlon et al. 2006). The algal membrane requires a specific pH for the proper functioning of the Se membrane transporters and maintains the overall charge of the surface. Thus, change in pH can affect the optimal efficiency of Se uptake. Riedel and Sanders (1996) reported that optimum uptake rate of SeO_3^{2-} and SeO_4^{2-} occurred at pH values lower than 5 and near, respectively, in the case of green algae *Chlamydomonas reinhardtii* (Riedel and Sanders 1996).

Uptake of Se is also affected by the competing ionic species such as phosphate, sulfate, and silicate (Lo et al. 2015; Vriens et al. 2016). SeO_4^{2-} and sulfate ions both share high chemical similarity and compete for the attachment on sites of sulfate permeases for the uptake by the cells (Cabannes et al. 2011; Schiavon et al. 2015; Schiavon and Pilon-Smits 2017). Various authors have reported the decrease in the selenate uptake by algal cells when grown under ambient sulfate concentration (Williams et al. 1994; Fournier et al. 2010; Simmons and Emery 2011; Lo et al. 2015; Vriens et al. 2016). In a study conducted by Vriens et al. (2016), when *C. reinhardtii* was grown at sulfate concentration of 600 μM, it led to the decrease in selenate uptake by 55% as compared to the control containing 150 μM sulfate. In the same study, the periodic increase in the selenium concentration decreased the sulfate concentration in the microalgae (Vriens et al. 2016).

As already mentioned, that Se shares chemical similarity with the sulfur. Therefore, once the Se is absorbed by the cells, it enters the assimilation pathway of S reduction and gets converted into selenide, which further gets converted into Se amino acid selenomethionine (SeMet) and selenocysteine (SeCys) (Neumann et al. 2003). These amino acids are then methylated by some microalgae species, resulting in the production of volatile compounds, and/or accumulated in the form of Se^0 (Elemental selenium) (Neumann et al. 2003; Vriens et al. 2016). Interesting to note that some microalgae species requires Se for their metabolism; therefore they hold the mechanism of Se incorporation in the amino acids as discovered in mammals and humans (Novoselov et al. 2002).

12.3.2 Treatment Technologies

12.3.2.1 Constructed Wetland

Constructed wetland are artificially designed systems that are effective in reducing Se concentration from various sources of wastewater such as agricultural wastewater (Gao et al. 2003) and oil refineries (Hansen et al. 1998). In the constructed wetland system, microbially mediated dissimilation process occurring within the sediments performs the anaerobic reduction of Se oxyanions into Se^0 (Masscheleyn and Patrick 1993). The maximum portion of the Se can be absorbed in the deposits, but the risk of Se ecotoxicity occurs due to Se buildup (Saiki and Lowe 1987; Hamilton 2004). This risk can be reduced by biologically volatilizing the selenium using the application of microalgae, as discussed in section 12.3.1. Therefore, based upon this fact that algae can volatize Se (Neumann et al. 2003), a system was build up by integrating pretreatment unit of algae with the constructed wetland for increasing the Se removal efficiency from the Salton Sea (Huang et al. 2013). Six different algal species were tested, out of which the *Chlorella vulgaris* strain showed the maximum efficiency in terms Se volatilization. *Ch. vulgaris* showed the removal efficiency of Se up to 96% within 72 hours in a microcosm water column, in which 61% was removed by volatilization. Moreover, it was revealed from the X-ray absorption spectroscopy, that Se accumulation in algal cells was in the form of selenomethionine which acts as the precursor of volatile and elemental Se formation. This study suggested that integration of algal pretreatment system with the constructed wetland can increase the overall process of Se removal with the added advantage of reducing the risk of toxicity arising due to the potential bioaccumulation of Se (Huang et al. 2013).

12.3.2.2 Algal-Bacterial Selenium Removal (ABSR) System

A high-rate algal pond was integrated with bacterial treatment to form an Algal-Bacterial Selenium Removal (ASBR) system (Gerhardt et al. 1991). This reactor was utilized for the treatment of wastewater generated from agriculture farms that composed of nitrate (NO_3^-) and Se oxyanions. In this system, microalgae are utilized as source of carbon for bacteria which in turns performs the Se bioreduction. Firstly, wastewater is treated by high-rate algal pond and allowing sufficient time for the partial assimilation of nitrate (NO_3^--N) by the metabolism performed by microalgae. After that, algal-treated wastewater is flowed into a reduction unit comprising of both a denitrifying unit and Se-reducing bacteria which feed upon the biomass of algae as a source of carbon. At last, Se^0 and residual biomass of algae and bacteria are removed in the solid–liquid separator before discharging. Up to 90% of NO_3^- removal and complete reduction of SeO_4^{2-} to SeO_3^{2-} was obtained. Ferric chloride was added in the anoxic unit for further reduction of SeO_3^{2-}. The remaining algal biomass was then used for methane production, recovered at the rate of 0.16 l/g of volatile solid (Gerhardt et al. 1991).

12.3.2.3 Biosorption

Microbial biomass is capable of adsorbing metals ions from aqueous solution on their surface. They can adsorb metals on to their surface through physicochemical and biological interactions (Mane et al. 2011). In a study performed, pretreated biomass of *Spirogyra* and *Nostoc* was used for the treatment of Se by adsorption. Biomass was firstly treated via physical and chemical pretreatment methods including heat, autoclaving, sodium hydroxide, and acetic acid. All of the mentioned methods increased the biosorption of Se on comparing with the live biomass. Maximum sorption of Se (*Spirogyra* – 50% and *Nostoc* – 52.4%) was obtained by drying the biomass for 12 hours at 60 °C in an oven (Mane et al. 2011).

In another experiment, Fe was used for the treatment of macroalgae biomass and converted into Fe-biochar through a slow process of pyrolysis, which showed a high affinity for oxyanions (Johansson et al. 2016). Two sources of algal biomass: (i) "*Gracilaria* waste" (organic portion remaining after the extraction of agar from *Gracilaria*); (ii) *Oedogonium* – freshwater macroalgae. With the increase in the temperature, the effectiveness of Fe-biochar decreases, but the effectiveness was increased by increasing the concentration of Fe. Fe^{3+} solution having composition of 12.5% w/v followed by slow pyrolysis produces the optimized Fe-biochar and showed higher efficiency of Se. Fe-biochar produced by using biomass *Oedogonium* shows higher removal efficiency than the *Gracilaria* waste (Johansson et al. 2016).

12.4 Phytoremediation

Pollution of soil and water with inorganic and inorganic waste results in significant health and environmental issues (Raskin et al. 1997). From the past few years, plant-based natural remediation (for example, phytoremediation) has been broadly studied as a positive remediation technology and is a region of concentrated logical examination (Rugh 2004). Phytoremediation is described as the utilization of a plant to expel contaminants from the surroundings (Cunningham et al. 1995), as shown in Figure 12.3. By utilizing hyperaccumulators, phytoremediation can be carried out. These plants are capable of absorbing pollutants by their roots. Plants are engaged with translocation, absorption, degradation, and sequestration of contaminations. Plants are harvested and processed for drying, composting, and ashing. After that, few metals may be recovered from the debris (Raskin et al. 1997). The hyperaccumulators have the potential of accumulating the metals by 50–100 times greater than normal plants.

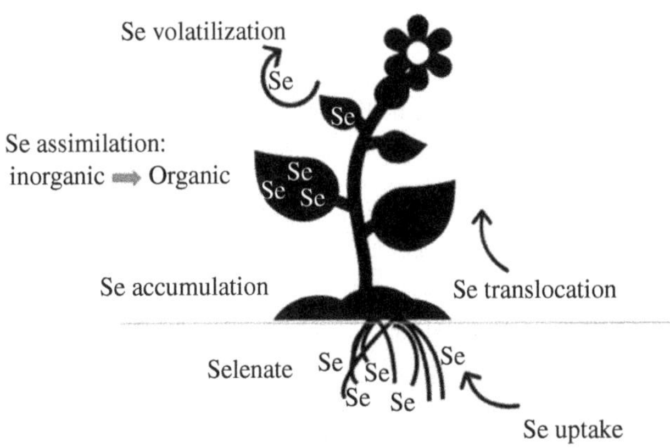

Figure 12.3 Selenium uptake, accumulation, translocation, assimilation, and volatilization – Major steps of phytoremediation.

These plants can accumulate 1–5% of their dry weight. *Thlaspi* species can accumulate more than 0.5% Pb, 0.1% Cd, and 3% Zn in their shoot (Pulford and Watson 2003). Till now, 400 species of plants which are capable hyperaccumulating the metals have been identified (Dobson et al. 1997). The best hyperaccumulators ought to have these qualities: capacity to accumulate high concentration of pollutants; higher rate of growth; large production of biomass; and protection from pest and diseases (Watanabe 1997). As metallic pollutants generally exist in insoluble and inaccessible forms, the utilization of plants to eliminate pollutants can be supported by the correct utilization of soil activation (Dobson et al. 1997). Some components, like soil conditions (for example pKa, pH, water and organic content) and plant physiology, influence the bioavailability and take-up of the target component (Salt et al. 1998). There are different transgenic plants like Indian Mustard, tobacco, Arabidopsis, etc. that have been employed for the treatment of Se from soil and water as shown in Table 12.2.

Phytoremediation has more advantages than conventional physicochemical methods. For instance:

- It is a standard technology (cost 50–80% less of the traditional methods).
- It is an eco-friendly technology.
- A broad range of environmental contamination can be treated.
- The recovery and reuse of valuable metals can be done after the harvesting process.
- With the help of microbial, thermal, or chemical means, the harvested biomass can be reduced in terms of weight or volume (Cunningham and Ow 1996).

Phytoremediation has various limitations too, such as:

- Phytoremediation is most efficient in the area with soils and water contamination <5 m in depth.
- It is a tedious procedure, and atmospheric conditions may hinder or suppress the growth of the plant.
- Inorganic and organic pollutants might be poisonous to plants, and in this way, plants growth might be influenced by the lethality of the pollutants.
- Plants in phytoremediation have a high risk of transferring pollutants to the higher level of food chain.

Despite having such limitations, phytoremediation is getting attention, and new technologies are developing that results in cheaper and effective remediation of the contaminated areas. A lot of phytoremediation is in practice in the United States and European countries (Boyajian and Carreira 1997; Watanabe 1997; Van Der Lelie et al. 2001).

12.4.1 Phytoremediation Processes

Phytoremediation involves various remediation processes for the treatment of Se which includes: (i) phytoextraction; and (ii) phytovolatilization.

12.4.1.1 Phytoextraction

For the most efficient phytoextraction, plants that can produce large amount of biomass which helps in accumulating a high concentration of Se, particularly in over-the-ground shoots tissues that can be harvested easily. Plants, like *Stanleya* and *Astragalus*, have a

Table 12.2 Transgenic plants for the phytoremediation of Se.

Source	Gene	Product	Target plant	Activity	Reference
E. coli	*GshI, gshII, and APS1*	γ-Glu-Cys synthetase, glutathione synthetase, and ATP sulfurylase	Indian mustard	Accumulated 4.3, 2.8, and 2.3× more Se, with *gshI*, *gshII*, and *APS1*, respectively, in comparison to wild type	Bańuelos et al. (2005)
A. thaliana	*OASTL*	Cysteine synthase	tobacco	Tolerance up to 250 mM Se, and production of higher biomass when agar medium was used for growth	Kawashima et al. (2004)
A. bisulcatus	SMT	selenocysteine methyltransferase	arabidopsis	Accumulation of MeSeCys increased to an average of 0.5 m mol/g dry weight	Ellis et al. (2004)
ATPsulfurylase	APS	*A. thaliana*	Indian mustard	2–3 times higher accumulation of Se in shoot and 1.5 times higher in roots than WT	Pilon-Smits et al. (1999)
ATP sulfurylase	APS	*A. thaliana*	Indian mustard	2.5 times higher concentration of different metals in shoot, both at seedling and mature stages	Wangeline et al. (2004)
ATPsulfurylase, cystathionine-γ-synthase	APS and CGS	*A. thaliana*	Indian mustard	2.5 times higher shoot Se in APS and 40% lower shoot Se levels in CGS plants than in WT	Huysen et al. (2004)

potential to absorb up to 10 000 mg/kg Se DM (Wu and Guo 2002), these species are not appropriate for use as they have very slow growth. Phytoextraction frameworks to be practically feasible for the accumulation of Se in Se-enhanced soils, chosen plants ought to be viewed as a feature of a regular yield and not bring about monetary misfortunes for the landowners. In central California, the potential crop used for the phytoextraction of Se is canola and Indian mustard (Bañuelos et al. 1993).

12.4.1.2 Phytovolatilization

Microorganisms and plants performs the biomethylation process for the formation of volatile compounds from inorganic Se (Frankenberger and Karlson 1994). These volatile compounds present fundamentally as dimethyl selenide (DMSe) (Karlson and Frankenberger 1988), alongside a limited extent of dimethyl diselenide (DMDSe) and other Se natural compounds (Doran and Alexander 1976; Karlson et al. 1994). Volatile Se compounds in the environment might undergo oxidation process and changed into an aerosol or particle forms (Rael et al. 1996). Airborne Se particulate matter might be shipped and stored on different regions where Se does not involve primary natural concern. The organic volatilization of Se represents a naturally stable and cheap innovation for the reduction of Se pollution from soil and water (Hill 1988; Watanabe 1997). In rats, volatile DMSe was found to be 500 times less toxic than inorganic selenite (Lin and Terry 1999). The change and passage of DMSe in the soil might be subject to the soil moisture (McConnell and Portman 1952). In some wetland areas, organic volatilization of Se was assessed to represent 10–30% of the Se reduced from wastewater (Hansen et al. 1998).

12.5 Remediation of Selenium by Fungi

The fast cell growth of fungi and yeast made them the most considered in the field of biotechnology. Easy genetic as well as morphological manipulation, and the capacity to grow in harsh environmental conditions like pH and temperature has made them a potential biosorbent. These organisms can be isolated or recovered free of cost by the industries as a by-product. The large-scale production of these organisms can be done very easy quickly without the utilization of complex growth medium and growth technology (Kuyucak 1990). Although a lot of studies has already been done on fungi and yeast for the reduction in concentration of different heavy metals, the remediation of selenium is yet less explored. For the biosorption of selenium, fungal species like *Candida utilis* (Zhil'tsova et al. 1996), *Saccharomyces cerevisiae* (Lazard et al. 2010; Khakpour et al. 2014), *Aspergillus sp.* (Li et al. 2013), *Lactobacillus Plantarum* (Calomme et al. 1995), *Rhodotorula Plantarum*-13B (Ruocco et al. 2014) have shown a potential effect. Selenium is bonded by the fungi and yeast in its inorganic or organic forms. Bioaccumulation of selenium is intracellularly and extracellularly by the fungi. During the extra binding of selenium phospholipids, polysaccharides, or protein perform an important part in the formation of an ionic bond or Se complexation. Intracellular accumulation occurs via active transport within the cells (Kieliszek et al. 2015). Pérez-Corona et al. (1997) utilized *Sa. cerevisiae* to isolate Se in its different oxidative forms. According to the report by the author, the bioaccumulation of Se(IV) depends on factors like biomass used and temperature. The low affinity of the Se(VI) was observed in high biomass, and no effect was observed on the accumulation of Se(IV) because of the pH (Pérez-Corona et al. 1997). According to a study by (Espinosa-Ortiz et al. 2015), the *Phanerochaete chrysosporium* (a white-rot fungus) was able to remediate SeO_3^{2-} from synthetic wastewater is continuously operated bioreactor. Observation stated that after 41 days of bioreactor operation, around 70% of total soluble Se was removed. At a continuous Se feeding rate of 10 mg Se/L/day, because of the intracellular production of elemental selenium nanoparticles, morphological changes were observed in fungus (Espinosa-Ortiz et al. 2015). A study performed by Gharieb et al. (1995) concluded that fungi like *Trichoderma reesei*, *Fusarium sp.*, *Rhizopus arrhizus*, *Aspergillus niger*, and *Mucor SK* were able to reduce SeO_3^{2-} to its Se^0 (Gharieb et al. 1995). Marinescu et al. (2011), with the help of *Saccharomyces uvarum*, demonstrated selenium adsorption by treating it with sodium selenite at different parameters like pH, temperature, biomass proportion, Se concentrations using malt wort and sparge water (Marinescu et al. 2011). Another study by Nettem and Almusallam (2013), using *Ganoderma lucidum* biomass, showed that by optimizing the different parameters like temperature, adsorbent dose, pH, contact time, and initial ion concentration, for the biosorption of selenium an adsorption capacity up to 127 mg/g can be achieved. The FTIR analysis results of the fungal cells showed that carboxyl, amino, carbonyl, and hydroxyl groups facilitate the biosorption process (Nettem and Almusallam 2013).

12.6 Conclusion

In comparison to the physical and chemical treatments, biological treatment offers the best treatment technology for the remediation of selenium. Various bioreactors discussed in the literature mentioned above were able to remove more than 90% of selenium from the wastewater, offering multiple advantages such as cost-efficiency, no formation of secondary contaminated sludge, scalability, and the ability to recover selenium in elemental form. They are also able to perform effectively in the varying conditions of incoming effluent. But despite various advantages, some disadvantages are also associated with biological treatment. This includes the requirement of large footprints by different reactors, carbon source competition by ions, and complex nature of set-up and wastewater. Bioreduction have the advantage of the formation of biogenic selenium, but its application has not been fully utilized and recovery potential from the biological reactor. Therefore, more investigations are needed in the field of biological treatment keeping the view of the complex nature of effluent, reducing the size of land requirement, more utilization of biogenic selenium. These improvements will undoubtedly scale up the process of biological treatment of selenium.

Acknowledgment

The authors of the current chapter are obliged to the School of Biochemical Engineering, Indian Institute of Technology (Banaras Hindu University), Varanasi and Department of Bioengineering, Integral University, Lucknow for extending their technical and financial help.

References

Araie, H., Sakamoto, K., Suzuki, I., and Shiraiwa, Y. (2011). Characterization of the selenite uptake mechanism in the Coccolithophore *Emiliania huxleyi* (Haptophyta). *Plant and Cell Physiology* 52 (7): 1204–1210. https://doi.org/10.1093/pcp/pcr070.

Bańuelos, G.S., Cardon, G.E., Phene, C.J. et al. (1993). Soil boron and selenium removal by three plant species. *Plant and Soil* 148 (2): 253–263.

Bańuelos, G., Terry, N., LeDuc, D.L. et al. (2005). Field trial of transgenic Indian mustard plants shows enhanced phytoremediation of selenium-contaminated sediment. *Environmental Science & Technology* 39 (6): 1771–1777.

Bhunia, P. and Ghangrekar, M.M. (2007). Required minimum granule size in UASB reactor and characteristics variation with size. *Bioresource Technology* 98 (5): 994–999. https://doi.org/10.1016/j.biortech.2006.04.019.

Boyajian, G.E. and Carreira, L.H. (1997). Phytoremediation: a clean transition from laboratory to marketplace? *Nature Biotechnology* 15 (2): 127–128.

Brown, K.M. and Arthur, J.R. (2001). Selenium, selenoproteins and human health: a review. *Public Health Nutrition* 4 (2b): 593–599.

Cabannes, E., Buchner, P., Broadley, M.R., and Hawkesford, M.J. (2011). A comparison of sulfate and selenium accumulation in relation to the expression of sulfate transporter genes in Astragalus species. *Plant Physiology* 157 (4): 2227–2239. https://doi.org/10.1104/pp.111.183897.

Calomme, M., Hu, J., Van Den Branden, K., and Berghe, D.V. (1995). Seleno-lactobacillus. *Biological Trace Element Research* 47 (1–3): 379–383.

Chan, Y.J., Chong, M.F., Law, C.L., and Hassell, D.G. (2009). A review on anaerobic–aerobic treatment of industrial and municipal wastewater. *Chemical Engineering Journal* 155 (1): 1–18. https://doi.org/10.1016/j.cej.2009.06.041.

Chapman, P.M., Adams, W.J., Brooks, M. et al. (2010). *Ecological Assessment of Selenium in the Aquatic Environment*. CRC Press.

Chung, J., Ryu, H., Abbaszadegan, M., and Rittmann, B.E. (2006a). Community structure and function in a H2-based membrane biofilm reactor capable of bioreduction of selenate and chromate. *Applied Microbiology and Biotechnology* 72 (6): 1330–1339. https://doi.org/10.1007/s00253-006-0439-x.

Chung, J., Rittmann, B.E., Wright, W.F., and Bowman, R.H. (2006b). Simultaneous bio-reduction of nitrate, perchlorate, selenate, chromate, arsenate, and Dibromochloropropane using a hydrogen-based membrane biofilm reactor. *Biodegradation* 18 (2): 199. https://doi.org/10.1007/s10532-006-9055-9.

Chung, J., Nerenberg, R., and Rittmann, B.E. (2006c). Bioreduction of selenate using a hydrogen-based membrane biofilm reactor. *Environmental Science & Technology* 40 (5): 1664–1671. https://doi.org/10.1021/es051251g.

Chung, J., Rittman, B.E., Her, N. et al. (2010). Integration of H2-based membrane biofilm reactor with RO and NF membranes for removal of chromate and selenate. *Water, Air, and Soil Pollution* 207 (1): 29–37. https://doi.org/10.1007/s11270-009-0116-7.

Cunningham, S.D. and Ow, D.W. (1996). Promises and prospects of phytoremediation. *Plant Physiology* 110 (3): 715.

Cunningham, S.D., Berti, W.R., and Huang, J.W. (1995). Phytoremediation of contaminated soils. *Trends in Biotechnology* 13 (9): 393–397.

Cutter, G.A. and Bruland, K.W. (1984). The marine biogeochemistry of selenium: a re-evaluation. *Limnology and Oceanography* 29 (6): 1179–1192. https://doi.org/10.4319/lo.1984.29.6.1179.

Dessì, P., Jain, R., Singh, S. et al. (2016). Effect of temperature on selenium removal from wastewater by UASB reactors. *Water Research* 94: 146–154. https://doi.org/10.1016/j.watres.2016.02.007.

Dobson, A.P., Bradshaw, A.D., and Baker, A.J.M. (1997). Hopes for the future: restoration ecology and conservation biology. *Science* 277 (5325): 515–522.

Doran, J.W. and Alexander, M. (1976). Microbial formation of volatile selenium compounds in soil 1. *Soil Science Society of America Journal* 40 (5): 687–690.

Ellis, D.R., Sor, T.G., Brunk, D.G. et al. (2004). Production of Se-methylselenocysteine in transgenic plants expressing selenocysteine methyltransferase. *BMC Plant Biology* 4 (1): 1.

El-Ramady, H., Abdalla, N., Alshaal, T. et al. (2014). Selenium in soils under climate change, implication for human health. *Environmental Chemistry Letters* 13 (1): 1–19.

Espinosa-Ortiz, E.J., Rene, E.R., van Hullebusch, E.D., and Lens, P.N. (2015). Removal of selenite from wastewater in a Phanerochaete chrysosporium pellet based fungal bioreactor. *International Biodeterioration & Biodegradation* 102: 361–369.

Fan, T.W.M., Teh, S.J., Hinton, D.E., and Higashi, R.M. (2002). Selenium biotransformations into proteinaceous forms by foodweb organisms of selenium-laden drainage waters in California. *Aquatic Toxicology* 57 (1–2): 65–84.

Fisher, J.C. and Hollibaugh, J.T. (2008). Selenate-dependent anaerobic arsenite oxidation by a bacterium from mono Lake, California. *Applied and Environmental Microbiology* 74 (9): 2588–2594. https://doi.org/10.1128/AEM.01995-07.

Fournier, E., Adam-Guillermin, C., Potin-Gautier, M., and Pannier, F. (2010). Selenate bioaccumulation and toxicity in *Chlamydomonas reinhardtii*: influence of ambient sulphate ion concentration. *Aquatic Toxicology* 97 (1): 51–57. https://doi.org/10.1016/j.aquatox.2009.12.003.

Frankenberger, W.T. Jr. and Karlson, U. (1994). Microbial volatilization of selenium from soils and sediments. In: *Selenium in the Environment* (eds. W.T. Frankenberger Jr. and S. Benson), 369–387. New York: Marcel Dekker, Inc.

Fujita, M., Ike, M., Nishimoto, S. et al. (1997). Isolation and characterization of a novel selenate-reducing bacterium, Bacillus sp. SF-1. *Journal of Fermentation and Bioengineering* 83 (6): 517–522. https://doi.org/10.1016/S0922-338X(97)81130-0.

Fujita, M., Ike, M., Kashiwa, M. et al. (2002). Laboratory-scale continuous reactor for soluble selenium removal using selenate-reducing bacterium, Bacillus sp. SF-1. *Biotechnology and Bioengineering*: 755–761. https://doi.org/10.1002/bit.10425.

Gao, S., Tanji, K.K., Lin, Z.Q. et al. (2003). Selenium removal and mass balance in a constructed flow-through wetland system. *Journal of Environmental Quality* 32 (4): 1557–1570.

Geering, H.R., Cary, E.E., Jones, L.H.P., and Allaway, W.H. (1968). Solubility and redox criteria for the possible forms of selenium in soils. *Soil Science Society of America Journal* 32 (1): 35–40. https://doi.org/10.2136/sssaj1968.03615995003200010009x.

Gerhardt, M.B., Green, F.B., Newman, R.D. et al. (1991). Removal of selenium using a novel algal-bacterial process. *Research Journal of the Water Pollution Control Federation* 63 (5): 799–805.

Gharieb, M.M., Wilkinson, S.C., and Gadd, G.M. (1995). Reduction of selenium oxyanions by unicellular, polymorphic and filamentous fungi: cellular location of reduced selenium and implications for tolerance. *Journal of Industrial Microbiology* 14 (3–4): 300–311.

Golder Associates Inc. (2009). *Literature Review of Treatment Technologies to Remove Selenium from Mining Influenced Water*. Calgary: Golder Associates Inc.

Gusek, J., Conroy, K., and Rutkowski, T. (2008). Past, present and future for treating selenium-impacted water. Tailings and Mine Waste. 08 – Proceedings of the 12th International Conference.

Hageman, S.P., van der Weijden, R.D., Weijma, J., and Buisman, C.J. (2013). Microbiological selenate to selenite conversion for selenium removal. *Water Research* 47 (7): 2118–2128. https://doi.org/10.1016/j.watres.2013.01.012.

Hamilton, S.J. (2004). Review of selenium toxicity in the aquatic food chain. *Science of the Total Environment* 326 (1–3): 1–31.

Hansen, D., Duda, P.J., Zayed, A., and Terry, N. (1998). Selenium removal by constructed wetlands: role of biological volatilization. *Environmental Science & Technology* 32 (5): 591–597.

Hill, C. (1988). *Air Quality Impacts of Enhanced Selenium Volatilization at Kesterson Reservoir*. CH2M Hill.

Hill, C. (2013). Technical addendum: Review of available technologies for the removal of selenium from water. *NAMC White Paper Report Addendum*.

Huang, J.C., Suárez, C., Yang, S.I. et al. (2013). Development of a constructed wetland water treatment system for selenium removal: incorporation of an algal treatment component. *Environmental Science and Technology* 47 (18): 10518–10525. https://doi.org/10.1021/es4015629.

Huysen, T., Terry, N., and Pilon-Smits, E.A.H. (2004). Exploring the selenium phytoremediation potential of transgenic Indian mustard overexpressing ATP sulfurylase or cystathionine-γ-synthase. *International Journal of Phytoremediation* 6 (2): 111–118.

Johansson, C.L., Paul, N.A., de Nys, R., and Roberts, D.A. (2016). Simultaneous biosorption of selenium, arsenic and molybdenum with modified algal-based biochars. *Journal of Environmental Management* 165: 117–123. https://doi.org/10.1016/j.jenvman.2015.09.021.

Kagami, T., Narita, T., Kuroda, M. et al. (2013). Effective selenium volatilization under aerobic conditions and recovery from the aqueous phase by *Pseudomonas stutzeri* NT-I. *Water Research* 47 (3): 1361–1368. https://doi.org/10.1016/j.watres.2012.12.001.

Karlson, U. and Frankenberger, W.T. Jr. (1988). Effects of carbon and trace element addition on alkylselenide production by soil. *Soil Science Society of America Journal* 52 (6): 1640–1644.

Karlson, U., Frankenberger, W.T. Jr., and Spencer, W.F. (1994). Physicochemical properties of dimethyl selenide and dimethyl diselenide. *Journal of Chemical and Engineering Data* 39 (3): 608–610.

Kashiwa, M., Nishimoto, S., Takuhashi, K. et al. (2000). Factors affecting soluble selenium removal by a selenate-reducing bacterium Bacillus sp. SF-1. *Journal of Bioscience and Bioengineering* 89 (6): 528–533. https://doi.org/10.1016/S1389-1723(00)80051-1.

Kashiwa, M., Ike, M., Mihara, H. et al. (2001). Removal of soluble selenium by a selenate-reducing bacterium Bacillus sp. SF-1. *BioFactors*: 261–265. https://doi.org/10.1002/biof.5520140132.

Kawashima, C.G., Noji, M., Nakamura, M. et al. (2004). Heavy metal tolerance of transgenic tobacco plants over-expressing cysteine synthase. *Biotechnology Letters* 26 (2): 153–157.

Khakpour, H., Younesi, H., and Mohammadhosseini, M. (2014). Two-stage biosorption of selenium from aqueous solution using dried biomass of the baker's yeast *Saccharomyces cerevisiae*. *Journal of Environmental Chemical Engineering* 2 (1): 532–542.

Kieliszek, M., Błażejak, S., Gientka, I. et al. (2015). Accumulation and metabolism of selenium by yeast cells. *Applied Microbiology and Biotechnology* 99 (13): 5373–5382.

Kuroda, M., Notaguchi, E., Sato, A. et al. (2011). Characterization of *Pseudomonas stutzeri* NT-I capable of removing soluble selenium from the aqueous phase under aerobic conditions. *Journal of Bioscience and Bioengineering* 112 (3): 259–264. https://doi.org/10.1016/j.jbiosc.2011.05.012.

Kuyucak, N. (1990). Feasibility of biosorbents application. *Biosorption of Heavy Metals* 4: 372–377.

Lazard, M., Blanquet, S., Fisicaro, P. et al. (2010). Uptake of selenite by Saccharomyces cerevisiae involves the high and low affinity orthophosphate transporters. *Journal of Biological Chemistry* 285 (42): 32029–32037.

Lenz, M. and Lens, P.N.L. (2009). The essential toxin: the changing perception of selenium in environmental sciences. *Science of the Total Environment* 407 (12): 3620–3633. https://doi.org/10.1016/j.scitotenv.2008.07.056.

Lenz, M., Smit, M., Binder, P. et al. (2008). Biological alkylation and colloid formation of selenium in methanogenic UASB reactors. *Journal of Environmental Quality* 37 (5): 1691–1700. https://doi.org/10.2134/jeq2007.0630.

Lenz, M., van Hullebusch, E.D., Hommes, G. et al. (2008). Selenate removal in methanogenic and sulfate-reducing upflow anaerobic sludge bed reactors. *Water Research* 42 (8): 2184–2194. https://doi.org/10.1016/j.watres.2007.11.031.

Li, Z., Li, H., Yang, X. et al. (2013). Characterization of Se (IV) removal from aqueous solution by Aspergillus sp. J2. *Chemical Engineering Journal* 220: 67–71.

Li, J., Shan, F., Xiong, G. et al. (2014). EGF-induced C/EBPβ participates in EMT by decreasing the expression of miR-203 in esophageal squamous cell carcinoma cells. *Journal of Cell Science* 127 (17): 3735–3744. https://doi.org/10.1242/jcs.148759.

Lin, Z.-Q. and Terry, N. (1999). 4 remediation of selenium-polluted soils and waters by phytovolatilization. In: *Phytoremediation of Contaminated Soil and Water* (eds. N. Terry and G.S. Bañuelos), 61. CRC Press.

Lo, B.P., Elphick, J.R., Bailey, H.C. et al. (2015). The effect of sulfate on selenate bioaccumulation in two freshwater primary producers: a duckweed (*Lemna minor*) and a green alga (*Pseudokirchneriella subcapitata*). *Environmental Toxicology and Chemistry* 34 (12): 2841–2845. https://doi.org/10.1002/etc.3127.

Losi, M.E. and Frankenberger, W.T. (1997). Reduction of selenium oxyanions by Enterobacter cloacae SLD1a-1: isolation and growth of the bacterium and its expulsion of selenium particles. *Applied and Environmental Microbiology* 63 (8): 3079–3084. https://doi.org/10.1128/aem.63.8.3079-3084.1997.

Maiers, D.T., Wichlacz, P.L., Thompson, D.L., and Bruhn, D.F. (1988). Selenate reduction by bacteria from a selenium-rich environment. *Applied and Environmental Microbiology* 54 (10): 2591–2593. http://aem.asm.org/content/54/10/2591.abstract.

Mal, J., Nancharaiah, Y.V., Van Hullebusch, E.D., and Lens, P.N.L. (2016). Effect of heavy metal co-contaminants on selenite bioreduction by anaerobic granular sludge. *Bioresource Technology* 206: 1–8. https://doi.org/10.1016/j.biortech.2016.01.064.

Mane, P.C., Bhosle, A.B., Jangam, C.M., and Vishwakarma, C.V. (2011). Bioadsorption of selenium by pretreated algal biomass. *Advances in Applied Science Research* 2 (2): 202–207.

Marinescu, G., Stoicescu, A.G., and Teodorof, L. (2011). Industrial nutrient medium use for yeast selenium preparation. *Annals of the University Dunarea de Jos of Galati Fascicle VI – Food Technology* 35 (1).

Masscheleyn, P.H. and Patrick, W.H. Jr. (1993). Biogeochemical processes affecting selenium cycling in wetlands. *Environmental Toxicology and Chemistry: An International Journal* 12 (12): 2235–2243.

McConnell, K.P. and Portman, O.W. (1952). Toxicity of dimethyl selenide in the rat and mouse. *Proceedings of the Society for Experimental Biology and Medicine* 79 (2): 230–231.

Morlon, H., Fortin, C., Adam, C., and Garnier-Laplace, J. (2006). Selenite transport and its inhibition in the unicellular green alga *Chlamydomonas reinhardtii*. *Environmental Toxicology and Chemistry* 25 (5): 1408–1417. https://doi.org/10.1897/2512039.1.

Nancharaiah, Y.V. and Lens, P.N.L. (2015). Ecology and biotechnology of selenium-respiring bacteria. *Microbiology and Molecular Biology Reviews* 79 (1): 61–80. https://doi.org/10.1128/mmbr.00037-14.

Nettem, K. and Almusallam, A.S. (2013). Equilibrium, kinetic, and thermodynamic studies on the biosorption of selenium (IV) ions onto Ganoderma lucidum biomass. *Separation Science and Technology* 48 (15): 2293–2301.

Neumann, P.M., De Souza, M.P., Pickering, I.J., and Terry, N. (2003). Rapid microalgal metabolism of selenate to volatile dimethylselenide. *Plant, Cell & Environment* 26 (6): 897–905.

Nnaji, C.C. (2014). A review of the upflow anaerobic sludge blanket reactor. *Desalination and Water Treatment* 52 (22–24): 4122–4143. https://doi.org/10.1080/19443994.2013.800809.

Nogueira, C.W. and Rocha, J.B.T. (2011). Toxicology and pharmacology of selenium: emphasis on synthetic organoselenium compounds. *Archives of Toxicology* 85 (11): 1313–1359. https://doi.org/10.1007/s00204-011-0720-3.

Novoselov, S.V., Rao, M., Onoshko, N.V. et al. (2002). Selenoproteins and selenocysteine insertion system in the model plant cell system, *Chlamydomonas reinhardtii*. *The EMBO Journal* 21 (14): 3681–3693.

Obata, T., Araie, H., and Shiraiwa, Y. (2004). Bioconcentration mechanism of selenium by a Coccolithophorid, *Emiliania huxleyi*. *Plant and Cell Physiology* 45 (10): 1434–1441. https://doi.org/10.1093/pcp/pch164.

Oremland, R.S., Blum, J.S., Culbertson, C.W. et al. (1994). Isolation, growth, and metabolism of an obligately anaerobic, selenate-respiring bacterium, strain SES-3. *Applied and Environmental Microbiology* 60 (8): 3011–3019. http://aem.asm.org/content/60/8/3011.abstract.

Oremland, R.S., Herbel, M.J., Blum, J.S. et al. (2004). Structural and spectral features of selenium nanospheres produced by Se-respiring bacteria. *Applied and Environmental Microbiology* 70 (1): 52–60. https://doi.org/10.1128/aem.70.1.52-60.2004.

Pérez-Corona, T., Madrid, Y., and Cámara, C. (1997). Evaluation of selective uptake of selenium (Se (IV) and Se (VI)) and antimony (Sb (III) and Sb (V)) species by baker's yeast cells (*Saccharomyces cerevisiae*). *Analytica Chimica Acta* 345 (1–3): 249–255.

Pettine, M., Gennari, F., Campanella, L. et al. (2012). The reduction of selenium(IV) by hydrogen sulfide in aqueous solutions. *Geochimica et Cosmochimica Acta* 83: 37–47. https://doi.org/10.1016/j.gca.2011.12.024.

Pilon-Smits, E.A.H., Hwang, S., Lytle, C.M. et al. (1999). Overexpression of ATP sulfurylase in Indian mustard leads to increased selenate uptake, reduction, and tolerance. *Plant Physiology American Society of Plant Biologists* 119 (1): 123–132.

Plant, J.A., Bone, J., Volvoulis, N. et al. (2004). *Arsenic and Selenium*. Elsevier.

Presser, T.S. (1994). The Kesterson effect. *Environmental Management* 18 (3): 437–454. https://doi.org/10.1007/BF02393872.

Pulford, I.D. and Watson, C. (2003). Phytoremediation of heavy metal-contaminated land by trees—a review. *Environment International* 29 (4): 529–540.

Rael, R.M., Tuzaon, E.C., and Frankenberger, W.T. Jr. (1996). Gas-phase reactions of dimethyl selenide with ozone and the hydroxyl and nitrate radicals. *Atmospheric Environment* 30 (8): 1221–1232.

Raskin, I., Smith, R.D., and Salt, D.E. (1997). Phytoremediation of metals: using plants to remove pollutants from the environment. *Current Opinion in Biotechnology* 8 (2): 221–226.

Rege, M.A., Yonge, D.R., Mendoza, D.P. et al. (1999). Selenium reduction by a denitrifying consortium. *Biotechnology and Bioengineering* 62 (4): 479–484. https://doi.org/10.1002/(SICI)1097-0290(19990220)62:4<479:AID-BIT11>3.0.CO;2-G.

Riedel, G.F. and Sanders, J.G. (1996). The influence of pH and media composition on the uptake of inorganic selenium by *Chlamydomonas reinhardtii*. *Environmental Toxicology and Chemistry: An International Journal* 15 (9): 1577–1583.

Rugh, C.L. (2004). Genetically engineered phytoremediation: one man's trash is another man's transgene. *Trends in Biotechnology* 22 (10): 496–498.

Ruocco, M.H., Chan, C.S., Hanson, T.E., and Church, T.M. (2014). Characterization and distribution of selenite reduction products in cultures of the marine yeast *Rhodotorula mucilaginosa*-13B. *Geomicrobiology Journal* 31 (9): 769–778.

Rutkowski, T., Hanson, R., and Conroy, K. (2013). Mine water treatment options for meeting stringent selenium regulatory limits. *Reliable Mine Water Technology*. Vol I, 711–717. International Mine Water Association.

Saiki, M.K. and Lowe, T.P. (1987). Selenium in aquatic organisms from subsurface agricultural drainage water, San Joaquin Valley, California. *Archives of Environmental Contamination and Toxicology* 16 (6): 657–670.

Salt, D.E., Smith, R.D., and Raskin, I. (1998). Phytoremediation. *Annual Review of Plant Physiology* 49: 643–668.

Schiavon, M. and Pilon-Smits, E.A.H. (2017). The fascinating facets of plant selenium accumulation – biochemistry, physiology, evolution and ecology. *New Phytologist* 213 (4): 1582–1596. https://doi.org/10.1111/nph.14378.

Schiavon, M., Pilon, M., Malagoli, M., and Pilon-Smits, E.A. (2015). Exploring the importance of sulfate transporters and ATP sulphurylases for selenium hyperaccumulation-a comparison of *Stanleya pinnata* and *Brassica juncea* (Brassicaceae). *Frontiers in Plant Science* 6: 2. https://doi.org/10.3389/fpls.2015.00002.

Schiavon, M., Ertani, A., Parrasia, S., and Dalla Vecchia, F. (2017). Selenium accumulation and metabolism in algae. *Aquatic Toxicology* 189: 1–8. https://doi.org/10.1016/j.aquatox.2017.05.011.

Schröder, I., Rech, S., Krafft, T., and Macy, J.M. (1997). Purification and characterization of the selenate reductase from Thauera selenatis. *Journal of Biological Chemistry* 272 (38): 23765–23768.

Seghezzo, L., Zeeman, G., van Lier, J.B. et al. (1998). A review: the anaerobic treatment of sewage in UASB and EGSB reactors. *Bioresource Technology* 65 (3): 175–190. https://doi.org/10.1016/S0960-8524(98)00046-7.

Simmons, D.B.D. and Emery, R.J.N. (2011). Phytochelatin induction by selenate in *Chlorella vulgaris*, and regulation of effect by sulfate levels. *Environmental Toxicology and Chemistry* 30 (2): 469–476. https://doi.org/10.1002/etc.392.

Singh, V. and Mishra, V. (2019). Bioremediation of nutrients and heavy metals from wastewater by microalgal cells. In: *Microbial Genomics in Sustainable Agroecosystems* (eds. V. Tripathi et al.), 319–357. Singapore: Springer Singapore https://doi.org/10.1007/978-981-32-9860-6_16.

Soda, S., Kashiwa, M., Kagami, T. et al. (2011). Laboratory-scale bioreactors for soluble selenium removal from selenium refinery wastewater using anaerobic sludge. *Desalination* 279 (1–3): 433–438. https://doi.org/10.1016/j.desal.2011.06.031.

Sonstegard, J., Pickett, T., Harwood, J., and Johnson, D. (2008). Full scale operation of GE ABMet® biological technology for the removal of selenium from FGD wastewaters. *IWC* 8: 31.

Sonstegard, J., Harwood, J., Pickett, T., and Kennedy, W. (2010). ABMeT: setting the standard for selenium removal. In: *Proceedings of 71st International Water Conference; Texas, USA*, 216.

Sors, T.G., Ellis, D.R., and Salt, D.E. (2005). Selenium uptake, translocation, assimilation and metabolic fate in plants. *Photosynthesis Research* 86 (3): 373–389. https://doi.org/10.1007/s11120-005-5222-9.

Staicu, L.C., van Hullebusch, E.D., and Lens, P.N.L. (2015). Production, recovery and reuse of biogenic elemental selenium. *Environmental Chemistry Letters* 13 (1): 89–96. https://doi.org/10.1007/s10311-015-0492-8.

Stolz, J.F., Basu, P., Santini, J.M., and Oremland, R.S. (2006). Arsenic and selenium in microbial metabolism. *Annual Review of Microbiology* 60 (1): 107–130. https://doi.org/10.1146/annurev.micro.60.080805.142053.

Tan, L.C., Nancharaiah, Y.V., van Hullebusch, E.D., and Lens, P.N. (2016). Selenium: environmental significance, pollution, and biological treatment technologies. *Biotechnology Advances* 34 (5): 886–907. https://doi.org/10.1016/j.biotechadv.2016.05.005.

Turner, R.J., Borghese, R., and Zannoni, D. (2012). Microbial processing of tellurium as a tool in biotechnology. *Biotechnology Advances* 30 (5): 954–963. https://doi.org/10.1016/j.biotechadv.2011.08.018.

Tuzen, M. and Sari, A. (2010). Biosorption of selenium from aqueous solution by green algae (Cladophora hutchinsiae) biomass: equilibrium, thermodynamic and kinetic studies. *Chemical Engineering Journal* 158 (2): 200–206. https://doi.org/10.1016/j.cej.2009.12.041.

Twidwell, L., McCloskey, J.M., Joyce, H. et al. (2005). Removal of selenium oxyanions from mine waters utilizing elemental iron and galvanically coupled metals. In: *Proceedings of the Jan D. Mill Symposium—Innovations in Natural Resource, Salt Lake City, UT, USA*, 299–313.

Umysová, D., Vítová, M., Doušková, I. et al. (2009). Bioaccumulation and toxicity of selenium compounds in the green alga Scenedesmus quadricauda. *BMC Plant Biology* 9 (1): 58.

Van Der Lelie, D., Schwitzguébel, J.-P., Glass, D.J. et al. (2001). *Peer Reviewed: Assessing phytoremediation's Progress in the United States and Europe*. ACS Publications.

Van Ginkel, S.W., Yang, Z., Kim, B.-O. et al. (2011). The removal of selenate to low ppb levels from flue gas desulfurization brine using the H2-based membrane biofilm reactor (MBfR). *Bioresource Technology* 102 (10): 6360–6364. https://doi.org/10.1016/j.biortech.2011.03.010.

Viamajala, S., Bereded-Samuel, Y., Apel, W.A., and Petersen, J.N. (2006). Selenite reduction by a denitrifying culture: batch- and packed-bed reactor studies. *Applied Microbiology and Biotechnology* 71 (6): 953–962. https://doi.org/10.1007/s00253-005-0276-3.

Vriens, B. et al. (2016). Selenium uptake and methylation by the microalga *Chlamydomonas reinhardtii*. *Environmental Science & Technology* 50 (2): 711–720. https://doi.org/10.1021/acs.est.5b04169.

Wangeline, A.L., Burkhead, J.L., Hale, K.L. et al. (2004). Overexpression of ATP sulfurylase in Indian mustard. *Journal of Environmental Quality* 33 (1): 54–60.

Watanabe, M.E. (1997). Phytoremediation on the brink of commericialization. *Environmental Science & Technology* 31 (4): 182A–186A.

Williams, M.J., Ogle, R.S., Knight, A.W., and Burau, R.G. (1994). Effects of sulfate on selenate uptake and toxicity in the green alga *Selenastrum capricornutum*. *Archives of Environmental Contamination and Toxicology* 27 (4): 449–453. https://doi.org/10.1007/BF00214834.

Wu, L. and Guo, X. (2002). Selenium accumulation in submerged aquatic macrophytes *Potamogeton pectinatus* L. and *Ruppia maritima* L. from water with elevated chloride and sulfate salinity. *Ecotoxicology and Environmental Safety* 51 (1): 22–27.

Yudovich, Y.E. and Ketris, M.P. (2006). Selenium in coal: a review. *International Journal of Coal Geology* 67 (1): 112–126. https://doi.org/10.1016/j.coal.2005.09.003.

Zannoni, D., Borsetti, F., Harrison, J.J., and Turner, R.J. (2007). The bacterial response to the chalcogen metalloids Se and Te. In: *Advances in Microbial Biology* (ed. R.A. Poole), 1–71. Academic Press. doi: https://doi.org/10.1016/S0065-2911(07)53001-8.

Zhang, Y., Amrhein, C., Chang, A., and Frankenberger, W.T. Jr. (2008). Effect of zero-valent iron and a redox mediator on removal of selenium in agricultural drainage water. *Science of the Total Environment* 407 (1): 89–96. https://doi.org/10.1016/j.scitotenv.2008.09.009.

Zhil'tsova, T.S., Gradova, N.B., Shagova, M.V., and Golubkina, N.A. (1996). Resistance to selenium compounds in yeasts of the genus Candida. *Applied Biochemistry and Microbiology* 32 (5): 511–513.

13

Nanomaterials for the Remediation of Selenium in Water

Ketki Kulkarni, Ketki Nagpure, Amit Bansiwal, and Nitin Labhsetwar

CSIR-National Environmental Engineering Research Institute (CSIR-NEERI), Nagpur, Maharashtra, India

13.1 Introduction

Selenium is a metalloid with the notation Se. Selenium (Se) in the Earth's crust is often combined with sulfur-containing minerals. The generally observed range for selenium in nature is found to be 50–90 μg/kg. Sedimentary and carbonate rocks and some volcanic activities are responsible for higher concentrations of selenium in the environment. Various forms of selenium that are found in nature are selenide, selenite, and Selenate. Selenite refers to +4 oxidation state of selenium with formula SeO_3^{2-} whereas selenate refers SeO_4^{2-} with the oxidation state +6 (Holmes and Gu 2016). Of the three forms mentioned above, the most dominant forms that are found in water and which can cause contamination are selenite and selenate. Both these forms are not odiferous and hence are difficult to identify as a contaminant. Both the selenite and selenate species are soluble in water. Selenite and selenate can be found as selenous and selenic acid, respectively, as diprotic acids. It is observed that the presence of selenium species is affected by the pH of water. The generally observed pH of most of natural waters ranges from 5 to 9, for this pH range dominant selenate species would be SeO_4^{2-} compared to the selenite form (Sharma et al. 2019). Selenium species are found to be highly influenced by redox potential. Various acidic and reducing conditions lead to conversion of inorganic selenite to constituent selenium, whereas basic and oxidizing conditions lead to the generation of selenate (WHO 2011). If water is strong oxidizing then the dominating species will be selenate and for a mild oxidizing environment selenite is the dominating species.

Selenium is one of the essential micro-nutrients for both plants' and animals' nutrition. There is a very fine boundary between deficiency and toxicity in humans, which is one of the important causes of concern (Sharma et al. 2019). The entry of selenium can be through both food and water due to the presence of selenium in natural waters and soil. According to FAO and WHO 1998 guidelines, the recommended intake of selenium varies according

Selenium Contamination in Water, First Edition. Edited by Pooja Devi, Pardeep Singh, Arindam Malakar, and Daniel Snow.

to age and gender. For the infants and children, it is 6–21 μg/day, for minor females and males it is observed to be 26 and 30 μg/day, respectively, and for grown-up females and males it is 26 and 35 μg/day, respectively (WHO 2011). Intake of selenium above these concentrations is harmful for human beings. Selenium is found to be toxic to humans, causing teratogenic and mutagenic defects. Also, its anthropogenic nature and toxicity for marine life is well reported.

Drinking water is a primary source for the entry of selenium into the body. Water-soluble selenium species are found over the gastrointestinal tract which paves its way towards the liver and then transported and distributed to all the organs. As mentioned earlier, both the inorganic forms of selenium, selenite and selenate, exist predominantly in water.

Due to the high toxicity of selenium various researchers have made extensive efforts to develop techniques to remove selenium from the water. The available technologies include physical/mechanical removal, oxidation/reduction removal, biological removal, reverse osmosis etc. However, most of these techniques are not cost-effective, require highly skilled manpower to operate, are inefficient, and require hazardous chemicals and high energy. These drawbacks are overcome by one of the most effective techniques: adsorption. Remediation of selenium in water by adsorption is one of the most competent and cost-effective techniques. A variety of adsorbents, namely activated carbon, metal-modified zeolites, and bio-polymeric materials, have been noted for the elimination of selenium in water.

As an adsorbent, various magnetic materials have been reported for the selenium removal. Adsorptivity of the material depends upon various factors like size of the particle, number of active sites present and also magnetic strength in the case of magnetic adsorbents. In such cases nanomaterials suit best for the adsorption phenomena. Nanoparticles are materials with sizes in the nano scale, which act as a bridge between bulky materials and atomic compositions with sizes ranging from 1 to 100 nm (Siddiqui et al. 2019). There are many studies where nanomaterials proved to be the finest of all other adsorbents. Different numbers of nanoparticles have been developed and studied for the removal of different categories of contaminants from the water. Many nanomaterials with metals such as Fe, Cu, and Mn have been reported as among the most promising adsorbents for the removal of selenium. The unique properties of nanoparticles play an important role in reducing various difficulties associated with the other technologies for selenium removal. This chapter broadly explains the uses of various nanoparticles for the removal of selenium from the water.

13.2 Various Selenium Remediation Techniques

It is well known that selenium toxicity in water even at low concentration brings out serious health issues to animals and human beings, therefore identification and application of different removal methods is crucial. Presently available various remediation technologies include reduction, oxidation, coagulation, phytoremediation, adsorption, etc. Several removal techniques are briefly described in the following sections.

13.2.1 Chemical Precipitation

Chemical precipitation deals with the separation of solid substance from a solution by changing the state of substance into an insoluble form by chemical reaction. According to the available literature, sulfur dioxide is the most commonly used chemical to precipitate selenous ions from aqueous solution. Many literatures show effective selenium removal with concentrated hydrochloric acid and sulfuric acid solutions; this is because of the high reduction capacities of these acids to convert all selenium into an elemental state. Several other reagents have been discovered so far for removal of selenium by this technique. Among other reagents sodium dithionite has wide industrial uses for the removal of selenium. However, sodium dithionite has disadvantages like high cost, subsequent redissolution of the selenium, and relatively poor efficiency for the removal.

Removal of selenium using sodium sulfide is one of the oldest methods, initially reported almost two centuries ago. The chemistry of this phenomenon is still yet to be optimized to treat industrial water. Selenous ions are precipitated effectively using sodium sulfide at weak acidic conditions. This chemical reaction gives optimum results at pH less than 7, molar ratio of sulfide to selenium is found to be 1.7–11, with reaction time of 10 minutes at room temperature. This reaction explains the loss of electron from sulfides and the gain of electron by selenium, resulting in the conversion of selenium ions into constituent selenium (He et al. 2018).

13.2.2 Phytoremediation

This technique makes use of green plants for the elimination, degradation, or stabilization of unwanted materials in ground water and soil. This is an efficient green technology for selenium removal from contaminated soil and water. This method uses phytoextraction and phytovolatilization for the selenium removal and is proven to be a powerful method to solve the selenium contamination problem in the environment. This can be achieved with the help of plants which shows high tolerance for the selenium. The *True Se* hyper accumulator shows 1000–15 000 mg Se/kg without showing any toxicity symptoms. The antioxidant responses of plants determine the mechanism of accumulation of selenium by that particular species. According to recent research the *Chinese Brake fern* species can accumulate high selenium without showing any adverse effect in its biomass. In this case fronds absorbed less selenium compared to the roots. Various enzymes play an active role in the selenium accumulation in this plant. The *Chinese Brake fern* is very useful species for the removal of selenium in hydroponic conditions, which is beneficial to treat contaminated water. A drawback to this technique lies in the stringent conditions required to carry out the process. More research is required in this field to achieve high efficiency (He et al. 2018).

13.2.3 Bioremediation

This technique makes use of various microorganisms or other minute life forms to break down or consume the pollutant present in the environment. For selenium remediation various microorganisms are used. The role of these microorganisms is to convert selenium ions to elemental form, which is insoluble in water. This insoluble form obtained has specific nanometer size and can directly be separated from the water. This method makes use of different bio-films or microorganism to transform toxic selenium to non-toxic selenium, which is referred as biotransformation of selenium. These bio-films after interaction with selenium have also been characterized by various instruments like Confocal laser scanning microscopy (CLSM), X-ray absorption spectroscopy (XAS), and extended X-ray absorption fine structure (EXAFS), which confirm the removal mechanism and oxidation states of selenium at various stages of treatment. Proof for the accumulation of selenium on to the bio-film is given by micro X-ray fluorescence tuned with CLSM. Bacterial species named *Alishewanella sp.* WH16-1 which is discovered in mining soils, reduces selenium efficiently under aerobic conditions. So far, several bacterial species have been identified that are capable of reducing Se (4+) to Se (0) leading to detoxification. This reaction can occur in both ways, i.e. enzymatically or non-enzymatically. The non-enzymatic reaction includes chemicals like glutathione, siderophore, and H_2S. In the enzymatic condition, reduction of selenium occurs during the respiration of anaerobic bacteria. Some enzymes, namely hydrogenase, fumarate reductase, arsenate reductase, and sulfite reductase, have been reported that can reduce selenium. Although some aerobic bacteria have also been reported for selenium decontamination, still more research is required to establish this approach (He et al. 2018).

13.2.4 Coagulation-Flocculation

This technique deals with the removal of colloidal particles. So far very few reports are available on the removal of selenium from water through coagulation. According to the research reports there are various parameters responsible for the efficient selenium removal by this method. These parameters include pH of the medium, Co-ions present in water, amount of coagulant added etc. Fe-based coagulants and Al-based coagulants are mostly compared and it is observed that Fe-based coagulants are more effective than Al analogue. Selenate was less removed compared to selenite by coagulation technique. Data reported so far shows that coagulation with Fe salt is effective only for removal of selenite, however, it is always essential to remove both selenite and selenate simultaneously. Chitosan is another coagulant which has been reported for selenate removal by coagulation (He et al. 2018).

13.2.5 Electrocoagulation (EC)

Electrocoagulation is a broad-spectrum treatment technology that destabilizes the suspended or dissolved contaminants in liquids with the help of electric current in the medium. This technology makes use of electric current to bring out a chemical reaction. The electrocoagulation (EC) process has been reported for eradication of selenium from the water. Iron electrodes were run at 200 mA, which shows 97% removal. Aluminum electrode runs at high current intensity of 300 mA and showed 96% of removal. In the same working conditions the best selenium removal was observed with Fe electrodes with about 93% removal efficiency whereas Al electrodes showed only 54% removal. When leaching tests were performed Fe-Se sediment showed the selenium leaching within regulatory conditions whereas for Al it was 20 times higher. Complete analysis suggests that Fe electrodes shows greater efficiency compared to Al electrodes for selenium removal by this method (He et al. 2018).

13.2.6 Co-Precipitation Method

Co-precipitation mainly occurs through three main types: adsorption, inclusion, and occlusion. Occlusion is the method used for the elimination of Se (IV) and Se (VI) from the polluted water system. The co-precipitation technique has many advantages over other methods. Advantages include short treatment time and cost-effectiveness; it has the capability to retain substituent ions in the crystal lattice for a longer period, etc. Earlier research was carried out to construct and optimize this technique using barite ($BaSO_4$) and analyze efficiencies in inserting Se (VI) and Se (IV). Barite is one form of barium sulfate commonly present in nature which is commonly used to remove toxic or radioactive elements from the polluted water. Barite has the characteristic properties which show greater efficiency for contaminant removal. It has large ionic radii, extremely low solubility, high density, and, most importantly, rapid sedimentation during the co-precipitation process. Barite also has high crystal stability at wide range for temperature, pressure. All these properties make barite useful for selenium removal from the water via this method (He et al. 2018).

13.3 Selenium Removal Using Adsorption

Numerous techniques are known for the removal of selenium from water; broadly, lime softening, coagulation/filtration, anion exchange, RO, phytoremediation, and electro dialysis, but the adsorption method is the most cost-saving, simple, and effective technique. The method of adsorption is always the preferred technique for eradication of heavy metals and metalloid anions from water.

Selenium removal has been reported using a well-identified cationic layered rare earth hydroxide (LRDH). The chemical formula of cation is $Y_2(OH)_5Cl1.5H_2O$. The batch parameters like adsorption efficiency, selectivity, sorption kinetics desorption, and isotherms of both selenite and selenate on the cation LRDH were investigated, broadly maintaining pH between 7 and 8.5. For selenite and selenate the maximum binding capacities are found to be 124 and 207 mg/g, respectively. These are among the best efficiencies reported for variety of adsorbents reported for selenium uptake. It is also found that in low amounts LRDH can remove the selenium almost completely, even in the presence of co-ions like HPO_4, CO_3, NO_3, and Cl^-. The final concentration of selenium obtained is below 10 μg/l, which is within the permissible limit for the drinking water given by WHO. The LDH can be reused after desorption to remove selenite and selenate from it. Eventually, this study confirms that the cationic LRDH, $Y_2(OH)_5Cl\,1.5H_2O$ is a promising and spurting material that proves to be effective for removal of selenate and selenite.

If the adsorption capacities of Se (IV) and Se (VI) are compared, the adsorbents like silica gel, activated aluminum oxide, clay, activated carbon, ferric oxyhydroxide, zeolite, and molecular sieve show efficient adsorption of Se (IV) while for Se (VI) most of the adsorbents shows very weak removal performance. The most probable reason for high affinity of adsorbents is that Se (IV) forms an inner-sphere layer with the adsorbent active surfaces whereas Se (VI) gets adsorbed in the outer sphere, which is not as strong as inner-sphere adsorption, hence removal of Se (VI) is relatively lesser by common adsorbents. Thus, it is a very essential requirement to find a recyclable and durable adsorbent material to remove both the Se (IV) and the Se (VI) from water. Recent studies on graphene oxide (GO) and graphene have gathered much recognition. Some studies show their promising performance for the elimination of heavy metal ions, organic components (Holmes and Gu 2016), and microbial community (He et al. 2018).

One researcher (Ghosh et al. 1994) has studied the adsorption of Se (IV) and Se (VI) on hydrous alumina. Batch studies are performed. It is observed that the only significant parameter that affects both Se (IV) and Se (VI) is pH, and the extent of adsorption of oxy-anions was determined by the presence of surface hydroxy groups on alumina. The Triple Layer Model (TLM) can satisfactorily describe the adsorption of both the anions of selenium. Co-ions study suggests that the adsorption of selenium decreases in presence of SO_4^{2-}. The optimum pH for this experiment observed to be 5.5. Fixed-bed adsorption setup was used for the selenium removal. Contact time of two minutes found to be sufficient for adsorption of Se (IV) from the water at pH ranging between 5 and 8. It is also found that the adsorption of selenium in hydrous alumina will increase if Empty Bed Contact Time (EBCT) is longer. They have also tried to regenerate spent alumina bed using 0.25% sodium hydroxide solution and rinsing with DI water and 0.05 M HCl and this confirms almost complete desorption with very minute loss of alumina (Ghosh et al. 1994).

13.4 Nano Materials for Remediation of Selenium in Water

More recently nanomaterials are catching attention for their vast and efficient applications in every field; and this is also the case in water treatment technologies. The exceptional characteristics of having nano-scale size provide nanomaterials with improved catalytic and adsorption properties with greater reactivity. Multitudinous studies have been conducted that describe the capability of nanomaterials to remove various pollutants and heavy metals from the water. Selenium is considered a difficult contaminant to separate from waste water because of its range of solubility, toxic nature, and state of nature of different oxidation states. Recently, the wide uses of nanomaterial for removal of selenium have gained huge interest. Some groups of nanomaterials, such as nano-scale adsorbents, catalysts, and reactants, show high capacity in removing selenium. Various nanomaterials used as adsorbents to be highly effective. Several nano-adsorbents reported for removal of selenium are summarized in Table 13.1 along with removal efficiencies, conditions, etc. Brief details of various nanomaterials for removal of selenium are also discussed in subsequent sections (He et al. 2018).

13.4.1 Single Metallic Nanoparticles as Adsorbents

(Gonzaleza 2011a) has investigated the adsorption of selenite and selenate by nano crystalline iron (III) oxide. The author has studied the comparison of removal of selenium through Fe_3O_4 material developed using non-microwave and microwave assisted techniques. Both the synthesized nanomaterials are characterized by XRD and the observed grain size is 27 nm for non-microwave assisted and 25 nm for microwave assisted Fe_3O_4.

Table 13.1 Nanomaterials reported for selenium removal.

Sr. No	Nanoparticle	Experimental Condition	Amount of Adsorbent	Maximum Selenium Removal	Remarks	Ref.
I Single metallic nanoparticles						
i	Fe_3O_4 -Non-microwave assisted	Efficient pH range is 2–4. Grain size is 27 nm. Langmuir isotherm	–	Se (IV) – 1923 mg Se/kg Se (VI) – 1423 mg Se/kg	25% larger volume and 10% larger surface area than microwave assisted Fe_3O_4	Gonzalez et al. (2010)
ii	Fe_3O_4 -Microwave assisted	Efficient pH range is 2–4. Grain size 25 nm. Langmuir isotherm	–	Se (IV) – 2380 mg Se/kg Se (VI) – 2369 mg Se/kg	Smaller volume and surface area still Se binding efficiency is more	Gonzalez et al. (2010)
iii	Mn_3O_4 – non microwave assisted	Optimum pH is 4. Langmuir isotherm	–	Se (IV)- 507 mg Se/kg Se (VI) – 800 mg Se/kg	The presence of co-ions like Cl^- and NO_3^- has no considerable effect on selenite binding, while selenate bonding was reduced to less than 15%	Gonzalez et al. (2011b)
iv	Mn_3O_4 – microwave assisted	Optimum pH is 4. Langmuir isotherm	–	Se (IV) – 1000 mg Se/kg Se (VI)- 934.5 mg Se/kg		
II Mixed-metal-based nanoparticles						
i	Nano-$MnFe_2O_4$	Efficient pH is 2–6	2.5 g/l	6.6 mg/g	Has greater adsorption capacity for Se oxyanions than natural magnetite	Holmes and Gu (2016)
ii	Nano-$CuFe_2O_4$	Efficient pH is 7.4	0.4 g/l	14.4 mg/g	Copper nanoparticle has lowest adsorption capacity of selenite and selenate as compared to cobalt and manganese	
III Magnetic nanoparticles						
i	Nano scale zero valent iron (n-ZVI)	Effective removal time 3 min. Selenite has single core-shell nano structure.	5 g/l n-ZVI for 3 mM of selenite	Not mentioned	The reaction mechanisms studied by Cs-STEM integrated with XEDS	Gonzalez et al. (2011a)

IV Nanocomposites						
A	Nanocomposites using inorganic frameworks					
i	n-Al_2O_3 impregnated chitosan beads (AICB)	Efficient pH range is 6.5–7. Langmuir isotherm	70 mg of original AICB (8% n-Al_2O_3, 92% chitosan by mass or 0.113 mm of Al)	Se (IV)Binding constant: n-Al_2O_3 = 1.21×10^{-2} Chitosan = 3.048×10^{-3} Se (VI)Binding constant: n-Al_2O_3 = 1.349×10^{-2} Chitosan = 1.990×10^{-3}	For coinciding systems of selenite and selenate, AICB is the most robust option	He et al. (2018)
ii	Al/Si coprecipitate	Optimum pH is 5. Langmuir isotherm	3 g/l of Al/Si	Not mentioned	Adsorption follows pseudo second-order kinetic model	Yamani et al. (2014)
B	Nanocomposites using porous polymeric matrix					
i	PPN-150-MoS_4	In presence of oxo ion Hg^{2+}. The parallel removal of Hg/Se species in aqueous solutions.	303 mg/g	51.5 mg/g for Se (VI) 124 mg/g for Se (IV)	In absence of Hg^{2+} the adsorption is very low	Feng et al. (2019)
C	Carbon/graphene-based nanocomposites					
i	Carbon-supported magnetic nanoparticle (C-MNA)	Efficient pH is 3 and 5	20 mg in 5 ml	48 ± 5 mg/g binding capacity with >99% efficiency	C-MNA shows greater adsorption as compared to MNA	Evans et al. (2019)
ii	Magnetic nanoparticle-graphene oxide (MGO)	Optimum pH is 7	1 g/l	4.7 mg/g	MGO is reusable and has efficient binding capacity toward Se (IV) and Se (VI)	He et al. (2018)
V Photocatalytic nanoparticles/nanocomposites						
i	n-TiO_2 impregnated chitosan beads (TICB)	Efficient pH range is 1–4	Not mentioned	55.56 mg/g	Adsorption of selenite on n-TiO_2 is more effective and quicker	He et al. (2018)

The effective pH range for removal of selenite is 2–6 while for selenate it is 2–4. The rise in pH shows a reduction in the binding capacity of the adsorbent, thus selenite shows a reduction in binding at pH 5 and 6. The adsorption capacities obtained from Langmuir isotherm were found to be 1923 and 1423 mg Se/kg for non-microwave assisted synthesis and 2380 and 2369 mg Se/kg for microwave assisted prepared iron (III) oxide for selenite and selenate, respectively. The comparative study shows that the microwave assisted iron (III) oxide nanoparticles are more efficient than non-microwave assisted synthesized material. The author has also studied the pH effects, kinetics, co-ions, and adsorption capacity of manganese oxide NPs on removal of selenium from water. The mechanism follows Langmuir isotherm and shows most active binding on pH 4. The effective adsorption rate is 10 minutes for both selenite and selenate. Author also investigated the effect of other anions but not found increased binding effect on both selenite and selenate.

The use of iron (Fe^{3+}) oxides or hydroxides nanoparticles sol, called nano Fe, as adsorbents has also been reported. The described technique has several advantages: it is cost-effective and one of the most efficient methods for selenium removal. The concentration and the pH level of the polluted water were noticed to strongly affect the removal efficiency of iron (Fe^{3+}) oxides/hydroxides-based nanomaterial. The concentration of residual selenium obtained by this technique is below 0.01 mg/l, which are permissible by water quality regulations. The regeneration efficiency of selenium is found to be between 95 and 98% with the synthesized adsorbent (Gonzalez et al. 2010).

13.4.2 Mixed Metal-Based Nanoparticles

In recent years, mixed metal nano particles have gained attention for the removal of selenium from the water. Some mixed metal nanoparticles reported for the selenium removal includes Nano-$MnFe_2O_4$, Nano-$CuFe_2O_4$, Fe–Mn binary oxide. $MnFe_2O_4$ synthesized via hydrothermal method shows effective adsorption for selenium oxyanions. Nano Fe_2O_4 has proved to have more selenium removal than its naturally occurring magnetite material. In this literature selenium adsorption was found to be in the pH range 2–6 with about five minutes of contact time. Presence of co-ions like chlorides and nitrates has no as such effects on adsorption efficiency of selenium (Holmes and Gu 2016).

Gonzalez et al. (2010) explains the removal of selenium species from aqueous medium using nano synthesized $MnFe_2O_4$. The adsorption studies of selenium on $MnFe_2O_4$ were conducted using batch experiments and DRC-ICP-MS. This adsorbent was synthesized by titrating Fe^{2+} and Mn^{2+} ions together with NaOH solution. The $MnFe_2O_4$ were characterized using XRD and size of the nanomaterial was determined using Scherrer's equation. It is found that selenium binding on $MnFe_2O_4$ was independent in the pH range 2–6. The kinetics studies show maximum adsorption within first five minutes of the experiment, moreover not much appreciable increase in adsorption was observed after 60 minutes. Co-ions study was performed with Cl^-, NO_3^-, SO_4^{2-}, and PO_4^{3-} ions. Ions like Cl^-, NO_3^- show no impact on adsorption but the presence of SO_4^{2-}, PO_4^{3-} ions shows the change in adsorption of both selenite and selenate to different extent. Binding capacity was determined using Langmuir isotherm (Gonzalez et al. 2011b).

13.4.3 Magnetic Nanoparticles

(Ling et al. 2015) has vastly studied the nature of nano-scale zero valent iron (n-ZVI) for the removal of selenium from waste water. The n-ZVI was used as adsorbent and falls in the magnetic category due to the presence of iron nanoparticles. A vessel containing 100 ml of 1.3–6.3 mM of diluted Se (IV) liquid is kept for shaking, along with 1–5 g/l of known n-ZVI solution. The vessels were air-sealed and the temperature of shaking was continuously maintained at 22 °C. High-purity nitrogen played an important role in this experiment as it acts as potential oxidant and removes dissolved oxygen from the batch. High-purity nitrogen was purged into each batch, 30 minutes before introducing the n-ZVI. The results obtained for adsorption show that 5 g/l of n-ZVI has successfully removed at initial concentration of 1.3 mM of selenite quickly within three minutes. This work concludes that n-ZVI is a good adsorbent for the removal and treatment of selenium-contaminated water.

The work establishes the reaction mechanisms which were studied by spherical aberration corrected scanning transmission electron microscopy (Cs-STEM) integrated with X-ray energy dispersive spectroscopy (XEDS). It shows that the selenite has single core-shell nano structure on the synthesized n-ZVI. The structure of fresh n-ZVI and used n-ZVI is also studied. It is observed that the thickness of smooth core-shell particle (having shaking time of 48 hours) is increased from 2 to 4 nm and even more as compare to original n-ZVI. This method was also used to determine the solid phase transformation and translocation of selenium, includes deposition, reduction, diffusion, and to detect the surface defects on single nanoparticle. It was observed that Se (IV) was reduced to Se (II) and Se (0) oxidation states which forms a layer of selenium having thickness of 0.5 nm on the iron oxide interface of thickness 6 mm.

This broad study proves that the change in the structure of the outer shell of n-ZVI particle enhances the selenium diffusion and improves the selenium removal capacity of n-ZVI. Additionally, structure studies indicate that STEM and XEDS are useful tools for demonstrating solid phase reactions and elemental mapping (Ling et al. 2015).

13.4.4 Nanocomposites

It has been found that every nanoparticle has its own advantages and disadvantages. One of the most commonly reported nanoparticles, namely nZVI, suffers from drawbacks like oxidation and aggregation and even faces difficulty in separation of degraded systems. Nanoparticles such as TiO_2 NP or ZnO NPs have the limitation that they work in UV region because of their large band gaps. To overcome the problems related to presence of only nanoparticles and in order to achieve efficient removal of the contaminants, use of nanocomposites proved to be beneficial (Lu et al. 2016).

Nanocomposites can be defined as the materials that embed the nanoparticle into the matrix of support material which are heterogeneous in nature. In past few years, different nanocomposites have been reported in the field of nanomaterials and have been shown to be very much efficient for the removal of different contaminants. For example, a novel adsorbent has been reported by a chemical deposition of nZVI onto carbon nanotubes, which demonstrated high efficiency for the removal of various contaminants including selenium removal. In the same context thin film nanocomposites membranes have also been reported by incorporation of TiO_2 NPs. Nanocomposites in selenium removal, though, is a less researched topic. Advantages of nanocomposites includes that they are highly useful for waste water treatment, as they possess long-term stability, and are cost-efficient and exhibit properties of both nanoparticles and porous support (Lu et al. 2016).

13.4.4.1 Nanocomposites Using Inorganic Frameworks

Different zeolites and other clay minerals have also been reported as porous media for the incorporation of various nanomaterials. And such materials are observed to be effective for the removal of selenium from water.

Chan et al. (2018) studied the adsorption of selenium on Al/Si coprecipitates. They studied the adsorption of selenate and selenite on Al/Si as adsorbent. According to this paper, adsorption of both SeO_3 and SeO_4 species was greatly related to surface characteristic property of Al/Si. Optimum adsorption occurs at pH 5 and at dose of 3 g/l of adsorbent. Surface Al present in Al/Si coprecipitates shields a section of negative charges from the inside SiO_2, giving higher point zero charge and thus providing the adsorption of SeO_3/SeO_4 on this adsorbent. It is also mentioned that adsorption of selenium species onto this material is an exothermic process. The experiment follows pseudo second-order kinetic model and Langmuir isotherm model (Yamani et al. 2014). Removal of both selenium species SeO_4 and SeO_3 were highly affected by the presence of PO_4 ligand Whereas SO_4 being weak ligand has only affected SeO_4 adsorption.

Yamani et al. (2014) reported the synthesis of n-Al_2O_3 impregnated chitosan beads (AICB). It is the combination of nanocrystalline metal oxide impregnated chitosan beads (MICB) with nanocrystalline aluminum oxide (n-Al_2O_3). Both n-Al_2O_3 and chitosan shows the adsorption for selenium species, but their combination AICB shows the simultaneous adsorption of both selenite and selenate in aqueous medium more efficiently. The adsorption efficiency of AICB was compared with n-TiO_2 and n-TiO_2 impregnated chitosan beads (TICB) which are already well known for their effectiveness in selenium adsorption. n-Al_2O_3 shows high affinity for selenite system whereas chitosan shows good affinity for selenate. This hypothesis is proved with the help of adsorbent/adsorbate association constants values. AICB is potentially the strong option when it comes to the removal of both the selenium species from same system as it shows most rational performance irrespectively the number of selenium present in one system. This experiment follows Langmuir adsorption isotherm and order of kinetics is found to be pseudo-second order kinetics. Adsorption found to be highly dependent on pH and redox conditions of the systems. High concentration of sulfates in the removal system shows a decrease in the adsorption of selenium and therefore pre-procedure is suggested to reduce sulfates (Chan et al. 2018).

13.4.4.2 Nanocomposites Using Porous Polymeric Matrix

Encapsulation of different nanomaterials within the macro molecules (polymer) has been an interesting choice for the preparation of nanocomposites. Different polymers have their own different advantages, which contribute to the greater efficiency. The presence of various functional groups in the polymers increases the adsorption capacity. Most commonly used and highly cost-effective nanomaterials for the removal of selenium are iron and iron oxide nanoparticles. These Fe-based adsorbents bring out a redox reaction to convert a selenium species into elemental selenium, which is easily removed.

Gui et al. (2015) developed iron-functionalized membranes to adsorb selenium from industrial waste water. These membranes have advantages over suspended Fe in that they prevent particle aggregation and dissolution. The method of preparation can be stated as large membranes of polyvinylidene difluoride (PVDF) first coated with polyacrylic acid (PAA). To this coated membrane, ion exchange treatment is given to introduce ferrous ions into it, which is followed by reduction of Fe ions to zero-valent nanoparticle. With the increasing coating of the polymer by PAA, water permeability of membranes decreases. These membranes were characterized by scanning electron microscopy (SEM) and transmission emission microscopy (TEM). Selenium concentration below 10 mg/L was achieved in waste water, which follows pseudo-first-order law (Gui et al. 2015).

13.4.4.3 Carbon/Graphene-Based Nanocomposites

Many carbon/graphene nanocomposites have shown to be effective for the removal of selenium from the water. The oxides magnetite and hematite are very effective adsorbents in the removal of selenite. The continuous research for the advancement of a noble adsorbent for the removal of selenium from water took an interesting turn when the surface traits of graphene and iron oxide were combined through modified GO sheets by magnetic iron oxide nanoparticles.

The concept of magnetic GO is categorized as nano-composites. For Se (IV) and Se (VI) magnetic graphene oxide (MGO) shows high efficiency and binding capacity, having removal percentage >90 and 80%, respectively; and MGO dosage of 1 g/l with initial concentration of selenium was 300 µg/l. The removal was achieved in a few seconds. It is also observed that the acidic pH increases the adsorption of selenium ion on MGO, which boosts the removal percentage of Se (VI) to more than 95% at pH 2.0 (WHO 2011). Thus it can be concluded that MGO has an outstanding removal capacity of selenite and selenate in water. MGO also has tremendous potential of practical applications for selenium adsorption from waste water (He et al. 2018).

Li et al. (2018) has demonstrated the preparation of nanocomposites beads from Mg-Al-CO_3 LDH nanoparticles and chitosan. This composition is achieved by the direct mixing of LDH nanoparticles into chitosan matrix. Removal kinetics were studied and compared to LDH nano powder for the removal of selenium in water. Development of LDH into the beads showed a characteristic property difference from the original and in favor of removal. This combination is found to be effective for a wide pH range. Good results were obtained, which showed effective alternative to granulated media. These materials were thoroughly characterized by SEM, FTIR, and XRD (Li et al. 2018).

Evans et al. (2019) introduced novel mesoporous carbon framework as a base for MNA (magnetic iron oxide NP), which helps to selectively adsorb selenite from the waste water. C-MNA showed greater removal as compared to only MNA from 5 to 50 mg/l selenite conc at 48 ± 5 mg/g capacity with almost 99% removal at pH 3 and 5. SEM and XRD were used for the confirmation (Evans et al. 2019).

13.4.5 Photocatalytic Nanoparticles/Nanocomposites (Titanium-Based Nanoparticles)

The most commonly used photocatalyst is TiO_2, which has been extensively studied for the elimination of various contaminants in water. Various nanocomposites are reported in combination with TiO_2. Some of the combinations are mixed oxide of Al_2O_3 and TiO_2-impregnated chitosan beads, TiO_2-pillared montmorillonite, hydrous TiO_2, etc. Extensive research has been undertaken to study nanocrystalline titanium dioxide (n-TiO_2) for selenium adsorption.

Scientists have described the adsorption of both selenite and selenate by n-TiO_2. n-TiO_2 removes both the selenium species but removal efficiency for selenite is greater than for selenate. As n-TiO_2 is insufficient to remove selenate, selenate can be reduced to selenite with help of photolysis and thus can be removed easily. Carbon nanotubes in combination with titanium showed good selenium removal from waste water. The reaction mechanism of photocatalysis of TiO_2 for selenium removal can be explained as follows: when the UV light falls on the TiO_2 catalyst it releases the electron which is taken by both selenite and selenate to convert into elemental selenium, which is then easily removed. Research has showed that a photocatalyst like zirconia mixed with titanium has higher photocatalytic nature compared to standard n-TiO_2 for further reduction of selenium ions species (Holmes and Gu 2016).

13.5 Conclusions and Future Trends

Selenium toxicity from consumption of selenium-contaminated water is a major problem with respect to environment and human health. Although different remediation techniques are available for selenium removal, each of its technique has its own pros and cons. Most of the technologies are complex or have a high cost. Adsorption has proved to be one of the simplest and most cost-effective techniques for selenium removal. Nanoparticles are emerging materials in the area of adsorption and have proved to be highly effective for selenium removal. Selenium removal using nanomaterial as an adsorbent has proved to be an attractive technique owing to their unusual properties like large specific surface area, easily accessible pores, desorption processes, etc.

This work has given a brief overview of various technologies in use for selenium removal, worldwide. These technologies include: reduction, bioremediation, electrochemical, phytoremediation, coagulation flocculation, EC, and adsorption. Adsorption using various adsorbents such as activated alumina, activated carbon, and nanomaterials have been explained. Activated alumina is modified by impregnation on chitosan beads to form the MICB and used as adsorbent to remove selenium in water.

Different nanoparticles are categorized depending upon the composition and functionality they possess. In this chapter nanomaterials are broadly classified into five main categories. The first category belongs to metal-based nanoparticles; they are named so because metal plays an active role in the adsorption of selenium onto the matrix. For example, nano crystalline iron (III) oxide shows removal in both microwave and non-microwave assisted nanoparticles but the binding capacity of microwave assisted nanoparticles is more efficient. The second category belongs to mixed metal-based nanoparticles. As the name suggests, these nanoparticle are composed of two or more metals working together to bring about efficient adsorption of selenium in aqueous medium. For example, Nano Fe_2O_4 has proved to have more selenium removal than its naturally occurring magnetite material. Magnetic nanoparticles fall under the third category, which contains ferrous. For example, research has been done to establish the properties of nano zero-valent iron in removing selenium from water in very less time. n-ZVI is one of the most studied nanomaterials for the adsorption of selenium. The fourth category is nano composites, defined as those materials that incorporate the nanoparticle into the matrix of standard material to increase the efficiency of the materials. These are further divided into three sub-categories depending upon the nature of the base material used for binding the nanomaterial. Nanocomposites can be sub-categorized into composites with inorganic frameworks, composites with a polymeric matrix, and carbon/ graphene-based nanocomposites. Al/Si by co-precipitation belongs to nanocomposites with inorganic framework; iron grafted PAA and PVDF belongs to nanocomposites with polymeric matrix, which shows high efficiency for selenium removal as compared to only iron-based nanomaterials. LDH with chitosan belongs to the third sub-category of nanocomposites, and the combination of these two proved to be more effective compared to single LDH. The fifth category is photocatalytic nanoparticles. TiO_2, being an established photocatalyst, is the most commonly used photocatalytic nanoparticle. The combination of light and TiO_2 catalyst brings about the reduction of selenium to its non-toxic and easily removable form and hence this is the most commonly used combination. The chapter summarizes that the nanoparticles and nanocomposites are important class of nanomaterials for selenium removal.

Although adsorption with the help of nanoparticles is an effective method for selenium removal it suffers from certain drawbacks which need to be studied. Several challenges are still to be addressed in terms of selenium adsorption with the help of nanomaterials. Though several nanoparticles are already been reported, more work can be done in terms of green synthesis and reducing the cost. More optimization is needed in terms of selenium removal in all the forms present in nature. Moreover, regeneration and the recovery of selenium from saturated adsorbents is not well studied. The disposal of spent adsorbents is still an area of environmental concern. Finally, the environmental fate and impacts of nanoparticles need to be studied in detail.

References

Chan, Y.T., Liu, Y.T., Tzou, Y.M. et al. (2018). Kinetics and equilibrium adsorption study of selenium oxyanions onto Al/Si and Fe/Si coprecipitates. *Chemosphere* 198: 59–67. https://doi.org/10.1016/j.chemosphere.2018.01.110.

Evans, S.F., Ivancevic, M.R., Yan, J. et al. (2019). Magnetic adsorbents for selective removal of selenite from contaminated water. *Separation Science and Technology* 54 (13). https://doi.org/10.1080/01496395.2019.1617742.

Feng, L., Chen, W.M., Li, J.L. et al. (2019). Biological antagonism inspired detoxification: removal of toxic elements by porous polymer networks. *ACS Applied Materials & Interfaces* 11: 14383–14390. https://doi.org/10.1021/acsami.9b02826.

Ghosh, M.M., Cox, C.D., and Yuan-Pan, J.R. (1994). Adsorption of selenium on hydrous alumina. *Environmental Progress* 13 (2): 79–88. https://doi.org/10.1002/ep.670130210.

Gonzalez, C.M., Hernandez, J., Parsons, J.G., and Gardea-Torresdey, J.L. (2010). A study of removal of selenite and selenate from aqueous solutions using a magnetic iron/manganese oxide nanomaterial and ICP-MS. *Microchemical Journal* 96: 324–329. https://doi.org/10.1016/j.microc.2010.05.005.

Gonzalez, C.M., Hernandez, J., Peralta-Videa, J.R. et al. (2011a). Sorption kinetic study of selenite and selenate onto a high- and low-pressure aged iron oxide nanomaterial. *Journal of Hazardous Materials* 211–212: 138–145. https://doi.org/10.1016/j.jhazmat.2011.08.023.

Gonzalez, C.M., Hernandez, J., Parsons, J.G., and Gardea-Torresdey, J.L. (2011b). Adsorption of selenite and selenate by a high- and low-pressure aged manganese oxide nanomaterial. *Instrumentation Science and Technology* 39 (1): 1–19. https://doi.org/10.1080/10739149.2010.537721.

Gui, M., Papp, J.K., Colburn, A.S. et al. (2015). Engineered iron/iron oxide functionalized membranes for selenium and other toxic metal removal from power plant scrubber water. *Journal of Membrane Science* 488: 79–91. https://doi.org/10.1016/j.memsci.2015.03.089.

He, Y., Xiang, Y., Zhou, Y. et al. (2018). Selenium contamination, consequences and remediation techniques in water and soils: a review. *Environmental Research* 164: 288–301. https://doi.org/10.1016/j.envres.2018.02.037.

Holmes, A.B. and Gu, F.X. (2016). Emerging nanomaterials for the application of selenium removal for wastewater treatment. *Environmental Science: Nano* 3: 982. https://doi.org/10.1039/c6en00144k.

Li, M., Dopilka, A., Kraetz, A.N. et al. (2018). Layered double hydroxide/chitosan nanocomposite beads as sorbents for selenium oxoanions. *Industrial & Engineering Chemistry Research* 57 (14): 4978–4987. https://doi.org/10.1021/acs.iecr.8b00466.

Ling, L., Pan, B., and Zhang, W. (2015). Removal of selenium from water with nanoscale zero-valent iron: mechanisms of intraparticle reduction of Se (IV). *Water Research* 71: 271–281. https://doi.org/10.1016/j.watres.2015.01.002.

Lu, H., Wang, J., Stoller, M. et al. (2016). An overview of nanomaterials for water and wastewater treatment. *Advances in Materials Science and Engineering* 2016: 4964828. https://doi.org/10.1155/2016/4964828.

Sharma, V.K., Sohn, M., and McDonald, T.J. (2019). Remediation of selenium in water: a review. *Advances in Water Purification Techniques* 2019: 203–218. https://doi.org/10.1016/B978-0-12-814790-0.00008-9.

Siddiqui, S.I., Naushad, M., and Chaudhry, S.A. (2019). Promising prospects of nanomaterials for arsenic water remediation: a comprehensive review. *Process Safety and Environmental Protection* 126: 60–97. https://doi.org/10.1016/j.psep.2019.03.037.

WHO (2011). *World Health Organization, Guidelines for Drinking-Water Quality*. Geneva, Switzerland: WHO Press.

Yamani, J.S., Lounsbury, A.W., and Zimmerman, J.B. (2014). Adsorption of selenite and selenate by nanocrystalline aluminum oxide, neat and impregnated in chitosan beads. *Water Research* 50: 373–381. https://doi.org/10.1016/j.watres.2013.10.054.

14

Harnessing Biogeochemical Principals for Remediation of Selenium-Contaminated Soils

Jennifer Cooper

Department of Agronomy and Horticulture, University of Nebraska, Lincoln, Nebraska, USA

14.1 Introduction

Selenium (Se) was identified in 1817 and has been described as “red sulfur.” Jöns Jacob Berzelius was the first scientist to characterize the selenium isolate, although he was probably not the first to discover it (El-Ramady et al. 2015). The name for selenium was derived from Selene, the moon goddess. This was due to the similar behavior between Se and tellurium, who was named after “Tellus” (the word for earth in Latin) (Reilly 2006). Selenium is found throughout the Earth, and known to be in the atmosphere, lithosphere, hydrosphere, and biosphere of our planet (Sharma et al. 2015). Selenium is found on most of the Earth’s crust, averaging approximately 90 μg Se/kg, with areas of higher and lower concentration (James and Shupe 1984). Selenium is commonly found in the vicinity of sulfur; common biological interaction occur with sulfur, arsenic, and phosphorus (Frost and Olson 1972). Because the inner shells of Se are filled $3d^{10}4s^{2}4p^{4}$, its behavior falls between that of tellurium and sulfur (Se is positioned between these elements in the periodic table in column 16), generally in the crystalline metalloid behavioral category (Reilly 2006; He et al. 2018). Selenium allotropes include the amorphous form of Se (black color), metallic Se (gray color), monoclinic Se (red color), and two amorphous powdered Se forms. Common inorganic species of Se in the environment include oxidation states of −2 (organic, primary following biological methylation), 0 (elemental Se), and +4 (selenite, Se(IV)) and +6 (selenate, Se(VI)). The +4 and +6 oxidation states are not stable alone, and will generally adsorb to metal hydroxides or oxides in the soil. There are approximately 50 known mineral compounds that include Se, often in combination with mercury, tellurium, copper, sulfur, iron, lead, silver, and manganese (El-Ramady et al. 2015).

Both humans and animals require Se as an essential nutrient, and it is linked to prevention of oxidative damage during cellular respiration, and involved in fertility, metabolism of the thyroid, and links to other organ functions (Reilly 2006). However, consumption of

Selenium Contamination in Water, First Edition. Edited by Pooja Devi, Pardeep Singh, Arindam Malakar, and Daniel Snow.

small amounts of Se in excess of essential requirements can lead to toxicity, and in severe cases humans can develop nerve damage, kidney damage, respiratory problems, cardiac distress, tremors, or even die (Nuttall 2006). There are countries and regions worldwide that have Se contamination of either soil or water sources. Locations with excess Se in groundwater or aquifers include the San Joaquin Valley in California USA, Japan, the Amman Zarqa Basin of Jordan, Barwa in Jainpur, Hisar, Chandigarh, and Chennai and Punjab regions of India (Kuisi and Abdel-Fattah 2010; Bajaj et al. 2011; Kumar and Riyazuddin 2011; Soda et al. 2011; Zelmanov and Semiat 2013). Contamination of soils include multiple locations of the USA, Ireland, China, and India (Bajaj et al. 2011). With exception of the extreme cases listed above, the general low solubility of Se in water makes soil a more important medium of research for control of Se as both a nutrient and a toxin (Vinceti et al. 2013a).

14.2 Selenium as a Nutrient for Humans and Animals

Humans, animals, bacteria, fungi, algae, and most organisms require low levels of Se in their diets to maintain health for proper function of thyroid glands, to fight off infection, to maintain reproductive health, and for protection from oxidative stress (Rayman 2000; Navarro-Alarcon and Cabrera-Vique 2008; El-Ramady et al. 2015; Sharma et al. 2015). The dietary source of Se is primarily from proteins and amino acids. The recommended daily Se nutrient requirement for humans is at least 60–70 μg Se/d (Waegeneers et al. 2013). Livestock require higher levels of Se to prevent deficiency than humans, with requirements ranging 100–1000 μg Se/kg (Ekermans and Schneider 1982). Humans with Se deficiencies in their diet have been known to develop fatigue, muscle weakness, mental fog, weakened immune systems, hair loss, and infertility in both sexes (Rayman 2012). Between 500 million and 1 billion people have a major health problem that is caused by Se deficiency (Haug et al. 2007). In livestock, deficiencies in Se are exhibited as lameness or stiffening of the legs, lethargic behavior, diarrhea, poor growth rates, low fertility rates, and high mortality of young offspring (Fordyce 2013). For these reasons, Se is added to the diets of humans and animals through a variety of mechanisms.

While plants do not require Se as a micronutrient, some plants are fortified with Se using commercial fertilizers to make up for deficiencies in the human and animal diets, however, plant uptake can be minimal and large portions are lost to the environment (Rayman 2000; Haug et al. 2007). Direct fortification of foods with Se in processing plants is another method for increasing dietary consumption, with less loss to the environment, however production methods need to be controlled to preserve optimally available forms (Haug et al. 2007; Gharibzahedi and Jafari 2017). Selenium salt supplements are available to humans whose diets are Se deficient, while veterinarians also supply Se through supplements or processed foods to livestock and domestic animals as needed (Reilly 2006; El-Ramady et al. 2015). While large regions of the world require supplementation of dietary Se, it can also be found in many forms in environment, which vary in their toxicity to humans.

14.3 Selenium Toxicity to Humans

The range of Se that humans need to consume daily lies between the minimum threshold (~60 μg Se/d) and the toxic level <300 μg Se/d. Food chain source of Se are predominantly through plants, which uptake Se from soils, whereas drinking water sources are considered negligible (Reilly 2006). When concentrations of Se reach toxic levels, there are negative consequences for human, animal, and environmental health (Sakamoto et al. 2012). When considered from the perspective of Se concentrations associated with various levels of toxicity, clinicians usually only have access to total Se concentrations, even though Se toxicity varies by chemical form. Toxicity of Se is greatly influenced by its chemical form, with Se(IV) the most toxic, Se(VI) generally less toxic, and the organic form the least toxic (Mehdi et al. 2013; Awual et al. 2015). Nuttall (2006) found the following associations between total serum (clear liquid separated following blood clotting) Se and toxicity: acute toxicity was associated with 400–30 000 μg Se/L serum, chronic toxicity was associated with 500–1400 μg Se/l, and <1400 μg Se/l serum were found to be free of toxicity. While the highest levels of Se ingestion and death have been reported from industrial accidents and murder (Glover 1970; Ruta and Haider 1989), consumption of crops and products of livestock receiving toxic levels of Se can also harm humans. Some typical Se concentrations in foods produced on non-seleniferous soils are as follows: fruits (<100 μg Se/kg), dairy (100–300 μg Se/kg), grains (100–800 μg Se/kg), meat (100–400 μg Se/kg), and seafood (400–1500 μg Se/kg) (World Health Organization 2004).

Humans experiencing Se toxicity are observed to have thicker nails with stratified ridges that are often brittle, hair loss, and a garlic odor to both skin and breath (Whanger et al. 1996). The suggested Se consumption limit for adults varies by region, with the recommendation for adults set at 30 μg Se/d (Europe), 55 μg Se/d (USA), and 50–250 μg Se/d (China) (He et al. 2018). The Tolerable Upper Intake Level is set at 60 μg Se/d for children 1–3 years, 250 μg Se/d for teenagers 15–17, and 300 μg Se/d for adult by the 2006 European Food Safety Authority guidelines (Mehdi et al. 2013). During 1961–1964 in the Hubei Province of China, an outbreak of selenosis affected approximately 50% of the population in small villages characterized by subsistence farming in remote areas (Fordyce 2013). Crops were grown in soils with Se contents exceeding 300 μg Se/g soil (Yang et al. 1983). Selenium accumulation in the Hubei Province soil was determined to be from weathering of coal, which was compounded by the addition to lime, resulting in consumption of 4900 μg of Se per person on a daily basis, and inducing symptoms of selenosis from retention of 3200 μg Se/l blood into sickened people of the region (Yang et al. 1983).

Selenium can produce reactive oxygen species, and is therefore recognized as an environmental pollutant and an inorganic carcinogen. This is exacerbated by the relatively high environmental levels and accessibility to both humans and animals, as well as measured toxicological effects (El-Ramady et al. 2015; He et al. 2018). Neurodegenerative effects can be induced by Se at high concentrations (Vinceti et al. 2013a). Lower levels of blood glutathione peroxidase activity were found following exposure to drinking water with elevated levels of Se, with greater amounts of Se producing reduced enzyme activity (Lee and Jeong 2012). Prolonged consumption of Se leads to selenosis, which causes impairment of neurological functions (Sharma et al. 2015). Cardiovascular diseases and other degenerative diseases, such as amyotrophic lateral sclerosis have been observed in connection with elevated Se in the diet (Vinceti et al. 2013b). Increased tumor formation has been linked with high Se consumption from increased expression of regulatory T cells (TR1), as well as association of high Se consumption with increased incidence of diabetes (Sun et al. 2014). Recognition of Se toxicity prompted governments to establish limits for Se in drinking water. Despite epidemiological studies suggesting the potential for adverse health effect for Se concentrations above 1 μg Se/l, most regulations and guideline for Se are set above this level. Vinceti et al. (2009) found risk of neurological diseases, endocrine disruption, and neoplasm issue increased from chronic exposure to >1 μg Se/l. Drinking water Se concentrations <40 μg Se/l are recommended guidelines from the World Health Agency. Most countries (e.g. Australia, New Zealand, Thailand, Europe, Canada, Japan) established drinking water concentration of 10 μg Se/l as the regulatory limit for safety, while others countries set guidelines as low as 1 μg Se/l (Russia) and as high as 50 μg Se/l set by the US (Vinceti et al. 2009).

A large part of the toxicity of Se is due to its ability to become methylated and volatile. Volatility provides movement and terrestrial deposition pathways that may cause repeated accumulation of Se that reach toxic levels. Some regions are known to have toxic levels of Se primarily due to the buildup of atmospheric deposition (Kabata-Pendias 2010). The microbially driven methylation process for Se forms products that are less toxic than the reactants, which is unusual for most toxic substances. Volatile methylation products of dimethyl selenide and dimethyl diselenide ($(CH_3)_2Se$) and $(CH_3)_2Se_2$, respectively) are less toxic than precursors that were not volatile (Spallholz 1994). Large natural deposits of Se in California and China put crops at risk of taking up Se concentrations close to toxicity levels, causing crops to be unacceptable for human consumption (El-Ramady et al. 2015). The following sections will outline Se toxicity to plants and animals, and how they may transfer excess Se to humans upon consumption.

14.4 Selenium Toxicity in Plants

Plants need many trace elements in the parts per billion (ppb, μg/kg) concentration to achieve healthy growth, many of which are heavy metals, e.g. copper, iron, manganese, zinc (El-Ramady et al. 2015). Plants do not require trace Se. However, humans and animals do require it in their diet. At high concentrations, Se is toxic to plants, animals, and humans. Selenium is present in soil from a variety of sources that include natural and human-derived. If the soil parent material contains Se, then it will be transferred to the soil as it undergoes pedogenesis (the process of soil formation). Selenium can also be added to the soil through human activities, during additions of fertilizers, liming products, pesticides, herbicides, fungicides, compost, animal manure, slurry, or organic carbon waste additions that may contain trace or large amounts of Se (Senesi et al. 1999; He et al. 2005). Intentional addition of Se does occur to most agricultural soils, not because it represents a plant essential nutrient, but to ensure that humans and animals have sufficient Se levels in their diets. Unfortunately, the range of safe Se consumption (minimum essential amount required to the lower limit of toxicity) is a narrow range of concentrations; this makes it possible that repeated Se addition to meet essential nutrient requirements may eventually reaching toxic levels (Kabata-Pendias 2010). Use of recycled wastewater for irrigation or the land application of sewage sludge are other sources of Se additions because humans excrete Se in their waste. Particulate or aerosol forms of Se in the atmosphere can undergo deposition to soil surfaces through settling, or in conjunction with precipitation events.

Humans primarily obtain Se from their diet, suggesting that soil Se concentration, and the amount accumulated it the plant, are important for human health (WHO 1987). Inhibitory factors for plant uptake of Se include the combination of high concentrations of iron oxide or hydroxide minerals and high organic matter (Fordyce et al. 2000; Rayman 2008); both of these factors are known to strongly retain soil Se, making it unavailable for plant uptake. The parts of plants (i.e. root, leaf, stem) that will accumulate higher amounts of Se are dependent on the species of the plant, physiological conditions, and the development phase (Terry et al. 2000). Selenium hyper-accumulating plant species have several similarities through development; specifically during early vegetative stages, Se will collect in leaves; however, concentration of Se will occur in the seed when the plant transitions to the reproductive stage of development (Terry et al. 2000; Rayman 2008). Conversely, in plants that do not accumulate large amounts of Se (e.g. rye, barley, or oats), very little Se will be present in leaves and stems, while larger amounts are present in root and grain portion of crops (Terry et al. 2000). Hyper-accumulating plants are capable of accruing 100 000–100 000,000 μg Se/kg air dry biomass, non Se-accumulating agricultural crops will accumulate much less, even if crops are grown on seleniferous soils (Barceloux 1999; Whanger 2002). Root vegetables (beets, potatoes, carrots, turnips), peas, cucumbers, tomatoes, and beans will amass upward of 6 000 000 μg Se/kg when planted in high Se-containing soil (Mehdi et al. 2013).

The relationship with plant health and exposure is not a clear picture. Plants primarily assimilate Se in the selenite form, rather than the selenite form. Selenate uptake occurs via plant sulfur transportation pathways, whereas the small amounts of selenite assimilated in plant tissues are primarily taken up via the phosphate transportation pathway (Terry et al. 2000; White 2004; Li et al. 2008). While Se is not a required micronutrient for plants, at low levels it has beneficial effects; above concentrations of 1000 μg Se/kg toxicity can be observed in most agricultural crops (Kaur et al. 2014). Low levels of Se may benefit plants by acting as an antioxidant, regulating growth, serving as an abiotic stress modulator, or defending the plant from pathogenic attack (Kaur et al. 2014). Classic toxicity stress symptoms appear in plants in higher Se concentration soils, including drying, wilting, or yellowing of the leaves, slow growth rates, reduced protein synthesis rate, and in extreme cases, plant death (Lawlor et al. 2004). The degree of plant sensitivity to Se toxicity depends if the plant is a Se accumulator. *Neptunia amplexicaulis* (a herbaceous legume) and *Morinda reticulata* (Mapoon Bush) are both Se accumulators and did not show toxicity after accumulation of up to 4000 μg Se/g dry tissue (Shrift 1969). *Stanleya pinnata* (prince's plume) is known to be a Se hyper-accumulator, as is *Astragalus bisulcatus* (two-grooved milkvetch) (Quinn et al. 2010; de Jong et al. 2015). Unfortunately, rice experience a 10% loss in yields from growth in soil with 2000 μg Se/kg soil (Girling 1984), whereas *Lolium perenne* (rye grass) exhibited signs of oxidative stress and a reduction in yields when grown in soil with 10 000 μg Se/kg soil (Hartikainen et al. 2000).

14.5 Selenium Toxicity in Animals

Selenium has also demonstrated toxicity to animals (terrestrial and aquatic), in large part because it can contaminate freshwater sources through leaching from elemental earth deposits, and from the previously mentioned volatility and deposition issues. Large natural deposits of Se in California put crops at risk of taking up Se concentrations close to toxicity levels, causing crops to be unacceptable for human and animal consumption. Plants that are hyper-accumulators can take up such large amounts of Se that their tissues contain toxic levels if ingested by animals (El-Ramady et al. 2015). In order for gazing animals to become symptomatic of selenosis, they need to consume Se >5000000 μg Se/kg dry weight (Moreno Rodriguez et al. 2005). Toxins from Se are responsible for alkali disease that has been observed the northern great plains in USA for grazing livestock (El-Ramady et al. 2015). Lameness and a stiff gate are common in animals experiencing Se toxicity because Se will generally target the musculoskeletal system (He et al. 2018). Cattle have been observed to lose their wool and form lesions on their hooves and base of their horns from chronic exposure to toxic levels of Se (Aitken 2001). Feeding sheep toxic concentrations of Se during a five-month study resulted in death of 57% of the animals, although they were asymptomatic of illness (Aitken 2001). Within hours of receiving a vitamin shot prior to a polo match, six horses died after displaying acute onset of tachycardia and disruption of central nervous system functions in addition to other symptoms; liver and kidneys were found to contain 6130–6250 μg/kg (wet weight basis) of Se in the postmortem (Desta et al. 2011).

In aquatic birds, when the Se-containing amino acid selenomethionine reached concentrations higher than 3000 μg/kg on a wet weight basis, embryo deformity increased, as well as overall hatchability (Spallholz and Hoffman 2002). Dolphins were found to have increasing Se concentrations in their bodily tissues with age, suggesting a constant source and uptake during their lifespan (Itano et al. 1984). Fathead minnow, eastern mosquitofish, black bullhead, and common carp have demonstrated tolerance to high Se levels, whereas pumpkinseed and largemouth bass were more affected by high Se exposure (He et al. 2018). In smaller aquatic organisms, Janz et al. (2010) found higher Se tolerance algae, with indications of some bioaccumulation possible. Furthermore, dinoflagellates, prymnesiophytes, and prasinophytes were also capable of significant Se enrichment, while chlorophytes were below par Se accumulators (Janz et al. 2010). Now that we have discussed the effects of Se on humans, plants, and animals, we will outline the sources and movement of Se in the environment in the following sections.

14.6 Sources of Selenium: Natural and Anthropogenic

Earth's average crustal abundance of Se is 90 μg Se/kg crust (National Research Council (US) Subcommittee on selenium 1983), however, there are many areas with higher or lower Se concentrations, from both solid and volatile origins. Natural processes that produce solid forms of Se on the Earth's surface include weathering of rocks and soil parent materials that contain Se, as well as volcanic eruptions that disperse solid Se downslope and in gaseous forms to the atmosphere (El-Ramady et al. 2015) (Figure 14.1). Anthropogenic

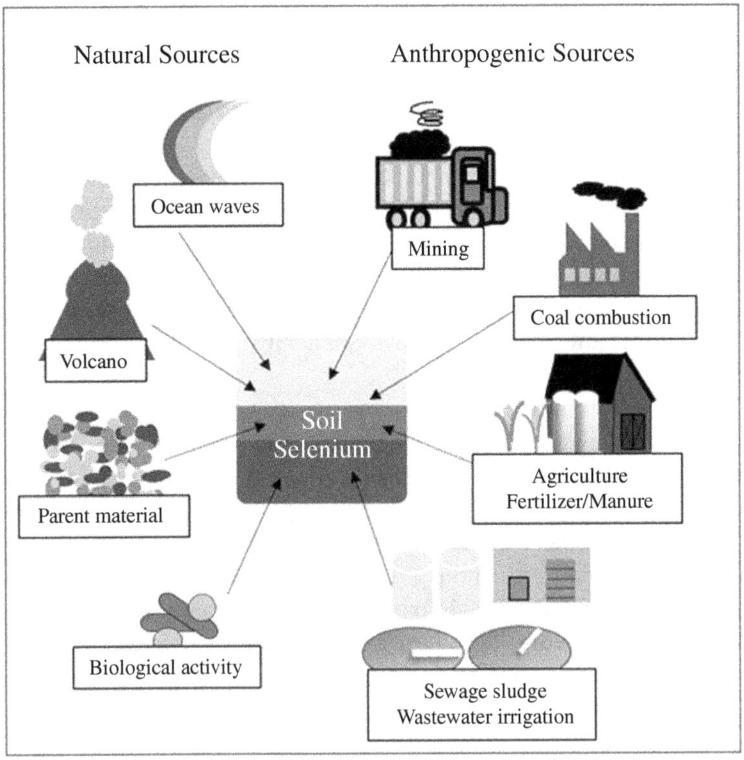

Figure 14.1 Natural and anthropogenic sources of selenium in soil.

processes that can cause Se accumulation in solid forms include oil refining, metal smelting, addition to some fertilizers and other agricultural products, irrigation water, introduction from sewage sludge application, and unintentional release from landfills (El-Ramady et al. 2015) (Figure 14.1). Frequently, movement of Se and accumulation in different regions of the world is due to atmospheric deposition of volatile Se forms. Atmospheric Se consists of about 50–65% natural origin Se, while the remaining portion comes from anthropogenic sources (Buchs et al. 2013; Sharma et al. 2015). Atmospheric Se that originates from natural sources includes volcanic gases, aerosols sea spray particulates, volatile organoselenium molecules from biological activity, and dust particles from the Earth's crust undergoing weathering and erosion (Nriagu and Pacyna 1988; Nriagu 1989) (Figure 14.1). Atmospheric Se originating from anthropogenic processes encompass dust particles from burning of coal, pharmaceuticals, glass, and ceramics manufacturing, coal combustion, and metal refining factory emissions (El-Ramady et al. 2015; Sharma et al. 2015) (Figure 14.1).

It is estimated that approximately 1890 tons of Se is produced annually through industrial processes (Foster and Sumar 1997). Selenium is used for many industrial purposes, and can serve as a source for contamination during production, use, movement in the environment, and disposal. Due to the ability of Se to conduct electricity, large amounts of Se are used in manufacturing of electrical devices. Boiling of elemental Se during production of electronic devices leads to atmospheric pollution, as well as release from improperly sealed landfills (Diplock and Kurzer 1974). Selenium is also used in the production of photovoltaic batteries and photometers; recycling of these components is far from complete, and there remains uncertainty about leaching of Se and other metals from landfills (Cyrs et al. 2014). Compounds formed from selenides can be used as metal lubricants, to enhance stainless steel machinability, to increase the abrasive resistance in rubber, and as a print photography toner. One of the largest uses of Se is for pigmentation in the industrial glass and ceramics factories, as well as serving as galvanizing agents (El-Ramady et al. 2015). Because Se is common to so many products, loss to the environment through chipped paint blowing away, dissolution of lubricants while washing machinery, and release from slow abrasion of moving machinery parts that contain Se over time.

Selenium is commonly used in trace amounts in cosmetics, or prescribed in medicines (Kabata-Pendias 2010). Humans and animals consume supplements and foods that contain Se to meet dietary requirements, where after fecal excretions and urine from animals and humans can return or concentrate Se in the soil. Cleaning of wastewater for irrigation purposes is mainly focused on biological contaminants, with trace amounts of dissolved Se in reclaimed irrigation water building up in the soil over time. Sewage sludge application to land has become a regular disposal method, frequently with elevated heavy metal concentrations strongly absorbed to the sludge and capable of causing toxic levels of Se to be land applied through repeated applications. The variety of sources of Se outlined in this section will move in the environment and accumulate in various regions, as outlined in the following section.

14.7 Concentrations of Selenium in Terrestrial, Aquatic, and Atmospheric Environments

The world average soil Se concentration is 400 μg Se/kg, while various individual soil Se concentrations range between 10 and 2000 μg Se/kg (Fairweather-Tait et al. 2011). Selenium enrichment in some soils have resulted in concentrations as high as 1 200 000 μg Se/kg in areas described as seleniferous regions (Johnson 2004; Fairweather-Tait et al. 2011; König et al. 2012; Schmidt et al. 2013). High Se regions are often associated with marine sediments, volcanic soils, mine waste, or are alkaline regions with little rainfall (Lakin 1972; Floor and Román-Ross 2012; Schmidt et al. 2013). The Se content of parent material rocks is highly correlated with the concentration of Se in soil. Shales and clay average 500 μg Se/kg, ultramafic rocks average 130 μg Se/kg, and granite and mafic rocks average 50 μg Se/kg (Alloway 1995). The concentration of Se is typically lower in igneous rocks, due to the lower clay portion, than in sedimentary rocks; therefore, sandstone, and limestone have less Se (10–100 μg Se/kg) than argillic (high clay) sediments (300–600 μg Se/kg) (Chipinda et al. 2008; El-Ramady et al. 2014). Cretaceous period rocks are particularly high in Se content (approximately 100 000 μg Se/kg) because of the high rate of volcanic eruption during that period the deposited volcanic dust and glass (El-Ramady et al. 2015). Phosphatic rocks have elevated Se concentrations because selenate and phosphate ions have many similarities. For example, sandstones and limestones will have <100 μg Se/kg, whereas phosphatic rock contain Se concentrations up to 300 000 μg Se/kg (Sharma et al. 2015). Elevated Se deposits were also found in oxidized sandstones, frequently in combination with uranium (Plant et al. 2013).

Volcanic soils are high in Se content because eruptions deposit volcanic glass and dust sourced from Earth's mantle and crust (El-Ramady et al. 2015). In volcanic soils, which are generally high in oxides of aluminum and iron and low in organic matter content, mobility of Se is typically limited due to strong adsorption of Se onto the oxides, hydroxides, and oxy-hydroxides of iron and aluminum (Tolu et al. 2014a; Tolu et al. 2014b). Selenium resembles sulfur, and is often found with high sulfur, and in coal deposits (Dauphas 2013; Wang and Becker 2013). Up to 43 000 μg Se/kg have been found in coal from China, Russia, and USA, however, more commonly coal Se content ranges from 1000 to 1600 μg Se/kg (Lenz and Lens 2009).

Mining waste soils are frequently observed to have high Se contents. Selenium is not the primary mineral mining companies are seeking, but it is commonly associated with mine waste regions. The primary damage to soil from mining is due to deposition of mine waste directly on the soil, complete removal of vegetation, or removal of upper horizons of soil; this creates a region with great susceptibility to erosion, and the potential for large releases of unfriendly chemicals the environment growth (Liu and Lal 2012; Charmbury 2019). Selenium is most often found in mining operations targeting coal, phosphorus, or shale oil (Lemly 1997). In a Mexican river valley mine waste material contained an average of 4600 μg Se/kg, whereas surface soils with mine waste incorporated resulted in Se content ranging 300–20 000 μg Se/kg (Thornton et al. 1965).

Industrial regions can contaminate soils with Se, primarily through volatile releases. In Croatia, a coal-fired power plant produced Se pollution of soils within the vicinity of plant due to the combination of mining, combustion, refining, and presence of former untreated industrial operations in the town; this resulted in soil Se contamination of 10 300 μg Se/kg soil (Medunić et al. 2018). In the Swansea Valley in the United Kingdom, there is surface soil with 200 000 ug Se/kg near a nickel refinery that has persisted for 80 years (Perkins 2011). In Portugal, the soils in the region that contained oil-fired power stations and other industry did not have elevated soil Se; however, airborne Se particles were elevated (8.0×10^{-4} μg/m^3), as were populous areas with many industrial plants near the capital (9.2×10^{-4} μg/m^3) (Ventura et al. 2005).

Agricultural soils can also accumulate high levels of Se. As previously mentioned, the Hubei Province in China had agricultural soils that became contaminated with Se in the early 1960s. In the case of Hubei Province, Se accumulated due to low precipitation rates and lime addition to soils that contained weathered coal, resulting in 300 μg Se/g soil (Yang et al. 1983; Fordyce 2013). Norway observed high Se in farmland areas; however, this was not due to anthropogenic inputs, but rather atmospheric input from sea spray due to the close proximity to the ocean (Wu and Låg 1988). Former agricultural land in California, USA was contaminated with Se through irrigation with Se-enriched drainage water, leading to soil Se concentrations in excess of 1400 μg Se/kg (Fujii et al. 1988; Bañuelos et al. 2005).

In aquatic environments, Se is poorly soluble, with concentrations generally <0.1 μg Se/l in freshwater and <10 μg Se/l in natural waters (Wallschläger and Feldmann 2010; Sohrin and Bruland 2011). Groundwater tends to have higher concentrations of Se than surface waters, sometimes reaching up to 1000 μg Se/l (Fordyce 2007), likely due to more contact with Se-containing soil, sediments, and rocks. Groundwater source, amount of Se transported in solution, degree of environmental aridity, and the release and retention of Se at the soil-water interface will dictate the chemical concentration and behavior of Se in groundwater (Nicolli et al. 2012).

Atmospheric Se is primarily in three forms – organic Se with various forms of methylation ($-CH_3$), e.g. dimethyl selenide and dimethyl diselenide, inorganic forms of Se that are volatile (H_2Se, SeO_2), and particulate Se (Sharma et al. 2015). In Scandinavia, soil Se concentrations are low, with atmospheric deposition considered to be the primary pathway of terrestrial Se addition. Ross (1985) measured atmospheric urban, rural, remote, and oceanic concentrations of Se. Measurements of atmospheric Se in remote locations and marine locations have been noted to contain excessive concentrations of Se in particulate forms of sea salt and soil dust (Adams et al. 1977). Mean particulate atmospheric Se concentrations were found to be greatest in urban (3700 pg Se/m^3) and rural (1200 pg Se/m^3) environments, and lowest in oceanic (209 pg Se/m^3) and remote regions (Ross 1985). In Australia, Ellwood et al. (2016) measured volatile Se flux from a lake contaminated with anthropogenic Se from overflowing coal ash dams. Flux rates between 8 and 53 ng Se/m^2/h were observed from the lake, in the form of dimethylselenide (Ellwood et al. 2016). In central California, USA, volatile Se from fields vegetated with pickleweed (a phytoextractor of Se) removed a mean of 62 mg Se/m^2/year, whereas bare soil were observed to produce an average of 17 mg Se/m^2/year. The sources and concentrations of Se outlined in the previous two section will move in the environment based on the biogeochemical properties of Se discussed in the following section.

14.8 Selenium Chemistry and Movement in the Environment

Selenium undergoes oxidation and reduction processes depending on its environmental conditions, meaning that speciation will change with the oxidation–reduction potential of its environment. Common inorganic species of Se in in nature include the following oxidation states: −2, 0, +4, +6. The oxidized form of the Se ion is selenate (SeO_4^{2-}), with Se in the +6-oxidation state, or Se(VI) (Figure 14.2). The reduced form of the Se ion is selenite (SeO_3^{2-}), with Se in the +4-oxidation state, or Se(IV). The +4 and +6 oxidation states predominate in aquatic systems and in soil. $HSeO_3^-$ and SeO_3^{2-} are natural forms under aerobic conditions, with HSe^- common for anaerobic conditions with pH > 4 (Sharma et al. 2015) (Figure 14.2, Figure 14.3). Organic forms of Se include various forms of methylation ($-CH_3$), such as dimethyl selenide and dimethyl diselenide, which are volatile forms of Se. Inorganic forms of Se that are volatile include H_2Se, SeO_2, and particulate Se (Sharma et al. 2015) (Figure 14.2).

Mobility and Se speciation in soil and sediments depends on photochemical behavior, pH, oxidation–reduction conditions, microbial community composition, cation exchange capacity and competitive ions, and the presence of adsorption sites for Se, such as organic matter and metal oxides. There are a few studies of photochemical induced changes to selenium species. In a controlled aqueous environment, Chen et al. (2005) irradiated

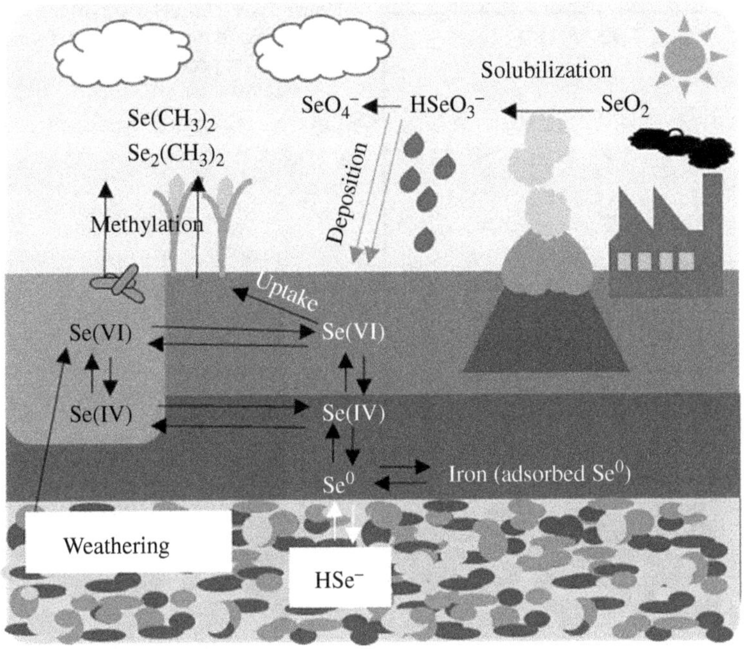

Figure 14.2 Speciation and mobility of selenium in the environment.

selenite with UV light in distilled water, and slow oxidation to selenate was observed, with addition of nitric acid found to speed the oxidation of selenite to selenate. Selenium in inorganic forms combined with carboxylic acids and irradiated with UV light produces methylated and ethylated Se species that were volatile (Guo et al. 2003).

The pH range for most natural soils is approximately pH 5.0–9.0; at this pH range, speciation for Se will be predominantly as SeO_4^{2-}, or the deprotonated form of Se(VI) (Figure 14.3). As conditions become more acidic in soil, the bioavailability of Se decreases. However, the overall mobility of Se increases (Torres et al. 2010). When strongly acidic conditions occur in a soil, the minerals holding Se immobile will be dissolved, thereby remobilizing Se (Tabelin et al. 2014). Adsorption of selenite in most soil types is more strongly influenced by pH than for the selenate form. Maximum selenite adsorption is achieved at approximately pH 3–4, while the lowest amount of adsorption is found around pH 8 (Goldberg and Glaubig 1988). At environmentally relevant pH ranges, selenite will strongly adsorb to soil particles, and is only sparingly susceptible to leaching (Sharma et al. 2015). Oxides of iron and aluminum are the primary minerals responsible for the strong sorption of selenite in soils (Torres et al. 2010; Nakamaru and Altansuvd 2014).

Oxidation–reduction potential and pH together influence the speciation and mobility of Se (Figure 14.3). Under strongly reducing conditions, species of H_2Se (volatile), HSe^-, and Se^0 (elemental) are possible, however at $pH<4$ H_2Se and Se^0 are more common, while $pH>4$ will be dominated by HSe^- (Figure 14.3). The Se (II–) form of Se generally will form a bond to metals, whereas elemental Se is an insoluble solid (Fernández-Martínez and Charlet 2009). Under sufficiently aerobic conditions, the surface horizon of soils will

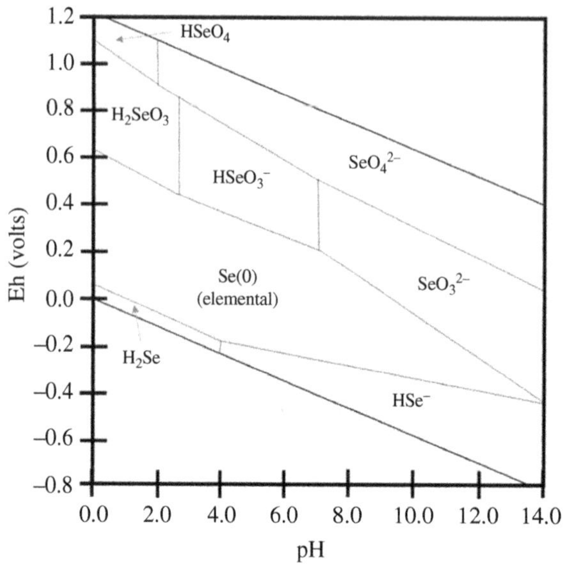

Figure 14.3 Eh-pH diagram for selenium and main speciation found under varying conditions.

predominantly contain Se in the +4 and +6 oxidation states. In locations lower down in the soil profile where oxygen is limited, or if conditions become anaerobic in the surface horizon, Se will be present in reduced forms, Se^0 and Se^{2-}, which are insoluble precipitates typically with iron (i.e. $FeSe_2$, FeSe) (Sharma et al. 2015).

The mineral composition of soil or sediment will strongly dictate the amount of Se that can potentially be adsorbed, as well as the strength of that interaction. Adsorption of Se readily occurs with the oxide, hydroxide, or oxy-hydroxide portion of iron, aluminum, or manganese minerals present in soil, however, the strength of the adsorption may be influenced by pH or oxidation–reduction conditions. For calcareous soils and those high in montmorillonite clay, sorption of selenate appears to be unaffected by pH between 2 and 9 (Goldberg and Glaubig 1988). Conversely, the sorption of selenate on goethitic soils was shown to be affected by pH, with poor adsorption at neutral pH, and higher adsorption observed at lower pH (Goldberg and Glaubig 1988). Adsorbed selenates on iron, aluminum, and manganese minerals are only held weakly by electrostatic attraction, and lack a strong bond (Torres et al. 2010).

Concentrations of soil selenium are well correlated with soil organic matter content, and selenium is known to form Se■C bonds (Abrams et al. 1990). Organic matter found in soil and sediments contains carbon, oxygen, nitrogen, and sometimes sulfur functional groups bound in large, amorphous macromolecules. When Gustafsson and Johnsson (1992) added selenite to a coniferous forest soil in Sweden, it was incorporation into soil organic matter following reduction by microorganisms. Adsorption sites on soil organic matter as well as its solubility are dictated by ionic strength and pH. The water-soluble form of soil organic matter is dissolved organic matter, and highly mobile. Mobility of Se from soil to water can occur through movement with dissolved organic matter. Selenium in the Atlantic Ocean consist primarily of organoselenium associated with soluble peptides (Cutter and Cutter 1995).

As conditions become more acidic, the bioavailability of Se decreases (Torres et al. 2010). The different mineral types present in the soil or sediment used to adsorb Se are also known to alter its bioavailability to microorganisms (Nakamaru and Altansuvd 2014). Selenite added to soil will undergo microbial reduction before integration into the soil organic matter (Gustafsson and Johnsson 1992). Immobilization of Se in organic matter complexes blocks Se from microorganisms and prevents assimilation or microbial transformation (Sharma et al. 2015). Adsorbed selenates on iron, aluminum, and manganese minerals are still plant available because they are held weakly by electrostatic attraction (Torres et al. 2010). At environmentally relevant pH ranges, selenite will strongly adsorb to soil particles, and is only sparingly available for plant uptake or susceptible to leaching (Sharma et al. 2015).

Soils with low pH generally hold Se more strongly, leaving less Se available for plant uptake than high pH soils, which can oxidize Se to more plant available forms (He et al. 2018). While Se is mobile between pH 2 and 9 (Figure 14.3), at the more acidic pH ranges, Se is adsorbed tightly to iron oxides, and therefore inaccessible to plants (Stadlober et al. 2001). Plants will generally contain low concentration of Se in leaf tissues, averaging 25 μg Se/kg, with a maximum of 100 μg Se/kg in most plants (El-Ramady et al. 2015). Low Se uptake is potentially because Se is not an essential plant nutrient. Hyper-accumulator plants may accumulate large masses of Se, and are capable to storing upwards of 1 000 000 μg Se/kg in leaf tissues (El-Ramady et al. 2015). In Punjab, India, excessive amounts of plant Se uptake (3000–670 000 μg Se/kg) have created plants too toxic for human or animal consumption (Bajaj et al. 2011). Plants capable of taking up large amounts of sulfur are also more susceptible to uptake of Se due to the ability to replace sulfur binding locations on proteins (He et al. 2018).

Selenium originating in the soil may take a variety of pathways, including entering food, undergoing plant uptake, be assimilated by microbes into tissue proteins, or become airborne through volatile methylated forms (e.g. dimethylselenide). Once in air, Se can be deposited onto terrestrial or aquatic environments with precipitation or settling. (El-Ramady et al. 2015). A variety of sources of Se and its behavior and chemical mobility outlined in this section have contributed to dispersal of Se to various forms that may accumulate in the soil. The following section will outline conventional and advanced remediation technologies for soils contaminated with high Se.

14.9 Conventional Remediation Techniques

While Se is an essential element at trace levels, we have seen in the previous sections that the consequences of excess Se in the soil traveling up the food chain to animals and humans can produce severe consequences. The narrow range of Se (concentrations between minimal essential nutrient and toxicity) that is suitable for the diet of humans and livestock is complicated by the high mobility of the element through many environmental pathways, including volatile aerosol and particulate forms in the atmosphere. The threshold for classification of a soil as seleniferous are those capable of producing vegetation with 45 000 μg Se/kg dry plant tissue (Kaur et al. 2014). For soils deemed either high in Se, or at the level of classification as seleniferous, a variety of remediation techniques exist. Because soil is the primary accumulator and distributor of Se to other forms (aquatic, atmospheric), as well the main mechanism of introduction to the food chain, it is important to focus remediation methods on soil. The current range of remediation methodologies for Se-contaminated soils are phytoremediation, selenium reduction to lock it into unavailable forms with the aid of iron, bioremediation, volatilization, and electrokinetics (Table 14.1).

Phytoremediation can operate in a variety of ways to remove contaminants from soil. These include phytovolatilization, during which root uptake of contaminants are lost as volatile products through plant leaves; phytoextraction, during which root uptake of accumulated contaminants are stored in plant above ground biomass and harvested; phytodegradation, during which root uptake of contaminant are broken down and assimilated into the plant tissues; phytostabilization, during which the contaminant is precipitated onto plant roots'; and rhizodegradation, during which root zone processes break apart the contaminant (Song et al. 2019) (Figure 14.4). Several forms of phytoremediation are frequently effective at remediating soils containing elevated levels of Se, including phytovolatilization, and phytoextraction (Figure 14.4). In general, phytoremediation may be an economical approach as well as an environmentally friendly option to some of the more intense treatments discussed below. Phytoremediation works because of the similarities between sulfur and Se, where plants can use the biochemical pathways designed for sulfur transport to take up Se, even though it is not an essential nutrient (Pilon-Smits and LeDuc 2009). Once plants assimilate Se through this sulfur pathway it may become methylated and volatilized because the pant cannot use Se for nutrient requirements, and as a means to prevent accumulation to toxic levels. Conversely, hyper-accumulators have the capability to amass up to 15 000 000 μg Se/kg dry biomass without exhibiting symptoms of Se toxicity (El Mehdawi and Pilon-Smits 2012; He et al. 2018). Plants designated as Se accumulators are those capable of amassing up to 1000 000 ug Se/kg dry biomass without exhibition of toxicity in the plant (He et al. 2018). Leeks, garlic and onions from *Allium* family, and cabbage broccoli and rapeseed from the *Brassica* family may accumulate large amounts of Se, but are not considered hyper-accumulators (Terry et al. 2000; Broadley et al. 2006). The Brazil nut tree is one of the few Se hyper-accumulating plants that is also used to produce a food for human consumption (Rayman 2008). The ability of flowering plants to assimilate Se varies greatly from hyper-accumulators to non-accumulators, with the Se uptake potential related to the plant's tolerance to the ratio of Se to sulfur in shoot and root parts of the plant (Rayman 2008). On a dry weight basis, and in high Se content soils, hyper-accumulating flowering plants can amass up to 40 000 000 μg Se/kg dry biomass, while the non-accumulators can only collect a maximum of 100 000 μg Se/kg dry biomass (Broadley et al. 2006). Two-grooved milkvetch and prince's plume (*Stanleya pinnata* and *Astragalus bisulcatus*, respectively) are examples of hyper-accumulators (Quinn et al. 2010; de Jong et al. 2015). Plants with the genera *Astragalus*, *Oonopsis*, *Stanleya*, and *Xlorhiza* are considered excellent hyper-accumulators, with no toxicity symptoms apparent at 15000000 μg Se/kg dry biomass uptake (El Mehdawi and Pilon-Smits 2012). In California, USA, phytoextraction resulted in a field planted with pickleweed to volatilize 62 000 μg Se/m^2/year, in comparison to only 17 μg Se/m^2/year volatilized from a nearby bare field (Lin 2002). Additionally, measurement of Se accumulate in pickleweed plant tissues was revealed to be 5.5 times less than the amount of volatilized Se, confirming successful phytoextraction and release of Se in a volatile form to the atmosphere (Lin 2002). Chinese brake fern (*Pteris vittata L.*), was previously used for arsenic hyper-accumulation, and was recently found to be effective at Se hyper-accumulation by Wu et al. (2015). However, Se was found to accumulate predominantly in the roots at rates of 1500–242 000 μg Se/kg dry biomass instead of the foliage (Feng and Wei 2012).

One of the methodologies for remediation of Se from soil is to convert the mobile Se form to an immobilized form by conversion of selenate (Se(VI)) to elemental Se (Se^0), often with the aid of iron minerals (Schellenger and Larese-Casanova 2013). Because elemental Se is of low solubility and low plant availability, this method of remediation will remove the element from cycling in the environment and limit potential pollution by reducing mobility (Reilly 2006), in effect sequestering Se into the soil in an unavailable form. Use of zero-valent iron at has demonstrated promise for remediation of Se not only in agricultural soils, but also from hazardous waste sites and groundwater. Confirmation of the reduction of selenite to Se^0 was by zero-valent iron was observed through instrumentation by Zhou et al. (2016). Zero-valent iron first reduces Se to Se(II), followed by encapsulation of elemental Se^0 by the iron (Zhou et al. 2016). An added benefit of using zero-valent iron is the possibility to remediate other toxic compounds if the soil contains multiple contaminants, including flame retardants, perchlorate, halogenated hydrocarbons, antibiotics, hexavalent chromium, nitroaromatic compounds, heavy radio nuclides, metal ions, and azo dyes

Table 14.1 Summary of conventional and innovative remediation methods.

Technique	Mechanism	Advantages	Disadvantages
Phytoremediation	Phytovolatilization Phytoextraction	Economical, passive, effective	Lose Se to atmosphere (phytovolatilization)
Reduction with iron	Se(VI)→ Se^0	Effective	Excess Se remains in soil
Bioremediation	Se(VI)→ Se^0	Effective	Excess Se remains in soil
Microbial methylation	Se(VI)→$Se(CH_3)_2$ Se(VI)→$Se_2(CH_3)_2$	Occurs naturally	Very slow
Electrokinetics	Low-level direct current transports Se to electrode	Good for fine-textured, moist, soil	Poor for coarse textured, dry soil
Nanoscale zero-valent iron	Se(VI)→ Se^0	Effective, more effective than bulk iron	Excess Se remains in soil
Nanoscale iron oxide	Se adsorption/precipitation on iron surface	Environmentally friendly	Not as effective as zero-valent iron
Soil washing	Separation of Se from soil, anaerobic digestion to Se^0	Can capture Se for reuse	Largely untested
Combined approaches	Electrokinetics/ zero-valent Fe Microbial reduction/ nanoparticle Biochar/zero-valent Fe/ polysulfide	Very effective	Very complex

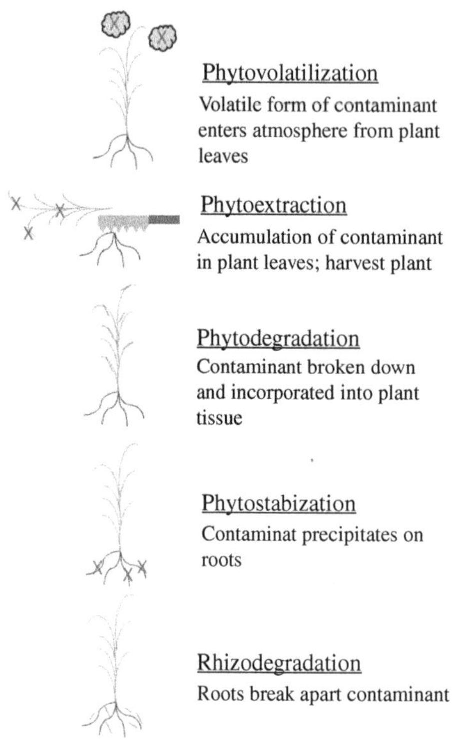

Figure 14.4 Varieties of phytoremediation possible for soil metal contaminants.

(Tang and Lo 2013; Raychoudhury et al. 2014; Liu et al. 2017). As previous mentioned in this chapter, there are strong associations between Se and the carbon in soil organic matter. Sharma et al. (2011) tested this relationship by adding organic matter – farmyard manure, poultry manure, and sugarcane press mud – to a seleniferous soil in India, and planted oilseed rape and wheat. The sugarcane press mud and poultry manure both reduced accumulation of Se in the grains of both oilseed rape and wheat by 75–95%, whereas the reduction from farmyard manure treatment was only 23% (Sharma et al. 2011). Lower Se uptake with farmyard manure treatment was suggested to be from lower water-soluble forms of Se following organic amendments, and the observation that more elemental Se was present, signifying that the organic matter had chemically reduced Se, similar to the effects of iron.

Bioremediation methods for high Se soils focus on the use of microorganism to reduce mobile forms of Se to unavailable elemental Se (Se^0), similar to the principle described above using zero-valent iron and organic matter. For this remediation method, microorganisms serve as the reducing agents. Specific sets of microbial communities or biofilms can be engineer for specific remediation functions, such as transformation of chemicals from a mobile or toxic form to an immobile or inert form (Fu et al. 2014). Yang et al. (2011) utilized a biofilm to transform mobile Se in the +4 and +6 oxidation states to immobile Se^0 primarily by four genera of bacteria discovered through 16S rRNA sequencing. Confirmation of Se species conversion was achieved with extended x-ray adsorption fine structure (EXAFS) and near-edge x-ray absorption spectroscopy (XAS) at the Se K near-edge (Yang et al. 2011). Under anoxic conditions, the facultative anaerobic bacterium *Alishewanella sp.*WH16-1 will reduce selenite (He et al. 2018). Isolation of the bacterium occurred from soil that contained mining waste, creating conditions for enrichment of a Se-reducer. The electron donor for the reduction process Nicotinamide Adenine Dinucleotide Phosphate (NADPH), and optimal conditions for reduction of Se(IV) to Se^0 were pH 7 and temperature between 30 and 37 °C (Xia et al. 2018). The reduction was catalyzed by a novel flavoprotein identified in the genome, named CsrF (Xia et al. 2018). Bacteria capable of this detoxification reaction can reduce Se from the +4 oxidation state to the elemental 0 oxidation state, either with or without enzymatic assistance, have been identified (Nancharaiah and Lens 2015). Siderophores, glutathione, and hydrogen sulfide gas are all capable of reducing Se(IV) without enzymes (He et al. 2018). During respiration, enzymatic reduction of Se(IV) can occur by the following enzymes: sulfite reductase, hydrogenase I, nitrate reductase, fumarate reductase, and arsenate reductase (Harrison et al. 1984; DeMoll-Decker and Macy 1993; Yanke et al. 1995; Afkar et al. 2003; Li et al. 2015). Electron acceptors for the reduction reaction can be arsenate, fumarate, nitrate, sulfate, or selenate (Zhou et al. 2014). Another microbial driven form of remediation is methylation and volatilization of Se by microorganisms. While this process occurs naturally, amendments to increase microbial community robustness may accelerate the process of volatile Se loss. Ellwood et al. (2016) measured the flux of Se from sediments of a 3000–6000 μg Se/kg sediment contaminated lake in Australia. From the natural volatile remediation by microorganisms process, approximately 1 300 000 μg of Se were lost per year through volatile pathways (Ellwood et al. 2016).

The electrokinetics method of remediation for Se-contaminated soils is particularly effective in fine-textured soils due to better connectivity of soil particles providing a higher ability to carry a current (Wang et al. 2010; Mahmoodi et al. 2013). The electrokinetic method to induce multiple mechanism simultaneously, utilizes a direct current at low levels. The mechanisms and reactions induced during electrokinetics are electroosmosis, electrophoresis, and electromigration (transport mechanism), and electrolysis and electrodeposition (electrochemical reactions) (He et al. 2018). Dissolved ions will undergo ionic migration directed toward electrodes in a wet contaminated soil in conjuncture with non-ionic pore liquids migrating through the porous media due to electroosmosis drivers (Tabelin et al. 2017). A benefit of this technique is the ability to remediate soils in situ for sites that excavation is not ideal; this method is especially effective on soils of low permeability which are difficult to treat with other methods, or when other inorganic and organic pollutants are present for simultaneous treatment in multiple contaminant soils (He et al. 2018). Clay textured soils that with high water content near saturation were observed to be the most effectively treated by this method, when compared to sandy textured, dry soils (Suzuki et al. 2013). The rate and effectiveness of electromigration in the subsurface is dependent on multiple variable, including soil pore spaces, soil texture, ionic mobility, concentration of the ionic solution, concentration of the contaminant of concern, the presence of interfering factors, and density of the current of water (Singh et al. 2015). Cations and anions will migrate to corresponding electrodes during the process, depending on the degree of hydraulic gradient and electroosmotic flow that cause advection currents (Singh et al. 2015). Due to the low rate of current applied, soil microorganisms are not adversely affected by electrokinetics, except for the area close to anode insertion, which becomes acidified (Lear et al. 2004). Below we discuss innovative and new remediation techniques, building on the conventional methods just outlined.

14.10 Nanomaterial-Based and Innovative Remediation Techniques

The use of nanotechnology for a variety of purposes has increased in recent decades. These unique materials can be engineered to be effective for a number of applications, and overcome some limitations found with conventional materials, or those of larger size (Banfield and Zhang 2001; Liu and Lal 2012). Materials with a diameter less than 100 nm are considered nanomaterials, with their small size providing unique properties such as high reactivity, quick dissolution, and strong sorption to other solids (Hochella 2002; Pradeep and Anshup, 2009; Khajeh et al. 2013; Gehrke et al. 2015; Jeevanandam et al. 2018). A high surface area to volume ratio provided by the small size of nanomaterials gives them novel properties for specific functions. Nanomaterials are specifically useful for enhancing remediation of Se-contaminated soils, including those classified as mine waste soils with higher concentrations of contaminants, due to these properties. Nanomaterial-enhanced remediation methodologies for Se-contaminated soils include the use of iron nanoparticles in either the iron oxide or zero-valent varieties (Liu and Lal 2012) (Table 14.1). Specific to soil, nanoscale remediation methods provide three primary benefits: the ease of delivery into the pores of soil, the ability to remediate in situ, and the higher reactivity of nanomaterials compared to their larger-sized counterparts (Liu and Lal 2012).

In the previous section on conventional remediation, we discussed the use of iron for conversion of Se from the mobile +6 and +4 oxidation states to immobile Se^0. A similar process is proposed using of zero-valent iron at the nanoscale, however the speed and effectiveness is increased if nanoparticulate forms of zero-valent iron are used (He et al. 2018). In nano-sized form, the reducing power of zero-valent iron is increased because the smaller-sized particles are more reactive, in addition to easy subsurface delivery and being an environmentally green treatment option (Liu and Lal 2012). Soluble forms of Se(VI) were observed to be removed by use of zero-valent iron nanoparticles by adsorbing up to 0.1 mol of Se/mol of iron, demonstrating the high efficiency of nanoscaled forms of zero-valent iron (Olegario et al. 2010). Confirmation of the Se reduction to bio unavailable form, was confirmed with advanced x-ray spectroscopy. Once Se is adsorbed onto iron, very little movement is expected under typical subsurface conditions. A strong interaction between the nanoparticles and soil particles, as well as agglomeration between nanoparticles, was observed to keep movement of nanoparticles to within several centimeters to several meters (Tratnyek and Johnson 2006; Saleh and Kim 2008; O'Carroll et al. 2013). Furthermore, zero-valent iron is susceptible to oxidation to immobile precipitates of magnetite or maghemite under oxic conditions, further reducing the possibility of particle migration (Reinsch et al. 2010).

Nanoparticles of iron oxide function by a different mechanism than zero-valent iron nanoparticles discussed previously. Whereas zero-valent iron is used to reduce Se to its immobile elemental form, iron oxide particles will use its oxide group surfaces to bind metals through adsorption or complexation/precipitation of the surface (Bigham et al. 2018). Due to the wide range of compounds that can bond to iron oxide and their environmental friendliness, iron oxides can be used on a large range of contaminated soils, including those with multiple contaminants (Liu and Lal 2012). Commonly used iron hydroxides for remediation include maghemite (γ-Fe_2O_3, 10 nm particles), magnetite (Fe_3O_4, approximately 10 nm particles), goethite (α-Fe-OOH, 50nm × 200 nm needles), hematite (α-Fe_2O_3, 75 nm granular), and amorphous iron oxides such as ferrihydrite (Hua et al. 2012; Liu and Lal 2012). From wastewater, Shipley et al. (2011) demonstrated that Se in addition to 12 other heavy metals – cadmium, cobalt, chromium, lead, manganese, molybdenum, thorium, tin, thallium, and vanadium – could be removed simultaneously using iron oxides in a column study. Iron oxides are thought to have limited toxicity, with no clear enhanced toxicity due to the nanoscales size of the particle, compared to bulk material on its ability to cause damage to mitochondria, DNA, or to cause death in cells (Karlsson et al. 2009).

In a novel approach, Wadgaonkar et al. (2018) proposed ex situ washing of seleniferous soils as an approach to both reduce soil Se content, and to capture Se for fertilization on soils deficient in the essential nutrient. Given the limited availability of Se on the planet, and that direct mining of Se does not occur, this study proposes a potential solution for Se capture. This approach constitutes the use of benign leaching agents to separate Se from soil particles, followed by bioreactor treatment of the Se-containing washings from the soil (Wadgaonkar et al. 2018). While soil washing has been previously applied to other heavy metals, including lead, arsenic, zinc, mercury, it has yet to be tested for Se (Dermont et al. 2008; Mohanty and Mahindrakar 2011; Kim et al. 2012). Soil washing for reclamation of Se may be more effective in alkaline soils, where a larger portion of Se will be in a water-soluble form (Wadgaonkar et al. 2018). Using potassium chloride as the reagent, and multiple extractions, the authors were able to remove 90% of Se from a seleniferous soil (Dhillon and Dhillon 2003). The authors further found that mobilization of Se was increased 12–20 fold following addition of 7.5 mmol/kg ethylene diamine-N,N′-disuccinic acid and 1.0 mmom/kg diethylenetriamine pentaacetate, which may aid in greater Se recovery with less washing repetitions (Bolan et al. 2014). Wadgaonkar et al. (2018) advocates for a sequential extraction procedure, which will remove each Se species in a sequential series of steps, which the authors believe will be effective in many soils with different properties. Following extraction, Se-laden leachate may undergo anaerobic digestion in a bioreactor to recover elemental Se (Tan et al. 2016; Vallini et al. 2005).

In another combined approach, Huang et al. (2019) coupled a nanoscale activated carbon-zero-valent iron reactive barrier with electrokinetics to treat Se in the +4 and +6 oxidation states in a contaminated soil. Results on initial testing of the coupled system were promising, with 90% removal efficiency achieved for the +6 oxidation state, and 95% achieved for the +4 oxidation state, and conversion to elemental Se and FeSe observed (Huang et al. 2019). In a different combined approach, Vogel et al. (2018) utilized *Azospirillum brasilense*, a bacterium the promotes plant growth, to interact with Se in the +6 and +4 oxidation states to form Se nanoparticles with sulfur, with Se reduced to the immobilized Se^0 form. Spectroscopic investigations found that the formed nanoparticles has the formula $Se_{8\text{-}n}S_n$, were spherical in shape, and averaged 400 nm in diameter (Vogel et al. 2018). In an alternative approach to Se soil pollution, Mandal et al. (2020) coated biochar with both polysulfide and zero-valent iron nanoparticles for use as a soil amendment for immobilization of Se through reduction to Se^0. Following 30 days of incubation of the engineered biochar in the soil, a 77% decrease in available Se was recorded (Mandal et al. 2020). The mechanism of reduction in Se availability was determined to be a combination of complexation, chemical reduction, and adsorption to biochar particle surface. Innovative studies such as these point to more creative solutions potentially producing better Se removal or immobilization in soils.

14.11 Conclusions

Selenium is a complicated element to manage in the soil because of the small range of concentrations that satisfy essential nutrient requirements and those that will cause toxicity in humans, animals, and organisms. Selenium is further complicated due to the many different possible oxidation state, which affect its toxicity, solubility, and volatility. Because Se can volatilize and move to the atmosphere in both particulate and volatile forms, this can further complicate remediation efforts as deposition events or settling during rain can slowly cause Se to accumulate (or re-accumulate) in soil to toxic levels. While natural sources of Se are abundant, and primarily include the weathering of Se-containing parent material into soil, anthropogenic sources from agricultural and industrial sectors are increasing. Humans and animals consume excess Se through food chain source, which are predominantly through plants – which take up Se from soils, or from animals – that consume high Se plants. Selenium toxicity in humans can be exhibit as neurological disorders, cardiovascular issues, or other degenerative diseases if exposure is chronic. Remediation of Se-contaminated soils is important for maintenance of a healthy human populace, and to maintain the health and economic function of livestock. Conventional remediation techniques for soils with elevated Se exploit the many Se forms, focusing on either locking Se in the soil in a biologically unavailable form, or converting Se to a volatile form through use of plants or microorganisms. More recent remediation strategies have started to focus on capture and recycling of Se to areas that are deficient in this essential nutrient. With an average concentration of Se on the Earth's crust of only approximately 90 μg Se/kg, these methods to capture and reuse Se become very important due to the limited quantities available. Composting of Se-loaded hyper-accumulating plants or Se recovered from anaerobic digestion of washings from Se-contaminated soils in the future will likely be more common practices to more effectively distribute the limited quantities of Se in a more equitable fashion across the planet.

References

Abrams, M.M., Burau, R.G., and Zasoski, R.J. (1990). Organic selenium distribution in selected California soils. *Soil Science Society of America Journal*. https://doi.org/10.2136/sssaj1990.03615995005400040007x.

Adams, F., Dams, R., Guzman, L., and Winchester, J.W. (1977). Background aerosol composition on Chacaltaya mountain, Bolivia. *Atmospheric Environment (1967)* 11 (7): 629–634. https://doi.org/10.1016/0004-6981(77)90116-0.

Afkar, E., Lisak, J., Saltikov, C. et al. (2003). The respiratory arsenate reductase from bacillus selenitireducens strain MLS10. *FEMS Microbiology Letters* 226 (1): 107–112. https://doi.org/10.1016/S0378-1097(03)00609-8.

Aitken, P. (2001). Selenium toxicity. *In Practice*. https://doi.org/10.1136/inpract.23.5.286.

Al Kuisi, M. and Abdel-Fattah, A. (2010). Groundwater vulnerability to selenium in semi-arid environments: Amman Zarqa Basin, Jordan. *Environmental Geochemistry and Health* https://doi.org/10.1007/s10653-009-9269-y.

Alloway, B. (1995). Heavy metals in soils – trace elements and metalloids in soils. *Heavy Metals in Soils* https://doi.org/10.1007/9789400744707.

Awual, M.R., Hasan, M.M., and Khaleque, M.A. (2015). Efficient selenium(IV) detection and removal from water by tailor-made novel conjugate adsorbent. *Sensors and Actuators B: Chemical* https://doi.org/10.1016/j.snb.2014.11.010.

Bajaj, M., Eiche, E., Neumann, T. et al. (2011). Hazardous concentrations of selenium in soil and groundwater in North-West India. *Journal of Hazardous Materials* 189 (3): 640–646. https://doi.org/10.1016/j.jhazmat(.2011.).01.086.

Banfield, J.F. and Zhang, H. (2001). Nanoparticles in the environment. *Reviews in Mineralogy and Geochemistry* 44 (1): 1–58. https://doi.org/10.2138/rmg.2001.44.01.

Bañuelos, G., Terry, N., Leduc, D.L. et al. (2005). Field trial of transgenic Indian mustard plants shows enhanced phytoremediation of selenium-contaminated sediment. *Environmental Science & Technology* https://doi.org/10.1021/es049035f.

Barceloux, D.G. (1999). Selenium. *Journal of Toxicology. Clinical Toxicology* https://doi.org/10.1081/CLT-100102417.

Bigham, J.M., Fitzpatrick, R.W., and Schulze, D.G. (2018). Iron Oxides. In: *Soil Mineralogy with Environmental Applications*, Vol. 7 (ed. J.B. Dixon and D.G. Schulze), 323–366. doi: 10.2136/sssabookser7.c10.

Bolan, N., Kunhikrishnan, A., Thangarajan, R. et al. (2014). Remediation of heavy metal(loid)s contaminated soils – to mobilize or to immobilize? *Journal of Hazardous Materials* https://doi.org/10.1016/j.jhazmat.2013.12.018.

Broadley, M.R., White, P.J., Bryson, R.J. et al. (2006). Biofortification of UK food crops with selenium. *Proceedings of the Nutrition Society* https://doi.org/10.1079/pns2006490.

Buchs, B., Evangelou, M.W.H., Winkel, L.H.E., and Lenz, M. (2013). Colloidal properties of nanoparticular biogenic selenium govern environmental fate and bioremediation effectiveness. *Environmental Science & Technology* 47 (5): 2401–2407. https://doi.org/10.1021/es304940s.

Charmbury, H.B. (2019). Coal and the environment. In: *Coal Surface Mining* (ed. J.E. Rowe), 71–75. Routledge https://doi.org/10.4324/9780429051128-6.

Chen, Y.-W., Zhou, X.-L., Tong, J. et al. (2005). Photochemical behavior of inorganic and organic selenium compounds in various aqueous solutions. *Analytica Chimica Acta* 545 (2): 149–157. https://doi.org/10.1016/j.aca.2005.03.033.

Chipinda, I., Zhang, X.-D., Simoyi, R.H., and Siegel, P.D. (2008). Mercaptobenzothiazole allergenicity – role of the thiol group. *Cutaneous and Ocular Toxicology* 27 (2): 103–116. https://doi.org/10.1080/15569520701713008.

Cutter, G.A. and Cutter, L.S. (1995). Behavior of dissolved antimony, arsenic, and selenium in the Atlantic Ocean. *Marine Chemistry* 49 (4): 295–306. https://doi.org/10.1016/0304-4203(95)00019-N.

Cyrs, W.D., Avens, H.J., Capshaw, Z.A. et al. (2014). Landfill waste and recycling: use of a screening-level risk assessment tool for end-of-life cadmium telluride (CdTe) thin-film photovoltaic (PV) panels. *Energy Policy* https://doi.org/10.1016/j.enpol.2014.01.025.

Dauphas, N. (2013). Sulphur from heaven and hell. *Nature* 501 (7466): 175–176. https://doi.org/10.1038/nature12554.

DeMoll-Decker, H. and Macy, J.M. (1993). The periplasmic nitrite reductase of Thauera selenatis may catalyze the reduction of selenite to elemental selenium. *Archives of Microbiology* https://doi.org/10.1007/BF00249131.

Dermont, G., Bergeron, M., Mercier, G., and Richer-Laflèche, M. (2008). Soil washing for metal removal: a review of physical/chemical technologies and field applications. *Journal of Hazardous Materials* https://doi.org/10.1016/j.jhazmat.2007.10.043.

Desta, B., Maldonado, G., Reid, H. et al. (2011). Acute selenium toxicosis in polo ponies. *Journal of Veterinary Diagnostic Investigation* https://doi.org/10.1177/1040638711404142.

Dhillon, K.S. and Dhillon, S.K. (2003). Distribution and management of seleniferous soils. *Advances in Agronomy* https://doi.org/10.1016/S0065-2113(02)79003-2.

Diplock, A.T. and Kurzer, F. (1974). Organic selenium compounds: their chemistry and biology. *Biochemical Society Transactions* https://doi.org/10.1042/bst0020571.

Ekermans, L.G. and Schneider, J.V. (1982). Selenium in livestock production: a review. *Journal of the South African Veterinary Association* 53 (4): 223–228.

El Mehdawi, A.F. and Pilon-Smits, E.A.H. (2012). Ecological aspects of plant selenium hyperaccumulation. *Plant Biology* 14 (1): 1–10. https://doi.org/10.1111/j.1438-8677.2011.00535.x.

Ellwood, M.J., Schneider, L., Potts, J. et al. (2016). Volatile selenium fluxes from selenium-contaminated sediments in an Australian coastal lake. *Environmental Chemistry* 13 (1): 68. https://doi.org/10.1071/EN14228.

El-Ramady, H.R., Domokos-Szabolcsy, É., Abdalla, N.A. et al. (2014). Selenium and nano-selenium in agroecosystems. *Environmental Chemistry Letters* 12 (4): 495–510. https://doi.org/10.1007/s10311-014-0476-0.

El-Ramady, H.R., Domokos-Szabolcsy, É., Shalaby, T.A. et al. (2015). Selenium in agriculture: water, air, soil, plants, food, animals, and nanoselenium. In: *CO2 Sequestration, Biofuels, and Depollution* (eds. E. Lichtfouse, J. Schwarzbauer and D. Robert), 1147–1150. London: Springer https://doi.org/10.1016/B978-008043018-8/50191-0.

Fairweather-Tait, S.J., Bao, Y., Broadley, M.R. et al. (2011). Selenium in human health and disease. *Antioxidants & Redox Signaling* 14 (7): 1337–1383. https://doi.org/10.1089/ars.2010.3275.

Feng, R.W. and Wei, C.Y. (2012). Antioxidative mechanisms on selenium accumulation in Pteris vittata L., a potential selenium phytoremediation plant. *Plant, Soil and Environment* 58 (3): 105–110. https://doi.org/10.17221/162/2011-PSE.

Fernández-Martínez, A. and Charlet, L. (2009). Selenium environmental cycling and bioavailability: a structural chemist point of view. *Reviews in Environmental Science and Bio/Technology* 8 (1): 81–110. https://doi.org/10.1007/s11157-009-9145-3.

Floor, G.H. and Román-Ross, G. (2012). Selenium in volcanic environments: a review. *Applied Geochemistry* 27 (3): 517–531. https://doi.org/10.1016/j.apgeochem.2011.11.010.

Fordyce, F. (2007). Selenium geochemistry and health. *Ambio* 36: 94–97. https://doi.org/10.1579/0044-7447(2007)36[94:SGAH]2.0.CO;2.

Fordyce, F.M. (2013). Selenium deficiency and toxicity in the environment. In: *Essentials of Medical Geology*. Rev. Ed. (eds. O. Selinus, B. Alloway, J.A. Centeno, et al.), 375–416. Dordrecht: Springer Netherlands https://doi.org/10.1007/978-94-007-4375-5_16.

Fordyce, F.M., Guangdi, Z., Green, K., and Xinping, L. (2000). Soil, grain and water chemistry in relation to human selenium-responsive diseases in Enshi District, China. *Applied Geochemistry* 15 (1): 117–132. https://doi.org/10.1016/S0883-2927(99)00035-9.

Foster, L.H. and Sumar, S. (1997). Selenium in health and disease: a review. *Critical Reviews in Food Science and Nutrition* https://doi.org/10.1080/10408399709527773.

Frost, D.V. and Olson, O.E. (1972). The two faces of selenium – can selenophobia be cured? *Critical Reviews in Toxicology* https://doi.org/10.3109/10408447209103467.

Fu, Y., Wang, J., Liu, Q., and Zeng, H. (2014). Water-dispersible magnetic nanoparticle–graphene oxide composites for selenium removal. *Carbon* 77: 710–721. https://doi.org/10.1016/j.carbon.2014.05.076.

Fujii, R., Deverel, S.J., and Hatfield, D.B. (1988). Distribution of selenium in soils of agricultural fields, Western San Joaquin Valley, California. *Soil Science Society of America Journal* https://doi.org/10.2136/sssaj1988.03615995005200050011x.

Gehrke, I., Geiser, A., and Somborn-Schulz, A. (2015). Innovations in nanotechnology for water treatment. *Nanotechnology, Science and Applications* 1. https://doi.org/10.2147/NSA.S43773.

Gharibzahedi, S.M.T. and Jafari, S.M. (2017). The importance of minerals in human nutrition: bioavailability, food fortification, processing effects and nanoencapsulation. *Trends in Food Science and Technology* https://doi.org/10.1016/j.tifs.2017.02.017.

Girling, C.A. (1984). Selenium in agriculture and the environment. *Agriculture, Ecosystems and Environment* https://doi.org/10.1016/0167-8809(84)90047-1.

Glover, J.R. (1970). Selenium and its industrial toxicology. *IMS, Industrial Medicine and Surgery* 39 (1): 50–54.

Goldberg, S. and Glaubig, R.A. (1988). Anion sorption on a calcareous, Montmorillonitic soil-arsenic. *Soil Science Society of America Journal* https://doi.org/10.2136/sssaj1988.03615995005200050015x.

Guo, X., Sturgeon, R.E., Mester, Z., and Gardner, G.J. (2003). Photochemical alkylation of inorganic selenium in the presence of low molecular weight organic acids. *Environmental Science & Technology* 37 (24): 5645–5650. https://doi.org/10.1021/es034418j.

Gustafsson, J.P. and Johnsson, L. (1992). Selenium retention in the organic matter of Swedish forest soils. *Journal of Soil Science* https://doi.org/10.1111/j.1365-2389.1992.tb00152.x.

Harrison, G., Curle, C., and Laishley, E.J. (1984). Purification and characterization of an inducible dissimilatory type sulfite reductase from Clostridium pasteurianum. *Archives of Microbiology* 138 (1): 72–78. https://doi.org/10.1007/BF00425411.

Hartikainen, H., Xue, T., and Piironen, V. (2000). Selenium as an anti-oxidant and pro-oxidant in ryegrass. *Plant and Soil* https://doi.org/10.1023/A:1026512921026.

Haug, A., Graham, R.D., Christophersen, O.A., and Lyons, G.H. (2007). How to use the world. scarce selenium resources efficiently to increase the selenium concentration in food. *Microbial Ecology in Health and Disease* https://doi.org/10.1080/08910600701698986.

He, Z.L., Yang, X.E., and Stoffella, P.J. (2005). Trace elements in agroecosystems and impacts on the environment. *Journal of Trace Elements in Medicine and Biology* https://doi.org/10.1016/j.jtemb.2005.02.010.

He, Y., Xiang, Y., Zhou, Y. et al. (2018). Selenium contamination, consequences and remediation techniques in water and soils: a review. *Environmental Research* 164: 288–301. https://doi.org/10.1016/j.envres.2018.02.037.

Hochella, M.F. (2002). There's plenty of room at the bottom: nanoscience in geochemistry. *Geochimica et Cosmochimica Acta* 66 (5): 735–743. https://doi.org/10.1016/S0016-7037(01)00868-7.

Hua, M., Zhang, S., Pan, B. et al. (2012). Heavy metal removal from water/wastewater by nanosized metal oxides: a review. *Journal of Hazardous Materials* https://doi.org/10.1016/j.jhazmat.2011.10.016.

Huang, T., Liu, L., Zhang, S., and Xu, J. (2019). Evaluation of electrokinetics coupled with a reactive barrier of activated carbon loaded with a nanoscale zero-valent iron for selenite removal from contaminated soils. *Journal of Hazardous Materials* https://doi.org/10.1016/j.jhazmat.2019.01.036.

Itano, K., Kawai, S., Miyazaki, N. et al. (1984). Mercury and selenium levels in striped dolphins caught off the Pacific coast of Japan. *Agricultural and Biological Chemistry* https://doi.org/10.1271/bbb1961.48.1109.

James, L.F. and Shupe, J.L. (1984). Selenium poisoning in livestock. *Rangelands* 6 (2): 64–67.

Janz, D., DeForest, D., Brooks, M. et al. (2010). Selenium toxicity to aquatic organisms. In: *Ecological Assessment of Selenium in the Aquatic Environment* (eds. P.M. Chapman, W.J. Adams, M. Brooks, et al.), 141–231. CRC Press https://doi.org/10.1201/EBK1439826775-c6.

Jeevanandam, J., Barhoum, A., Chan, Y.S. et al. (2018). Review on nanoparticles and nanostructured materials: history, sources, toxicity and regulations. *Beilstein Journal of Nanotechnology* 9: 1050–1074. https://doi.org/10.3762/bjnano.9.98.

Johnson, T.M. (2004). A review of mass-dependent fractionation of selenium isotopes and implications for other heavy stable isotopes. *Chemical Geology* 204 (3–4): 201–214. https://doi.org/10.1016/j.chemgeo.2003.11.015.

de Jong, M.S., Reynolds, R.J.B., Richterova, K. et al. (2015). Selenium hyperaccumulators harbor a diverse endophytic bacterial community characterized by high selenium resistance and plant growth promoting properties. *Frontiers in Plant Science* https://doi.org/10.3389/fpls.2015.00113.

Kabata-Pendias, A. (2010). *Trace Elements in Soils and Plants*, 4e. CRC Press https://doi.org/10.1201/b10158.

Karlsson, H.L., Gustafsson, J., Cronholm, P., and Möller, L. (2009). Size-dependent toxicity of metal oxide particles – a comparison between nano- and micrometer size. *Toxicology Letters* https://doi.org/10.1016/j.toxlet.2009.03.014.

Kaur, N., Sharma, S., Kaur, S., and Nayyar, H. (2014). Selenium in agriculture: a nutrient or contaminant for crops? *Archives of Agronomy and Soil Science* 60 (12): 1593–1624. https://doi.org/10.1080/03650340.2014.918258.

Khajeh, M., Laurent, S., and Dastafkan, K. (2013). Nanoadsorbents: classification, preparation, and applications (with emphasis on aqueous media). *Chemical Reviews* 113 (10): 7728–7768. https://doi.org/10.1021/cr400086v.

Kim, K., Cheong, J., Kang, W. et al. (2012). Field study on application of soil washing system to arsenic- contaminated site adjacent to J. refinery in Korea. In: *International Conference on Environmental Science and Technology*.

König, S., Luguet, A., Lorand, J.-P. et al. (2012). Selenium and tellurium systematics of the Earth's mantle from high precision analyses of ultra-depleted orogenic peridotites. *Geochimica et Cosmochimica Acta* 86: 354–366. https://doi.org/10.1016/j.gca.2012.03.014.

Kumar, A.R. and Riyazuddin, P. (2011). Speciation of selenium in groundwater: seasonal variations and redox transformations. *Journal of Hazardous Materials* https://doi.org/10.1016/j.jhazmat.2011.05.013.

Lakin, H.W. (1972). Selenium accumulation in soils and its absorption by plants and animals. *Bulletin of the Geological Society of America* 83 (1): 181–190. https://doi.org/10.1130/0016-7606(1972)83[181:SAISAI]2.0.CO;2.

Lawlor, D.W., Mengel, K., and Kirkby, E.A. (2004). Principles of plant nutrition. *Annals of Botany* 93 (4): 479–480. https://doi.org/10.1093/aob/mch063.

Lear, G., Harbottle, M.J., van der Gast, C.J. et al. (2004). The effect of electrokinetics on soil microbial communities. *Soil Biology and Biochemistry* 36 (11): 1751–1760. https://doi.org/10.1016/j.soilbio.2004.04.032.

Lee, K.H. and Jeong, D. (2012). Bimodal actions of selenium essential for antioxidant and toxic pro-oxidant activities: the selenium paradox (review). *Molecular Medicine Reports* https://doi.org/10.3892/mmr.2011.651.

Lemly, A.D. (1997). Environmental implications of excessive selenium: a review. *Biomedical and Environmental Sciences* 10 (4): 415–435.

Lenz, M. and Lens, P.N.L. (2009). The essential toxin: the changing perception of selenium in environmental sciences. *Science of the Total Environment* 407 (12): 3620–3633. https://doi.org/10.1016/j.scitotenv.2008.07.056.

Li, H.-F., McGrath, S.P., and Zhao, F.-J. (2008). Selenium uptake, translocation and speciation in wheat supplied with selenate or selenite. *New Phytologist* 178 (1): 92–102. https://doi.org/10.1111/j.1469-8137.2007.02343.x.

Li, D.-B., Cheng, Y.-Y., Wu, C. et al. (2015). Selenite reduction by Shewanella oneidensis MR-1 is mediated by fumarate reductase in periplasm. *Scientific Reports* 4 (1): 3735. https://doi.org/10.1038/srep03735.

Lin, Z. (2002). Managing selenium-contaminated agricultural drainage water by the integrated on-farm drainage management system: role of selenium volatilization. *Water Research* 36 (12): 3150–3160. https://doi.org/10.1016/S0043-1354(02)00007-6.

Liu, R. and Lal, R. (2012). Nanoenhanced materials for reclamation of mine lands and other degraded soils: a review. *Journal of Nanotechnology* 2012: 1–18. https://doi.org/10.1155/2012/461468.

Liu, X., Yang, D., Zhou, Y. et al. (2017). Electrocatalytic properties of N-doped graphite felt in electro-Fenton process and degradation mechanism of levofloxacin. *Chemosphere* 182: 306–315. https://doi.org/10.1016/j.chemosphere.2017.05.035.

Mahmoodi, S.R., Bayati, M., Hosseinirad, S. et al. (2013). AC electrokinetic manipulation of selenium nanoparticles for potential nanosensor applications. *Materials Research Bulletin* 48 (3): 1262–1267. https://doi.org/10.1016/j.materresbull.2012.12.026.

Mandal, S., Pu, S., Wang, X. et al. (2020). Hierarchical porous structured polysulfide supported nZVI/biochar and efficient immobilization of selenium in the soil. *The Science of the Total Environment* https://doi.org/10.1016/j.scitotenv.2019.134831.

Medunić, G., Kuharić, Ž., Krivohlavek, A. et al. (2018). Selenium, Sulphur, trace metal, and BTEX levels in soil, water, and lettuce from the croatian raša bay contaminated by superhigh-organic-Sulphur coal. *Geosciences* https://doi.org/10.3390/geosciences8110408.

Mehdi, Y., Hornick, J.L., Istasse, L., and Dufrasne, I. (2013). Selenium in the environment, metabolism and involvement in body functions. *Molecules* https://doi.org/10.3390/molecules18033292.

Mohanty, B. and Mahindrakar, A.B. (2011). Removal of heavy metal by screening followed by soil washing from contaminated soil. *International Journal of Technology And Engineering System* 2 (3): 290–293.

Moreno Rodriguez, M.J., Cala Rivero, V., and Jiménez Ballesta, R. (2005). Selenium distribution in topsoils and plants of a semi-arid Mediterranean environment. *Environmental Geochemistry and Health* https://doi.org/10.1007/s10653-005-8625-9.

Nakamaru, Y.M. and Altansuvd, J. (2014). Speciation and bioavailability of selenium and antimony in non-flooded and wetland soils: a review. *Chemosphere* 111: 366–371. https://doi.org/10.1016/j.chemosphere.2014.04.024.

Nancharaiah, Y.V. and Lens, P.N.L. (2015). Ecology and biotechnology of selenium-respiring bacteria. *Microbiology and Molecular Biology Reviews* 79 (1): 61–80. https://doi.org/10.1128/MMBR.00037-14.

National Research Council (US) Subcommittee on Selenium (1983). *Selenium in Nutrition*. 2nd rev. ed. Washington, DC: National Academy of Sciences https://doi.org/10.17226/40.

Navarro-Alarcon, M. and Cabrera-Vique, C. (2008). Selenium in food and the human body: a review. *Science of the Total Environment* 400 (1–3): 115–141. https://doi.org/10.1016/j.scitotenv.2008.06.024.

Nicolli, H.B., Bundschuh, J., Blanco, M. et al. (2012). Arsenic and associated trace-elements in groundwater from the Chaco-Pampean plain, Argentina: results from 100 years of research. *Science of the Total Environment* https://doi.org/10.1016/j.scitotenv.2012.04.048.

Nriagu, J.O. (1989). A global assessment of natural sources of atmospheric trace metals. *Nature* 338 (6210): 47–49. https://doi.org/10.1038/338047a0.

Nriagu, J.O. and Pacyna, J.M. (1988). Quantitative assessment of worldwide contamination of air, water and soils by trace metals. *Nature* 333 (6169): 134–139. https://doi.org/10.1038/333134a0.

Nuttall, K.L. (2006). Evaluating selenium poisoning. *Annals of Clinical and Laboratory Science* 36 (4): 409–420.

O'Carroll, D., Sleep, B., Krol, M. et al. (2013). Nanoscale zero valent iron and bimetallic particles for contaminated site remediation. *Advances in Water Resources* https://doi.org/10.1016/j.advwatres.2012.02.005.

Olegario, J.T., Yee, N., Miller, M. et al. (2010). Reduction of Se(VI) to Se(−II) by zerovalent iron nanoparticle suspensions. *Journal of Nanoparticle Research* https://doi.org/10.1007/s11051-009-9764-1.

Perkins, W.T. (2011). Extreme selenium and tellurium contamination in soils – an eighty year-old industrial legacy surrounding a Ni refinery in the Swansea Valley. *The Science of the Total Environment* https://doi.org/10.1016/j.scitotenv.2011.09.056.

Pilon-Smits, E.A. and LeDuc, D.L. (2009). Phytoremediation of selenium using transgenic plants. *Current Opinion in Biotechnology* https://doi.org/10.1016/j.copbio.2009.02.001.

Plant, J.A., Bone, J., Voulvoulis, N. et al. (2013). Arsenic and selenium. In: *Treatise on Geochemistry*, 2ee, vol. 11 (eds. H.D. Holland and K.K. Turekian), 13–57. Elsevier Science https://doi.org/10.1016/B978-0-08-095975-7.00902-5.

Pradeep, T. and Anshup (2009). Noble metal nanoparticles for water purification: a critical review. *Thin Solid Films* 517 (24): 6441–6478. https://doi.org/10.1016/j.tsf(.2009.).03.195.

Quinn, C.F., Freeman, J.L., Reynolds, R.J.B. et al. (2010). Selenium hyperaccumulation offers protection from cell disruptor herbivores. *BMC Ecology* https://doi.org/10.1186/1472-6785-10-19.

Raychoudhury, T., Tufenkji, N., and Ghoshal, S. (2014). Straining of polyelectrolyte-stabilized nanoscale zero valent iron particles during transport through granular porous media. *Water Research* 50: 80–89. https://doi.org/10.1016/j.watres.2013.11.038.

Rayman, M.P. (2000). The importance of selenium to human health. *Lancet* https://doi.org/10.1016/S0140-6736(00)02490-9.

Rayman, M.P. (2008). Food-chain selenium and human health: emphasis on intake. *British Journal of Nutrition* https://doi.org/10.1017/S0007114508939830.

Rayman, M.P. (2012). Selenium and human health. *The Lancet* 379 (9822): 1256–1268. https://doi.org/10.1016/S0140-6736(11)61452-9.

Reilly, C. (2006). *Selenium in Food and Health*. Springer https://doi.org/10.1007/978-0-387-33244-4.

Reinsch, B.C., Forsberg, B., Penn, R.L. et al. (2010). Chemical transformations during aging of zerovalent iron nanoparticles in the presence of common groundwater dissolved constituents. *Environmental Science and Technology* https://doi.org/10.1021/es902924h.

Ross, H.B. (1985). An atmospheric selenium budget for the region 30° N to 90° N. *Tellus B* https://doi.org/10.1111/j.1600-0889.1985.tb00057.x.

Ruta, D.A. and Haider, S. (1989). Attempted murder by selenium poisoning. *British Medical Journal* https://doi.org/10.1136/bmj.299.6694.316.

Sakamoto, M., Man Chan, H., Domingo, J.L. et al. (2012). Changes in body burden of mercury, lead, arsenic, cadmium and selenium in infants during early lactation in comparison with placental transfer. *Ecotoxicology and Environmental Safety* https://doi.org/10.1016/j.ecoenv.2012.07.014.

Saleh, N. and Kim, H.J., Phenrat, T., Matyjaszewski, K., Tilton, R.D. and Lowry, G.V. (2008). Ionic strength and composition affect the mobility of surface-modified Fe0 nanoparticles in water-saturated sand columns. *Environmental Science & Technology* 42(9):3349–3355. https://doi.org/10.1021/es071936b.

Schellenger, A.E.P. and Larese-Casanova, P. (2013). Oxygen isotope indicators of selenate reaction with Fe(II) and Fe(III) hydroxides. *Environmental Science and Technology* https://doi.org/10.1021/es4000033.

Schmidt, R., Tantoyotai, P., Fakra, S.C. et al. (2013). Selenium biotransformations in an engineered aquatic ecosystem for bioremediation of agricultural wastewater via brine shrimp production. *Environmental Science & Technology* 47 (10): 5057–5065. https://doi.org/10.1021/es305001n.

Senesi, G.S., Baldassarre, G., Senesi, N., and Radina, B. (1999). Trace element inputs into soils by anthropogenic activities and implications for human health. *Chemosphere* 38 (2): 343–377. https://doi.org/10.1016/S0045-6535(99)00115-0.

Sharma, S., Bansal, A., Dogra, R. et al. (2011). Effect of organic amendments on uptake of selenium and biochemical grain composition of wheat and rape grown on seleniferous soils in northwestern India. *Journal of Plant Nutrition and Soil Science* https://doi.org/10.1002/jpln.200900265.

Sharma, V.K., McDonald, T.J., Sohn, M. et al. (2015). Biogeochemistry of selenium. A review. *Environmental Chemistry Letters* 13 (1): 49–58. https://doi.org/10.1007/s10311-014-0487-x.

Shipley, H.J., Engates, K.E., and Guettner, A.M. (2011). Study of iron oxide nanoparticles in soil for remediation of arsenic. *Journal of Nanoparticle Research* https://doi.org/10.1007/s11051-010-9999-x.

Shrift, A. (1969). Aspects of selenium metabolism in higher plants. *Annual Review of Plant Physiology* 20 (1): 475–494. https://doi.org/10.1146/annurev.pp.20.060169.002355.

Singh, R., Singh, S., Parihar, P. et al. (2015). Arsenic contamination, consequences and remediation techniques: a review. *Ecotoxicology and Environmental Safety* 112: 247–270. https://doi.org/10.1016/j.ecoenv.2014.10.009.

Soda, S., Kashiwa, M., Kagami, T. et al. (2011). Laboratory-scale bioreactors for soluble selenium removal from selenium refinery wastewater using anaerobic sludge. *Desalination* https://doi.org/10.1016/j.desal.2011.06.031.

Sohrin, Y. and Bruland, K.W. (2011). Global status of trace elements in the ocean. *TrAC Trends in Analytical Chemistry* 30 (8): 1291–1307. https://doi.org/10.1016/j.trac.2011.03.006.

Song, B., Xu, P., Chen, M. et al. (2019). Using nanomaterials to facilitate the phytoremediation of contaminated soil. *Critical Reviews in Environmental Science and Technology* https://doi.org/10.1080/10643389.2018.1558891.

Spallholz, J.E. (1994). On the nature of selenium toxicity and carcinostatic activity. *Free Radical Biology and Medicine* https://doi.org/10.1016/0891-5849(94)90007-8.

Spallholz, J.E. and Hoffman, D.J. (2002). Selenium toxicity: cause and effects in aquatic birds. *Aquatic Toxicology* 57 (1–2): 27–37. https://doi.org/10.1016/S0166-445X(01)00268-5.

Stadlober, M., Sager, M., and Irgolic, K.J. (2001). Effects of selenate supplemented fertilisation on the selenium level of cereals – identification and quantification of selenium compounds by HPLC–ICP–MS. *Food Chemistry* 73 (3): 357–366. https://doi.org/10.1016/S0308-8146(01)00115-7.

Sun, H.J., Rathinasabapathi, B., Wu, B. et al. (2014). Arsenic and selenium toxicity and their interactive effects in humans. *Environment International* https://doi.org/10.1016/j.envint.2014.04.019.

Suzuki, T., Moribe, M., Okabe, Y. et al. (2013). A mechanistic study of arsenate removal from artificially contaminated clay soils by electrokinetic remediation. *Journal of Hazardous Materials* 254–255: 310–317. https://doi.org/10.1016/j.jhazmat.2013.04.013.

Tabelin, C.B., Hashimoto, A., Igarashi, T. et al. (2014). Leaching of boron, arsenic and selenium from sedimentary rocks: II. pH dependence, speciation and mechanisms of release. *The Science of the Total Environment* https://doi.org/10.1016/j.scitotenv.2013.12.029.

Tabelin, C.B., Sasaki, R., Igarashi, T. et al. (2017). Simultaneous leaching of arsenite, arsenate, selenite and selenate, and their migration in tunnel-excavated sedimentary rocks: II. Kinetic and reactive transport modeling. *Chemosphere* 188: 444–454. https://doi.org/10.1016/j.chemosphere.2017.08.088.

Tan, L.C., Nancharaiah, Y.V., van Hullebusch, E.D., and Lens, P.N.L. (2016). Selenium: environmental significance, pollution, and biological treatment technologies. *Biotechnology Advances* https://doi.org/10.1016/j.biotechadv.2016.05.005.

Tang, S.C.N. and Lo, I.M.C. (2013). Magnetic nanoparticles: Essential factors for sustainable environmental applications. *Water Research* 47 (8): 2613–2632. https://doi.org/10.1016/j.watres.2013.02.039.

Terry, N., Zayed, A.M., De Souza, M.P., and Tarun, A.S. (2000). Selenium in higher plants. *Annual Review of Plant Biology* 51: 401–432.

Thornton, R.M., Rosenfeld, I., and Beath, O.A. (1965). Selenium: Geobotany, biochemistry, toxicity and nutrition. *Bulletin of the Torrey Botanical Club* https://doi.org/10.2307/2483845.

Tolu, J., Di Tullo, P., Le Hécho, I. et al. (2014a). A new methodology involving stable isotope tracer to compare simultaneously short- and long-term selenium mobility in soils. *Analytical and Bioanalytical Chemistry* 406 (4): 1221–1231. https://doi.org/10.1007/s00216-013-7323-1.

Tolu, J., Thiry, Y., Bueno, M. et al. (2014b). Distribution and speciation of ambient selenium in contrasted soils, from mineral to organic rich. *Science of the Total Environment* 479–480 (1): 93–101. https://doi.org/10.1016/j.scitotenv.2014.01.079.

Torres, J., Pintos, V., Domínguez, S. et al. (2010). Selenite and selenate speciation in natural waters: interaction with divalent metal ions. *Journal of Solution Chemistry* https://doi.org/10.1007/s10953-009-9491-3.

Tratnyek, P.G. and Johnson, R.L. (2006). Nanotechnologies for environmental cleanup. *Nano Today* https://doi.org/10.1016/S1748-0132(06)70048-2.

Vallini, G., Di Gregorio, S., and Lampis, S. (2005). Rhizosphere-induced selenium precipitation for possible applications in phytoremediation of se polluted effluents. *Zeitschrift fur Naturforschung – Section C* https://doi.org/10.1515/znc-2005-3-419.

Ventura, M.G., Freitas, M.D.C., and Pacheco, A.M.G. (2005). Selenium levels in mainland Portugal. *Water, Air, and Soil Pollution* https://doi.org/10.1007/s11270-005-7322-8.

Vinceti, M., Maraldi, T., Bergomi, M., and Malagoli, C. (2009). Risk of chronic low-dose selenium overexposure in humans: insights from epidemiology and biochemistry. *Reviews on Environmental Health* https://doi.org/10.1515/REVEH.2009.24.3.231.

Vinceti, M., Crespi, C.M., Bonvicini, F. et al. (2013a). The need for a reassessment of the safe upper limit of selenium in drinking water. *The Science of the Total Environment* https://doi.org/10.1016/j.scitotenv.2012.11.025.

Vinceti, M., Solovyev, N., Mandrioli, J. et al. (2013b). Cerebrospinal fluid of newly diagnosed amyotrophic lateral sclerosis patients exhibits abnormal levels of selenium species including elevated selenite. *Neurotoxicology* 38: 25–32. https://doi.org/10.1016/j.neuro.2013.05.016.

Vogel, M., Fischer, S., Maffert, A. et al. (2018). Biotransformation and detoxification of selenite by microbial biogenesis of selenium-sulfur nanoparticles. *Journal of Hazardous Materials* https://doi.org/10.1016/j.jhazmat.2017.10.034.

Wadgaonkar, S.L., Nancharaiah, Y.V., Esposito, G. et al. (2018). Environmental impact and bioremediation of seleniferous soils and sediments. *Critical Reviews in Biotechnology* https://doi.org/10.1080/07388551.2017.1420623.

Waegeneers, N., Thiry, C., De Temmerman, L. et al. (2013). Predicted dietary intake of selenium by the general adult population in Belgium. *Food Additives and Contaminants – Part A Chemistry, Analysis, Control, Exposure and Risk Assessment* https://doi.org/10.1080/19440049.2012.746474.

Wallschläger, D. and Feldmann, J. (2010). Formation, occurrence, significance, and analysis of Organoselenium and Organotellurium compounds in the environment. *Metal Ions in Life Sciences*(ed.D.WallschlägerandJ.Feldmann):319–364.https://doi.org/10.1039/9781849730822-00319.

Wang, Z. and Becker, H. (2013). Ratios of S, se and Te in the silicate Earth require a volatile-rich late veneer. *Nature* 499 (7458): 328–331. https://doi.org/10.1038/nature12285.

Wang, M.C.P., Zhang, X., Majidi, E. et al. (2010). Electrokinetic assembly of selenium and silver nanowires into macroscopic fibers. *ACS Nano* 4 (5): 2607–2614. https://doi.org/10.1021/nn901923z.

Whanger, P.D. (2002). Selenocompounds in plants and animals and their biological significance. *Journal of the American College of Nutrition* 21 (3): 223–232. https://doi.org/10.1080/07315724.2002.10719214.

Whanger, P., Vendeland, S., Park, Y.C., and Xia, Y. (1996). Metabolism of subtoxic levels of selenium in animals and humans. *Annals of Clinical and Laboratory Science* 26 (2): 99–113. http://www.ncbi.nlm.nih.gov/pubmed/8852418.

White, P.J. (2004). Interactions between selenium and Sulphur nutrition in Arabidopsis thaliana. *Journal of Experimental Botany* 55 (404): 1927–1937. https://doi.org/10.1093/jxb/erh192.

WHO (1987). Guidelines for drinking-water quality, volume 2: health criteria and other supporting information. *Science of the Total Environment* https://doi.org/10.1016/0048-9697(87)90388-3.

World Health Organization (2004). Selenium. In: *Vitamin and Mineral Requirements in Human Nutrition*, 2e. Geneva, Switzerland: WHO Press.

Wu, X. and Låg, J. (1988). Selenium in Norwegian farmland soils. *Acta Agriculturae Scandinavica* https://doi.org/10.1080/00015128809437988.

Wu, Z., Bañuelos, G.S., Lin, Z.-Q. et al. (2015). Biofortification and phytoremediation of selenium in China. *Frontiers in Plant Science* 6 https://doi.org/10.3389/fpls.2015.00136.

Xia, X., Wu, S., Li, N. et al. (2018). Novel bacterial selenite reductase CsrF responsible for Se(IV) and Cr(VI) reduction that produces nanoparticles in Alishewanella sp. WH16-1. *Journal of Hazardous Materials* 342: 499–509. https://doi.org/10.1016/j.jhazmat.2017.08.051.

Yang, G.Q., Wang, S.Z., Zhou, R.H., and Sun, S.Z. (1983). Endemic selenium intoxication of humans in China. *The American Journal of Clinical Nutrition* 37 (5): 872–881. https://doi.org/10.1093/ajcn/37.5.872.

Yang, S.I., Lawrence, J.R., Swerhone, G.D.W. et al. (2011). Biotransformation of selenium and arsenic in multi-species biofilm. *Environmental Chemistry* 8 (6): 543. https://doi.org/10.1071/EN11062.

Yanke, L.J., Bryant, R.D., and Laishley, E.J. (1995). Hydrogenase I of Clostridium pasteurianum functions as a novel selenite reductase. *Anaerobe* 1 (1): 61–67. https://doi.org/10.1016/S1075-9964(95)80457-9.

Zelmanov, G. and Semiat, R. (2013). Selenium removal from water and its recovery using iron (Fe3+) oxide/hydroxide-based nanoparticles sol (NanoFe) as an adsorbent. *Separation and Purification Technology* https://doi.org/10.1016/j.seppur.2012.10.037.

Zhou, Y., Tang, L., Zeng, G. et al. (2014). Mesoporous carbon nitride based biosensor for highly sensitive and selective analysis of phenol and catechol in compost bioremediation. *Biosensors and Bioelectronics* 61: 519–525. https://doi.org/10.1016/j.bios.2014.05.063.

Zhou, Y., Tang, L., Zeng, G. et al. (2016). Phosphorus-doped ordered mesoporous carbons embedded with Pd/Fe bimetal nanoparticles for the dechlorination of 2,4-dichlorophenol. *Catalysis Science & Technology* 6 (6): 1930–1939. https://doi.org/10.1039/C5CY01514F.

15

Membrane Separation Technologies for Selenium

Majid Peyravi and Abolfazl Arjmandi

Department of Chemical Engineering, Babol Noshirvani University of Technology, Babol, Iran

15.1 Introduction

Selenium (Se) discharge into the environment is becoming a matter of increasing concern because it induces toxic effects to biota at low concentrations (several micrograms per liter). Industrial activities that include energy generation, metal and oil refining, mining, and agriculture irrigation generate effluents contaminated with Se (Winkel et al. 2012; Staicu et al. 2017). It is obvious that the environmental pollution has direct and indirect impacts on human health. As some of the main examples of how environmental pollution by Se affects human health, the pollution of drinking water, the contamination of irrigation water of edible plants and crops, the pollution of the respiratory air, the pollution of water in which aquatics grow, and the contamination of animal feed can be mentioned (Bailey 2017; He et al. 2018c). Therefore, due to the human health concerns and environmental issues, separation and removal of Se is an essential work. Until now, different methods including reduction techniques, phytoremediation, bioremediation, electrocoagulation, electrochemical methods, adsorption, co-precipitation, electrokinetics, membrane technology, and chemical precipitation have been used for Se reduction in water and wastewaters (Shi et al. 2009; Eswayah et al. 2016; Holmes and Gu 2016; Adio et al. 2017; Lu et al. 2017; Schiavon and Pilon-Smits 2017; Hansen et al. 2019; Huang et al. 2019; Qiu et al. 2020).

Membrane technology as one of the major and most efficient separation methods has been broadly considered for water, and wastewater treatment and studies are intensely going on in this field (Dabaghian et al. 2018; Arjmandi et al. 2019a, b). Based on the type of material the membrane is made of, there are different kinds of membranes, including polymer, ceramic, glass, and liquid membranes (Noble 1987). Due to the high flexibility and selectivity (ultra-small pores), overall, and especially in aqueous systems, polymer membranes are the most commonly used membranes (Bruggen et al. 2003). Low cost, low energy demand, feasibility, environmental-friendly, and the ability to treat on temperature-sensitive materials are the main advantages of membrane systems (Kaya et al. 2018; Guo et al. 2019).

Selenium Contamination in Water, First Edition. Edited by Pooja Devi, Pardeep Singh, Arindam Malakar, and Daniel Snow.

Due to the variety of driving forces in membranes including gradients in hydrostatic pressure, chemical potential, and electrical voltage and also the variety of membranes based on their pore size, they have widespread applications in different fields (Noble 1987). Based on the membrane pore size, four general membrane types including microfiltration (MF, 0.08–2 μm), ultrafiltration (UF, 0.005–2 μm), nanofiltration (1–10 nm), and reverse osmosis (RO, 0.1–5000 nm) membranes can be mentioned (Ulbricht 2006). Permeation flux (F), rejection efficiency (R), and water recovery (WR) are the most important parameters defining membranes performance and can be calculated by the following equations (Reverberi et al. 2014; Arjmandi et al. 2020b):

$$R(\%)=\left(1-\frac{C_P}{C_F}\right)\times 100 \tag{15.1}$$

$$F=\frac{V}{A\times t} \tag{15.2}$$

$$WR(\%)=\frac{\textit{volume of permeate}}{\textit{volume of feed}} \tag{15.3}$$

where C_P and C_F are the concentrations of target solute in the permeate and feed side, V is the permeate volume over a predestined time (t), and A is the effective membrane area (encountering the solution). All that passes through the membrane (mostly water in water or wastewater treatment) is called the permeate, and all that remains behind the membrane and rejected is called the retentate. Based on our target material, one of these two streams, or even both of them, can be our desired products.

To get acceptable results, especially for commercial applications, the fabricated membranes should have suitable performance in terms of water flux, rejection, antifouling, concentration polarization, durability, etc. (Arjmandi et al. 2018; Arjmandi et al. 2020a). These are the main challenges ahead in membrane processes and have to be modified until they get to the point of commercialization. Also, the more cost-effective and eco-friendlier these methods are, the more attractive they are for utilization at an industrial scale.

What we are going to find out in this chapter is how to recognize the main Se resources that contaminate the water or wastewater and the danger of this contamination, to investigate the application of membranes to separate Se from the rest of the water and to have a look at the valuable research done in this area and, finally, to understand the commercial status of membrane applications in the treatment of Se-containing wastewater and the challenges ahead.

15.2 Se Resources

All Se content that presents in the environment is produced by natural and human actions. Naturally, Se element is randomly distributed in the Earth's crust and has a low concentration in the range of 0.05–0.5 mg/kg (Kabata-Pendias 2000). All raw and processed materials

Table 15.1 Main selenium resources and its average concentration.

Material	Se concentration (mg/kg)
Seleniferous soils	1–5000
Coal (fly ash)	73–440
Phosphate	1–300
Carbon-shale China	206–280
Copper ore	20–82
Ash-settling ponds	0.087–27 mg/l
USA coal	0.46–10.7
Crude oil	0.01–1.4
Fly ash leachate	0.04–0.61 mg/l
Earth's crust	0.05–0.5
Soil (adequate)	0.4
Volcanic rocks	0.35
Refined oils	0.005–0.26 mg/l

which contain Se and the concentration of Se are illustrated in Table 15.1 (Lemly 2004; Plant et al. 2014).

The major natural processes that result in Se production are as follows: weathering of rocks and soils, evaporation from water bodies, vegetation and soils, wildfires, and igneous activities (Haygarth 1994; Haug et al. 2007). But these days human industrial activities have to be widely considered due to their breadth and the significant impact they can have on the Se production and consumption cycle (Wen and Carignan 2007). In nature, pure Se element is scarce, and it is mainly present in other materials. Because Se can be exchanged with sulfur, coal, and metal-sulfide minerals such as chalcopyrite and pyrite are prone to increase the existence of this element in the environment (Kyle et al. 2011). For example, common selenide minerals include tiemannite (HgSe), stilleite (ZnSe), berzelianite (CuSe), penroseite (NiSe), klockmannite (CuSe), and clausthalite (PbSe). But in general, Se sources can be classified as soil, waters, plants, air, and food, that each of these will be briefly described below.

15.2.1 Soil

Worldwide, the total amount of Se in soil is about 0.4 mg/kg and ranges from 0.01 to 2.0 mg/kg with an average concentration of 0.4 mg/kg (Fairweather-Tait et al. 2011). Soils containing high concentrations of Se (more than 1200 mg/kg) are commonly originated from seleniferous materials such as limestones, shales, rocks, sandstones, coal series, and boulders as parent materials (Fairweather-Tait et al. 2011). Soil is the main resource of Se and can transfer it to the leaves and accumulate Se in them, through water stream from solum. Some sulfur-rich vegetables such as brassica and some types of allium, due to their ability to replace Se in proteins, have more potential to accumulate Se compounds. But even in this case, the level of accumulation is dependent on Se concentration in their surrounding soil (De et al. 2014). The level of Se content in soil depends on the texture and the amount of organic compounds of soil and the rate of local rainfall. In general, Se has different forms, including ferric selenite, selenate salts, and organic forms. SeO_3^{2-} and SeO_4^{2-} can be found in most soils and in large extent. The level of Se uptake in plants is controlled by the physicochemical factors, such as redox status and pH, of soil and microbiological activities. The mean concentration of Se content in soils is in the range of 100–700 ppb. In the case of clay, this amount is in the range of 800–2000 ppb, while in tropical soils, this value is in the range of 2000–4500 ppb (He et al. 2018c).

15.2.2 Waters

Results obtained from laboratory and experiments performed on humans in terms of the toxicity of Se for human health made the world reconsider the safety of Se existing in drinking water and rearrange the health standards relating to Se content (Vinceti et al. 2009). In this regard, since 1970s, the toxicity of Se in drinking water was widely discussed in health organizations worldwide. According to the standard of the World Health Organization (WHO), the maximum permitted amount of Se content in drinking water is 10 ppb. According to the United States Environmental Protection Agency (US EPA) this value is 50 ppb. In 2003, results obtained from a valuable piece of research recommended a maximum permitted concentration of 40 μg/L for Se content in drinking water, which has been considered by WHO for standard updates in 2011 (Toft 2003; WHO 2011). In some areas, Se concentration in ground water (e.g. well water) can reach 10–20 μg/l and sometimes it may reach to several hundred μg/l of Se content (Emmanuelle et al. 2012). It is important to note that generally, mineral waters contain negligible levels of Se concentrations and that drinking water is not considered a source of Se (Vinceti et al. 2013a).

15.2.3 Plants

In some countries, such as Egypt and China, Se deficiency in people's diet is a widespread problem. To overcome this problem, solutions such as spraying Se on crops and incorporation of Se in cord used in agriculture were employed by researchers (Feng et al. 2013). Plants can take up Se in plants occurs via sulfur (S) transportation and through biochemical pathways. They can also evaporate methylated Se, which is harmful to eat. Plants can be used both to compensate for Se deficiency in diet and to remove Se pollution in seleniferous areas (Pilon-Smits and Quinn 2010). In general, Se concentration in plants depends on the amount of Se in the surrounding soil. Se uptake in roots is a fast process but it also depends on the plant species and the physiological conditions. It is notable that the danger of poisoning livestock by Se is high, and the standard concentration of Se in forage is 0.1–0.5 ppm (Ma et al. 2017).

15.2.4 Air

The main source of Se emissions in the air is the burning of oil and coal. In general, the amount of Se in the surrounding air is low. Because of the impact on the transport and transformation, the atmosphere plays a significant role in Se cycle on Earth (Huang et al. 2012). The existence of Se in the air depends on natural events such as burning fossil fuels, tires, garbage, and papers (Marcocci et al. 2011). Also, natural events such as volcanoes, soil erosion, and forest fires can also affect the level of Se in the air.

15.2.5 Food

Major foods with Se content in people's diet are milk and diary products generally, cereals and bread, fish, and eggs. Moreover, some of the Brazilian nuts have also a considerable amount of Se content, with a concentration of 0.03–512 mg/kg (Silva et al. 2017). The level of Se content in grains and vegetables depends on the surrounding soil. Tomatoes, turnips, beets, carrots, cucumbers, beans, and potatoes, if grown in seleniferous soil, will have the most Se concentration (6 mg/g). Vegetables like asparagus and onions can also accumulate Se up to a concentration of 17 μg/g in some soils. Also, the amount of Se that animals consume depends on the amount of Se in their diet (Mehdi et al. 2013).

15.3 Health Hazards

Selenium plays an important role as an antioxidant in the human body. Considering the importance of Se for humans, the recommended amounts of Se in daily diet are as follows: 30 and 55 μg/day for adults in Europe and America, and based on the standard value recommended by Chinese Nutrition Society, 50–250 μg/day for adults in China. The recommended daily consumption of Se reported by the Council of Health (CSS) in Belgium is in the range of 60 μg/day for women and 70 μg/day for men (older than 14). This recommended value can increase to 65 μg/day for pregnant women and 75 μg/day during breastfeeding. The European Food Safety Authority 2006 (EFSA) has determined the maximum allowed values of Se in daily diet. Based on this standard, the maximum values of 30, 60, and 250 μg/day were recommended for adults, children (1–3 years), and teenagers (15–17 years), respectively (Sun et al. 2014). Se, as an essential compound from selenoproteins, plays an important role in several biological functions: fertility and reproduction, antioxidant defense, the formation of thyroid hormones, and DNA synthesis.

In organisms, Se can convert to different types of metabolites such as methylselenol, which plays an important role in preventing cancers. Just like vitamin E, Se can impact on muscle function by improving endurance and can also provide relief and improvement to the aging process (Mehdi et al. 2013). The scarce Se element played an important role in humans' health over the years, because of the various compounds with low molecular weight and its presence in 25 selenoproteins of selenocysteine amino acid. Although selenoproteins are not completely explored in term of performance, but general functions such as thyroid hormone metabolism, redox homeostasis, and redox regulation of intercellular signaling, were completely developed. It is important to note that there is a narrow band between the amount of Se necessary to avoid deficiency and a level of consumption that is toxic for the human body. However, it is almost impossible to determine a specific diet that causes toxicity and harms the human body as a result of producing selenol, because toxicity depends on the form of Se in the food, the type of food compounds coupling with Se, and also probably the reactions of Se with genotype (Fairweather-Tait et al. 2011). It is well believed that a sufficient supply of Se, and even a large amount of it, has an antiviral effect which is a major factor in successful reproduction and decreasing the danger of thyroid autoimmune disease. But an excess amount of Se, due to the production of reactive oxygen species (ROS) which have a direct relation with carcinogenesis, can increase the risk of getting cancer. Also, some studies showed that high concentrations of Se can exacerbate the risk of diabetes and can also cause tumorigenesis by increasing the production of TR1 cells (Sun et al. 2014). In addition to the carcinogenic effect, the excess amount of Se has correlation with the development of degenerative diseases such as amyotrophic lateral sclerosis and cardiovascular disease (Vinceti et al. 2009). Moreover, Se can have a destructive effect of nerve degeneration and subsequent amyotrophic lateral sclerosis. In general, Se is neurotoxic, but organic Se is four times less neurotoxic than inorganic Se (Vinceti et al. 2013b). However, the toxicity of Se depends on its concentration, the type of Se, and the compounds coupled with Se. These compounds can also have significant impacts on the target organism, because of the antagonistic or synergistic effects. Therefore, these compounds have to be measured completely, especially in the case of inorganic Se (Sharma et al. 2017).

Some disorders and diseases originating from Se deficiency in humans and animals are well known. Se deficiency can has negative outcomes such as cognitive decline, increased mortality, and poor immune function. Se deficiency in humans, can cause disorders including anemia, hair thinning, peripheral myopathy, opacification of the nails, cardiomyopathy, and decreased muscle endurance. Moreover, it was indicated that in peoples with low Se in their plasma (<115 ng/ml), the risk of cancer was higher (Fairweather-Tait et al. 2011). A disease called Keshan has been reported in China, which is related to Se deficiency in children and also known as cardiomyopathy in children (Koyama et al. 2013). Se deficiency in animals is common, and it can manifest in different forms in adult and young animals. First organisms to be affected by se deficiency, including skeletal muscle, heart, and liver. The most common disease related to the Se deficiency, especially in young animals, is white muscle disease (Sordillo 2013). Considering all the health concerns related to Se pollution, it seems necessary to remove Se from water and wastewater to prevent lots of problems related to human health and environmental issues.

15.4 Membrane Applications

15.4.1 Hybrid System

Using membrane processes in combination with conventional separation methods can increase the efficiency of separation systems (Lee and Hyun 2020; Rastgar et al. 2020). Mavrov et al. have experimentally investigated the combination of membrane process with electrocoagulation process to remove Se from artificial and real wastewaters containing Se (Mavrov et al. 2006). For this purpose, a hybrid electrocoagulation/microfiltration (MF) system, as shown in Figure 15.1, was used to treat on artificial wastewater containing Se and arsenic (As) and industrial copper (Cu) production wastewater containing Se and other dissolved elements. First, the artificial wastewater containing 5 mg/l Se and 100 mg/l As was used to optimize the system operation conditions. Then, to evaluate the real

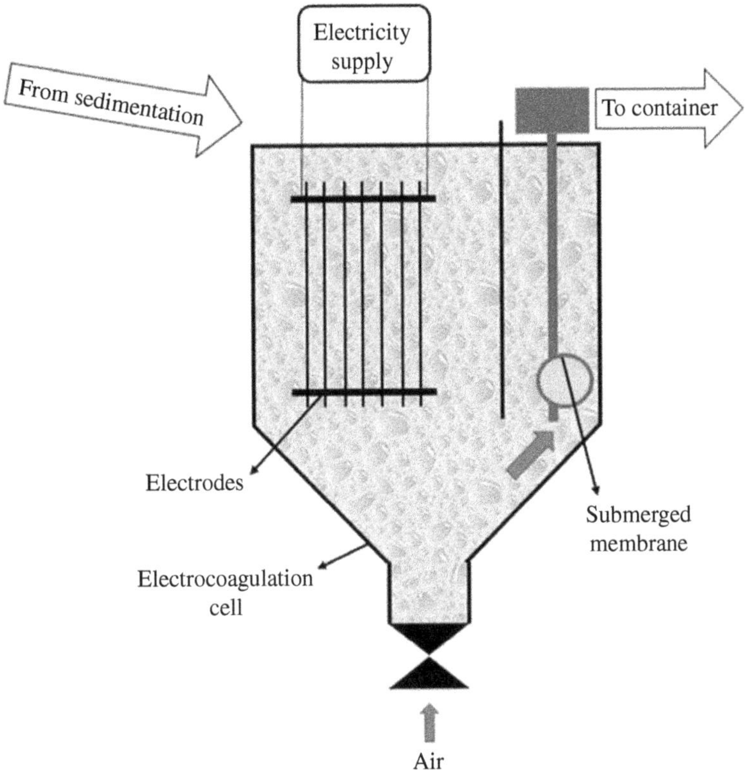

Figure 15.1 Schematic view of a hybrid electrocoagulation/microfiltration (MF) system.

performance of, the Cu production wastewater was treated by the electrocoagulation/MF membrane system. The membrane process used in this combination was a submerged flat-sheet multi-channel MF ceramic membrane in a dead-end system. Also, the real wastewater used in this study includes As, iron (Fe), Cu, Se, lead (Pb), cadmium (Cd), and zinc (Zn). The submerged membrane exhibited the permeate fluxes of 150–500 l/m^2.h under the vacuum pressures of 60–680 mbar. Adsorption and co-precipitation of Se occur by Fe hydroxide produced by electrocoagulation process. Separation of Fe hydroxide from water is challenging, and it was done by MF membrane efficiently. Through this membrane separation, a Fe hydroxide cake forms on the membrane, which also helps the further adsorption of remained Se particles from water permeating through the membrane. Separation performance of this hybrid system was significant in treating on Cu production wastewater with 20 minutes operation time and anode current density of 4.8 mA/cm^2, by the rejection values of 98.7% for Se, more than 98% for Cu and Pb, more than 99.9% for As, more and than 99.9% for Zn and Cd. The waste produced in all parts of this system, including sediment and cake from neutralization and electrocoagulation/MF sections, respectively, was 40% less than other conventional methods. Also, the chemicals used in this system was 15% less than other conventional methods. It is important to note that the submerged MF ceramic membrane showed a stable performance and removed filtration cake efficiently.

15.4.2 Adsorbent Membrane

Due to the ionic nature of Se compounds in the water, adsorbent membranes can be considered for Se removal. Gui et al. investigated the performance of iron/iron oxide functionalized polyacrylic acid (PAA) coated polyvinylidene fluoride (PVDF) MF membrane to remove toxic Se metal from power plant scrubber water (Gui et al. 2015). Scrubber water results from the flue gas desulfurization (FGD) process in power plants burning coal to remove sulfuric dioxide (SO_2) from the plant exhaust (Meikap et al. 2002). Usually, in this process, a forced-oxidation of SO_2 to sulfates using excess air performs. Next, by adding limestone ($CaCO_3$), it will react with sulfates and gypsum will precipitate and remove. These waters usually have TDS values in the range of 15000 to 50000 mg/l (Huang et al. 2013). In general, ions existing in the scrubber water include sulfate, calcium, chloride, magnesium, and volatile metals such as mercury, arsenic, and Se (Blázquez et al. 2019). In this study, commercial PVDF400HE MF membrane with 420 nm pore size and full-scale PVDF400-PAA (functionalized) MF membrane with 124 nm pore size were used. Also, the spiral-wound module was used in all experiments. The PVDF membranes were coated by PAA using in-situ polymerization (mostly formed on the pore surfaces) and were functionalized with iron through ion exchange process. The main predicted advantages of these iron-functionalized membranes compared with the adsorption method using iron nanoparticles were the inhibition of agglomeration and the dissolving of iron nanoparticles. The impacts of modification of membrane pores on the pH-responsive behavior, ion exchange capacity, and PAA gel permeability were evaluated. It was observed that membranes with about 20% weight gain due to the functionalization, exhibited the best pH-responsive pores. Also, the PAA network inhibited the agglomeration and dissolving of iron nanoparticles, unlike what happens in adsorption, which increased se removal. In high salt concentrations, because of the corrosion of the iron surface by chloride and sulfate and/or competitive adsorption, se removal was intensely inhabited. It is important to note that by increasing the temperature, the destructive impacts of salt were moderated and the iron oxidation was facilitated. Selenate rejection from DI water using functionalized membranes with a residence time of 10 seconds, was about 91.4% at 25.2 l/m^2.h.bar, while the conversion was only 21.8% at 15.3 l/m^2.h.bar for scrubber water with a residence time of 8.4 s. This low conversion is attributed to the presence of dissolved metal ions and salt in scrubber water, which decreases the iron reactivity. To reduce the TDS for solving this problem, a NF membrane was used as a pretreatment which rejected the chloride and sulfate with values of about 93.02 and 99.42%, respectively. By subsequent treating using functionalized membranes through ion reduction mechanism, the conversion of Se was about 77.5% with a residence time of 8 s.

15.4.3 Ultrafiltration and Nanofiltration

To compare the effects of using symmetric ultrafiltration (UF) and nanofiltration (NF) membranes to remove Se from water, an experimental study was performed by Chabane et al. (Chabane and Dahmani 2011). The removal and transportation of ionized and non-ionized Se including selenate (Se (VI)) and selenite (Se (IV)) were evaluated in this study. Also, two different membrane modules, including flat-sheet and spiral-wound modules, were used in bench-scale and under different hydrodynamic conditions for this experiment. These different conditions were obtained by altering the main operational factor of the ratio of J_0/k which J_0 is the initial pure water flux and k is the back-diffusion mass transfer coefficient. The J_0 and k can be controlled by use of applied pressure and cross-flow velocity, respectively. Measuring Se removal in this study, was obtained using three different commercial membranes including two UF membranes (PW (Desal.) and GM (Desal.)) and one NF membrane (ESNA (Hydranautics)). The main characteristics of these membranes are illustrated in Table 15.2. It is well known that electrostatic exclusion can affect Se removal by membranes. It was predicted by researchers that selenite and selenate, exhibit different transport performances which are correlated with the oxidation state and pH. Also, membrane characteristics such as pore size and surface charge can affect the transport performance (Teixeira et al. 2005; Qu et al. 2014). In general, different membranes and different J_0/k values, obtain different operational performances. In detail, three systems were used in this study including Se removal using spiral-wound module NF membrane, flat-sheet module UF (GM) and NF (ESNA) membranes, and flat-sheet module UF membranes (PW & GM). In the same condition, NF membrane shows better removal efficiency than UF membranes, and in general spiral-wound module shows better results than flat-sheet module in term of Se rejection. Moreover, in all cases, Se removal has increased by increasing the pH and the ratio of J_0/k. Selenate removal was also higher than selenite removal in all cases. The high removal was related to selenate removal using spiral-wound module NF membrane with a value of about 92%, due to the suitable electrostatic interaction between ionized selenate ions and negatively-charged membrane surface. According to the obtained results, it was indicated that the ratio of J_0/k was the most dominant factor controlling the membrane performance. So that the rejection of highly charged Se ions using UF membranes was competitive with the rejection of non-charged Se ions using NF membrane. The high Se removal using UF membranes obtained in this study was about 87%.

Kharaka et al. used nanofiltration (NF) to remove Se from contaminated agricultural drainage wastewater in an area in California, USA (Kharaka et al. 1996). Se concentration in drainage water, soils, and groundwater was high in this area. The wastewater was filtered (0.45 μm) before being pumped into a FILMTEC NF membrane module. First, a portion of wastewater was added into the feed container and pumping continued to reach 75% water recovery (WR). At this point, a significant amount of $CaSO_4.2H_2O$ (gypsum) was rapidly precipitated in the feed stream. Next, a portion of wastewater was diluted by distilled DI water to a volume ratio of 4:1 and added to the feed container, and system ran. In about 44% WR, the remained solution (in feed container) was concentrated to the values of 2.5 mg/l Se^{+4} (as $HSeO_3^-$) and 4.5 mg/l Se^{+6} (as SeO_4^{-2}) and until 80% WR, no precipitation was observed. By these dilution and Se spiking, different concentrations of Se and other dissolved chemicals were tested in this study. The chemicals in the agricultural drainage wastewater used in this study include Se^{+4}, Se^{+6}, SO_4, Cl, Ca, Sr, K, Na, B, Mg, Br, SiO_2, NO_3, and HCO_3. Based on the obtained data, the NF membrane selectively rejected more than 95% of Se in almost all wastewaters (Se < 1000 μg/l) tested. However, the efficiency of membrane filtration of such wastewaters depends on the type of chemicals in them. As an important and related example, membrane application can be limited by fouling phenomenon as a result of the precipitation of minerals (gypsum in this case) on the membrane which negatively impacts on the water flux and membrane selectivity (Cifuentes-Araya et al. 2013). Moreover, it was indicated by geochemical modeling that gypsum is the most likely precipitate in undiluted wastewater due to the saturation of this mineral (Brusilovsky et al. 1992). Wastewater dilution and using mineral precipitation inhibitors are common solutions to this problem and can enhance the WR in which precipitation occurs, to a value of 90%. Also, it was reported that because of operating under lower pressures (about three times lower), less pretreatment requirement, and higher water output, the NF membranes can get better results than traditional reverse osmosis (RO) membranes in such applications.

15.4.4 Novel Thin-film Composite Membranes

15.4.4.1 What Is TFC?

After inventing and using conventional RO membranes for water treatment in past years, thin-film composite (TFC) membranes have attracted widespread attention for separation in aqueous systems (Rao et al. 2003; Lau et al. 2012; Li et al. 2020). These membranes usually consist of two parts: a symmetric or composite porous thick support layer (SL) to provide the mechanical strength, and an ultra-thin dense active layer (AL) to be responsible for separation (Lau et al. 2012). A schematic view of the structure of a TFC membrane is shown in Figure 15.2. Due to the presence of the AL with ultra-low pore diameters ranging from 0.5 to 2 nm on top of the SL in TFC membranes, almost every dissolved and suspended nanoparticle can be removed from the system using this kind of membranes (Al-Najar et al. 2020; Zhao et al. 2020).

15.4.4.2 Thin-Film Nanocomposite Membranes

He et al. investigated the application of UiO-66 as a water-stable metal-organic framework (MOF) incorporated in the active layer of the TFC membrane to improve the separation efficiency toward removing Se and As from aqueous system (He et al. 2017). In this study, the UiO-66 nanoparticles were synthesized with different sizes of 30, 100, and 500 nm to be encapsulated into the AL of the TFC membrane to form the thin-film nanocomposite (TFN) membranes. The impacts of UiO-66 on the hydrophilicity, surface charge, morphology, and mean pore diameter of the TFN membrane were evaluated. The SL in this study was made of polyethersulfone (PES) with 20% (w/w) polymeric solution via phase inversion (PI) method, and the polyamide layer was formed as the AL by interfacial polymerization (IP) method on top of the SL. The PI and IP methods are common TFC fabrication methods with suitable outcomes (Liu et al. 2020). In the IP method in this study, UiO-66 nanoparticles were added to the organic solution to create the TFN membrane.

Table 15.2 All membrane types used by Chabane et al.

Membrane	Material	Zeta potential (mV) at pH 7.0	Water contact angle (°)	MWCO (Dalton)	Pure water permeability (PWP) (l/day.m^2.kPa)
ESNA (NF)	Polyamide TFC	−9.90	63.0	250	2.35
PW (UF)	Polyethersulfonate	−29.11	73.9	10 000	28.22
GM (UF)	Polyamide TFC	−45.09	61.0	8000	4.34

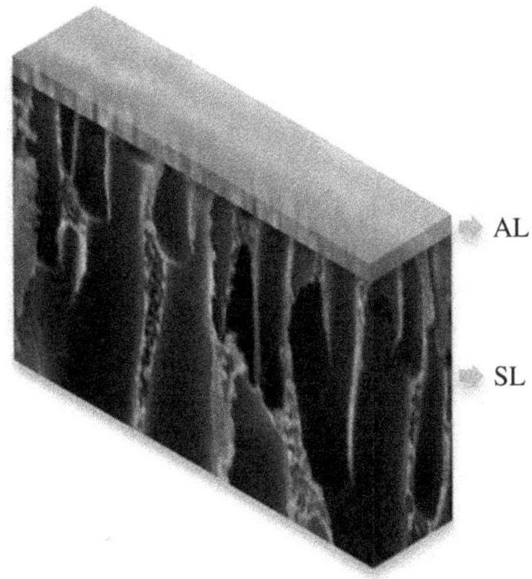

Figure 15.2 The structure of a TFC FO membrane made of polyethersulfone (as SL) and polyamide (as AL).

It was predicted that high rejection would be obtained for Se (IV), Se (VI), and As (V) especially, due to the similarity of the size of these particles and the UiO-66 pore diameter. According to zeta-potential analysis, in a wide range of pH values (2–12) fabricated TFN membranes showed negative surface charge, which proved the ability of these membranes to reject the negative charged ions through the Donnan effect (Sun et al. 2020). Because of the greater hydrophilicity and smaller pore size of the TFN membranes compared to the corresponding TFC membranes, these membranes showed better results in terms of pure water permeability (PWP) and rejection for both contaminants. Due to the better dispersion and thinner AL (50–100 nm), the 30 nm UiO-66 incorporated TFN membrane showed the best results with the highest PWP of 11.5 l/m^2.h.bar and the highest rejections of 96.5, 97.4, and 98.6% for SeO_3^{2-}, SeO_4^{2-}, and $HAsO_4^{2-}$, respectively. Also, the removal of $MgCl_2$, Na_2SO_4, $MgSO_4$, and NaCl was performed and, for example, these values were in the range of 94–97% for the TFN membrane containing 30 nm UiO-66. It is important to note that the fabricated TFN membranes showed acceptable long-term stability.

Carbon Quantum dots (CQDs), due to the special properties such as biocompatibility and non-toxicity, have many applications in bioimaging, biosensors, and drug delivery systems (Shankar et al. 2020; Venerando et al. 2020). CQDs have nanoscale size and consist of a carbon core surrounded by passive carboxyl groups. He et al. utilized sodium ion (Na^+) functionalized CQD (Na-CQD) for incorporating in the AL of TFC membrane to remove Se and As using this membrane (Figure 15.3) (He et al. 2018b). Therefore Na-CQD was added to the PIP aqueous solution in IP process with concentrations of 0, 0.02, 0.05, 0.08, 0.10, and 0.12 wt% to form TFN membranes. Also, the PES SL was used in these membranes because of outstanding chemical and mechanical properties (Misdan et al. 2014). The PWP and rejection efficiency in this study were measured using DI water and fed with different salt concentrations, respectively. With morphological analyses, it was found that the molecular size of synthesized Na-CQD was in the range of 2.5–4 nm with an average size of 3.2 nm. Also, due to the presence of Na containing functional groups and oxygen, high hydrophilicity was observed for the Na-CQD. By incorporating the Na-CQD, the surface structure of the TFN membranes was altered compared with the pristine TFC membrane. So, the TFN membrane surface was more granular, more convex, and rougher, in contrast with the TFC membrane containing globules and nodules on the surface. Also, the AL cross-linking degrees of the TFN membranes were higher than the pristine TFC membrane. The reason of all differences between the TFC and the TFN membranes can be explained in two statements: (i) The solubility and diffusion rate of the PIP molecules were increased due to the hydrogen bond with Na-CQD and (ii) In low loading, the hydrophilic Na-CQD made the organic–inorganic interface larger, due to its hydrophilic functional groups, which increased the repulsion of the organic phase. According to the zeta potential, all modified TFN membranes passed negatively-charged surfaces in a wide range of pH values (2–12), which is suitable, as mentioned earlier. The PWP was increased by increasing the loading percentage up to 0.05 wt% (10.5 l/m^2.h.bar) and decreased by further loading increment, but still higher than the pristine TFC membrane. The rejection of $MgCl_2$, $MgSO_4$, NaCl, and Na_2SO_4 salts was increased from a mean value of about 90% to a mean value of about 95%. By increasing the loading percentage, the removal of SeO_3^{2-}, SeO_4^{2-}, and $HAsO_4^{2-}$ was significantly increased from a mean value of about 90% to a mean value of about 99%. Also, the anti-fouling performance of the TFN membranes was measured using the bovine serum albumin (BSA) as a protein and organic foulant. The TFN-0.12 (0.12 wt% Na-CQD) showed the best performance with the values of 25.2% for flux decline rate (FDR) (50.5% for pristine TFC) and 93.4% for flux recovery ratios (FRR) (63.5% for pristine TFC).

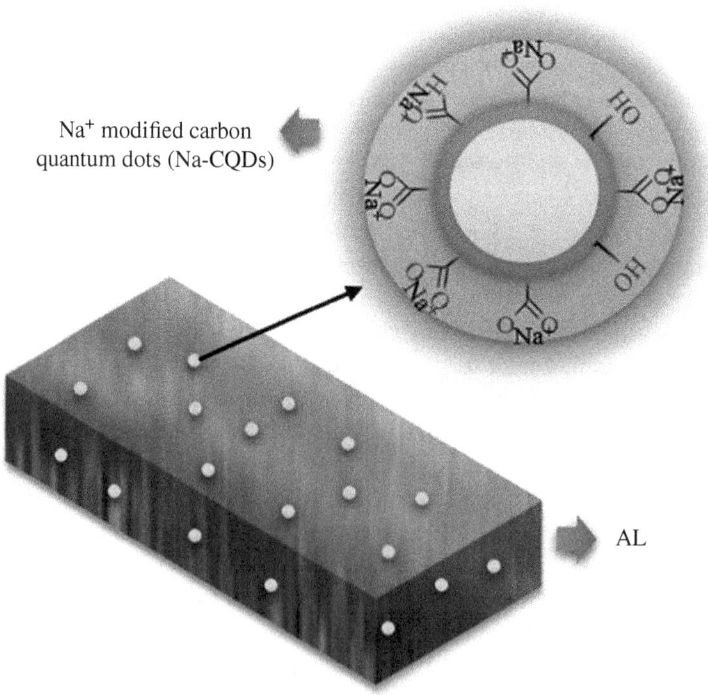

Figure 15.3 Schematic view of the Na-CQD modified TFN membrane.

15.4.4.3 TFC & Zwitterionic Polymer

He et al. synthesized a water-soluble zwitterionic co-polymer, via one-pot free-radical polymerization between 2-methacryloyloxyethyl phosphorylcholine (MPC) monomer and 2-aminoethyl methacrylate hydrochloride (AEMA) monomer named P[MPC-co-AEMA] for incorporation in the AL of TFC membrane to remove Se and As from aqueous medium (He et al. 2018a). The zwitterionic part of poly(MPC) contains negatively-charged phosphate anions and positively-charged ammonium cations and poly(AEMA) part of the co-polymer contains terminal amino functional groups. So, because of using trimesoyl chloride (TMC) in the organic phase in the IP method, AEMA acts as the linker between co-polymer and TMC and of course MPC induces the hydrophilicity to the co-polymer. It should be noted that piperazine (PIP) was used in the aqueous phase for IP reaction in this study. The co-polymer was incorporated into the aqueous phase by the values of 0, 10, 50, 70, and 100 wt%, and 100 wt% means that the PIP was completely replaced by the co-polymer in the aqueous solution. The impact of this modification on separation performance and the membrane composition was investigated. Due to the lower diffusion rate, higher molecular volume, and the hydrogen bond between the co-polymer, PIP, and water molecules, the diffusion of PIP from the aqueous phase to the organic phase was disturbed, resulting in more nodule-like points on the TFC membranes (10–70 wt%) compared with the pristine one. All modified TFC membranes passed negatively-charged surfaces in the wide range of pH values (2-12) which is suitable for rejection of negatively-charged particles like what is targeted in this study. The TFC membranes with 10 and 50 wt% co-polymer (TFC-10 & TFC-50) have the smallest pores with the values of 0.65 and 0.63 nm, respectively. Due to the lower amino functional groups existing in the co-polymer than in the PIP, the AL of the TFC membrane with 100 wt% co-polymer (TFC-100) has the lowest cross-linking degree and subsequently largest pores, which is unsuitable for separation mechanism. Because of the presence of the zwitterionic ions which have a good interaction with water molecules and form hydration layer, all modified TFC membranes showed higher PWP than the pristine one. Although the TFC-100 membrane has the highest PWP (85.5 l/m^2.h.bar), but the lowest rejection values for $MgCl_2$, $MgSO_4$, NaCl, and Na_2SO_4 salts were also attributed to this membrane due to the lowest cross-linking degree. The TFC-50 membrane has the highest rejection (about 94%) with acceptable PWP of 40 l/m^2.h.bar. Also, due to the formation of hydration layer with both negative and positive charges and filling the empty pores that can be seen on the pristine TFC membrane with the co-polymer and reducing the pore size, the TFC-50 was the best membrane, with the removal of 98.2% and 99.8% for SeO_3^{2-} and $HAsO_4^{2-}$, respectively.

15.4.5 Functionalized Nanowire Membrane

Qiu et al. used manganese dioxide (MnO_2) ultra-long nanowires coated with nanoscale zero valent-iron (NZVI) via self-assembly, as a hybrid membrane to separate Se (Qiu et al. 2020). The high specific surface of the NZVI and excellent mechanical properties of the MnO_2 were considered for this membrane. Batch experiments were performed in a san core funnel on this dynamic zero-valent iron/MnO_2 nanowire hybrid membrane (ZVINM) with a thickness of around 4 cm and 100 ml of the solution to be treated with different concentrations. The impact of several factors, including the NZVI content on the ZVINM surface, Se concentration, interfering ions, solution pH, operation time, and applied pressure, were evaluated. The reusability test was also performed on the ZVINM in the same separation conditions of Se (concentration of 100 mg/l, atmospheric pressure, and pH = 6). By increasing the pressure in the range of 20–100 kPa at Se concentration of 100 mg/l, the permeation flux and the pure water flux were increased, but the rejection was decreased from 80.75 to 65.73%. Also, by increasing the operation time from 20 to 140 minutes at the concentrations of 50, 100, and 150 mg/l, the rejection was increased and gradually reached to a constant value. In the first one hour of the test, almost 60% of Se was trapped on the membrane which is related to the unoccupied active sites on the surface of NZVI and the redox reaction between Se and zero-valent iron. Over time, due to the formation of a Se layer on the membrane, the rejection efficiency was gradually decreased. By increasing Se concentration ranging from 60–200 mg/l, the rejection rate was decreased from 90.81 to 58.73%, which is also related to Se accumulation on the membrane surface. Also, by increasing Se concentration, the permeation flux was decreased simultaneously, due to the increased solution polarity and Se accumulation on the membrane surface which in turn increased the diffusion resistivity. By increasing the NZVI content on the nanowire ranging from 20.25–133.67 mg/m^2, the rejection and flux were increased and decreased by the values of 70.83–94.26% and 48.98-15 l/m^2.h, respectively. The flux and rejection were decreased and increased by increasing the membrane thickness, respectively. The impact of interfering ions, including Cl^-, NO_3^-, CO_3^{2-}, and SO_4^{2-} on separation efficiency was investigated. Because of forming complex between Se and these ions and subsequent inhibition of redox reaction between Se and zero-valent iron, the rejection was dramatically decreased in the order of $NO_3^- \approx Cl^- < SO_4^{2-} < CO_3^{2-}$. Due to the high reactivity of NZVI, the impact of acidic and alkaline environments on separation efficiency was also tested. So, by increasing the pH values from 2 to 10, the rejection rate was decreased from 95.62 to 76.35%. It is worth mentioning the rejection rate of 95.62% was the best result in this study. Also, the rejection efficiency in the durability test and after three cycles, was steel more than 60%.

By reviewing some of the most valid studies based on Se separation via membrane technology, it is obvious that at least at the laboratory scale, high Se separation efficiency using different membranes has been obtained. Also, some experiments were performed at larger scales. To commercialize novel technologies and methods in order to improve the conventional technologies, laboratory-scale studies are much needed. These pieces of research can improve each other step by step until getting to the point of commercializing. This is the reason that the pieces of research were discussed in this chapter in order of year published, to follow the evolutionary path of using membrane technology to remove Se.

15.5 Commercial Aspect

As described earlier, it is well believed that only the NF and RO membranes can remove Se from water with efficient results. But what we have to consider is that the artificial wastewaters which are usually used in scientific research are very diluted compared to real industrial wastewaters. So the long-term performance of the commercial NF and RO membranes can be limited by the large amount of dissolved solids (such as Se) contained in industrial wastewaters (Azizighannad et al. 2020; Wang et al. 2020a). As one of the problems caused by high concentrations of contaminants, the membrane fouling can be mentioned, which is the most important drawback of commercial membranes resulting in low water flux and non-efficient separation over time. But it is not the whole story, in wastewater treatment plants, pretreatment and post-treatment (final) stages can have significant positive impacts on the durability of the whole system and the quality of the output stream (Samaei et al. 2020). Because of the high rejection of Se ions by the NF and RO membranes, using these membranes as the post-treatment or pretreatment of the plant can be seriously considered. Using other treatment methods, as mentioned in Section 15.4.1, as the pretreatment for membrane method, not only can somewhat solve the membrane fouling problem, but also can facilitate the using of another kind of membranes, such as MF which is cheaper. With this attitude, the combination of membrane processes with other conventional treatment methods such as thermal processes and sedimentation and other novel treatment methods such as activated carbon adsorption, biological treatment, and ion exchange can increase the output quality, improve the durability and performance of all parts, and decrease the energy and chemical consumptions (Morón-López and Molina 2020; Wang et al. 2020b; Zhang et al. 2020).

One prominent option in membrane-based separations is that because of physical separation in such processes, the residual Se in the concentrated retentate is not chemically changed and can be recovered and be effectively used. One important commercial aspect that can somewhat justify Se separation costs is the broad application of Se in several industries and its considerable price in the market. The major applications of Se are listed below (Waitkins et al. 1942; Ullrey 1992; McKenzie et al. 1998; Guin et al. 2004; Grant et al. 2007; Trenner et al. 2013; Lu et al. 2014; Peng et al. 2015; Skalickova et al. 2017):

1) In drug supplements.
2) In fertilizers containing Se.
3) In the glass industry to decolor the glasses or produce the red-color glasses.
4) As catalysis in lots of chemical reactions.
5) In photocells and solar cells (first solar cell was made of Se).
6) As photographic toner.
7) As an additive in stainless steel and brass (along with bismuth).
8) To enhance the machinability of the iron and copper.
9) In anti-dandruff shampoos (Se sulfide).
10) In manganese electrowinning process, adding Se dioxide can decrease the power needed for electrolysis cells operation.
11) In lithium-Se batteries.

These and more probable applications of Se along with its considerable price in market, can justify the costs of Se separation and recovery by membrane processes. It is also worth mentioning that the membrane processes are in many aspects such as cost, energy consumption, and simple equipment more efficient than other conventional and many novel separation processes. Therefore, the economic concerns, which unfortunately are prior to the environmental aspects at the industrial scale, can be answered by the widespread applications of Se recovered using membrane technology. Moreover, exploring and developing the novel cost-effective membranes that can be used for Se removal and recovery, can light up the future of this field even more.

15.6 Conclusion

Although Se is an essential element for the humans and animals, its deficiency can result in several diseases and disorders. However, the presence of Se in the body with more than the permitted values can also cause irreparable harms for humans and animals. Se element can originate from natural resources, but human activities also can produce a massive amount of this compound. Human activities such as agriculture irrigation, mining, metal and oil refining, and energy generation can produce effluents containing Se. Discharge these wastewaters into the environment can pollute the soils, plants, rivers, and etc., which in turn can give rise to human and animal health concerns, because of the inseparable correlation between the environment, humans, and animals. So, finding a reliable, eco-friendly, and cost-effective method to separate Se from these wastewaters can have huge justification. One of the most important separation methods that provide all the properties mentioned is membrane technology. According to the research orientated to this field, RO and NF membranes can efficiently remove Se by more than 98% of rejection values, and acceptable PWP values, at least at laboratory scale. Also, because of the fouling problem that negatively impacts the membrane separation performance, more research is needed on using these membranes at an industrial scale. But it is important to note that, according to the research done overall on membrane technology, via coupling membrane processes and other feasible processes such as sedimentation and large-pore filtration, the durability and efficiency of the whole system can be enhanced in a commercially viable way. Also, through this attitude, other cheaper membranes such as MF can be utilized for this order. Beside separation efficiency and by considering separation costs, due to the huge applications of Se in industrial activities and its suitable price in market, separation and recovery costs by membrane processes can be justified. For the future, developing the latest technologies in this field, e.g. TFC membranes, and moving on the frontier of science on to use novel materials, structures, and mechanisms, can justify the use of membrane technology to remove Se from water and wastewaters even more.

References

Adio, S.O., Omar, M.H., Asif, M., and Saleh, T.A. (2017). Arsenic and selenium removal from water using biosynthesized nanoscale zero-valent iron: a factorial design analysis. *Process Safety and Environmental Protection* 107: 518–527.

Al-Najar, B., Peters, C.D., Albuflasa, H., and Hankins, N.P. (2020). Pressure and osmotically driven membrane processes: a review of the benefits and production of nano-enhanced membranes for desalination. *Desalination* 479: 114323.

Arjmandi, M., Pakizeh, M., Saghi, M., and Arjmandi, A. (2018). Study of separation behavior of activated and non-activated MOF-5 as diller on MOF-based mixed-matrix membranes in H2/CO2 separation. *Petroleum Chemistry* 58 (4): 317–329.

Arjmandi, M., Peyravi, M., Pourafshari Chenar, M., and Jahanshahi, M. (2019a). A new concept of MOF-based PMM by modification of conventional dense film casting method: significant impact on the performance of FO process. *Journal of Membrane Science* 579: 253–265.

Arjmandi, M., Peyravi, M., Pourafshari Chenar, M., and Jahanshahi, M. (2019b). Channelization of water pathway and encapsulation of DS in the SL of the TFC FO membrane as a novel approach for controlling dilutive internal concentration polarization. *Environmental Science: Water Research & Technology* 5 (8): 1436–1452.

Arjmandi, M., Pourafshari Chenar, M., Peyravi, M., and Jahanshahi, M. (2020a). Physical modification of polymeric support layer for thin film composite forward osmosis membranes by metal–organic framework-based porous matrix membrane strategy. *Journal of Applied Polymer Science* 137 (19): 48672.

Arjmandi, M., Peyravi, M., Altaee, A. et al. (2020b). A state-of-the-art protocol to minimize the internal concentration polarization in forward osmosis membranes. *Desalination* 480: 114355.

Azizighannad, S., Intrchom, W., and Mitra, S. (2020). Raman imaging of membrane fouling. *Separation and Purification Technology* 242: 116763.

Bailey, T.T. (2017). Review: selenium contamination, fate, and reactive transport in groundwater in relation to human health. *Hydrogeology Journal* 25: 1191–1217.

Blázquez, E., Baeza, J.A., Gabriel, D., and Guisasola, A. (2019). Treatment of real flue gas desulfurization wastewater in an autotrophic biocathode in view of elemental sulfur recovery: microbial communities involved. *Science of the Total Environment* 657: 945–952.

Bruggen, B.V.D., Vandecasteele, C., Gestel, T.V. et al. (2003). A review of pressure-driven membrane processes in wastewater treatment and drinking water production. *Environmental Progress* 22 (1): 46–56.

Brusilovsky, M., Borden, J., and Hasson, D. (1992). Flux decline due to gypsum precipitation on RO membranes. *Desalination* 86 (2): 187–222.

Chabane, M. and Dahmani, B. (2011). Study the remove and transport mechanisms of Seleniums in membrane (UF and NF) processes. *Journal of Chemical Engineering & Process Technology* 2 (5): 116.

Cifuentes-Araya, N., Pourcelly, G., and Bazinet, L. (2013). Water splitting proton-barriers for mineral membrane fouling control and their optimization by accurate pulsed modes of electrodialysis. *Journal of Membrane Science* 447: 433–441.

Dabaghian, Z., Peyravi, M., Jahanshahi, M., and Shokuhi Rad, A. (2018). Potential of advanced nano-structured membranes for landfill leachate treatment: a review. *ChemBioEng Reviews* 5 (2): 119–138.

De, T.L., Waegeneers, N., Thiry, C. et al. (2014). Selenium content of Belgian cultivated soils and its uptake by field crops and vegetables. *Science of the Total Environment* 468–469: 77–82.

Emmanuelle, B., Virginie, M., Fabienne, S. et al. (2012). Selenium exposure in subjects living in areas with high selenium concentrated drinking water: results of a French integrated exposure assessment survey. *Environment International* 40: 155–161.

Eswayah, A.S., Smith, T.J., and Gardiner, P.H.E. (2016). Microbial transformations of selenium species of relevance to bioremediation. *Applied and Environmental Microbiology* 82 (16): 4848–4859.

Fairweather-Tait, S.J., Bao, Y., Broadley, M.R. et al. (2011). Selenium in human health and disease. *Antioxidants & Redox Signaling* 14: 1337–1383.

Feng, R., Wei, C., and Tu, S. (2013). The roles of selenium in protecting plants against abiotic stresses. *Environmental and Experimental Botany* 87: 58–68.

Grant, C.A., Buckley, W.T., and Wu, R. (2007). Effect of selenium fertilizer source and rate on grain yield and selenium and cadmium concentration of durum wheat. *Canadian Journal of Plant Science* 87 (4): 703–708.

Gui, M., Papp, J.K., Colburn, A.S. et al. (2015). Engineered iron/iron oxide functionalized membranes for selenium and other toxic metal removal from power plant scrubber water. *Journal of Membrane Science* 488: 79–91.

Guin, J.-P., Rouxel, T., Sanglebœuf, J.-C. et al. (2004). Hardness, toughness, and scratchability of germanium–selenium chalcogenide glasses. *Journal of the American Ceramic Society* 85 (6): 1545–1552.

Guo, Y.-S., Mi, Y.-F., Ji, Y.-L. et al. (2019). One-step surface grafting method for preparing zwitterionic nanofiltration membrane via in situ introduction of initiator in interfacial polymerization. *ACS Applied Polymer Materials* 1 (5): 1022–1033.

Hansen, H.K., Peña, S.F., Gutiérrez, C. et al. (2019). Selenium removal from petroleum refinery wastewater using an electrocoagulation technique. *Journal of Hazardous Materials* 364: 78–81.

Haug, A., Graham, R.B., Christophersen, O.A., and Lyons, G.H. (2007). How to use the world's scarce selenium resources efficiently to increase the selenium concentration in food. *Microbial Ecology in Health and Disease* 19 (4): 209–228.

Haygarth, P.M. (1994). Global importance and global cycling of selenium. In: *Selenium in the Environment* (eds. W.T. Frankenberger Jr. and S. Benson), 1–27. New York, USA: Marcel Dekker, Inc.

He, Y., Tang, Y.P., Ma, D., and Chung, T.-S. (2017). UiO-66 incorporated thin-film nanocomposite membranes for efficient selenium and arsenic removal. *Journal of Membrane Science* 541: 262–270.

He, Y., Liu, J., Han, G., and Chung, T.-S. (2018a). Novel thin-film composite nanofiltration membranes consisting of a zwitterionic co-polymer for selenium and arsenic removal. *Journal of Membrane Science* 555: 299–306.

He, Y., Zhao, D.L., and Chung, T.-S. (2018b). Na^+ functionalized carbon quantum dot incorporated thin-film nanocomposite membranes for selenium and arsenic removal. *Journal of Membrane Science* 564: 483–491.

He, Y., Xiang, Y., Zhou, Y. et al. (2018c). Selenium contamination, consequences and remediation techniques in water and soils: a review. *Environmental Research* 164: 288–301.

Holmes, A.B. and Gu, F.X. (2016). Emerging nanomaterials for the application of selenium removal for wastewater treatment. *Environmental Science: Nano* 3: 982–996.

Huang, J.-C., Passeport, E., and Terry, N. (2012). Development of a constructed wetland water treatment system for selenium removal: use of mesocosms to evaluate design parameters. *Environmental Science & Technology* 46 (21): 12021–12029.

Huang, Y.H., Peddi, P.K., Tang, C. et al. (2013). Hybrid zero-valent iron process for removing heavy metals and nitrate from flue-gas-desulfurization wastewater. *Separation and Purification Technology* 118: 690–698.

Huang, T., Liu, L., Zhang, S., and Xu, J. (2019). Evaluation of electrokinetics coupled with a reactive barrier of activated carbon loaded with a nanoscale zero-valent iron for selenite removal from contaminated soils. *Journal of Hazardous Materials* 368: 104–114.

Kabata-Pendias, A. (2000). *Trace Elements in Soil and Plants*, 3e. Boca Raton, FL: CRC Press.

Kaya, H., Bulut, O., Kamali, A.R., and Ege, D. (2018). L-Arginine modified multi-walled carbon nanotube/sulfonated poly(ether ether ketone) nanocomposite films for biomedical applications. *Applied Surface Science* 444: 168–176.

Kharaka, Y.K., Ambats, G., Presser, T.S., and Davis, R.A. (1996). Removal of selenium from contaminated agricultural drainage water by nanofiltration membranes. *Applied Geochemistry* 11 (6): 797–802.

Koyama, H., Abdulah, R., Yamazaki, C., and Kameo, S. (2013). Selenium supplementation trials for cancer prevention and the subsequent risk of type 2 diabetes mellitus: selenium and vitamin E cancer prevention trial and after. *Nihon Eiseigaku Zasshi. Japanese Journal of Hygiene* 68 (1): 1–10.

Kyle, J.H., Breuer, P.L., Bunney, K.G. et al. (2011). Review of trace toxic elements (Pb, Cd, Hg, As, Sb, Bi, Se, Te) and their deportment in gold processing. Part 1: mineralogy, aqueous solution chemistry and toxicity. *Hydrometallurgy* 107: 91–100.

Lau, W.J., Ismail, A.F., Misdan, N., and Kassim, M.A. (2012). A recent progress in thin film composite membrane: a review. *Desalination* 287: 190–199.

Lee, H. and Hyun, K. (2020). Effect of sequencing batch reactor (SBR)/granular activated carbon (GAC) bed and membrane hybrid system for simultaneous water reuse and membrane fouling mitigation. *Environmental Engineering Research* 26 (1): 190500.

Lemly, A.D. (2004). Aquatic selenium pollution is a global environmental safety issue. *Ecotoxicology and Environmental Safety* 59 (1): 44–56.

Li, Y., Xu, Z., Xie, M. et al. (2020). Resource recovery from digested manure centrate: comparison between conventional and aquaporin thin-film composite forward osmosis membranes. *Journal of Membrane Science* 593: 117436.

Liu, Y., Zhu, J., Zheng, J. et al. (2020). A facile and scalable fabrication procedure for thin-film composite membranes: integration of phase inversion and interfacial polymerization. *Environmental Science & Technology* 54 (3): 1946–1954.

Lu, J., Dreisinger, D., and Glück, T. (2014). Manganese electrodeposition — a literature review. *Hydrometallurgy* 141: 105–116.

Lu, Z., Yu, J., Zeng, H., and Liu, Q. (2017). Polyamine-modified magnetic graphene oxide nanocomposite for enhanced selenium removal. *Separation and Purification Technology* 183: 249–257.

Ma, B., Fernandez-Martinez, A., Grangeon, S. et al. (2017). Selenite uptake by Ca-Al LDH: a description of intercalated anion coordination geometries. *Environmental Science & Technology* 52 (3): 1624–1632.

Marcocci, C., Kahaly, G.J., Krassas, G.E. et al. (2011). Selenium and the course of mild Graves' orbitopathy. *The New England Journal of Medicine* 364 (20): 1920–1931.

Mavrov, V., Stamenov, S., Todorova, E. et al. (2006). New hybrid electrocoagulation membrane process for removing selenium from industrial wastewater. *Desalination* 201 (1–3): 290–296.

McKenzie, R.C., Rafferty, T.S., and Beckett, G.J. (1998). Selenium: An essential element for immune function. *Immunology Today* 19 (8): 342–345.

Mehdi, Y., Hornick, J.L., Istasse, L., and Dufrasne, I. (2013). Selenium in the environment, metabolism and involvement in body functions. *Molecules* 18 (3): 3292–3311.

Meikap, B.C., Kundu, G., and Biswas, M.N. (2002). Modeling of a novel multi-stage bubble column scrubber for flue gas desulfurization. *Chemical Engineering Journal* 86 (3): 331–342.

Misdan, N., Lau, W.J., Ismail, A.F. et al. (2014). Study on the thin film composite poly(piperazine-amide) nanofiltration membrane: impacts of physicochemical properties of substrate on interfacial polymerization formation. *Desalination* 344: 198–205.

Morón-López, J. and Molina, S. (2020). Optimization of Recycled-Membrane Biofilm Reactor (R-MBfR) as a sustainable biological treatment for microcystins removal. *Biochemical Engineering Journal* 153: 107422.

Noble, R.D. (1987). An overview of membrane separations. *Separation Science and Technology* 22 (2–3): 731–743.

Peng, X., Wang, L., Zhang, X. et al. (2015). Reduced graphene oxide encapsulated selenium nanoparticles for high-power lithium–selenium battery cathode. *Journal of Power Sources* 288: 214–220.

Pilon-Smits, E.A.H. and Quinn, C.F. (2010). Selenium metabolism in plants. *Plant Cell Monographs* 17: 225–241.

Plant, J.A., Bone, J., Vouvoulis, N. et al. (2014). Treatise on geochemistry. In: *Arsenic and Selenium*, vol. 11 (eds. H.D. Holland and K.K. Turekian), 13–57. Amsterdam, The Netherlands: Elsevier.

Qiu, Z., Tian, Q., Zhang, T. et al. (2020). Fabrication of dynamic zero-valent iron/ MnO2 nanowire membrane for efficient and recyclable selenium separation. *Separation and Purification Technology* 230: 115847.

Qu, F., Liang, H., Zhou, J. et al. (2014). Ultrafiltration membrane fouling caused by extracellular organic matter (EOM) from Microcystis aeruginosa: effects of membrane pore size and surface hydrophobicity. *Journal of Membrane Science* 449: 58–66.

Rao, A.P., Joshi, S.V., Trivedi, J.J. et al. (2003). Structure–performance correlation of polyamide thin film composite membranes: effect of coating conditions on film formation. *Journal of Membrane Science* 211 (1): 13–24.

Rastgar, M., Karkooti, A., Sohrabi, A. et al. (2020). Osmotic dewatering accelerates inherent sluggish kinetics of electro-Fenton process: toward sustainable removal of organic contaminants. *Chemical Engineering Journal* 394: 125043.

Reverberi, A.P., Maga, L., Cerrato, C., and Fabiano, B. (2014). Membrane processes for water recovery and decontamination. *Current Opinion in Chemical Engineering* 6: 75–82.

Samaei, S.M., Gato-Trinidad, S., and Altaee, A. (2020). Performance evaluation of reverse osmosis process in the post-treatment of mining wastewaters: case study of Costerfield mining operations, Victoria, Australia. *Journal of Water Process Engineering* 34: 101116.

Schiavon, M. and Pilon-Smits, E.A.H. (2017). Selenium biofortification and phytoremediation phytotechnologies: a review. *Journal of Environmental Quality* 46 (1): 10–19.

Shankar, S.S., Ramachandran, V., Raj, R.P. et al. (2020). Carbon quantum dots: a potential candidate for diagnostic and therapeutic application. *Nanobiomaterial Engineering*: 49–70.

Sharma, V.K., Mcdonald, T.J., Sohn, M. et al. (2017). Assessment of toxicity of selenium and cadmium selenium quantum dots: a review. *Chemosphere* 188: 403–413.

Shi, Y., Wang, L., Chen, Y., and Xu, D. (2009). Research on removing selenium from raw water by using Fe/Se co-precipitation system. *Journal of Water Supply: Research and Technology – AQUA* 58 (1): 51–56.

Silva, E.J., Wadt, L., Silva, K.E. et al. (2017). Natural variation of selenium in Brazil nuts and soils from the Amazon region. *Chemosphere* 188: 650–658.

Skalickova, S., Milosavljevic, V., Cihalova, K. et al. (2017). Selenium nanoparticles as a nutritional supplement. *Nutrition* 33: 83–90.

Sordillo, L.M. (2013). Selenium-dependent regulation of oxidative stress and immunity in periparturient dairy cattle. *Veterinary Medicine International* 2013: 154045.

Staicu, L.C., Hullebusch, E.D.V., Rittmann, B.E., and Lens, P.N.L. (2017). Industrial selenium pollution: sources and biological treatment technologies. In: *Bioremediation of Selenium Contaminated Wastewater* (ed. E.D.V. Hullebusch), 75–101. Cham: Springer International Publishing.

Sun, H.-J., Rathinasabapathi, B., Wu, B. et al. (2014). Arsenic and selenium toxicity and their interactive effects in humans. *Environment International* 69: 148–158.

Sun, J., Hu, C., Wu, B. et al. (2020). Improving ion rejection of graphene oxide conductive membranes by applying electric field. *Journal of Membrane Science* 604: 118077.

Teixeira, M.R., Rosa, M.J., and Nyström, M. (2005). The role of membrane charge on nanofiltration performance. *Journal of Membrane Science* 265 (1–2): 160–166.

Toft, P. (2003). *Background Document for Preparation of WHO Guidelines for Drinking water Quality.*

Trenner, J., Depken, C., Weber, T., and Breder, A. (2013). Direct oxidative allylic and vinylic amination of alkenes through selenium catalysis. *Angewandte Chemie International Edition* 52 (34): 8952–8956.

Ulbricht, M. (2006). Advanced functional polymer membranes. *Polymer* 47 (7): 2217–2262.

Ullrey, D.E. (1992). Basis for regulation of selenium supplements in animal diets. *Journal of Animal Science* 70 (12): 3922–3927.

Venerando, A., Magro, M., Baratella, D. et al. (2020). Biotechnological applications of nanostructured hybrids of polyamine carbon quantum dots and iron oxide nanoparticles. *Amino Acids* 52: 301–311.

Vinceti, M., Maraldi, T., Bergomi, M., and Malagoli, C. (2009). Risk of chronic low-dose selenium overexposure in humans: insights from epidemiology and biochemistry. *Reviews on Environmental Health* 24 (3): 231–248.

Vinceti, M., Solovyev, N., Mandrioli, J. et al. (2013a). Cerebrospinal fluid of newly diagnosed amyotrophic lateral sclerosis patients exhibits abnormal levels of selenium species including elevated selenite. *Neurotoxicology* 38: 25–32.

Vinceti, M., Crespi, C.M., Bonvicini, F. et al. (2013b). The need for a reassessment of the safe upper limit of selenium in drinking water. *The Science of the Total Environment* 443: 633–642.

Waitkins, G.R., Bearse, A.E., and Shutt, R. (1942). Industrial utilization of selenium and tellurium. *Industrial & Engineering Chemistry* 34 (8): 899–910.

Wang, S., Huang, X., and Elimelech, M. (2020a). Complexation between dissolved silica and alginate molecules: implications for reverse osmosis membrane fouling. *Journal of Membrane Science* 605: 118109.

Wang, L., Deng, N., Liang, Y. et al. (2020b). Metal-organic framework anchored sulfonated poly(ether sulfone) nanofibers as highly conductive channels for hybrid proton exchange membranes. *Journal of Power Sources* 450: 227592.

Wen, H. and Carignan, J. (2007). Reviews on atmospheric selenium: emissions speciation and fate. *Atmospheric Environment* 41: 7151–7165.

Winkel, L.H.E., Johnson, C.A., Lenz, M. et al. (2012). Environmental selenium research: from microscopic processes to global understanding. *Environmental Science & Technology* 46 (2): 571–579.

WHO (2011). *World Health Organization, Guidelines for Drinking-Water Quality*. Geneva, Switzerland: WHO Press.

Zhang, J., Nguyen, M.N., Li, Y. et al. (2020). Steroid hormone micropollutant removal from water with activated carbon fiber-ultrafiltration composite membranes. *Journal of Hazardous Materials* 391: 122020.

Zhao, D.L., Japip, S., Zhang, Y. et al. (2020). Emerging thin-film nanocomposite (TFN) membranes for reverse osmosis: a review. *Water Research* 173: 115557.

16

Intensifying Approaches for Removal of Selenium

Kailas L. Wasewar

Advance Separation and Analytical Laboratory (ASAL), Department of Chemical Engineering, Visvesvaraya National Institute of Technology (VNIT), Nagpur, Maharashtra, India

16.1 Introduction

Selenium is originated in metal sulfide ores and commercially it is obtained as a byproduct in the ores refining. Minerals with compounds having pure selenate or selenide are very rarely identified (Wikipedia 2019). Selenium (Se) is also observed as selenate (SeO_4^{2-}) and/or selenite (SeO_3^{2-}) in thermal power plant effluents, waste streams from smelting industry, and also from oil refineries' effluent streams (Lawson and Macy 1995). The major sources of Se are started with the basic mineral industries including mining, power industries (thermal, oil, gas) to agriculture sectors. Se has wide application in various industrial sectors, such as: catalysts, solar batteries, pigments, glass, and semiconductors (Kashiwa et al. 2000).

Selenium is one of the essential trace elements needed by the human body but it is toxic when consumed in excess. The intake of more than the upper tolerable limit of selenium of 400 µg/day may cause selenosis (NIH 2009). The limit of 400 µg/day (tolerable intake level) was decided based on the study performed in 1986 and further follow-up was taken by 1992 (PDARC 2000). The maximum safe limit for dietary Se intake was found to be ~800 µg/day against 400 µg/day which is to avoid an imbalance of nutrients (Wikipedia 2019; Yang and Zhou 1994). The chronic and acute toxicity was observed for the high amount of soluble selenium, selenate, and selenite and hence an excess amount of selenium must be removed (Haygarth 1994; Kashiwa et al. 2000).

In past few decades, due to health and environmental concerns, more attention has been given to contaminants' impact on the environment. High-value treatments of waste water are essential due to severe environmental strategies that compel treatment and recovery of more and more water for reuse. The major concern in developing the efficient water and waste water technologies are the requirement of energy and also space. Process intensification has been considered to tackle waste water technologies' issues. Process intensification is the development which can incorporate the sustainability, safer, cleaner,

Selenium Contamination in Water, First Edition. Edited by Pooja Devi, Pardeep Singh, Arindam Malakar, and Daniel Snow.

and energy-efficient manufacturing and processing units. The present chapter deals with the process intensification approaches for the removal of selenium. Conventional methods are also discussed and compared with new methods. The conventional and intensified methods include physical, chemical, and biological methods. A few of the conventional and intensified equipments and methods are: adsorption, advanced reduction processes (ARPs) assisted with ultraviolet (UV), aerobic wetland cells, algal – bacterial, algal assimilation, algal volatilization, biochemical reactor, biosorption, capacitive deionization, catalytic reduction, catalyzed reduction (cementation), cementation, chemical oxidation/reduction, chemical precipitation, coagulation, constructed wetlands, co-precipitation, elector-biochemical reactors, electrocoagulation, emulsion liquid membrane, enhanced evaporation system, enzymatic reduction, fluidized bed reactor, hydrogen-based membrane biofilm reactor, ion exchange, lime softening, mechanical evaporation/crystallization, media filtration, membrane biofilm reactor, membrane filtration, microbial treatment, microwave-assisted, moving bed biofilm reactor, moving bed biological reactors, nanofiltration, nanotechnology, packed bed reactor, permeable reactive barriers, photoreduction, phytoremediation, reverse osmosis, reverse osmosis (VSEP, vibratory shear enhanced process), upflow anaerobic sludge blanket (UASB) bioreactor, and zero-valent iron, etc. The basic concepts and their merits and demerits have been discussed with respect to application for selenium removal in various forms from a variety of streams.

16.2 Selenium

Selenium belongs to the essential metalloids and chalcogen group; conversely, it has a narrow range for essentiality and toxicity. In 1817, Swedish chemist Jöns Jacob Berzilius recognized selenium for the first time. Selenium falls in the category of a metalloid which has properties somewhere of the nonmetals and the metals. The selenium has been widely scattered in small quantities all over the world (CGWB 2014). Naturally selenium occurs in the Earth crust, in rocks, soils, and with the sulfides as the selenides in the metal ores of silver, iron, copper, and lead which are shown in Figure 16.1 (CGWB 2014; Fernandez-Martinez and Charlet 2009). Inorganic forms and organic forms are the naturally occurring forms of the selenium and these are mentioned in Figure 16.2. The selenite (SeO_3^{2-}), selenate (SeO_4^{2-}), selenide (Se^{2-}), and elementary selenium (Se0) in oxidation states are

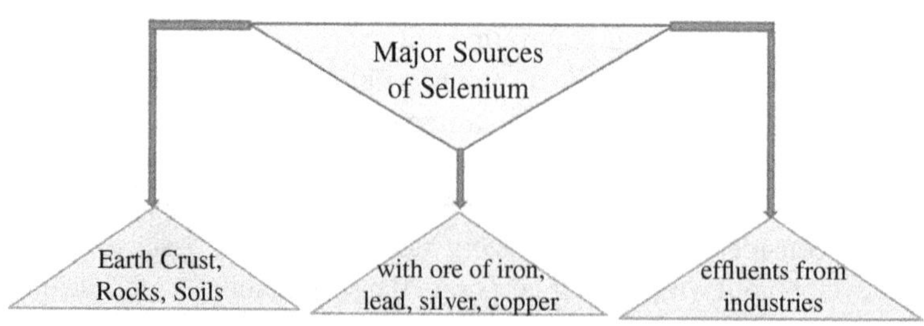

Figure 16.1 Major sources of selenium.

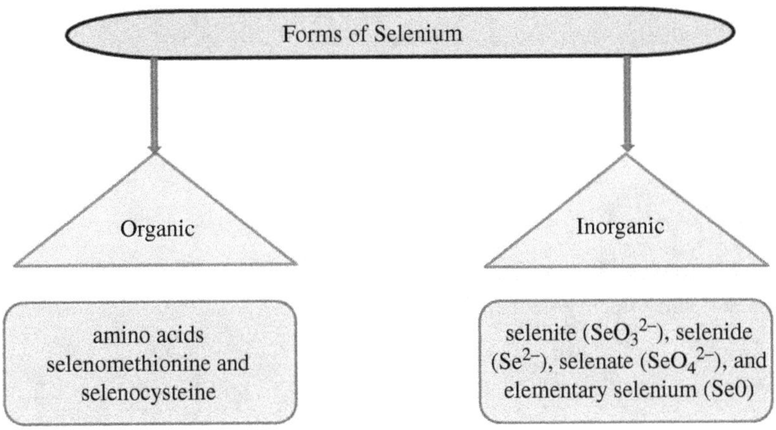

Figure 16.2 Forms of selenium.

the forms of inorganic selenium. The selenium occurs in organic forms as in selenocysteine and selenomethionine of amino acids.

In its compounds selenium is present in the form of oxidation states of −2, +2, +4, and +6. Elemental selenium is not poisonous but many compounds of selenium are enormously toxic (e.g. hydrogen selenide). Selenium forms selenides with most of the metals, e.g. cadmium selenide, aluminum selenide, and sodium selenide. Selenite makes complexes with components in common soil, such as aluminum and ferric sesquioxides. On the other side, selenium as selenate does not make complexes and it leaches to ground water from soil (CGWB 2014; Kapoor et al. 1995; Wikipedia 2019).

Reduction of inorganic selenites to elemental selenium can be taken place in reducing and acidic conditions and while selenates are formed at oxidizing and alkaline conditions. The In case of aerated and alkaline soils, selenium leaches in water in the oxidation form as selenates and selenites. Whereas, in the case of poorly aerated and reducing conditions soils, selenium and selenite are retained. Hence, alkaline soils are favorable for the uptake of selenium by plants. (CGWB 2014). Selenium subsist in various allotropic structures, having chemical properties similar to sulfur, as are selenides to sulfides Selenium in its all forms plays a very significant role for biological processes in human body. These includes DNA synthesis, maintaining antioxidant homeostasis, formation of thyroid hormones, reproductive cycles, for muscles to improve endurance and to slow down the aging process (Koudelkova et al. 2019). Selenium is not only essential to human beings but also for many living organisms. Controlling the sufficient amount of selenium in the human body is difficult because of the narrow range of selenium deficiency (<0.1 mg Se/kg of body weight/day) and toxicity (>mg Se/kg of body weight/day) (Pinochet et al. 1999).

Commercially, selenium is produced as a byproduct in the refining of ores. Selenium has wide application globally in industries including electronics, fungicides, fertilizers, anti-dandruff glass, shampoo, pigments, catalyst, semiconductors, and solar batteries (Figure 16.3). Therefore, selenium is found in a significant amount as soluble Se in the effluent of related industries (Kashiwa et al. 2000; Moore and Mahmoudkhani 2011).

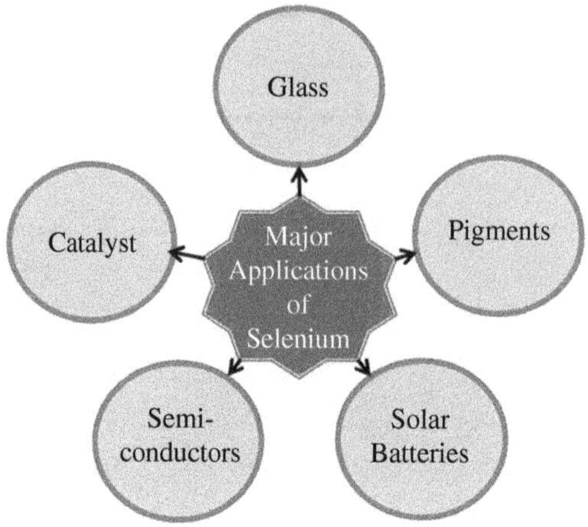

Figure 16.3 Major applications of selenium.

16.3 Selenium and Wastewater

Pure or elemental selenium (Se) and various forms of selenium are rising contaminants which are major concerns for environmental and human health. The contamination to the environment due to selenium and its compounds can be originated from smelting plants, mine drainage, thermoelectric power plants, radioactive waste, agricultural drainage water, oil refineries, residues from fossil fuel, coal products, thermal power stations, and metal ores (Dewberry et al. 2000; Lawson and Macy 1995; Lemly 2004; Stefaniak et al. 2018; Wasewar et al. 2009a). These major sources of selenium wastewater are presented in Figure 16.4.

Drinking water is a primary source of selenium entering the human body, with the maximum contaminant level in drinking water being 0.05 mg Se/l (Mandal et al. 2009; USEPA 2006). Ground water is contaminated by selenium due to various reasons and it has been given greater attention worldwide. A toxicity problem in livestock and wildlife arises at higher concentrations of selenium. For inland surface water, public sewers, and marine/coastal areas, the Central Pollution Control Board (CPCB), Delhi, India has also set a standard of 0.05 mg/l (CPCB 2004). According to EU and WHO standards, 0.01 ppm is the maximum permissible limit of selenium in drinking waters (EU and WHO 2008).

Mammals needed selenium essentially as a micronutrient in a very narrow concentration range. It naturally occurs in the environment, and beyond permissible limits its toxicity can be observed as carcinogenic. Intake of more than 5 mg/kg-day as per the US EPA may lead to gastrointestinal, cardiac, skin, and neurological disorders. The US EPA mentioned 50 ppb selenium in drinking water as a maximum contaminant level (Yamani et al. 2014). Higher intake (>400 μg/day) may lead to selenosis. Higher amounts of selenium as selenate and selenite can cause acute toxicity (Hamilton 2004). This prompts the development of suitable methods to control the selenium pollution in the environment for the protection of public health and wildlife. In general, selenium is found in two forms for bioavailability and bioaccumulation. Untreated selenium is bioaccumulated in soil and water, which poses an alarming problem for aquatic life as hydrogen selenide gas, which is extremely corrosive and toxic, is generated at acidic conditions from selenium (Moore and

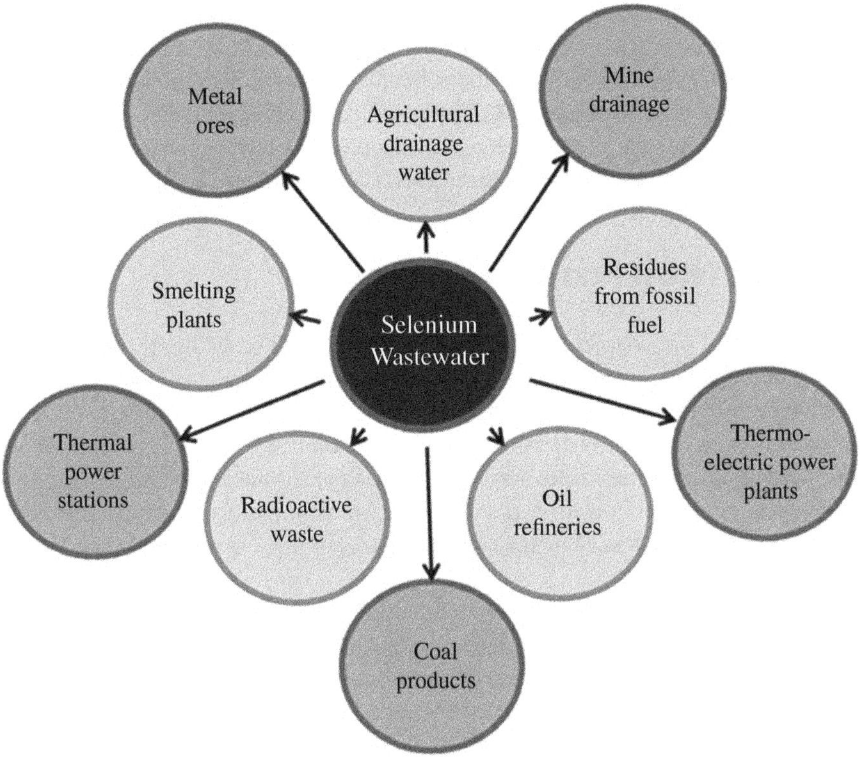

Figure 16.4 Major sources of selenium wastewater.

Mahmoudkhani 2011). Selenium is essential and also toxic to animals and human beings, having a narrow range for essentiality and toxicity (Burau 1985; Christensen et al. 1989; Conde and Alaejos 1997; Fernandez-Martinez and Charlet 2009; Fordyce 2007; Gissel-Nielsen et al. 1984; Jordan et al. 2013).

The intake or exposure of excess amounts of selenium in human body is mainly through food and water. The various health issues associated with selenium are: headache, nausea, teeth, and nail loss (CGWB 2014). The selenium toxicity scenario in India was found in northwestern regions on sugar cane and wheat plant as chlorosis and selenosis for humans and animals (CGWB 2014). The risks to wild animals are also observed for high selenium concentrations in water bodies in forest areas. These high concentrations of selenium are embryotoxic and teratogenic (Gonzalez et al. 2012; Hoffman 2002; Strawn et al. 2002).

Several occurrences of chronic toxicity of selenium in animals were observed in India. In Karnel, a northwest region of India, domestic buffaloes were found with selenium (Reill 1996). The alkaline soils responsible for keeping the high selenium levels are likely the region where leached selenium reached rice plants. The rice plant straw is the main fodder of animals, which may be the cause of the selenium toxicity. A similar kind of problem with selenium was observed in at least nine western US states (EPA 2001). Hence it is necessary to remove the selenium and its compounds from waste water, due to the acute and chronic toxicity of selenium toward living things (Haygarth 1994; Kashiwa et al. 2000).

Many methods are capable of removing selenium from wastewater, but still the reliable removal of selenium remains a challenge. Treatment of selenium by traditional/conventional methods requires primary, tertiary, and residuals treatment. The efficacy of selenium treatment is highly dependent on the type of selenium compounds and their concentration in wastewater and other parameters. Hence, a detailed wastewater characterization is important to select potential treatment alternatives (Dale and Liner 2016).

16.4 Process Intensification

In terms of GDP and employment, the process industry contributes significantly toward the economy of many countries. The manufacturing and process industries are major consumers of energy and resources (Górak and Stankiewicz 2011). In the process industry, for new and existing processes, further improvement in economical factors and environmental issues are needed. Process intensification is extensively accountable for process improvements and sustainable processes. Process intensification has directly related to society benefits such as energy saving, less CO_2 emission, clean processing, human health, and safety, etc. (de Vries 2008).

Process intensification has been gaining in attention and importance due to the strong potential of process intensification to develop innovative and more sustainable process design alternatives (Babi et al. 2016). Many definitions are available and summarized in the literature (Babi et al. 2016; Keil 2007; Stankiewicz and Moulijn 2000; Wasewar et al. 2020). Typically process intensification may be the improvement and/or development of a new technology/process which may be technically feasible, eco-friendly, and economical when compared to the existing one. Process intensification must be safer, cleaner, smaller, and cheaper. The number of scientist have defined the process intensification and few of these are summarized in Figure 16.5 (Ramshaw, 1983, 1995; Heggs, 1983; Stankiewicz and Moulijin, 2000; Dautzenberg and Mukherjee, 2001; Tsouris and Porcelli, 2003; Grunewald and Agar, 2004; Croscuoli and Drioli, 2007; Becht et al., 2009; Lutze etal., 2010).

Process intensification is of great interest not only in process industry but also in various allied industries for various reasons: novel or enhanced product, improved chemistry, enhanced safety, improved processing, energy and environment benefit, capital cost reduction, low inventories, enhanced corporate image, and value to customers (Tsouris and Porcelli 2003). The typical taxonomy is given in Figure 16.6. Processes intensification is mainly classified as equipments and methods. Sustainable development targeting process intensification must consider the know-how to develop the technologies to increase the renewable sources, lessen the waste generation, use less material and energy, and use recyclable or biodegradable products (Jenck et al. 2004).

Lutze et al. (2010) stated the four principles which are associated with process intensification: enhancements accomplished by the integration of functions, integration of phenomena, and integration of operations. Reay et al. (2008, 2013) discussed process intensification with various equipments and methods in detail. Process intensification is classified broadly into equipments and methods. Process-intensifying equipments are typified by optimizing the designs by considering mass, heat, and momentum transfer (e.g. spinning disk reactors, centrifugal contactors, static mixers, micro-channel reactors). Process-intensifying methods

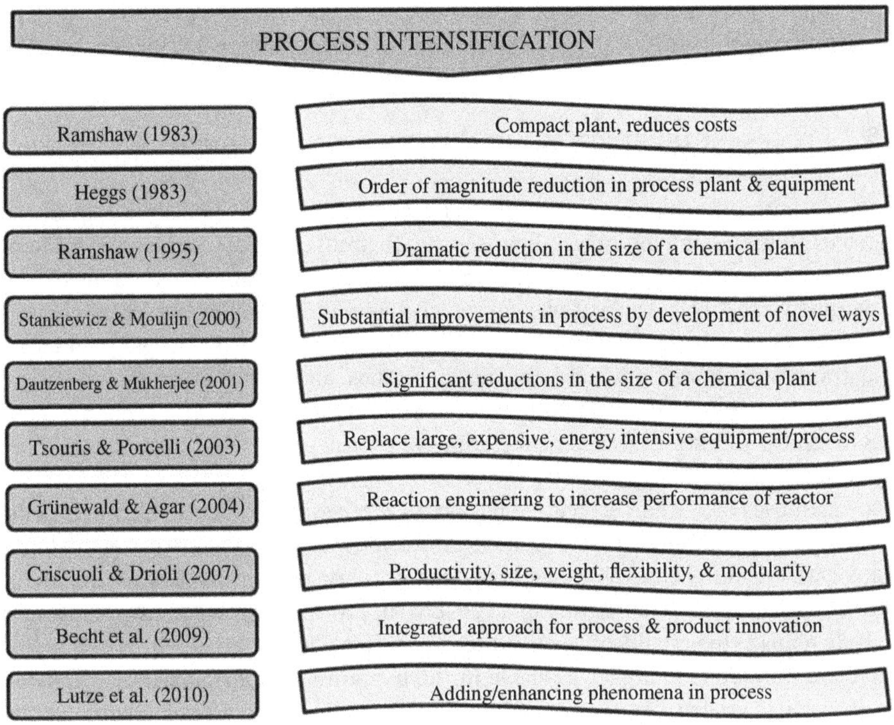

Figure 16.5 Glimpses of defining process intensification.

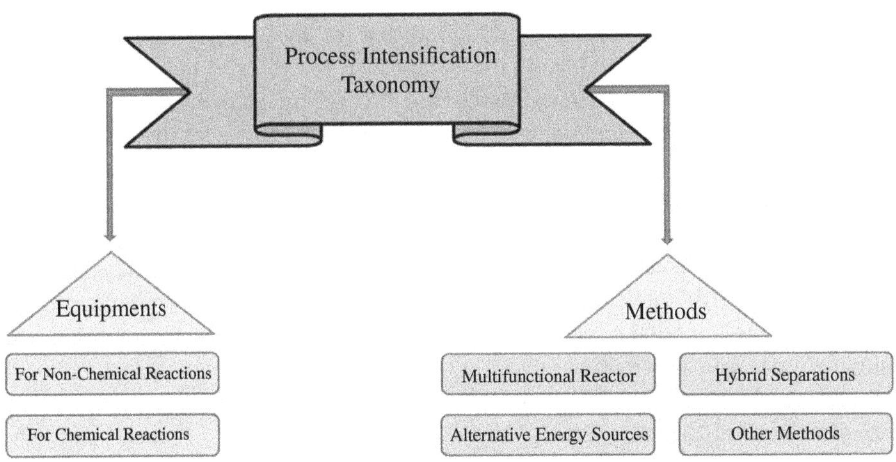

Figure 16.6 Classification of process intensification.

combine or integrate various operations/steps of process (e.g. reactive separations, fuel cells, adsorptive distillation, membrane absorption, reactive distillation) and use alternative energy sources (e.g. electric fields, centrifugal fields, ultrasound, microwaves) (QTR 2015). Process-intensifying approaches for equipment and methods can be applied to various industrial sectors, including biofuel production, oil and gas extraction, power generation, chemical manufacturing, petroleum refining, and water and waste water treatment (Kim et al. 2017).

16.5 Process Intensification in Wastewater Treatment

The process intensification in water and wastewater treatment technologies requires an integrated multiscale and multidisciplinary approach, from chemical engineering to chemistry, material science, biology, electronics, microbiology, and advanced computing (Gόrak and Stankiewicz 2011). Applications for process intensification technologies crosscut energy-intensive industries, giving the opportunity to have space in oil and gas production, petroleum refining, forest products, chemicals, plastics, and food industries, etc. Process intensification may provide the solutions to environmental, energy, and economic challenges in almost all industrial sectors, including water and waste water treatment and carbon capture (QTR 2015).

Many countries face water scarcity issues because of severe droughts and water pollution. Process intensification can be employed in various water and waste water treatment technologies, including anaerobic membrane digesters, wastewater and sewage treatment, and desalination, to more efficiently and reliably supply drinking water, as well as more effectively manage the quality of water sources (Kim et al. 2017).

There are numerous examples available in the literature for the application of process intensification in wastewater treatment. High gravity rotating packed bed has been used for the production of HCIO, where almost a one-third reduction has taken place in wastewater and hence further load on the wastewater treatment as compared to same capacity traditional plant (de Vries 2008).

Leyva-Díaz et al. (2015) studied a hybrid bioprocess reactor by combining moving bed biofilm reactor with advanced oxidation for waste water treatment. A membrane bioreactor and a moving bed biofilm reactor-advance oxidation are intensified unit operation and hybrid process respectively. Membrane technologies are listed as one of the milestones in process intensification for clean water from various water sources, including wastewater, targeting health, living, food and agriculture (Gόrak and Stankiewicz 2011). Vibratory shear enhanced membrane filtration can be used for waste water treatment (Drewes 2009). Reactive media filtration can be used for nutrient removal from waste waters (EPA 2013). Anderson et al. (2010) studied an electrochemical means to save energy by capacitive deionization for desalination of water. Reverse osmosis is used for cleaning of groundwater, surface water, and waste water (Keil 2018). Waste water from textile production can also be cleaned by a combination of ultratiltration/microfiltration and nanofiltration or reverse osmosis (Keil 2018).

Anaerobic membrane bioreactor technology is attractive for wastewater treatment which requires a smaller amount of energy, and produces effluent with less pollutants compared to other processes (QTR 2015). Commercially successful membrane reactor applications are membrane bioreactors (MBR). The variety of membrane bioreactors are considered largely for waste water treatment of various industries (Fontananova and Drioli 2014; Kemmere and Keurentjes 2001). In membrane bioreactors, treatment of waste water performed by enzymatic or cell catalyzed reactions facilitates the purification and further concentration of the product streams in a single unit (Tian et al. 2018). Apart from waste water treatment, MBRs have also application in various industries, such as in the production of amino acids, plant metabolites, antibiotics, foods, anticancer drugs, optically pure enantiomers, vitamins, proteins, anti-inflammatories, fine chemicals, isomers, biofuels, etc. (Drioli et al. 2011). Pertraction (membrane extraction based on membrane and extractant) can be used to treat the wastewater or fermentation broths by removing pollutants or products respectively (Tian et al. 2018).

For the extraction of copper and a wide range of metals from wastewater, chelating agent solutions are effective (Yu and Klarup 1994). These scavenging agents extract metal by forming a metal-chelate complex which can further be removed from the solution (Goel et al. 2009). An electric field in an electrohydrodynamic micromixing reactor is used for water treatment (Tsouris et al. 2003). Oscillatory flow reactors are used for the oxidation of waste water with ozone (Reis and Li 2014) and ethylene oligomerization (Shaik and Xu 2017). Fischer and Behn (2016) patented reactive distillation for waste water treatment. For waste water and water treatments, the various techniques of ultrasonics have been employed (Gogate and Pandit 2004). Microwaves have been used for waste water cleanup (Keil 2018). Ultrasound in environmental area can be used for waste water treatment through degradation of bacteria, treatment of waste water by sonophotocatalytic, and ultrasound and UV radiation assisted advanced oxidation processes for waste water treatment (Gogate and Pandit 2004; Mahamuni and Adenuyi 2010; Neis 2015).

16.6 Conventional and Intensified Ways for Selenium Removal

In further sections, various methods including conventional with intensified approaches used in these methods have been discussed. The majority of the studies are on biological methods and membrane technologies. These methods are summarized in Figure 16.7.

16.6.1 Biological Treatment

Biological treatment methods are the most popular technologies for waste water treatment. Biological methods are developing technologies for selenium removal from water and waste water. In these technologies bacteria is used to transfer selenium and its compounds from aqueous solutions to elemental selenium (Holmes and Gu 2016).

Biological reduction of selenium can be achieved by many autotrophic and heterotrophic bacteria. Biological transformation by most of these bacteria is slow. In the case of mining wastewaters, these biological transformations may be even slower. Bacteria must be selective to transform selenite and selenate in non-toxic forms. Facultative anaerobes or heterotrophic anaerobes consume carbon source materials such as methanol, sugars, ethanol, or acetic acid for the conversion of selenite and/or selenate to selenium in elemental form.

Selenium removal can be carried out from various wastewaters by means of various biological methods, which include microbial and fungal volatilization, and phytoremediation,

Figure 16.7 Various methods for treatment of selenium wastewater.

bioreactors, constructed wetlands, microbial reduction, fixed-film bioreactors, packed bed, fluidized bed, moving bed biofilm reactors, etc. (Dale and Liner 2016; Holmes and Gu 2016).

A few of the studies on biological treatment using various systems for selenium removal discussed in literature are: an alga-bacterial, fluidized bed reactor as a sludge blanket reactor and a bio-augmented reactor with bacteria to reduce selenate (Cantafio et al. 1996; Gerhardt et al. 1991; Macy et al. 1993; Squires et al. 1989).

16.6.1.1 Microbial Reduction

Various bacteria have been tested, such as *Pseudomonas stutzeri*, *Bacterial inoculum*, and *Clostridium bacteria* (Adams et al. 1996; Koren et al. 1992). He and Yao (2010) performed the experiments using two *Anaeromyxobacter dehalogenans* strains to transform toxic Se(IV) to less toxic insoluble Se(0). Most of the studies are on the development of waste water treatment methods for low concentrations (0.3–0.5 mg/l) of soluble Se (Kashiwa et al. 2000). On the other hand, in some cases, waste water with high concentrations of soluble Se was studied. Kashiwa et al. (2000) performed the experimental study to treat 4.76 mg Se/kg using *Bacillus sp. SF-l* to reduce selenate (Fujita et al. 1997). These can efficiently transform 20 mM of selenate to selenite and 2 mM of selenite to elemental selenium with suitable carbon source material anaerobically.

Development of new bacteria having more compatibility for the transformation of both selenite and selenate to elemental selenium with comparable reduction rates is the challenge. This may be considered as a process intensification of the development of new bacteria giving a higher removal rate and a better survival rate.

16.6.1.2 Microbial Volatilization

Selenium volatilization by microbes is another biological treatment method for selenium removal from wastewaters (Holmes and Gu 2016). SeO_4^{2-}, SeO_3^{2-}, and Se0 are converted aerobically into $(CH_3)_2Se_2$ and $(CH_3)_2Se$; various volatile compounds of selenium in gaseous phase by various bacteria strains (i.e. *P. stutzeri NT-I*) (Kagami et al. 2013; Kuroda et al. 2011; Sors et al. 2005). This method has the potential to recapture the selenium for sustainable utilization, which is an added advantage compared to other biological methods (de Souza et al. 1999; Holmes and Gu 2016). The removal or separation of selenium from various waste water streams by this method and production of selenium for utilization is the sustainable process intensification. This intensifying approach gives relative savings in many respects, including energy and operating cost.

16.6.1.3 Enzymatic Reduction

Enzymes can also be considered as a biological treatment for waste water. These are proteins which are used to digest the molecules into smaller units and significantly increase the rate of biochemical reactions (NAMC 2010). Experiments were performed on the enzymatic reduction of selenium (MSE 2001). Selenium content of 23 100 μg/l in mining wastewater was reduced to <100 μg/l by enzymatic reduction (MSE 2001). Enzymatic reduction for selenium removal from wastewater is an intensified approach as it has the added advantage that it does not need to add any nutrients.

16.6.1.4 Enhanced In-Situ Microbial Reduction

Organic amendments, nutrients, or inoculated micro-organisms may be used to enhance the microbial reduction as an intensified approach. The addition of nutrients, organic amendments, or inoculated micro-organisms can enhance the microbial reduction of oxidized forms of selenium (NAMC 2010). It can take place naturally or be improved with addition of amendments to enhance the removal rate of selenium. Martin et al. (2009) treated the dissolved from 460 to <10 μg/l.

16.6.1.5 Algal Treatment

Algal bacterial treatment can be employed for selenium removal, volatilization, and assimilation (NAMC 2010). In algal bacterial selenium removal, the growth of bacteria is infused with a supply of algae as a renewable food source material (NSMP 2007). The growth of the algae is enhanced by addition of carbon dioxide in ponds (Quinn et al. 2000). This will intensify the production of algae, which may then further be used for the reduction of selenium. There was no effect of salinity on selenium removal for any concentration of salinity (NAMC 2010; Takata 2003). Algal volatilization can be observed by biomethylation and conversion of selenium from waste water into the form of gases (Frankenberger et al. 2004). Golder (2009a, b). Microalgae are accountable for transformation of selenite into various volatile compounds: selenium ions, alkylselenides, and proteinaceous selenomethionine variety. The dimethylselenide is the main product of selenium volatilization (Frankenberger et al. 2004).

16.6.1.6 Microbial Fuel Cell

Multi-functioning in equipment is the process intensification which can also be achieved in a microbial fuel cell for selenium removal from waste water. Microbial fuel cells have been used for the removal of selenite from different waste water, producing electricity and removing selenium in a single unit simultaneously (Holmes and Gu 2016). In microbial fuel cells, microbes oxidize the organic substrates (in this case glucose), and this reduces selenite to elemental selenium with simultaneous generation of electricity. Selenium removal was obtained as 99% for 50–200 mg/l Se, indicating the prospective of microbial fuel cell for removal of Se from waste water (Catal et al. 2009).

16.6.1.7 Packed Bioreactor

Packed bioreactor is common equipment used for biological treatment. Intensification can be achieved in this type of bioreactor by using improved and novel packing material for bacterial attachment. Ji and Wang (2019) used an alum-impregnated activated alumina for packing in bioreactor for bacterial attachment and growth. The packing material is cultured with *Shigella fergusonii* strain TB42616 to remove Se(IV) during Se(VI) reduction. Approximately, 70% selenium has been removed with a hydraulic retention time of 3.1 days with no Se(IV) accumulation observed, indicating that Se(IV) removal is enhanced by the adsorption of the alum-impregnated activated alumina. By using novel packing material, the removal rate is intensified and 10% more selenium is removed by alum-impregnated activated alumina instead of unmodified packing (Ji and Wang 2019).

16.6.1.8 Fluidized Bed Reactor

In a fluidized bed reactor, the waste water passes through the fluidized bed having solid media (sand or granular activated carbon) with enough velocity to maintain fluidization for attached biological growth or biofilm. Fluidized bed reactors can be operated aerobically or anaerobically (USEPA 2014).

In a fluidized bed reactor, biological treatment of waste water is biological process of fixed-film or attached growth. The biological treatment in fluidized bed reactor under anaerobic conditions for removal of selenium is the transformation of selenate and selenite to elemental selenium by reduction (NAMC 2010). Brienne et al. (2009) successfully demonstrated removal of selenium using a fluidized bed reactor operating for 239 days in continuous operation. In this reactor, intensification is in the flow pattern achieved in the fluidized bed reactor, which is more favorable for attached growth of microbes and hence more removal of selenium as compared to conventional bioreactors for biological treatment of wastewater.

16.6.1.9 Hydrogen-Based Membrane Biofilm Reactor

Hydrogen may be supplied to a contaminant reducing biofilm by the membrane biofilm reactor. It is an effective and simple way compared to a conventional bioreactor. Supply of H_2 via membranes directly is more capable than using an organic compound as hydrogen donor. The biofilm utilizes the stoichiometric H_2 for reduction of the contaminants in the waste water (Van Ginkel et al. 2011). This may achieve the target of less cost and prevent the leakage of oxygen to the waste water, which achieves the objective of process intensification.

Chung et al. (2006) performed the selenium removal by hydrogen-based membrane biofilm reactor. The removal of selenium was sufficient in effluent to acceptable level (NAMC 2010). The hydrogen-based membrane biofilm reactor has bundles of membrane for bubble-less hydrogen transfer to micro-organisms biofilm which is cultivated by oxidizing the H_2 and reducing soluble electron acceptors, including $NO^-{}_3$, $NO^-{}_2$, and $SeO^{2-}{}_4$ (Chung et al. 2007; Van Ginkel et al. 2008, 2010). Van Ginkel et al. (2011) performed experimental investigation for selenate removal from flue gas desulfurization brine using a hydrogen-based membrane biofilm reactor. Selenate is removed to <50 ppb, which is acceptable for release into the environment.

16.6.1.10 Moving Bed Biofilm Reactor

A moving bed biofilm reactor is a biological method for selenium removal which operates for 20 years with the same support material, unlike other biological methods where granular activated carbon is replaced regularly (Dale and Liner 2016). In this reactor, media (prepared from polyethylene) having a large surface area for the growth of biofilm can be employed. In a moving bed biofilm reactor, back washing is not required and this also tolerates the high concentrations of suspended material and hence is not subject to blockage. A moving bed biofilm reactor is generally energy-efficient, which is a feature of intensifying approach. Also, a moving bed biofilm reactor can be designed for higher capacity of the same size compared to other bioreactors (Dale and Liner 2016). Dale and Liner (2016) performed a study on a moving bed biofilm reactor for removal of selenium from waste water of coal mines and power plant flue gas desulfurization. Less than 5 ppb of total selenium were obtained after chemical treatment (Dale and Liner 2016).

16.6.1.11 Biochemical Reactor

Biochemical reactors employ micro-organisms for the removal of contaminants from mining industry wastewater. It can be obtained in different designs such as open or buried ponds, in trenches between mine waste and a surface water body or in tanks (ITRC 2010). A biochemical reactor effectively raises pH and separate metals, sulfate, and selenium from waste water (USEPA 2014). Different media for gravel pit seepage (biochemical reactor) resulted in 98% selenium removal with <5 mg/l (CH2M Hill 2010).

16.6.2 Constructed Wetlands

Constructed wetlands are a kind of complex ecosystem where the biological and physical components interact and allow to remove many contaminants from water by means of mechanical and biogeochemical filter. Constructed wetlands are nothing but a biological and physical complex. In constructed wetlands, the chemical nature of contaminants is changed to less toxic by immobilizing and/or transforming contaminants. The type of vegetation and micro-organisms plays significant role in the performance of constructed wetlands. The efficacy of constructed wetlands varies from place to place. The use of constructed wetlands has been continuously increasing for removal of contaminant from wastewater (Brown and Reed 1994; Cooper and Findlater 1990; Kadlec and Knight 1996).

Hansen et al. (1998) performed the removal of Se from oil refinery wastewater, mainly Se(IV), by constructed wetlands and almost 89% of the Se was removed for a period of 16 weeks. Biological volatilization accounted for 10–30% of the total Se removed and immobilization into sediments and plant tissues removed the remainder. Se was transformed by reduction to insoluble Se(0) in constructed wetlands and incorporated into organic materials (Tokunaga et al. 1996; Zawislanski and Zavarin 1996; Zhang and Moore 1996, 1997, 1998).

Gao et al. (2003) performed experimental investigation for a period of three years 1996–1999 by flow through experimental wetland system with 10 cells (15×76 m) for removal of selenium (Se) from agricultural drainage water in the Tulare Lake at Corcoran, California, USA. In 1999, the selenium was reduced to 65% from 32% using these wetland cells. Lin and Terry (2003) studied the quantitative significance of biological volatilization in selenium removal from agricultural drainage water using constructed wetlands. They observed that the vegetated constructed wetlands reduced Se significantly from the inflow drainage water to an average of 69.2% of the total Se.

Constructed wetlands can be successfully employed to remove selenium from wastewater. Constructed wetlands can generate a toxic hazard for wildlife due to bioaccumulation of selenium in wetland plants and animals (Lemley and Ohlendorf 2002). Further, intensification in constructed wetlands may be done by sustainable design and working toward increasing in performance in less land space and a more cost-effective way.

16.6.3 Phytoremediation

The removal of selenium may be performed by phytoremediation, which is a passive method in constructed wetlands (Holmes and Gu 2016). In phytoremediation, plant uptake of selenium and its volatilization by cellular pathways is observed as microbial volatilization (Lin and Terry 2003). Selenium and its forms both selenite and selenate have been significantly removed in the wetland, which needs two years for the completion of treatment and remediation (Gao et al. 2000). Phytoremediation is not a suitable option for waste water treatment as there was significant investment. The similar results for effectiveness were observed for fungal systems to remove the selenium for low pH wastewaters (Espinosa-Ortiz et al. 2015). In phytoremediation intensification is achieved as removal and volatilization selenium simultaneously.

16.6.4 Adsorption

Various treatment methods have been developed for the removal of selenium, such as bacterial reduction, membrane filtration, chemical reduction, reverse osmosis, and solar ponds, etc. (Gonzalez et al. 2012). Adsorption is relatively cost-effective compared to other methods. Adsorption has been widely studied as one of the most cost-effective methods compared to other techniques due to the sustainability of the regeneration and the reuse of the adsorbents (Shafey, 2007; Wang and Lo 2009). A lot of studies have been available on adsorption as it removes multi-pollutants simultaneously (De et al. 2018a, b; Dhongde et al. 2017; Kumar et al. 2012, 2015; Madan et al. 2016a, b, c; Madan and Wasewar 2017; Madan et al. 2017; Madan and Wasewar 2018; Madan et al. 2019; Rai et al. 2013; Rajoriya et al. 2007; Singh et al. 2016, 2020; Uslu et al. 2014; Wasewar et al. 2007, 2008a, b, 2009a, b, c, d; Wasewar 2010; Wasewar et al. 2010). Various metals and other contaminants can be removed from water. Absorbents may include ferric hydroxides, aluminum hydroxides, alumina, activated carbon, and clays and other low-cost and novel adsorbents. Adsorbent must be cost competitive and have ease for regeneration and reuse. Therefore, novel adsorbents and approaches are required for the selenium containing waste water.

Removal of selenium from water by adsorption is cost-effective. The selenium in forms such as selenate, selenite, and selenocyanate, etc. from waste water get condensed and concentrated on the surface of the adsorbent, which represents the adsorption having a mechanism of physical, chemical, and ion exchange (NAMC 2010).

Process intensification in adsorptive removal of selenium may be achieved at equipment level and adsorbent level. Development of novel equipment for better contacting which reduce the operating cost due to higher adsorption. Also new materials may be low-cost or having higher surface area with more active sites and further it will be easily regenerated.

16.6.4.1 Metal Oxides

Laurie and Chao (1990) investigated the adsorptions of selenium using amorphous manganese dioxide and iron oxyhydroxide. They have compared and modeled the adsorption of selenium and other anions on a neutral to alkaline surface (amorphous iron oxyhydroxide) and an acidic surface (manganese dioxide). The performance of aluminum oxide adsorbent for the adsorption of selenium was investigated by Derek (2006). Trussell et al. (1980) studied aluminum hydroxide for selenium adsorption. Selenium (IV) adsorbed effectively over the pH range of 3–7.

Another absorbent, ferric oxyhydroxide, used to impregnate peat was employed for selenium removal (Chamberlin 1996; Corwin et al. 1994; Jeffers et al. 1991). To prepare this adsorbent, ferric oxyhydroxide and peat were mixed into beads of polysulfone resin and the resin along with Se of 1700 mg/l were loaded together with solutions containing both Se(VI) and Se(IV). The removal rate of Se(IV) can be >95%, while that of Se(VI) is about 80%. Absorbing selenium onto lanthanum oxide substrates is also investigated by Adutwum (1995). The adsorptive removal of selenium was effective for both Se(VI) and Se(IV).

16.6.4.2 Activated Adsorbents

Activated materials as adsorbents (carbon and alumina) are commonly used for the removal of organic and inorganic compounds from aqueous solutions (Yamani et al. 2014). Adsorption of selenium by activated alumina was performed by Hu et al. (2006). The poor affinity for selenium was observed for adsorptive separation using activated carbon compared to activated alumina for removal of selenium (Sharma et al. 2009).

Activated carbon for adsorption is the most universal absorbent in water treatment and also for removal of selenium from waste water of mining industries. Activated carbon is not effective as <4% removal of Se(IV) or Se(VI) at 30–100 μg/l was observed for up to 100 mg/l dosages (Sorg and Logsdon 1978). Wasewar et al. (2009a, b) used granular and powdered activated carbon for adsorption of selenium with >90% removal.

16.6.4.3 Low-Cost Adsorbent

Low-cost adsorbents can be prepared from the available biomass. El-Shafey (2007) studied the efficacy of peanut shell-based carbonaceous adsorbent prepared by sulfuric acid treatment for Se(IV) removal from simulated aqueous solution. The effect of various parameters was investigated, including adsorption time, solution pH, Se(IV) concentration in feed, operating temperature, adsorbent dosage, and whether its condition was wet or dry. The selenium removal of 63% was reported for 25 mg/l selenide solution.

Wasewar et al. (2009a, b) performed an experimental investigation on selenium removal using bagasse fly ash. They studied kinetics, equilibrium, thermodynamics, and the effect of various parameters and more than 95% removal rate was observed.

16.6.4.4 Soil as Adsorbent

Dhillon (1999) performed experiments using Indian soils for the removal of selenium. It was observed that more Se was adsorbed by acidic soils as compared to alkaline soils. The exception for the soil was the alkaline seleniferous soil from Barwa in northwestern India. The adsorption data were well fitted in Freundlich and Langmuir isotherm which represented the multilayer adsorption.

16.6.4.5 Layered Double Hydroxides

Layered double hydroxides (LDHs) are promising adsorbents for many oxoanions, anions, certain cations, and radionuclides (Bo et al. 2016; Cavani et al. 1991; Goh et al. 2008; Liang et al. 2013). LDHs are also identified as compounds like hydrotalcite. The structure of these compounds are: $[M^{II}{}_{1-x}M^{III}{}_{x}(OH)_2]^{x+}[A^{n-}{}_{x/n}]^{x-}\cdot mH_2O$. It has divalent and trivalent metals as $M^{II} = (Mg^{2+}, Zn^{2+}, Co^{2+}, Cu^{2+})$ and $M^{III} = (Al^{3+}, Fe^{3+}, Ni^{3+}$, etc.) respectively. These metals form brucite-type layers and also have the A^{n-} (an exchangeable anion with a valence of n) with $0.20 \leq x \leq 0.33$ (Goh et al. 2008; Mills et al. 2012).

Based on several studies it was observed that LDH-based adsorbents are intensified for removal of selenium (Chubar 2014; Constantino et al. 2017; Koilraj et al. 2018; Tian et al. 2017). The adsorption mechanism for adsorption of selenium using LDH is ion exchange and chemical sorption. In the case of selenite, it is a combination of ion exchange and some chemisorptions. For selenate, it is by ion exchange only (Constantino et al. 2017).

Youwen et al. (2001) studied adsorption of selenium using Mg■Al and Zn■Al LDHs. Li et al. (2019) evaluated granular mixed metal oxide of 61.7%w/w MgO and 38.3%w/w Al_2O_3 LDH for removal of selenium from natural gas power plant wastewaters.

16.6.4.6 Ferrihydrite Adsorbent

Ferrihydrite-based adsorbent may also be a promising material for adsorptive removal of selenium. Ferrihydrite adsorption has two chemical steps: one is addition of ferric chloride or ferric sulfate in the waste water with rapid mixing and pH adjustment, and the second is the addition of polymer or coagulant and flocculants to obtain a ferrihydrite solid. These solids are purely soluble crystalline form of ferric hydroxide (NAMC 2010; Twidwell et al. 1999). Adsorption of selenium may result on ferrihydrite floc or solids matrix, depending on the oxidation state of the selenium and pH. The selenium in the form selenite is adsorbed on the ferrihydrite floc strongly and selenate is adsorbed loosely (Merrill et al. 1986).

16.6.4.7 Maghemite Adsorbent

Jordan et al. (2013) studied the sorption properties of maghemite (c-Fe_2O_3) toward selenium(VI) on the macroscopic and the molecular level. The adsorption process is very fast and sensitive for higher pH and ionic strength, which can lead to a decrease in removal of selenium(VI) by adsorption.

16.6.4.8 Conjugate Adsorbent

The conjugate material is the combination of organic and inorganic materials which can be the promising adsorbent for the removal of selenium. The conjugate adsorbents have higher surface area, large pore volumes, and are free of resistance to internal diffusion, hence these are suitable for adsorption. The ligand immobilized as a functional group for conjugate adsorbent was employed for Se(IV) removal from wastewater (Awual et al. 2015). The organic ligand of (3-(3-(methoxycarbonyl) benzylidene) hydrazinyl) benzoic acid was synthesized by Awual et al. (2015). It was then immobilized onto the mesoporous silica, obtaining a high ratio of surface area to volume with pores of uniform nanostructures, and used as an adsorbent for removal of selenium.

16.6.4.9 Graphene Oxides

Koudelkova et al. (2019) employed iron oxide, graphene oxides, and their composites for the removal of Se(IV) from aqueous phase. The graphene oxides are synthesized by simplified Hummer's method. The composite material with modification in graphene oxide particles by iron oxide particles was found as the most capable material to remove Se(IV) with >90% removal efficiency and maximum adsorption capacity of 18.69 mg/g at 2.0 pH.

16.6.4.10 Ion Exchange

Ion exchange resins are considered for contaminants removal from water and wastewater. In this process, an undesirable ion present in waste waters is exchanged with desirable ions by passing water through ion exchange resins. In waste water treatment, ion exchange plays a significant role for desalination and for removal of hardness, alkalinity, ammonia, radioactive waste, and metal (NAMC 2010). Anionic resins of weak base and strong base are used for removal of selenite and selenate ions from waste water (Patterson 1985; Twidwell et al. 1999).

16.6.4.11 In-Situ Solidification and Chemisorption

Moore and Mahmoudkhani (2011) presented an intensifying and innovative approach for removal of selenium from wastewater using in-situ solidification and chemisorptions. This method used a combination of physical and chemical treatment methods. An adsorbent was prepared by chemical transformation of silica-based polymeric material providing active sites for the chemisorption of selenate. The total selenium was removed was >99% from aqueous phase of waste water (Moore and Mahmoudkhani 2011).

16.6.4.12 Nanoadsorbents

Nanomaterials of various types can be employed as adsorbent for removal of selenium (Holmes and Gu 2016). Nanomaterials are generally <100 nm at least in one dimension. Nanomaterials have distinctive properties depending on size, and many nanomaterials can be employed for the removal of selenium from wastewater (Holmes and Gu 2016). The most important characteristics of these nanoparticles are: large surface area, great catalytic potential, high reactivity, and high active sites to remove the oxyanions of selenium (Ali 2012).

Numerous nano-phenomena are accountable for improvements in nanoparticles and hence in preparing the catalysts, adsorbents, and reductants, etc. as compared to micro and macro particles. The nanomaterials used as an adsorbent are capable of carrying out selenium removal to very low level which may not be possible using any other material or method. The major factors are the high surface area available for adsorption, the activity of adsorbent, high reactivity, selectivity, and the sustainable treatment capability for selenium removal from wastewater (Holmes and Gu 2016).

In this section, many promising nanotechnologies for the removal of selenium from various streams of waste water are discussed. The two main ways of removing selenium using nanomaterials are non-destructive and destructive – adsorption and reduction, respectively. Holmes and Gu (2016) presented a detailed review on nanomaterials for the removal of selenium from wastewater. A glimpse of their work is highlighted in following subsections.

16.6.4.12.1 Metal Oxide/Hydroxide Nanomaterial For the last few decades, metal oxides have been attracting increasing interest for a wide range of applications due to their being benign, less expensive, and easily available (Holmes and Gu 2016; Miller and Zimmerman 2010; Sharma et al. 2009; Yamani et al. 2014). Nano metal oxides have large number of active sites due to their high surface area (Holmes and Gu 2016; Yamani et al. 2014). Nano-form adsorbents of iron oxide (n-Fe_2O_3), titanium dioxide (n-TiO_2), and aluminum oxide (n-Al_2O_3) were observed as very effective for removal of various inorganic and also selenium (Elzinga et al. 2009; Jordan et al. 2011; Peak 2006; Shi et al. 2009; Yamani et al. 2014; Zhang et al. 2009).

The adsorbent iron oxide magnetite (Fe_3O_4) has magnetic properties. Gonzalez et al. (2012) carried out the adsorption study of selenite and selenate onto a high and low pressure aged iron oxide nanomaterial. Microwave-assisted methods can be used for the synthesis of Fe_3O_4 nanomaterials. The average particle sizes of Fe_3O_4 nanoparticles were obtained as 27 and 25 nm by non-microwave and microwave-assisted synthesis respectively. One study developed iron (Fe^{3+}) oxide/hydroxide nanoparticles sol (NanoFe) by an intensifying approach and removed selenium (Se^{4+} and Se^{6+}) from waste water.

Nanocrystalline metal oxide impregnated chitosan beads are developed with nanocrystalline aluminum oxide (n-Al_2O_3) to form n-Al_2O_3 impregnated chitosan beads (Yamani et al. 2014). These adsorb inorganic selenite and selenate simultaneously more effectively than n-Al_2O_3 or chitosan alone and also compared with n-TiO_2 and n-TiO_2 impregnated chitosan beads.

16.6.4.12.2 Ligand Immobilized Nanocomposites Nanocomposites of immobilized ligand were synthesized by immobilizing Schiff base ligands on a nanostructure to remove selenium oxyanions (Holmes and Gu 2016). The Schiff base ligands ($R_2C = NR'$ where $R' \neq H$) are regular complexing agents for transition metals. In ligand, the imine nitrogen is basic in nature and have pi-acceptor and sigma-donor properties which are responsible for complexation with Se(IV) and transition metals.

Ligand immobilized nanocomposites prepared by immobilizing 3-(2-amino ethylamino) –propyl trimethoxysilane on multi-wall nanotubes of carbon. This type of adsorbent selectively extracts Se(VI) at low pH values (Peng et al. 2015). Very few macro- or micro-size adsorbents are selective for selenium, hence nanoadsorbents are more attractive and promising (Holmes and Gu 2016).

16.6.4.12.3 Nanoparticle Graphene Oxide Graphene oxides are also getting attention as an emerging material with promising nano-phenomena properties (Holmes and Gu 2016). Hydrophobic monolayer graphene oxide exhibits promising results for removing heavy metals because of high surface area and ample functional groups of carboxyl and hydroxyl. Fu et al. (2014) synthesized the magnetic nanoparticles using graphene oxide and employed for the removal of selenium from waste water and also for regeneration on large scale.

The high binding ability of magnetic nanoparticle graphene oxide composite adsorbent was observed for both selenite and selenate. The magnetic nanoparticle graphene oxide has at least two types of hydroxyl binding sites for Se(IV) and Se(VI) on the iron oxide nanoparticles and the graphene oxide sheets. This provides the higher adsorption capacities as compared to pure Fe_3O_4 nanoparticles or pure graphene oxide (Holmes and Gu 2016). The magnetic nanoparticle graphene oxide composites removed more than 99.9% selenite and more than 80% of selenate for adsorbent dosage of 1 g/l (Fu et al. 2014).

16.6.4.12.4 Nanocomposites Layered Double Hydroxide Nanocomposites of layered double hydroxide are promising anion exchange and attractive nanomaterials for the removal of oxyanion of selenium (Holmes and Gu 2016). LDHs are anionic clays (e.g. hydrotalicite ($Mg–Al–CO_3$)). Layered double hydroxide nanocomposites can incorporate anions between hydroxide layers. A novel Mg–Al Layered double hydroxide is synthesized by an alkoxide free sol–gel process. This method increases surface area, functional groups, interlayer species, and adsorptive capacity for selenate and selenite (Chubar et al. 2013).

16.6.4.12.5 Polymer-Clay Nanocomposites Polymer-clay nanocomposites have been studied for the last few years as an adsorbent for removal anionic contaminants (Holmes and Gu 2016). Chitosan–montmorillonite nanocomposites have been investigated for the adsorption of various anionic contaminants and selectively adsorbed selenate from wastewater (Celis et al. 2012). Multiple polymer polycations such as poly (diallyldimethylammonium chloride), chitosan, and poly (vinylpyrrolidone-co-styrene) have been used for the removal of selenium. Chitosan was found to have the highest effectiveness, with more than 99% removal of selenate (Bleiman and Mishael 2010).

16.6.5 Membrane Technologies

Membrane technologies are widely used for the water and waste water treatment. Pressure-driven membrane technologies used precise membrane to facilitate the physical barrier to the way of colloidal, particulate, and soluble constituents, on the side of permeate or filtrate (NAMC 2010). The typical membrane separations based on pressure difference are ultrafiltration, microfiltration, nanofiltration, and reverse osmosis. Membrane technologies are useful to remove pollutants from aqueous solutions. Reverse osmosis, emulsion liquid membranes, and nanofiltration are among various separation methods which are capable of separating selenium from contaminated wastewater (Stefaniak et al. 2018). The various species of selenium have an average size of 0.3–0.4 nm, hence the very small size of the membrane pores is needed for successful separation of selenium (Holmes and Gu 2016).

16.6.5.1 Reverse Osmosis

The membrane separation, reverse osmosis can be used for the removal of selenium with >80% separation efficiencies. As compared to other membrane separation and other methods, reverse osmosis is relatively costly for implementation and operations. Further, pretreatment of feed is needed and treatment of generated waste stream is also required pre treatment (Holmes and Gu 2016). U.S. EPA listed reverse osmosis for removal of selenium (Pontius 1995). Pretreatments are needed to avoid fouling of the membranes. Retention up to 94% has been observed in the pilot scale study and the field study to obtain Se lower than 5 μg/l (Richards et al. 2011).

16.6.5.2 Emulsion Liquid Membranes

In emulsion liquid membranes, the transfer of selenium is taking place from aqueous to liquid organic droplets (Gleason et al. 1996). During the processing steps, selenium is transferred from the bulk waste water to organics. Emulsion liquid membranes have been shown to be an effective method and Se(VI) can be extracted sufficiently fast even when sulfate is present for pH over 2.

16.6.5.3 Nanofiltration

Nanofiltration is similar to reverse osmosis but has the membranes prepared using a porous inert layer of polysulfone and a hydrophobic rejection layer of negatively charged. The operating pressure needed in nanofiltration system is one third of the reverse osmosis system. The nanofiltration selectively removed around 95% of selenium from waste water (Kharaka et al. 1996). The hybrid process of ultrafiltration, nanofiltration/reverse osmosis assisted by solar energy removed 74–94% of selenium from waste water (Richards et al. 2008). These are the hybrid intensified systems for selenium removal.

16.6.5.4 Ceramic Microfiltration

In the pressure-driven membrane separation category, microfiltration using ceramic membrane is also a promising method for the removal of heavy metals from various waste water including acid mine drainage. This method needs frequent cleaning of the membrane to maintain permeability of the ceramic membrane. Also, ceramic microfiltration needs a substantial amount of energy and hence has high operating costs. This methods replaces numerous units, hence it may be a better alternative and more cost-effective for treatment of waste water (USEPA 2014). Microfiltration with ceramic membrane confines the precipitated selenium. Górak and Stankiewicz (2011) targeted various milestones in membrane technologies achievable by 2030 and reaping benefits by 2050. These focused developments can be summarized as: optimization and development for high performance, novel options (membrane crystallization), further development of novel material (aquaporin channels, carbon nanotube, and new protein-based membranes), coupling with renewable energy, compact devices, etc.

16.6.6 Oxidation/Reduction

Redox (oxidation – reduction) reactions can be employed for the removal of selenium and other metals. In this method, reaction, solubilization, and precipitation are carried out to remove the desired metals or contaminants or compounds. The extent of reduction – oxidation reactions is decided on the chemical equilibria, which depend on pH, temperature, and ionic strength (NAMC 2010).

16.6.6.1 Zero-Valent Iron (ZVI)

Zero-valent iron (ZVI) is a readily available, less expensive, non-toxic, and moderately strong reducing agent. ZVI has been extensively considered for various waste water treatment and other environmental remediation (Tang et al. 2014). ZVIs have been successfully used for the removal of numerous organic and inorganic compounds such as: nitroaromatics, azoaromatics, halogenated organics, chlorinated organics, nitrate, uranium, arsenic, and chromate from various waste water (Liang et al. 2014).

Selenium as SeO_4^{2-} and SeO_3^{2-} can be removed effectively by chemical reduction to Se0 because lower oxidation states have poor solubility in aqueous phase. Also ZVI reduces selenate and selenite to selenium. The iron is the best ZVI as it can be work as a catalyst and an electron donor, which is needed for the reduction reaction (Frankenberger et al. 2004; Holmes and Gu 2016). The oxidized surface of ZVI make available the active sites of both ferrous and ferric iron for selenium oxyanions adsorption for the removal of selenium. The green rust (the complex at iron surface) is the form of ZVI which reduces selenate and selenite chemically to elemental selenium (Murphy 1988; Myneni et al. 1997; Zhang et al. 2005a; Zingaro et al. 1997). Zhang et al. (2005a, b) suggested two mechanisms that contribute to the removal of Se(VI) from solution by ZVI. The reduction of Se(VI) to Se(IV), with the rapid adsorption of Se(IV) to oxyhydroxides and the adsorption of Se(VI) directly to oxyhydroxides before any reduction to Se(IV) (Murphy 1988; Zhang et al. 2005b; Zingaro et al. 1997). X-ray photoelectron spectroscopy (XPS) revealed that Se(VI) reduced to Se(IV)/Se(0) on an iron surface (Qiu et al. 2000). Meng et al. (2002) also used XPS to determine that selenocyanate was removed from wastewater by ZVI through the formation of Se(0)/Se(-II). Yoon et al. (2011) and Liang et al. (2014) investigated the iron ZVI for the removal and confirmed by several spectroscopy and other analytical methods.

16.6.6.2 Catalyzed ZVI

The activity of the ZVI can be increased by the addition of catalyst where the reduction is taking place in the presence of catalyst. In the case of selenium, copper or nickel was added ZVI for the reduction of selenium in the presence of these catalyst by providing a higher electrochemical potential between the elemental iron and soluble selenium (MSE 2001).

16.6.6.3 Advanced Reduction Process

A new group of water/waste water treatment technology, ARPs have been explored which is the intensification in waste water treatment. ARPs produce highly reactive reducing radicals by combining reducing reagents (e.g. dithionite, sulfite, ferrous ion) with activating methods (e.g. UV light, ultrasound, electron beam) (Duan et al. 2017a). Sodium dithionite is a strong reductant and is known to be able to reduce Se(IV) rapidly by itself (Duan et al. 2017b; Geoffroy and Demopoulos 2009). However, application of UV irradiation increases the removal rates due to more rapid formation of highly reactive reducing radicals through dithionite photolysis (Duan et al. 2017b). Duan et al. (2017a) employed the combination of UV irradiation and dithionite as ARP for selenium removal from waste water at various levels of pH and in an anaerobic environment.

16.6.6.4 Nanoscale Zero-Valent Iron (nZVI)

Nanoscale zero-valent iron (nZVI) has more potential compared to macro or micro ZVI due to the availability of highly active sites and surface area for the removal and treatment of various organic and inorganic compounds from waste water. nZVI can be promising for removal and reduction, using the same mechanism as ZVI of selenate and selenite from wastewater (Holmes and Gu 2016). As a process intensification, threefold increase in selenium oxyanions reduction and removal was observed due to high surface area and active sites (Ling et al. 2015). A more than 96% reduction of selenate from 4 ppm was obtained using nZVI (Olegario et al. 2009).

16.6.7 Evaporation

Evaporation is one of the more popular methods for waste water treatment. It has enhanced evaporation systems, evaporation ponds, evaporator/crystallizer technologies and salinity gradient solar ponds (NAMC 2010). Enhanced evaporation systems and salinity gradient solar ponds are the intensified approaches having less energy requirements compared to conventional evaporation technology. The conventional evaporation unit needs large space, large residence times, and risks of contamination (Holmes and Gu 2016; Manning and Burau 1995).

16.6.7.1 Evaporation Ponds

Evaporation ponds reduced selenate to selenite, and then absorption of selenite onto sediments and minerals (NSMP 2007). A typical influent having 91% selenate, 7% selenite, and 2% organic selenium reduced in the evaporation pond to 21%, 62%, and 17% respectively (NAMC 2010; NSMP 2007). The selenium remaining in the pond as adsorbed selenium is accumulated in sediments (Gao et al. 2007).

16.6.7.2 Enhanced Evaporation Systems

Enhanced evaporation systems are an option for conventional evaporation ponds. They produce brine concentrate followed by crystallization, drying, and solid waste disposal (NAMC 2010). Evaporation is intensified by spraying water in the air using blower. This improves the contact of water with air, which increases the rate of water evaporation and thus concentration. It needs less land area as compared to evaporation pond (NAMC 2010). Average enhanced evaporator system efficiency is observed as 67% by assuming 70% of running time of evaporator (NAMC 2010; SSERP 2005).

16.6.7.3 Salinity Gradient Solar Pond

A temperature gradient is developed from top to bottom in deep bodies of saline water in the case of a salinity gradient solar pond (NAMC 2010). It has the combination of solar energy collection and storage. It is a kind of reliable source of heat if it properly designed. Electricity may be generated using the temperature difference between the top and bottom of the salinity gradient solar pond (NAMC 2010). It has a surface layer, gradient zone, and storage zone. Visible solar radiation penetrated and heated the bottom and retained heat (Frankenberger et al. 2004). It has been not used for removal of selenium (NSMP 2007) but may be considered as an intensified way to be used for selenium removal and heat utilization.

16.6.7.4 Mechanical Evaporator/Crystallizer

Vapor compression evaporator and multiple effect evaporators are the main configuration for mechanical evaporator. It can overcome the limitations of other evaporators having improved thermal equilibrium, which increased TDS (NAMC 2010). It concentrates the dissolved solids in the brine and produced salt solids with 10–20% moisture (NAMC 2010).

16.6.8 Coagulation

Coagulation by salts of Fe and Al is one of the vital water and waste water treatment methods (Hu et al. 2006, 2015). Contaminant adsorption to Al or Fe hydroxide oxide can play a very important role in the coagulation process. Hu et al. (2015) studied the removal of Se using coagulation and investigated the effect of various parameters such as species of Se, type of coagulant, pH of aqueous phase, and ions. Fe-based coagulant was much more efficient than Al-based coagulants for the removal of selenium, where selenite was removed more than selenate. The selenite removal was observed as more than 98% using an Fe coagulant.

16.6.9 Electrocoagulation

Electrocoagulation is a promising method in which the functions and advantages of conventional coagulation, flotation, and electrochemistry are combined for the treatment of water and wastewater treatment. The numerous metals and metalloids including selenium from waste waters can be treated by electrocoagulation (Hansen et al. 2019; Holmes and Gu 2016). With electrocoagulation, 98.7% removal of selenium was observed by applying a direct electrical current to an electrochemical cell (Mavrov et al. 2006). The colloidal biogenic selenium was removed to 93% by electrocoagulation (Staicu et al. 2015). Hansen et al. (2019) used electrocoagulation for selenium treatment of waste water of petroleum refinery. A batch airlift reactor with air stirring was employed for selenium treatment, and 90% removal was observed in six hours.

16.6.10 Photoreduction

Photoreduction is a type of hybrid process which is the combination of chemical adsorption and reduction. In photoreduction of selenium, UV light in the presence of a photocatalyst was used for the treatment of selenate and selenite to form elemental selenium (Shamas et al. 2009). The process of photoreduction involves absorption of selenium compounds on the photocatalyst for chemical redox reactions created through the electrons and the holes due to the exposure of UV light. The photocatalyst was regenerated by desorption and surface treatment (Golder 2009a). Naguyen et al. (2005) successfully used photoreduction for effective treatment of a mixed solution of selenate and selenite (20–40 μg/l) by TiO_2 with <380 nm UV light at 3.5 pH. Removal of selenite from wastewater of wet flue gas desulfurization using photocatalytic reduction was performed by Nakajima et al. (2011). An electron hole scavenger, HCOOH in excess for efficient removal was used along with TiO_2 photocatalyst.

16.6.11 Precipitation

Precipitation can be used to remove various contaminants from wastewater. The chemical reagents added to wastewater may react and precipitate the contaminants. Selenium in the form of selenate and selenite does not react with any chemical reagent. Only selenocyanate can be removed by precipitation. Selenocyanate present in refining waste water may be precipitated with various metals: silver, mercury, cadmium, lead, thallium, copper, and forms of various metal complexes. Copper salts were used in experimental investigation for approximately 6 mg/l selenocynate, and copper thiocynate lattice was observed by XRD analysis (Manceau and Gallup 1997). Merrill et al. (1986) performed the successful experiments in pilot scale by iron co-precipitation to remove the selenite. The disposal of the large amount of sludge formed during co-precipitation is the major concern for this process.

16.6.12 Permeable Reactive Barriers

A reactive material can be placed in permeable treatment zone where the pollutant treated by reaction from waste water (USEPA 2014). The reactive materials of ZVI, compost, limestone, activated carbon, zeolites, and apatite have been considered as permeable reactive barriers. If a permeable reactive barrier is designed and implemented properly, it can remediate many contaminants having no operational costs (USEPA 1998). A permeable reactive barrier successfully treated selenium and various other contaminants (USEPA 2014).

16.7 Discussion

Selenium presents mostly in the form of selenite and selenate. In the presence of suitable conditions for treatment, selenate transforms to selenite and then to elemental selenium. The removal of elemental selenium is easier due to its insoluble nature in aqueous phase, and also it is non-toxic (Oremland et al. 1994). Hence the economical process can be developed for the removal of selenium and their forms. The basic challenges for removal of selenium and features of various treatment methods have been discussed in several critical reviews (Gingerich et al. 2018; Holmes and Gu 2014; NAMC 2010; Santos et al. 2015; Tan et al. 2016). As such there is no universal method for removal of selenium from water and waste water (Zelmanov and Semiat 2013).

The selenium removal methods mainly include chemical, physical, and biological methods. The various waste water treatment options to remove selenium are: iron co-precipitation, reverse osmosis, the use of constructed wetlands, coagulation, ion exchange, membrane filtration, chemical reduction, evaporation, coagulation/flocculation, extraction with chelating agents, nanofiltration, electrochemical operation, cementation, lime softening, solvent extraction, adsorption, solidification/stabilization, and biological selenium reduction (Awual et al. 2015; Holmes and Gu 2016; Lee 1989; Lottermoser 2007; Yamani et al. 2014). High operating and maintenance costs, sludge generation, and use of toxic chemicals severely limit their wide application (Yamani et al. 2014).

The bioreactors for selenium removal using micro-organisms are one of the well established technologies and have been studied at pilot and full-scale wastewater treatment. The performance of biological methods depends mainly on the type of microorganism used (EPRI 2017; Gingerich et al. 2018; Sonstegard et al. 2008).

Adsorption is one of the best options for the treatment of heavy metals from waste water and other streams for high strength and low volumes and also for selenium waste water (Wasewar et al. 2009a). Most of the physicochemical methods are costly. The biological treatment have more residence time as compared to physicochemical methods (NAMC 2010). Most of these techniques for selenium removal depend on the oxidation state of selenium and require low pH conditions. Competing anions are the most challenging limitation on removal of selenium with ion exchange resins, adsorption, and enzymatic reduction in biological treatment. Effectiveness of removal of selenium by ion exchange resins and coagulation-precipitation processes is different for selenite and selenate. Production of secondary wastes and high operating and maintenance costs inhibit the wide application of precipitation. Membrane processes are effective technologies for removing selenium, but their operational costs are high.

Process intensification approaches may reduce the industrial water treatment energy by more than 50% (AMO 2015). Improved technologies are needed for water treatment at the industrial scale to enable internal reuse, including determination of critical parameters in various industries, and level of cleaning required. Smaller, modular systems should be developed to enable water treatment and heat integration at industrial or agricultural sites (AMO 2015). Transposable wastewater treatment processes, including generic solutions based on first principles, and enhanced, less energy-intensive water extraction from gas, solid, and liquid effluents, including extraction/separation of very dilute concentrations, are needed. Mitigating the effect of impurities on the performance of separation technologies, such as fouling and degradation, should be considered during development of these wastewater treatment processes (AMO 2015).

The dramatic improvements create various benefits, including reduction of energy consumption, cost savings, safety enhancement, and minimization of environmental impacts. These benefits are the key drivers and could accelerate future development and application of modular chemical process intensification to the chemical industry, pharmaceutical industry, materials, pulp and paper, petroleum, water treatment and desalination, and environmental management, among other fields. Thus, process intensification is expected to include a wide range of academic and practical topics in the future. A universal definition of process intensification may further enhance its range of applications (Kim et al. 2017).

References

Adams, J., Pickett, T.M., and Montgomery, J. (1996). Biotechnologies for metal and toxic inorganic removal from mining process and waste solutions. *Randol Gold Forum* 96 (4): 143–146.

Adutwum, K.O. (1995). Adsorption mechanism of oxyanions of selenium on to lanthanum oxide and alumina. M.S. thesis. University Reno, Reno, NV.

Ali, I. (2012). New generation adsorbents for water treatment. *Chem. Rev.* 112: 5073–5091.

AMO, Advanced Manufacturing Office (2015). *Process intensification workshop, Energy Efficiency and Renewable Energy, US Department of Energy, September 29–30*. Alexandria, VA: US Department of Energy.

Anderson, M.A., Cudero, A.L., and Palma, J. (2010). Capacitive deionization as an electrochemical means of saving energy and delivering clean water. Comparison to present desalination practices: will it compete? *Electrochim. Acta* 55 (12): 3845–3856.

Awual, M.R., Hasan, M.M., and Khaleque, M.A. (2015). Efficient selenium(IV) detection and removal from water bytailor-made novel conjugate adsorbent. *Sens. Actuators B* 209: 194–202.

Babi, D.K., Cruz, M.S., and Gani, R. (2016). Fundamentals of process intensification: a process systems engineering view. In: *Process Intensification in Chemical Engineering: Design Optimization and Control* (eds. J.G. Segovia-Hernández and A. Bonilla-Petriciolet), 7–33. Cham: Springer International Publishing.

Becht, S., Franke, R., Geißelmann, A., and Hahn, H. (2009). An industrial view of process intensification. *Chem. Eng. Process.* 48 (1): 329–332.

Bleiman, N. and Mishael, Y.G. (2010). *J. Hazard. Mater.* 183: 590–595.

Bo, A., Sarina, S., Liu, H. et al. (2016). Efficient removal of cationic and anionic radioactive pollutants fromWater using Hydrotalcite-based getters. *ACS Appl. Mater. Interfaces* 8: 16503–16510.

Brienne, S.H., Jones, R.G., Jensen, S.E., and Tompkins, A.N. (2009). Selenium release from coal mines in the Elk Valley, and Treatment R & D Plans. B.C.'s 33rd Annual Mine Reclamation Symposium, Cranbrook, British Columbia.

Brown, D.S. and Reed, S.C. (1994). Inventory of constructed wetlands in the United States. *Water Sci. Technol.* 29: 309–318.

Burau, R.G. (1985). Environmental chemistry of selenium. *Calif. Agrc.* 39: 16–18.

Cantafio, A.W., Hagen, K.D., Lewis, G.E. et al. (1996). Pilot-scale selenium bioremediation of San Joquin drainage water with Thauera selenatis. *Appl. Environ. Microbial.* 62: 3298–3303.

Catal, T., Bermek, H., and Liu, H. (2009). *Biotechnol. Lett.* 31: 1211–1216.

Cavani, F., Trifirb, F., and Vaccari, A. (1991). Hydrotalcite-type anionic clays: preparation, properties and applications. *Catal. Today* 11: 172–301.

Celis, R., Adelino, M.A., Hermosín, M.C., and Cornejo, J. (2012). *J. Hazard. Mater.* 209–210: 67–76.

CGWB, Concept note on geogenic contamination of ground water in India, Central Ground Water Board, Ministry of Water Resources, Govt. of India, Faridabad, February 2014.

CH2M Hill. (2010). Review of Available Technologies for the Removal of Selenium from Water. http://namc.org/docs/00062756.PDF.

Chamberlin, P.D. (1996). Selenium Removal from Wastewaters – An Update. Denver, CO, Randol Gold Forum 96: 9.

Christensen, B.T., Bertelsen, F., and Gissel-Nielsen, G. (1989). Selenite fixation by soil particle-size separates. *J. Soil Sci.* 40: 641–647.

Chubar, N. (2014). EXAFS and FTIR studies of selenite and Selenate sorption by Alkoxide-free sol-gel generated mg-Al-CO_3 layered double hydroxide with very labile interlayer anions. *J. Mater. Chem. A* 2: 15995–16007.

Chubar, N., Gerda, V., Megantari, O., M. et al. (2013). *Chem. Eng. J.* 234: 284–299.

Chung, J., Nerenberg, R., and Rittmann, B.E. (2006). Bio-reduction of selenate using a hydrogen based membrane bio-film reactor. *Environ. Sci. Tech.* 40 (5): 1664–1671.

Chung, J., Nerenberg, R., Rittmann, B.E. et al. (2007). Evaluation for biological reduction of nitrate and perchlorate in brine water using the hydrogen-based membrane biofilm reactor. *J. Environ. Eng.* 133: 157–164.

Conde, J.E. and Alaejos, M.S. (1997). Selenium concentrations in natural and environmental waters. *Chem. Rev.* 97: 1979–2003.

Constantino, L.V., Nunes Quirino, J., Monteiro, A.M. et al. (2017). Sorption-desorption of selenite and Selenate on mg-Al layered double hydroxide in competition with nitrate, sulfate and phosphate. *Chemosphere* 181: 627–634.

Cooper, P.F. and Findlater, B.C. (1990). *Constructed Wetlands in Water Pollution Control.* London: Pergamon Press.

Corwin, R., Isaacson, A.E., and Jeffers, T. (1994). Oxyanion Removal from Waste Waters Using Immobilized Ferric Oxyhydroxide/Peat Moss. Park City, UT, 16th Annual Assoc. Abandoned Mine Land Programs Conf., September 1994.

CPCB (2004). *Status of Water Treatment Plants in India*. Delhi: Central Pollution Control Board, Ministry Of Environment and Forests.

Criscuoli, A. and Drioli, E. (2007). New metrics for evaluating the performance of membrane operations in the logic of process intensification. *Ind. Eng. Chem. Res.* 46 (8): 2268–2271.

Dale, C. and Liner, B. (2016). Selenium removal using a moving bed biofilm reactor. Industrial Wastewater March: 1–4.

Dautzenberg, F.M. and Mukherjee, M. (2001). Process intensification using multifunctional reactors. *Chem. Eng. Sci.* 56 (2): 251–267.

De, B.S., Wasewar, K.L., Dhongde, V.R., and Sontakke, P.B. (2018a). Recovery of acrylic acid using calcium peroxide nanoparticles: thermodynamics and continuous column study. *Chem. Biochem. Eng. Q.* 32 (1): 19–28.

De, B.S., Wasewar, K.L., Dhongde, V.R. et al. (2018b). Recovery of acrylic acid using calcium peroxide nanoparticles: synthesis, characterisation, batch study, equilibrium, and kinetics. *Chem. Biochem. Eng. Q.* 32 (1): 29–39.

Derek, P. (2006). Adsorption mechanisms of selenium oxyanions at the aluminum oxide/water Interface. *Colloid Interface Science* 303: 337.

Dewberry, R.A., Leyba, J.D., and Boyce, W.T. (2000). Observation and measurement of se-79 in Savannah River site high level waste tank fission product waste. *J. Radioanal. Nucl. Chem.* 245: 491–500.

Dhillon, S.K. (1999). Adsorption–desorption reactions of selenium in some soils of India. *Geoderma* 93: 19.

Dhongde, V., Wasewar, K.L., and De, B.S. (2017). Development of nanohybrid adsorbent for defluoridation from aqueous systems. *Chemosphere* 188: 354–366.

Drewes, J.E. (2009). *An Integrated Framework for Treatment and Management of Produced Water – Technical Assessment of Produced Water Treatment Technologies*, 1e. Golden, CO: Colorado School of Mines. RPSEA Project 07122-12.

Drioli, E., Stankiewicz, A.I., and Macedonio, F. (2011). Membrane engineering in process intensification – an overview. *J. Membr. Sci.* 380 (1–2): 1–8.

Duan, Y., Kaushik, V., Jung, B. et al. (2017a). Kinetic study of selenium removal using advanced reduction process with dithionite. *Environ. Eng. Sci.* 35: 169–175.

Duan, Y., Luo, G., Jung, B. et al. (2017b). Photochemical degradation of arsenic and selenium with advanced reduction processes (ARPs) – effects of reagents. *Environ. Eng. Sci.* 34: 481.

El-Shafey, E.I. (2007). Removal of Se(IV) from aqueous solution using sulphuric acid treated peanut shells. *J. Environ. Manage.* 84: 620–627.

Elzinga, E.J., Tang, Y., McDonald, J. et al. (2009). Macroscopic and spectroscopic characterization of selenate, selenite, and chromate adsorption at the solid-water interface of gamma-Al_2O_3. *J. Colloid Interface Sci.* 340 (2): 153–159.

EPA (2001). Selenium Treatment and Removal Alternatives: Demonstration Project. Mine Waste Technology Program Activity III, Project 20, National Risk Management Research Laboratory, US EPA, Cincinnati, OH.

EPA, Environmental Protection Agency (2013). Emerging Technologies for Wastewater Treatment and In-Plant Wet Weather Management. EPA 832-R-12-011, March.

EPRI (2017). Biological Treatment of Flue Gas Desulfurization Wastewater at a Power Plant Burning Powder River Basin Coal: Pilot Demonstration with the ABMet Technology; Final Report 3002006089; Electric Power Research Institute: Palo Alto, CA, USA.

Espinosa-Ortiz, E.J., Gonzalez-Gil, G., Saikaly, P.E. et al. (2015). *Appl. Microbiol. Biotechnol.* 99: 2405–2418.

Fernandez-Martinez, A. and Charlet, L. (2009). Selenium environmental cycling and bioavailability: a structural chemist point of view. *Rev. Environ. Sci. Biotechnol.* 8: 81–110.

Fischer, K.J. and Behn M. (2016). Apparatus and method for reactive distillation for waste water treatment. WO Patent Application 2016207272A1.

Fontananova, E. and Drioli, E. (2014). Membrane reactors: advanced systems for intensified chemical processes. *Chem. Ing. Tech.* 86 (12): 2039–2050.

Fordyce, F. (2007). Selenium geochemistry and health. *Ambio* 36: 94–97.

Frankenberger, W.T., Amrhein, C. Jr., Fan, T.W.-M. et al. (2004). Advanced treatment technologies in the remediation of seleniferous drainage waters and sediments. *Irrig. Drain. Syst.* 18: 19–41.

Fu, Y., Wang, J., Liu, Q., and Zeng, H. (2014). *Carbon* 77: 710–721.

Fujita, M., Ike, M., Nisbimoto, S. et al. (1997). Isolation and characterization of a novel selenate-reducing bacterium, batiks sp. SF-1. *J. Ferment. Bioeng.* 83: 517–522.

Gao, S., Tanji, K.K., Peters, D.W., and Herbel, M.J. (2000). *J. Environ. Qual.* 29: 1275–1283.

Gao, S., Tanji, K.K., Peters, D.W. et al. (2003). Selenium removal from irrigation drainage water flowing through constructed wetland cells with special attention to accumulation in sediments. *Water Air Soil Pollut.* 144: 263–284.

Gao, S., Tanji, K.K., Dahlgren, R.A. et al. (2007). Chemical status of selenium in evaporation basins for agricultural drainage. *Chemisphere* 69 (4): 585–594.

Geoffroy, N. and Demopoulos, G.P. (2009). Reductive precipitation of elemental selenium from selenious acidic solutions using sodium dithionite. *Ind. Eng. Chem. Res.* 48: 10240.

Gerhardt, M.B., Green, F.B., Newman, R.D. et al. (1991). Removal of selenium using a novel alga-bacterial process. *Res. J. WPCF* 63: 799–805.

Gingerich, D.B., Grol, E., and Mauter, M.S. (2018). Fundamental challenges and opportunities in flue gas desulfurization wastewater treatment at coal fired power plants. *Environ. Sci. Water Res. Technol.* 4: 909–925.

Gissel-Nielsen, G., Gupta, U.C., Lamand, M., and Westermarck, T. (1984). Selenium in soils and plants and its importance in livestock and human-nutrition. *Adv. Agron.* 37: 397–460.

Gleason, K.J., Yu, J., Bunge, A.L., and Wright, J.D. (1996). Removal of selenium from contaminated waters using emulsion liquid membranes. *Chem. Separations Liquid Membr.* 642: 18.

Goel, S., Pant, K.K., and Nigam, K.D.P. (2009). *J. Hazard. Mater.* 171: 253–261.

Gogate, P.R. and Pandit, A.B. (2004). Sonophotocatalytic reactors for wastewater treatment: a critical review. *AIChE J* 50: 1051–1079.

Goh, K.H., Lim, T.T., and Dong, Z. (2008). Application of layered double hydroxides for removal of oxyanions: a review. *Water Res.* 42: 1343–1368.

Golder Associates Inc. (Golder) (2009a). Literature review of treatment technologies to remove selenium from mining influenced water. Prepared for Teck Coal Limited, Calgary, AB. 0842-0034.

Golder Associates Inc. (Golder) (2009b). Algal simulation pilot study, Cardinal River operations, Edmonton AB, Lakewood, Co. USA.

Gonzalez, C.M., Hernandez, J., Peralta-Videa, J.R. et al. (2012). Sorption kinetic study of selenite and selenate onto a high and low pressure aged iron oxide nanomaterial. *J. Hazard. Mater.* 211–212: 138–145.

Górak, A. and Stankiewicz, A. (2011). Research Agenda for Process Intensification Towards a Sustainable World of 2050, Creative Energy, Energy Transmission, Dutch Institute for Sustainable Process Technology (ISPT), Ministry of Economic Affairs, Agriculture and Innovation Netherlands www.ispt.nl.

Grünewald, M. and Agar, D.W. (2004). Enhanced catalyst performance using integrated structured functionalities. *Chem. Eng. Sci.* 59 (22–23): 5519–5526.

Hamilton, S.J. (2004). Review of selenium toxicity in the aquatic food chain. *Sci. Total Environ.* 326: 1–31.

Hansen, D.P.J., Zayed, D.A., and Terry, N. (1998). Selenium removal by constructed wetlands: role of biological volatilization. *Environ. Sci. Technol.* 32: 591–597.

Hansen, H.K., Pena, S.F., Gutierrez, C. et al. (2019). Selenium removal from petroleum refinery wastewater using an electrocoagulation technique. *J. Hazard. Mater.* 364: 78–81.

Haygarth, P.M. (1994). Global importance and global cycling of selenium. In: *Selenium in the Environment* (eds. W.T. Frankenberger Jr. and S. Benson), 1–7. New York;: Marcel Dekker.

He, Q. and Yao, K. (2010). Microbial reduction of selenium oxyanions by Anaeromyxobacter dehalogenans. *Bioresour. Technol.* 101: 3760–3764.

Heggs, P.J. (1983). Experimental techniques and correlations for heat exchangers surfaces: packed beds. *Chem. Eng.* 394 (13): 183.

Hoffman, D.J. (2002). Role of selenium toxicity and oxidative stress in aquatic birds. *Aquat. Toxicol.* 57: 11–26.

Holmes, A.B. and Gu, F.X. (2014). Emerging nanomaterials for the application of selenium removal for wastewater treatment. *Environ. Sci. Nano* 3: 982–996.

Hu, J., Chen, G., Lo, I.M.C., and Asce, M. (2006). Selective removal of heavy metals from industrial wastewater using maghemite nanoparticle: performance and mechanisms. *J. Environ. Eng.* 132 (7): 709–715.

Hu, C., Chen, Q., Chen, G. et al. (2015). Removal of Se(IV) and Se(VI) from drinking water by coagulation. *Sep. Purif. Technol.* 142: 65–70.

ITRC, Interstate Technology Regulatory Council. (2010). Mining Waste Treatment Technology Selection Website. http://www.itrcweb.org/miningwaste-guidance

Jeffers, T., Ferguson, C.R., and Bennett, P.G. (1991). *Biosorption of Metal Contaminants Using Immobilized Biomass – A Laboratory Study*. Washington, DC: U.S. Bureau Mines, RI-9340: 9.

Jenck, J.F., Agterberg, F., and Droescher, M.J. (2004). Products and processes for a sustainable chemical industry: a review of achievements and prospects. *Green Chem.* 6: 544–556.

Ji, Y. and Wang, Y-T., Selenium removal using a bacteria cultured reactor packed with alum-impregnated activated alumina, *Proceedings of the 4th World Congress on Civil, Structural, and Environmental Engineering (CSEE'19) Rome, Italy – April* (2019). *Paper No. ICEPTP 116, 1-6*

Jordan, N., Foerstendorf, H., Weiß, S. et al. (2011). Sorption of selenium(VI) onto anatase: macroscopic and microscopic characterization. *Geochim. Cosmochim. Acta* 75 (6): 1519–1530.

Jordan, N., Ritter, A., Foerstendorf, H. et al. (2013). Adsorption mechanism of selenium(VI) onto maghemite. *Geochim. Cosmochim. Acta* 103: 63–75.

Kadlec, R.H. and Knight, R.L. (1996). *Treatment Wetlands*. Boca Raton, FL: Lewis Pub.

Kagami, T., Narita, T., Kuroda, M. et al. (2013). *Water Res.* 47: 1361–1368.

Kapoor, A., Tanjore, S., and Viraraghavan, T. (1995). Removal of selenium from water and wastewater. *Int. J. Environ. Stud.* 49: 137–147.

Kashiwa, M., Nishimoto, S., Takahashi, K. et al. (2000). Factors affecting soluble selenium removal by a Selenate-reducing *Bacterium Bacillus sp*. SF-1. *J. Biosci. Bioeng.* 89 (6): 528–533.

Keil, F.J. (2007). Modeling of process intensification – an introduction and overview. In: *Modeling of Process Intensification* (ed. F.J. Keil), 1–7. Weinheim: Wiley-VCH Verlag GmbH & Co. KGaA.

Keil, F.J. (2018). Process intensification. *Rev. Chem. Eng.* 34 (2): 135–200.

Kemmere, M. and Keurentjes, J. (2001). Industrial membrane reactors. In: *Membrane Technology: In the Chemical Industry* (eds. S.P. Nunes and K.V. Peinemann), 191–221. Wiley-VCH.

Kharaka, Y.K., Ambats, G., Presser, T.S., and Davis, R.A. (1996). Removal of selenium from contaminated agricultural drainage water by nanofiltration membranes. *Appl. Geochem.* 11 (6): 797–802.

Kim, Y.-H., Park, L.K., Yiacoumi, S., and Tsouris, C. (2017). Modular chemical process intensification: a review. *Annu. Rev. Chem. Biomol. Eng.* 8: 359–380.

Koilraj, P., Kamura, Y., and Sasaki, K. (2018). Cosorption characteristics of $SeO2^4$ and Sr^{2+} radioactive surrogates using 2D/2D Graphene oxide-layered double hydroxide Nanocomposites. *ACS Sustain. Chem. Eng.* 6: 13854–13866.

Koren, D.W., Gould, W.D., and Lortie, L. (1992). Selenium Removal from Waste Water,@ Waste Processing and Recycling in Min. and Met. Industries. Edmonton, Alberta, Can., CIM,: 5.

Koudelkova, Z., Bytesnikova, Z., Xhaxhiu, K. et al. (2019). Electrochemical evaluation of selenium (IV) removal from its aqueous solutions by unmodified and modified graphene oxide. *Molecules* 24 (1063): 1–13.

Kumar, P., Agnihotri, R., Wasewar, K.L. et al. (2012). Status of adsorptive removal of dye from textile industry effluent. *Desalination Water Treat.* 50 (1–3): 226–244.

Kumar, P., Rao, R., Chand, S. et al. (2015). Adsorption of lead from aqueous solution onto coir-pith activated carbon. *Desalination Water Treat.* 51 (13–15): 2529–2535.

Kuroda, M., Notaguchi, E., Sato, A. et al. (2011). *J. Biosci. Bioeng.* 112: 259–264.

Laurie, S.B. and Chao, T.T. (1990). Adsorption of selenium by amorphous iron Oxyhydroxide and manganese dioxide. *Geochim. Acta* 54: 739.

Lawson, S. and Macy, J.M. (1995). Bioremediation of selenite in oil refinery wastewater. *Appl. Microbial. Biotechnol.* 43: 762–765.

Lee, E.W. (1989). Current options in treatment of agricultural drainage water. In: *Biotreatment of Agricultural Wastewater* (ed. M.E. Huntley), 33–45. Boca Raton, FL: CRC Press.

Lemley, A.D. and Ohlendorf, H.M. (2002). Regulatory implications of using constructed wetlands to treat selenium-laden wastewater. *Ecotoxicol. Environ. Saf.* 52: 46–56.

Lemly, A.D. (2004). Aquatic selenium pollution is a global environmental safety issue. *Ecotoxical. Environ. Saf.* 59: 44–56.

Leyva-Díaz, J.C., Lopez-Lopez, C., Martín-Pascual, J. et al. (2015). Kinetic study of the combined processes of a membrane bioreactor and a hybrid moving bed biofilm reactor-membrane bioreactor with advanced oxidation processes as a post-treatment stage for wastewater treatment. *Chem. Eng. Process.* 91: 57–66.

Li, M., Chowdhury, T., Kraetz, A.N. et al. (2019). Layered double hydroxide sorbents for removal of selenium from power plant wastewaters. *Chem. Eng.* 3: 20. https://doi.org/10.3390/chemengineering3010020, 1-24.

Liang, X., Zang, Y., Xu, Y. et al. (2013). Sorption of metal Cations on layered double hydroxides. *Colloids Surf. A* 433: 122–131.

Liang, L., Sun, W., Guan, X. et al. (2014). Weak magnetic field significantly enhances selenite removal kinetics by zero valent iron. *Water Res.* 49: 371–380.

Lin, Z.-Q. and Terry, N. (2003). Selenium removal by constructed wetlands: quantitative importance of biological volatilization in the treatment of selenium-laden agricultural drainage water. *Environ. Sci. Technol.* 37: 606–615.

Ling, L., Pan, B., and Zhang, W. (2015). *Water Res.* 71: 274–281.

Lottermoser, B. (2007). *Mine Wastes: Characterization, Treatment and Environmental Impacts.* Springer Science & Business Media.

Lutze, P., Gani, R., and Woodley, J.M. (2010). Process intensification: a perspective on process synthesis. *Chem. Eng. Process.* 49: 547–558.

Macy, J.M., Lawson, S., and DeMoB-Decker, H. (1993). Bioremediation of selenium oxyanions in San Joaquin drainage water using Thauera selenatis in a biological reactor syste+m. *Appl. Microbiol. Biotechnol.* 40: 588–594.

Madan, S.S. and Wasewar, K.L. (2017). Optimization for benzeneacetic acid removal from aqueous solution using CaO_2 nanoparticles based on Taguchi method. *J. Appl. Res. Technol.* 15 (4): 332–339.

Madan, S.S. and Wasewar, K.L. (2018). *Synthesis of CaO2 Nanoparticles for Environmental Remediation, Energy and Environment*, 69–76. Singapore: Springer.

Madan, S.S., Wasewar, K.L., and Pandharipande, S.L. (2016a). Modeling the adsorption of benzeneacetic acid on CaO_2 nanoparticles using artificial neural network. *Resource-Efficient Technol.* 2: S53–S62.

Madan, S.S., Wasewar, K.L., and Kumar, C.R. (2016b). Adsorption kinetics, thermodynamics, and equilibrium of α-toluic acid onto calcium peroxide nanoparticles. *Adv. Powder Technol.* 27 (5): 2112–2120.

Madan, S.S., Upwanshi, A., and Wasewar, K.L. (2016c). Adsorption of α-toluic acid by calcium peroxide nanoparticles. *Desalination and Water Treatment* 57 (35): 16507–16513.

Madan, S.S., Wasewar, K.L., and Kumar, C.R. (2017). Optimization of adsorptive removal of α-toluic acid by CaO_2 nanoparticles using response surface methodology. *Resource-Efficient Technologies* 3 (3): 329–336.

Madan, S.S., De, B.S., and Wasewar, K.L. (2019). Adsorption performance of packed bed column for benzylformic acid removal using CaO_2 nanoparticles. *Chem. Data Collections* 23: 100267.

Mahamuni, N.N. and Adenuyi, Y.G. (2010). Advanced oxidation processes (AOP) involving ultrasound for waste water treatment: a review with emphasis on cost estimation. *Ultrason. Sonochem.* 17: 990–1003.

Manceau, A. and Gallup, D. (1997). Removal of selenocyanate in water by precipitation: characterization of copper-selenium precipitate by X-ray diffraction, infrared, and X-ray absorption. *Environ. Sci. Tech.* 31: 968–976.

Mandal, S., Mayadevi, S., and Kulkarni, B.D. (2009). Adsorption of aqueous selenite [Se(IV)] species on synthetic layered double hydroxide materials. *Ind. Eng. Chem. Res.* 48: 7893–7898.

Manning, B.A. and Burau, R.G. (1995). *Environ. Sci. Technol.* 29: 2639–2646.

Martin, A.J., Jones, R., and Buckwalter-Davis, M. (2009). Passive and semi-passive treatment alternatives for the bioremediation of selenium from mine waters. B. C.'s 33rd Annual Mine Reclamation Symposium, Cranbrook, British Columbia.

Mavrov, V., Stamenov, S., Todorova, E. et al. (2006). *Desalination* 201: 290–296.

Meng, X., Bang, S., and Korfiatis, G.P. (2002). *Water Res.* 36: 3867–3873.

Merrill, D.T., Manzione, M.A., Peterson, J.J. et al. (1986). Field evaluation of arsenic and selenium removal by iron coprecipitation. *J. Water Pollut. Control Fed.* 58 (1): 18–26.

Miller, S.M. and Zimmerman, J.B. (2010). Novel, bio-based, photoactive arsenic sorbent: TiO_2-impregnated chitosan bead. *Water Res.* 44 (19): 5722–5729.

Mills, S.J., Christy, A.G., Génin, J.-M.R. et al. Nomenclature of the hydrotalcite supergroup: natural layered double hydroxides. *Mineral. Mag.* 76: 1289–1336.

Moore, L. and Mahmoudkhani, A. (2011). Methods for removing selenium from aqueous systems. Proceedings Tailings and Mine Waste 2011 Vancouver, BC, November 6 to 9.

MSE Technology Applications Inc. (2001). Final Report – Selenium treatment/removal alternatives demonstration project. Mine Waste Technology Program Activity III, Project 20. Report prepared for US EPA, National Energy Technology Laboratory, Office of Research and Development, Cincinnati, OH, and US Department of Energy, Federal Energy Technology Center, Pittsburg, PA, USA. http://www.epa.gov/ORD/NRMRL/pubs/600r01077/600r01077.pdf

Murphy, A.P. (1988). *Ind. Eng. Chem. Res.* 27: 187–191.

Myneni, S.C.B., Tokunaga, T.K., and Brown, G.E. (1997). *Science* 278: 1106–1109.

Naguyen, V.M.H., Amal, R., Beydoun, D. et al. (2005). Photocatalytic reduction of selenium ions using different TiO_2 photocatalysts. *Chem. Eng. Sci.* 60 (21): 5759–5769.

Nakajima, T., Yamada, K., Idehara, H. et al. (2011). Removal of selenium (VI) from simulated wet flue gas desulfurization wastewater using photocatalytic reduction. *J. Water Environ. Technol.* 9 (1): 13–19.

NAMC (2010). Review of Available Technologies for the Removal of Selenium from Water; Final Report prepared for North American Metals Council. CH2MHill: Englewood, CO, USA. Available online: http://www.namc.org/docs/00062756.PDF (accessed March 2013).

Neis, U. (2015). The use of power ultrasound for waste and biomass treatment. In: *Power Ultrasonics – Applications of High-Intensity Ultrasound* (eds. J.A. Gallego-Juarez and K.F. Graff). Oxford/UK: Woodhead Publishing.

NIH (2009). Dietary Supplement Fact Sheet: Selenium (http://ods.od.nih.gov/factsheets/selenium.asp#h7). National Institutes of Health; Office of Dietary Supplements. Retrieved 1 May 2009.

NSMP, Nitrogen and Selenium Management Program (2007). Identification and assessment of selenium and nitrogen treatment technologies and best management practices. Orange County, CA. http://www.ocnsmp.com/team/NSMP%20Task%202.2%20Update%2030Mar07.pdf

Olegario, J.T., Yee, N., Miller, M. et al. (2009). *J. Nanopart. Res.* 12: 2057–2068.

Oremland, R.S., Blum, J.S., Culbertson, C.W. et al. (1994). Isolation, growth, and metabolism of an obligately anaerobic, selenate-respiring bacterium, strain SES-3. *Appl. Environ. Microbial.* 60: 3011–3019.

Patterson, J.W. (1985). Selenium. In: *Industrial Wastewater Treatment Technology* (ed. J.W. Patterson), 395–403. New York;, NY: Butterworth-Heinemann.

PDARC (2000). Panel on Dietary Antioxidants and Related Compounds, Subcommittees on Upper Reference Levels of Nutrients and Interpretation and Uses of DRIs, Standing Committee on the Scientific Evaluation of Dietary Reference Intakes, Food and Nutrition Board, Institute of Medicine (15 August 2000). In: *Dietary Reference Intakes for Vitamin C, Vitamin E, Selenium, and Carotenoids* (ed. Institute of Medicine), 314–315. Institute of Medicine. http://www.nap.edu/openbook.php?record_id=9810&page=315. doi:10.17226/9810.

Peak, D. (2006). Adsorption mechanisms of selenium oxyanions at the aluminum oxide/water interface. *J. Colloid Interface Sci.* 303 (2): 337–345.

Peng, H., Zhang, N., He, M. et al. (2015). *Talanta* 131: 266–272.

Pinochet, H., de Gregori, I., Lobos, M.G., and Fuentes, E. (1999). Selenium and copper in vegetables and fruits grown on long-term impacted soils from Valparaiso region, Chile. *Bull. Environ. Contam. Toxicol.* 63: 327–334.

Pontius, F.W. (1995). An update of the Federal Drinking Water Regs. *Am. Water Works Assoc.* 87 (2): 11.

Qiu, S.R., Lai, H.F., Roberson, M.J. et al. (2000). *Langmuir* 16: 2230–2236.

QTR, Quadrennial Technology Review (2015). Chapter 6 Innovating clean energy technologies in advanced manufacturing, Technology Assessments, US Department of Energy, USA.

Quinn, N.W.T., Lundquist, T.J., Green, F.B. et al. (2000). Algal bacterial treatment facility removes selenium from drainage water. *Calif. Agric.* 54: 50–55.

Rai, S., Wasewar, K.L., Lataye, D.H. et al. (2013). Seawater neutralized red mud as an adsorbent. *Res. J. Eng. Technol.* 4 (2): 57–61.

Rajoriya, R.K., Prasad, B., Mishra, I.M., and Wasewar, K.L. (2007). Adsorption of benzaldehyde on granular activated carbon: kinetics, equilibrium, and thermodynamic. *Chem. Biochem. Eng. Q.* 21 (3): 219–226.

Ramshaw, C. (1983). "Higee" distillation – an example of process intensification. *Chem. Eng.* 389: 13–14.

Ramshaw, C. (1995). The incentive for process intensification. In: 1st International Conference on proceedings of Intensification for Chemical Industry, London, 1995.

Reay, D., Ramshaw, C., and Harvey, A. (2008). *Process Intensification: Engineering for Efficiency, Sustainability and Flexibility*. Oxford: Butterworth-Heinemann.

Reay, D., Ramshaw, C., and Harvey, A. (2013). *Process Intensification, Engineering for Efficiency, Sustainability and Flexibility*, 2e. Oxford: Butterworth Heinemann.

Reill, C. (1996). *Selenium in Food and Health*, 308. London: Blackie Academic and Professionals.

Reis, N. and Li, P.G. (2014). Oscillatory Baffled Reactor and Gas-liquid Reaction Process, WO2016009177A1.

Richards, B., Capão, D., and Schäfer, A. (2008). Renewable energy powered membrane technology. 2. The effect of energy fluctuations on performance of a photovoltaic hybrid membrane system. *Environ. Sci. Technol.* 42: 4563–4569.

Richards, L.A., Richards, B.S., and Schäfer, A.I. (2011). Renewable energy powered membrane technology: Salt and inorganic contaminant removal by nanofiltration/reverse osmosis. *J. Membr. Sci.* 369: 188–195.

Santos, S., Ungureanu, G., Boaventura, R., and Botelho, C. (2015). Selenium contaminated waters: an overview of analytical methods, treatment options and recent advances in sorption methods. *Sci. Total Environ.* 521–522: 246–260.

Shafey, E.I. (2007). Removal of Se(IV) from aqueous solution using sulphuric acid-treated peanut shell. *J. Environ. Manage.* 84: 620.

Shaik, K.M. and Xu, W. (2017). Ethylene oligomerization process, U.S. Patent No. 9,598,329.

Shamas J., Wagener C., and Cooke T. (2009). Technologies and strategies for the treatment of selenium as a micro constituent in industrial wastewater. WEF Microconstituents and industrial water quality specialty conference, Baltimore, MD, 2009.

Sharma, Y.C., Srivastava, V., Singh, V.K. et al. (2009). Nanoadsorbents for the removal of metallic contaminants from wastewater. *Environ. Technol.* 30 (6): 583–609.

Shi, K., Wang, X., Guo, Z. et al. (2009). Se(IV) sorption on TiO_2: sorption kinetics and surface complexation modeling. *Colloids Surf. A Physicochem. Eng. Asp.* 349 (1–3): 90–95.

Singh, K., Lataye, D.H., and Wasewar, K.L. (2016). Removal of fluoride from aqueous solution by using low-cost sugarcane bagasse: kinetic study and equilibrium isotherm analyses. *J. Hazard., Toxic, Radioact. Waste* 20 (3): 04015024.

Singh, S., Wasewar, K.L., and Kansal, S.K. (2020). Low-cost adsorbents for removal of inorganic impurities from wastewater. In: *Inorganic Pollutants in Water* (eds. D. Pooja, P. Singh and S.K. Kansal), 173–203. Elsevier Inc.

Sonstegard, J., Pickett, T., Harwood, J., and Johnson, D. (2008). Full Scale Operation of GE ABMet®Biological Technology for the Removal of Selenium from FGD Wastewaters. In: 69th Annual International Water Conference; Curran Associates, Inc.: San Antonio, TX, USA.

Sorg, T.J. and Logsdon, G.S. (1978). Treatment technology to meet interim primary drinking water regulations for inorganics: part 2. *J. AWWA* 70: 379–393.

Sors, G., Ellis, D.R., and Salt, D.E. (2005). *Photosynth. Res.* 86: 373–389.

de Souza, M.P., Chu, D., Zhao, M. et al. (1999). *Plant Physiol.* 119: 565–574.

Squires, R.C., Groves, G.R., and Jhonston, W.R. (1989). Economics of selenium removal from drainage water. *J. Brig. Drain. Eng.* 115: 18–57.

SSERP, Salton Sea Ecosystem Restoration Program (2005). Final Technologies and management techniques to limit exposure of selenium. April 2005

Staicu, L.C., van Hullebusch, E.D., Oturan, M.A. et al. (2015). *Chemosphere* 125: 130–138.

Stankiewicz, A. and Moulijn, J. (2000). Process intensification: transforming chemical engineering. *Chem. Eng. Prog.* 96 (1): 22–34.

Stefaniak, J., Dutta, A., Verbinnen, B. et al. (2018). Selenium removal from mining and process wastewater: a systematic review of available technologies. *J. Water Supply: Res. Technol. – AQUA* 67 (8): 903–918.

Strawn, D., Doner, H., Zavarin, M., and McHugo, S. (2002). Microscale investigation into the geochemistry of arsenic, selenium, and iron in soil developed in pyritic shale materials. *Geoderma* 108: 237–257.

Takata, M. (2003). The effects of increased salinity on the algal bacterial reduction of selenium from excess wastewater in the Panoche Drainage District of the San Joaquin Valley. http://nature.berkeley.edu/classes/es196/projects/2003final/Takata.pdf

Tan, L.C., Nancharaiah, Y.V., van Hullebusch, E.D., and Lens, P.N.L. (2016). Selenium: environmental significance, pollution, and biological treatment technologies. *Biotechnol. Adv.* 34: 886–907.

Tang, C., Huang, Y.H., Zeng, H., and Zhang, Z. (2014). Reductive removal of selenate by zero-valent iron: the roles of aqueous Fe2þ and corrosion products, and selenate removal mechanisms. *Water Res.* 67: 166–174.

Tian, N., Zhou, Z., Tian, X. et al. (2017). Superior capability of MgAl2O4 for selenite removal from contaminated groundwater during its reconstruction of layered double hydroxides. *Sep. Purif. Technol.* 176: 66–72.

Tian, Y., Demirel, S.E., Hasan, M.M.F., and Pistikopoulos, E.N. (2018). An overview of process systems engineering approaches for process intensification: state of the art. *Chem. Eng. Process.* 133: 160–210.

Tokunaga, T.K., Pickering, I.J., and Brown, G.E. Jr. (1996). Selenium transformations in ponded sediments. *Soil Sci. Soc. Am. J.* 60: 781–790.

Trussell, R.A., Trussell, A., and Kraft, P. (1980). Selenium Removal from Groundwater Using Activated Alumina. EPA-600/12-80-153.

Tsouris, C. and Porcelli, J.V. (2003). Process intensification – has its time finally come? *Chem. Eng. Prog.* 99 (10): 50–55.

Tsouris, C., Culbertson, C.T., DePaoli, D.W. et al. (2003). Electrohydrodynamic mixing in microchannels. *AIChE J.* 49: 2181–2186.

Twidwell, L., McClosky, J., Miranda, P., and Gale, M. (1999). *Technologies and Potential Technologies for Removing Selenium from Process and Mine Wastewater*. Montana Tech, University of Montana.

USEPA, United States Environmental Protection Agency (2006). Edition of the Drinking Water Standards and Health Advisories, EPA-822-R-06-013, Office of Water, USEPA, Washington, DC.

USEPA, United States Environmental Protection Agency (2014). Reference guide to treatment technologies for mining-influenced water, EPA 542-R-14-001, Office of Superfund Remediation and Technology Innovation

USEPA. United States Environmental Protection Agency (1998). Permeable Reactive Barrier Technologies for Contaminant Remediation. http://clu-in.org/download/rtdf/prb/reactbar.pdf.

Uslu, H., Demir, G., Bayat, C. et al. (2014). Adsorptive separation of adipic acid from aqueous solutions by perlite or its composites by manganese or copper. *Membr. Water Treat., Int. J.* 5 (4): 295–304.

Van Ginkel, S.W., Ahn, C.H., Badruzzaman, M. et al. (2008). Kinetics of nitrate and perchlorate reduction in ionexchange brine using the membrane biofilm reactor (MBfR). *Water Res.* 42 (15): 4197–4205.

Van Ginkel, S.W., Zhou, C., Lien, M., and Rittmann, B.E. (2010). Hydrogen-based nitrate and selenate bioreductions in flue-gas desulfurization brine. *J. Environ. Eng.* 137 (1): 63–68.

Van Ginkel, S.W., Yang, Z., Kim, B.-O. et al. (2011). The removal of selenate to low ppb levels from flue gas desulfurization brine using hydrogen based membrane biofilm reactor (MBfR). *Bioresour. Technol.* 102: 6360–6364.

Vries, de W. (2008). Action plan for process intensification, Creative Energy, Energy Transmission, Ministry of Economic Affairs, Netherland, http://www.creative-energy.org.

Wang, P. and Lo, I.M.C. (2009). Synthesis of mesoporous magnetic gamma-Fe_2O_3 and its application to Cr(VI) removal from contaminated water. *Water Res.* 43 (15): 3727–3734.

Wasewar, K.L. (2010). Adsorption of metals onto tea factory waste: a review. *Int. J. Res. Rev. Appl. Sci.* 3 (3): 303.

Wasewar, K.L., Ravichandra, Y., Kumar, M.A., and Godbole, V. (2007). Adsorption mechanism for the adsorption of heavy metals using tea waste as an adsorbent, i-Manager's. *J. Future Eng. Technol.* 3 (1): 41.

Wasewar, K.L., Atif, M., Prasad, B., and Mishra, I.M. (2008a). Adsorption of zinc using tea factory waste: kinetics, equilibrium and thermodynamics. *Clean (Weinh)* 36 (3): 320–329.

Wasewar, K.L., Atif, M., and Prasad, B. (2008b). Characterization of factory tea waste as an adsorbent for removal of heavy metals. *J. Future Eng. Technonl.* 3 (3): 47–53.

Wasewar, K.L., Prasad, B., and Gulipalli, S. (2009a). Adsorption of selenium using bagasse fly ash. *Clean* 37 (7): 534–543.

Wasewar, K.L., Prasad, B., and Gulipalli, S. (2009b). Removal of selenium by adsorption onto granular activated carbon (GAC) and powdered activated carbon (PAC). *Clean* 37 (11): 872–883.

Wasewar, K.L., Atif, M., Prasad, B., and Mishra, I.M. (2009c). Batch adsorption of zinc on tea factory waste. *Desalination* 244 (1–3): 66–71.

Wasewar, K.L., Kumar, S., and Prasad, B. (2009d). Adsorption of tin using granular activated carbon. *J. Environ. Prot. Sci* 3: 41–52.

Wasewar, K.L., Kumar, P., Chand, S. et al. (2010). Adsorption of cadmium ions from aqueous solution using granular activated carbon and activated clay. *Clean (Weinh)* 38 (7): 649–656.

Wasewar, K.L., Singh, S., and Kansal, S.K. (2020). Process intensification of treatment of inorganic water pollutants. In: *Inorganic Pollutants in Water* (eds. D. Pooja, P. Singh and S.K. Kansal), 245–271. Elsevier Inc.

Wikipedia. (2019). https://en.wikipedia.org/wiki/Selenium referred on 10 December 2019.

Yamani, J.S., Lounsbury, A.W., and Zimmerman, J.B. (2014). Adsorption of selenite and selenate by nanocrystalline aluminum oxide, neat and impregnated in chitosan beads. *Water Res.* 50: 373–381.

Yang, G. and Zhou, R. (1994). Further observations on the human maximum safe dietary selenium intake in a Seleniferous area of China. *J. Trace Elem. Electrolytes Health Dis.* 8 (3–4): 159–165.

Yoon, I.-H., Kim, K.-W., Bang, S., and Kim, M.G. (2011). Reduction and adsorption mechanisms of selenate by zero-valent iron and related iron corrosion. *Appl. Catal. Environ.* 104: 185–192.

Youwen, Y., Vance, G.F., and Zhao, H. (2001). Selenium adsorption on mg–Al and Zn–Al layered double hydroxides. *Appl. Clay Sci.* 20: 13.

Yu, J. and Klarup, D. (1994). *Water Air Soil Pollut.* 75: 205–225.

Zawislanski, P.T. and Zavarin, M. (1996). Nature and rates of selenium transformations: a laboratory study of Kesterson reservoir soils. *Soil Sci. Soc. Am. J.* 60: 791–800.

Zelmanov, G. and Semiat, R. (2013). Selenium removal from water and its recovery using iron (Fe^{3+}) oxide/hydroxide-based nanoparticles sol (NanoFe) as an adsorbent. *Sep. Purif. Technol.* 103: 167–172.

Zhang, Y.Q. and Moore, J.N. (1996). Selenium fractionation and speciation in a wetland system. *Environ. Sci. Technol.* 30: 2613–2619.

Zhang, Y.Q. and Moore, J.N. (1997). Interaction of selenate with a wetland sediment. *Appl. Geochem.* 12: 685–691.

Zhang, Y.Q. and Moore, J.N. (1998). Selenium accumulation in a Wetland Channel, Benton Lake, Montana. In: *Environmental Chemistry of Selenium* (eds. W.T. Frankenberger Jr. and R.A. Engberg), 243–257. New York;: Marcel Dekker, Inc.

Zhang, Y., Amrhein, C., and Frankenberger, W.T. Jr. (2005a). *Sci. Total Environ.* 350: 1–11.

Zhang, Y., Wang, J., Amrhein, C., and Frankenberger, W.T. Jr. (2005b). *J. Environ. Qual.* 34: 487–495.

Zhang, L., Liu, N., Yang, L., and Lin, Q. (2009). Sorption behavior of nano-TiO_2 for the removal of selenium ions from aqueous solution. *J. Hazard. Mater.* 170 (2–3): 1197–1203.

Zingaro, R.A., Dufner, D.C., Murphy, A.P., and Moody, C.D. (1997). *Environ. Int.* 23: 299–304.

17

The Emerging Threat of Selenium Pollution

A Spatial Analysis of Its Sources and Vulnerable Areas in India

Ipsita Nandi[1], Rahul Harshwardhan[2], and Anuradha Kumari[3]

[1] *Institute of Environmental Science, Banaras Hindu University, Varanasi, Uttar Pradesh, India*
[2] *Office of Registrar General of India, Ministry of Home Affairs, New Delhi, India*
[3] *School of Environmental Science, Jawaharlal Nehru University, New Delhi, India*

17.1 Introduction

Selenium is a naturally occurring element which when released from geogenic and anthropogenic sources in excess amounts causes pollution of aquifers. At trace levels, selenium is a micronutrient while at moderate levels it is observed as a potent toxic pollutant (Reich and Hondal 2016). Selenium was earlier an unknown pollutant in developing countries, but recently selenium gain the attention of scientists owing to its detrimental effects on biota at concentration above 5 mg/kg of dry weight (Janz et al. 2010; Staicu et al. 2017). Selenium inhibits growth and causes gross deformities in wildlife, as well as killing bottom-feeding ducks and fish (Karthikeyan et al. 2019).

Selenium pollution is a wide-reaching phenomenon and its sources include mining, agricultural, petrochemical, and manufacturing industries as well as surface runoff (Lemly 2002). Anthropogenic activities contributing to selenium pollution include both basic (agriculture) and highly advanced activities (industries) (Lemly 2004). Specific industrial activities releasing selenium-rich effluents include metal and oil refining, mining and power generation (Staicu et al. 2017). In aquatic environments, rapid bioaccumulation of selenium results in toxicity of aquatic organisms (Lemly 2004). Selenium contamination of aquifers occurs in urban and rural landscape varying from high peaks, plains, and arid areas to rainforests, tropical, and snowcapped regions (Lemly 2002).

Selenite is the most toxic water-soluble form of selenium. Toxic effects of Se are the result of sequence of processes occurring in nature (Luoma and Rainbow 2008). Dilution in water and its redistribution determine the level of contaminants remaining after input from sources. The speciation of selenium affects its transformation from aquifer to living and nonliving organism in the food web. The concentration level at the bottom of the food web decides the amount of contaminant being transferred to lower trophic levels. Food web transfer determines the extent of exposure of animals from higher trophic levels.

Selenium Contamination in Water, First Edition. Edited by Pooja Devi, Pardeep Singh, Arindam Malakar, and Daniel Snow.

Manifestation of toxicity in organisms is determined by their degree of internal exposure. Selenium mainly acts as a reproductive toxicant (both a gonadotoxicant and a teratogen) and the degree of damage determines the negative impact on populations.

There exists an enormous contradiction about the ecological risks of selenium in aquatic sphere, more than any other chemical contaminant (Luoma and Presser 2009). Numerous variations are reported in regulations among the countries and environments for selenium leading to lack of proper management strategies (Luoma and Presser 2009). Though selenium contamination is caused at a regional scale, its bioaccumulation effect threatens transboundary flora and fauna. For instance, migratory birds are severely affected with bioaccumulation of selenium in their eggs in the form of altered reproductive system and progeny loss (Hamilton 1999; Presser et al. 2004). Lack of proper information regarding the sources of selenium and vulnerable areas increases the cost of managing Se risks. It is also expected that increased exploitation of coal and other mining activities as well as industrial developments will likely to enhance Se-based issues in the near future (Presser et al. 2004).

To develop a global consensus regarding regulation and management of selenium pollution, it is vital to have as much understanding and knowledge of selenium pollution as possible. This book chapter therefore tries to bring forth the details about selenium, its role as an essential element and toxicant, different sources of pollution, global and Indian pollution status and gradation of states based on their vulnerability to contamination by selenium. The entire study is divided into four broad sections: understanding of selenium; toxicity of selenium; selenium pollution; and identification of vulnerable areas.

17.2 Understanding of Selenium

17.2.1 Selenium Chemistry

Selenium, with an atomic weight and number of 78.94 and 34 respectively, is an essential nutritional element (Martens 2003). Being a group XVI member of the periodic table, it is a semi-metallic chemical occupying the position between the metal tellurium and the non-metal sulfur (Martens 2003). In 1817, Jöns Jacob Berzelius first discovered selenium (Boyd 2011). Owing to its semi-metallic nature it exhibits intermediate chemical and physical properties with more similarities to sulfur than tellurium (Reich and Hondal 2016). Selenium forms a range of compounds with hydrogen, metals, or oxygen (Haygarth 1994). The majority of the selenides except for selenite and selenate are insoluble in aquatic solvents (Martens 2003).

In 1957 the discovery of selenoproteins identified selenium as a required element in the human body preventing cell damage. This leads to the synthesis of small antioxidant molecules containing selenium for treatment of numerous ailments like cancer and heart diseases (Huang et al. 2011; Mugesh and Singh 2000). Selenium plays a role in ribosome-mediated protein synthesis as it forms an essential constituent of enzyme glutathione perioxidase. In the genetic code of humans, Selenocysteine forms the 21st amino acid. The selenolate of selenocysteine is a ligand for a number of coenzymes in bacteria (Reich and Hondal 2016). Selenomethionine forms an important dietary source for plants which is further converted into selenocysteine in animals via the transsulfuration pathway

(Reich and Hondal 2016). Selenocysteine thus produced is converted into selenophosphate used as a nucleophile by bacteria (Lacourciere and Stadtman 2001). Taking into consideration the nutritional forms of selenium, L(1)-selenomethionine form exists as the major constituent of total selenium in seleniferous soybeans, corn, and wheat (Schrauzer 2001; Yang et al. 1997). Selenium is accumulated in plants as Se-methylselenocysteine and selenocystathionine (Shrift 1973). Se-methylselenocysteine has been suggested as effective against cancer (Finley et al. 2000). Selenomethionine forms an important constituent of pancreas, liver, kidney, stomach, gastrointestinal mucosa, and muscular proteins (Hansson and Jacobsson 1966; Schrauzer 2001). Unused selenomethionine in protein synthesis is degraded to selenocysteine via the transsulfuration pathway, forming serine and selenide as byproducts (Schrauzer 2000). Selenomethionine forms the basis of bioactive selenium needed for selenoprotein biosynthesis (Schrauzer 2001). Selenomethionine also acts as cellular antioxidant, wherein selenomethionine reacts with peroxynitrite, forming selenomethionine oxide, which is again reduced back to selenomethionine by ascorbic acid (Arteel et al. 1999). Selenomethionine, one of the many forms of selenium, is a major protein constituent of higher plants and animals. Other forms such as G-glutamyl-Se-methylcysteine, Selenohomocysteine, Se-methylselenocysteine, Selenocysteine, Selenocystathionine are present in insignificant amount (Butler et al. 1989; Salbe and Levander 1990).

17.2.2 Uses of Selenium

Selenium with its different forms is used in diverse sectors of society. One of the pioneer use is in solar cells, namely copper indium (gallium) diselenide (CIS or CIGS) as a semiconductor. Previously, high-purity selenium was used as a semiconductor in photocopiers machines and rectifiers while it is now being used in solar cells (Kavlak and Graedel 2013). Selenium removes the green tint on glass imparted by iron impurities and thereby functions as a decolorizer in glass manufacturing. It is used as a coloring agent in ceramics and glass industries (Kavlak and Graedel 2013). Last decade witnesses electrolytic manganese production as the major use of selenium (Kavlak and Graedel 2013). Doping of cobalt electrodes with selenium has shown remarkable enhancement in electrochemical performance of cobalt-based alkaline rechargeable batteries (Du et al. 2011). Selenium lowers the antimony-lead levels of alloys in lead acid batteries and hence it is added to generate lead-free alloys and stainless steels with improved mechanical properties (Kavlak and Graedel 2013).

Being an essential micronutrient, selenium is used as a fertilizer in soils deficient in selenium (IFA 2011) as well as in animal feed (FDA 1994). Presence of selenium in soil increases the microbiome of beneficial microbes but reduces that of its pathogenic forms, thereby decreasing many soil-borne diseases in plants (Liu et al. 2019). Spraying of selenium increased productivity of certain flowering and grain crops (Mohtashami et al. 2020). It enhances productivity in plants by various means like enhancing production of bioactive compounds, alleviation of salinity effect and more (Velez et al. 2019; Zahedi et al. 2019). Selenium was also observed to alleviate toxicity of cadmium by reducing cadmium uptake. It was also used for improvement in concentration of manganese, iron, and zinc in tomato plants and in its fruits (Lima et al. 2019).

As a food supplement, selenium is used in animal feed, weight-loss products, multivitamin preparations, protein mixes and infant formulas (Schrauzer 2001). Selenium-enriched yeast is popular these days, deliberately used as a substitute for food sources of particular nutritional uses (Aguilar et al. 2008). One study reveals association of increased selenium intake with longer telomeres, thereby increasing longevity of cells (Shu et al. 2020). This is further substantiated by the fact that selenoprotein H plays a major role in senescence, genome maintenance, and redox regulation (Zhang et al. 2018). Selenium forms an indispensable constituent in fruit juices due to its role in metabolic activities (Welna et al. 2018). In conjugation with Vitamin E it forms component of some essential enzymes showing redox activity, thereby preventing certain types of cancer, peroxidation of lipids, and many heart diseases, as well as assisting in reducing the undesirable impact of heavy metals (Sullivan et al. 2013; Ulusoy et al. 2013; Welna et al. 2018). Selenium deficiency also leads to cardiomyopathy, muscular dystrophy, and liver and brain dysfunction (Gallignani et al. 2000).

Selenium was also found to have antibacterial properties and was used against *Staphylococcus aureus* (Zang et al. 2020). Here it enhances the autophagy of macrophage cells RAW264.7, thereby inhibiting proliferation of *S. aureus* (Zang et al. 2020). Se deficiency adversely affects viral immunity in humans through reduced proliferation of T cells, lymphocyte-mediated toxicity, and NK cell activity (Ivory et al. 2017). Administration of daily dosage of selenium supplements in patients infected with HIV helps in delaying progression of HIV infection (Muzembo et al. 2019). Selenium provides protection against severe side effects caused by lithium-based medicines like renal, gastrointestinal, neurological, endocrine, and metabolic disorders (Kiełczykowska et al. 2018). Selenium was also effective in reproductive dysfunction in male rats owing to arsenic exposure. It was observed to attenuate accumulation of biometal, oxido-inflammatory stress and activation of caspase-3 protein cascade (Adedara et al. 2019). Certain studies have also reported a decline in Total Cholesterol (TC) and Total Triglycerides (TG) in lipid profile with selenium supplements (Rad et al. 2019). Selenium forms an essential part of the human enzyme system that alters the effects of hydrogen peroxide, thereby preventing oxidative damages (Toppo et al. 2008). Different uses of selenium are classified and detailed in Table 17.1.

17.3 Toxicity of Selenium

17.3.1 Selenium as a Toxic Substance

Though selenium has many uses and is essential to all organisms yet it is also toxic above certain threshold levels (Reich and Hondal 2016). Chemical formation of the compound affects its toxicity and availability on the Earth's crust (Haygarth 1994). Based on this fact, Dihydrogen selenide (H_2Se) is the most poisonous form at room temperature (Martens 2003). Selenium also forms numerous other lethal toxic organic compounds such as selenomethionine, selenocystathionone, dimethylselenopropionic acid, methylselenomethionine, trimethylselenonium, dimethyselenides, and dimethyldiselenides.

These toxic selenium compounds impair protein structure and result in various forms of abnormalities, including reproductive impairment and death of organisms (Maier

Table 17.1 Various uses of selenium.

Category	Uses	References
Electrical and semiconductor devices	1) Thin-film solar cells 2) CIGS semiconductor 3) Photocopiers and rectifiers	Kavlak and Graedel (2013)
Industries	1) Decolorizer in glass manufacturing 2) Coloring agent in ceramics and glass industries 3) Cobalt-based alkaline rechargeable batteries 4) Steels and lead-free alloys	Kavlak and Graedel (2013), Du et al. (2011)
Fertilizers	1) Additive to fertilizers	IFA (2011), Mohtashami et al. (2020)
Pharmaceuticals	1) Alleviate cadmium toxicity 2) Constituent of multivitamin preparations, infant formulas, protein mixes, weight-loss products, and animal feed 3) Prevention of cancer, peroxidation of lipids, and several heart diseases, Reducing the adverse impact of heavy metals 4) Effective against *Staphylococcus aureus* 5) Delaying progression of HIV infection 6) Protective agent against lithium-based medicines 7) Reduced total cholesterol, total triglycerides (TG) levels in lipid profile	Lima et al. (2019), Schrauzer (2001), Welna et al. (2018), Sullivan et al. (2013), Ulusoy et al. (2013), Zang et al. (2020); Muzembo et al. (2019), Kiełczykowska et al. (2018), Rad et al. (2019)
Food and feed	1) Animal feed 2) Selenium-enriched yeast 3) Fruit juices	FDA (1994), Aguilar et al. (2008), Welna et al. (2018)
Miscellaneous	1) Protection against reproductive dysfunction in male rat from arsenic exposure 2) Prevents oxidative damages	Adedara et al. (2019), Toppo et al. (2008)

et al. 1993). Selenium in its inorganic forms is more toxic compared to its organic forms. For example, toxicity of selenomethionine (organic form) is less compared to its inorganic counterparts (Yücel et al. 2020). Toxicity of selenium is attributed to its pro-oxidant role as it causes the formation of Reactive Oxygen Species (ROS) (Mézes and Balogh 2009; Spallholz 1997; Yücel et al. 2020). Adverse impacts of selenium also include inhibition of tissue respiration by replacing selenium for sulfur in certain essential proteins as well as inhibiting the activity of some sulfhydrylated enzymes, thereby decreasing the glutathione (GSH) level in the tissues (Kaya et al. 1998; Yücel et al. 2020).

Selenium poisoning frequently occurs due to the consumption of huge amounts of seleniferous plants over a long time duration, or through contamination from industry and pharmaceuticals effluents (Goehring et al. 1984; Yücel et al. 2020). In the human system, selenium mainly causes neurotoxicity (Adani et al. 2020). One such example is Amyotrophic Lateral Sclerosis (ALS), a rare but fatal degenerative disease of the motor neurons caused by selenium toxicity (Gunnarsson and Bodin 2018). Laboratory studies further revealed inert forms of selenium to induce toxic effects to neurons concerned in ALS etiology (Filippini et al. 2018). Symptoms of selenium intoxication include bad breath, worm infestation, chest pain, hair and nail abnormalities, nausea and vomiting, breathlessness and bad breath, garlic odor, edema, spontaneous abortion, and overall selenosis and cardiac symptoms (Chawla et al. 2019). It ultimately leads to mortality in several cases. In humans, mortality occurs with ingested doses of 1–100 mg Se/kg or with blood/plasma levels between 300 and 30 000 μg/l (normal level: 100 μg/l). Mortality-based urinary levels of selenium include 170 and 30 000 μg/l (normal level: 20–90 μg/l) (Hadrup and Ravn-Haren 2020). A high selenium level in plasma was also observed to increase the estimated Glomerular Filtration Rate (eGFR) in chronic kidney patients (Wu et al. 2019). Similarly, a high level of selenium in blood increases the risk of diabetes (Zhang et al. 2019).

Excess selenium levels also severely affect aquatic biota. A study on sea cucumber, *Apostichopus japonicus*, for the effect of selenium was found to show different forms of adverse effects like inhibition in gain of body weight, explicit growth rate and protein efficiency ratio, lower levels of antioxidant enzymes, reduced lysozyme activity, and more (Hu et al. 2019). Similar studies have highlighted bioaccumulation and poisonous effect of selenium on algae, diatoms, rotifers, and fishes (Abdel-Hamid and Skulberg 1995; Karthikeyan et al. 2019; Ponce et al. 2018; Schiavon et al. 2017).

Plants are also adversely affected by selenium toxicity majorly targeting its proteome. Selenium stress leads to the formation of various proteins like selenoproteins, oxyproteins, and nitroproteins (Kolbert et al. 2019). Selenite and selenate compounds facilitate absorption of arsenic from soil by paddy plants (Pokhrel et al. 2020). Lesions development in plants as well as lipid peroxidation is some of the adverse effects of selenium toxicity (Silva et al. 2018). Synthesis of non-specific plant proteins hindering the synthesis of normal proteins involved in antioxidant activities are also results of high Se concentration (Gupta and Gupta 2017). Besides, the absorption relationship between Se and sulfur/phosphorus leads to disorder of sulfur/phosphorus metabolism in plants. High Se concentration interferes with sulfur mechanism, displacing sulfur which is found in amino acids and proteins and thereby altering plant growth and metabolism (Dai et al. 2020). Excess of selenium exposure decreases yield as well as adversely affecting physiological seed quality (dos Reis et al. 2020).

All the above-mentioned facts point toward the adverse impact of high selenium levels on every living form, thereby indicating its role as a major heavy metal pollutant.

17.3.2 Mechanism of Selenium Toxicity

The toxic compounds of selenium stated in the above section are formed through the process of biological transformation of selenium-rich diets. Biotransformation of selenium takes place in three steps, namely oxidation, reduction, and methylation (Alexander 2015). Selenium present in compounds are oxidized via glutathione (GSH) as oxidizing agent in the presence of enzyme glutathione peroxidase. In this step, GSH is oxidized to GSSeSG, releasing ROH and water as byproducts. GSSeSG product thus formed is reduced to hydrogen selenide. Herein GSSeSG is converted to GSH-selenopersulfides (GSSeH), which is further changed to hydrogen selenide. The entire step of reduction is catalyzed by enzyme GSH reductase. Formation of hydrogen selenide is accompanied by subsequent methylations using methionine as methyl donor in the presence of methyltransferase enzyme. The process of methylation ultimately leads to the generation of excretal selenium compounds, causing toxicity. The entire selenium toxicity and its mechanism is illustrated in Figure 17.1.

17.4 Selenium Pollution

17.4.1 State of Art of Selenium Pollution

Selenium is present at an estimated quantity of 0.05–0.5 mg Se/kg (Lemly 2004), with rocks like quartzite, limestone, and sandstone having 40% of its total concentrations (Wang and Gao 2001). World mean for selenium content in soils is observed to be 0.4 mg Se/kg with seleniferous soil regions such as in the USA, Canada, Colombia, UK, China, and Russia having concentrations up to 1200 mg Se/kg (Fordyce 2005).The safe concentration of selenium lies between 40 and 400 μg/day (El-Ramady et al. 2014). Selenium pollution widely impacts several ecosystems, agricultural activities as well as human beings. However, owing to its low environmental impact compared to other metal pollutants, it has drawn less global scientific attention than required (Tan et al. 2016).

Selenium pollution occurs by natural as well as anthropogenic sources. Natural sources of selenium pollution include extensively mined geochemical landforms such as volcanoes and different types of igneous and sedimentary rocks (Kessi et al. 1999). It is also associated with fossil fuel, pyrites, and sulfur-containing mineral sources contributing toward 50–65% of selenium emission (Mehdi et al. 2013; Sharma et al. 2015). Anthropogenic activities forms the major factor for selenium mobilization, its accumulation, as well as 37–40% of the total atmospheric emissions of selenium (Wen and Carignan 2007; Winkel et al. 2012).

Anthropogenic sources include metal refinery and their processing plants; coal extraction and combustion, glass, electronic and petroleum industry, agricultural application of fertilizer and sewage sludge, and surface run-offs (Lemly 1997; Nancharaiah and Lens 2015; Pyrzyńska 2002). In addition, selenium-rich effluents are also observed from the power industry. It is substantiated by the fact that ash obtained from the power industry after burning coal has 1250 times more selenium compared to the amount from coal mining (Lemly 2004). Apart from these sources, there are few other sources which are recognized as emerging threat areas for selenium pollution. The wetlands which are constructed as a measure for abatement of heavy metal pollution of water are in turn serving as potent selenium pollution sources. Bioaccumulation of the selenium released from these areas leads to severe hazardous impacts downstream (Blanco 2019; Lemly 2004). Seleniferous ash released as byproduct from airborne particulate treatment plant involving burning of excess coal further add toward selenium pollution (Lemly 2004). Feedlot waste of cattle or

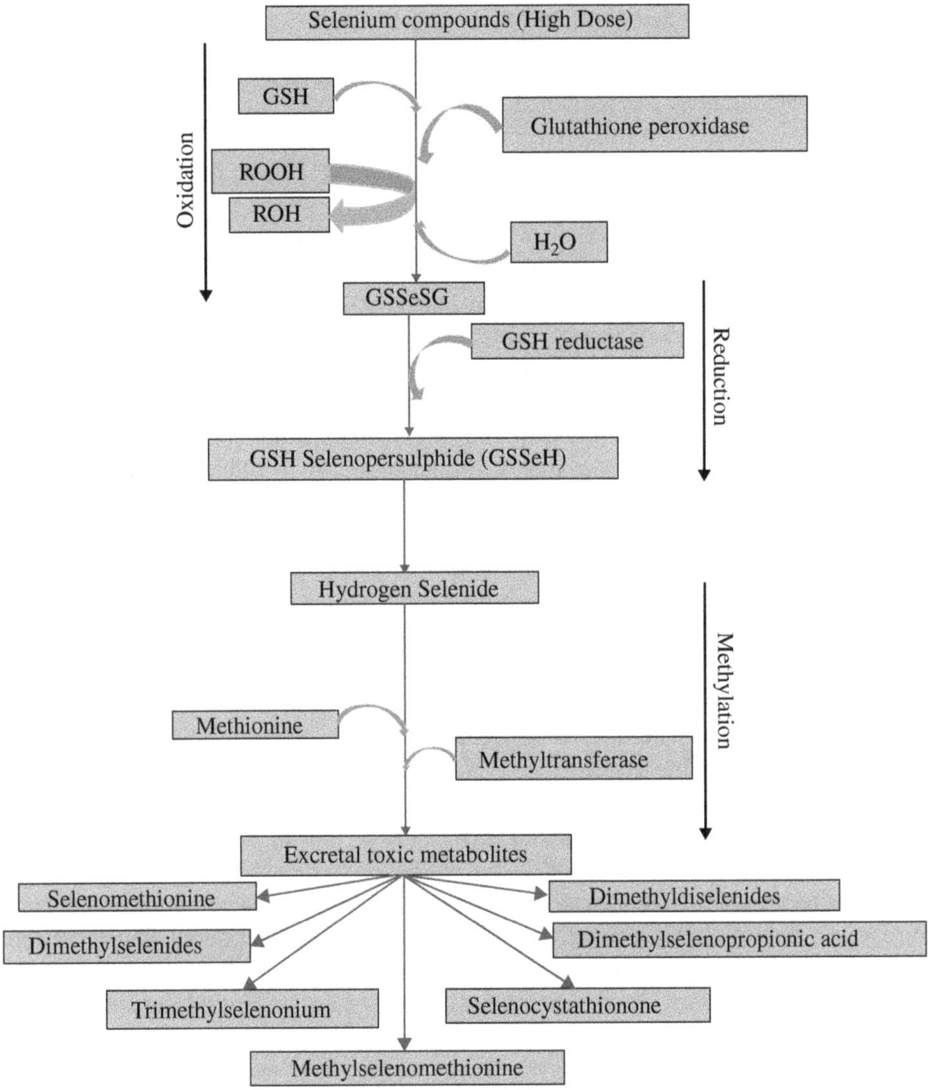

Figure 17.1 Flowchart showing mechanism of selenium toxicity.

swine also serves as a source of selenium pollution ranging from 50 to 150 μg Se/l. Feedlot waste releases a high amount of selenium (50–150 μg Se/l) owing to use of selenium as their nutritional supplement (Beyzi et al. 2019; Lemly 2004). Selenium is also associated with mineral matrix of phosphate deposits and hence is released in large amounts (2–20 μg Se/g) during its open pit mining (Khamkhash et al. 2017; Lemly 2004). Electrolytic production of manganese metal releases huge amounts of selenium due to its use as an additive in the process (Huang et al. 2014; Lemly 2004). Such and many more examples are obtained in literature regarding selenium pollution, highlighting it as an emergent pollutant.

17.4.2 Mechanism of Selenium Pollution

Selenium pollution occurs either by food chain transfer or through roots modification of plants causing long-term environmental concerns (Brown and Arthur 2001; Staicu et al. 2015). Leachates from various selenium-based industries are dispersed freely in the environment, reaching nearby rivers and other water bodies downstream (Etteieb et al. 2020; Sharma and Vance 2007).

Selenium lead oxidative stress is known to substitute sulfur in proteins formed during organogenesis in embryo development. Substituted proteins undergo improper folding leading to teratogenicity and embryotoxicity (Janz et al. 2010).

In the aquatic setting, selenium occurs in four oxidation states, i.e. VI, IV, 0, and −II depending on pH, redox potential, major elements, organic matter, and microbiological activity (Donner et al. 2018; Etteieb et al. 2020). Selenium species thus dominating the water body are further redistributed to different microenvironments by circulation and mixing of water (Lemly 2004). Variation in level of selenium in different aquatic microenvironments is further affected by its abilities of dissolution and desorption from solid phase into liquid phase (Etteieb et al. 2020). Selenite predominates under reducing conditions while alkaline and oxidizing conditions are predominated by selenate (Belzile et al. 2000; Etteieb et al. 2020). Oxygenated surface waters are dominated by species of selenate, selenite, and dissolved organic selenide while sediments and organic matter are attached with organic and inorganic species (Etteieb et al. 2020; Meseck and Cutter 2006). Low flowing rivers in contrast to high flowing ones were observed to have higher selenium content (Doblin et al. 2006). Selenium toxicity occurs either by bioconcentration (direct uptake from water) or through bioaccumulation (via food chain) (Maier et al. 1988).

Uptake and accumulation of selenium by algae is of crucial importance as they form potential vectors of selenium in aquatic environments (Etteieb et al. 2020). They mainly absorb selenium either as selenate or selenite while some species convert selenium to less toxic volatile form to combat its toxicity (Etteieb et al. 2020). Among different phytoplankton species, dinoflagellates accumulate selenium in higher concentration, while least concentration was observed in diatoms, chlorophytes, and cryptophytes (Doblin et al. 2006). Selenium in various forms are transferred by different food chain cycles in aquatic environment. Selenium-methionine is transferred from phytoplankton to zooplankton, followed by fishes, while selenite is primarily taken by zooplankton and transferred to further trophic levels (Sandholm et al. 1973).

Elevated ovarian selenium concentration causes reproductive failure in many fishes (Ogle et al. 1988). In fishes, selenium bound with the yolk precursor protein vitellogenin (VTG) disrupts the stages of egg development resulting in reproductive toxicity (Janz et al. 2010; Mo et al. 2019). Aquatic birds are also severely impacted from selenium pollution. Major impacts include mortality, impaired reproduction with teratogenesis, reduced growth, histopathological lesions, and alterations in hepatic glutathione metabolism (Hoffman 2002). This effect occurs due to oxidative pressure cause by intake and bioaccumulation of selenium in various lipotic tissues of aquatic birds (Spallholz and Hoffman 2002).

Apart from aquatic systems, selenium also advances from prey to predator in the terrestrial food chain. This was substantiated by a study wherein lizards feeding on selenium-enriched crickets showed high bioaccumulation of selenium in their gonads (Hopkins et al. 2005).

Bacteria are also known to uptake selenium and incorporate it into proteins and amino acid within minutes of exposure (Ogle et al. 1988). This may have severe implications in consumer organisms, as gut microbes may lead to bioaccumulation of selenium leading to toxicity (Ogle et al. 1988). Moreover, mineralization, i.e. conversion of inorganic to organic form of an element, becomes an essential step for selenium cycle. Microbial mineralization forms the major mechanism for regenerating and cycling of selenium in both terrestrial and aquatic systems (Maier et al. 1988).

Selenium uptake in plants occurs in the form of Se (IV), Se (VI) and organic Se (selenocysteine and selenomethionine) (Li et al. 2015; Gupta and Gupta 2017). Plants mainly utilize organic forms of selenium (Shahid et al. 2018). Selenium enters plants via sulfate transporters into the plasma membrane of root cells and is transported by phosphate transporters in the entire plant body (Shahid et al. 2018). Excess selenium causes toxic impact on photosynthetic apparatus of plants, leading to chlorosis and necrosis (Naz et al. 2015).

Humans are susceptible to selenium pollution transferred from agricultural soil contaminated food chain (Shahid et al. 2018). Other sources for selenium contamination affecting humans are polluted air, contaminated dairy/milk products or vegetables and drinking of selenium-enriched ground or surface waters (He et al. 2018). The upper tolerance limit set by the European Food Safety Authority (EFSA) guidelines in 2006 was 300 μg/day for adults and 60 μg/day (aged 1–3 years) to 250 μg/day (aged 15–17 years) for children (Mehdi et al. 2013; Sun et al. 2014). Excess selenium levels leads to carcinogenesis, diabetics, neurodegenerative disorders, cardiovascular disease, and severe diseases such as ALS (He et al. 2018).

17.4.3 Pollution at Global Level

The major portion of the selenium present in the Earth's crust is found in rocks, with concentration in soils accounting for only 0.4 mg Se/kg. Sulfide minerals, being the primary source of selenium, range from 1000 to 10 000 mg/kg, especially those associated with metal- and heavy metal-containing sedimentary rocks (Fishbein 1983). Next to rocks, "seleniferous" (Se-rich) coals, which could contain from 0.1 to 4 mg/kg of Se, form another natural source of selenium pollution. Higher concentration of selenium, i.e. approximately 1200 mg Se/kg is observed in seleniferous soils of Colombia, China, USA, UK, Canada, and Russia (Tan et al. 2016). Fossil fuels like black crude oil and shale rocks oil form another major natural source for selenium pollution (Lemly 1997; Shamberger 1981). One such example is the poisoning of Kesterson Wildlife Reservoir situated in California due to selenium pollution from shale-based rocks (Tan et al. 2016). It severely affected the existing wildlife and ultimately led to its closure. Toxic effects included physical deformities, reproductive failures, mutations, and ultimately death of organisms (Ellis et al. 2003). Other areas like Arkansas River, St. Charles River, and Belews Lake were also affected owing to excess release of selenium from marine shale and its derived soils (Tan et al. 2016).

Anthropogenic sources for selenium pollution include coal mining, power plants, selenium-based industries, agricultural activities, and more. Among these coal mining, power plants, and industries are the chief sources of selenium pollution. Table 17.2 shows the

Table 17.2 Significance of different sources for selenium pollution.

Source	Selenium concentration (mg/kg) (range)
Rocks and soil	0.35–0.5
Petroleum and oil refineries	0.005–1.4
Shale rocks	206–280
Electroplating industries	1–300
Coal mining and power plants	0.46–440
Seleniferous soil	1–5000

Source: Compiled from Staicu et al. (2017) and Lemly (2004).

severity of sources based on concentration of selenium. Various examples from the globe further substantiate the above-mention anthropogenic activities as chief sources of selenium pollution.

A study conducted by Young et al. (2010) revealed contamination of various aqueous reservoirs of North America from the discharges of various electroplating and selenium-based industries. Selenium poisoning owing to discharge from a power plant has destroyed many fish and wildlife reservoirs, causing mass deaths in the USA (Lemly 2002). Heavy coal combustion has led to severe particulate pollution of PM10 containing high selenium levels in Taiyuan, China. This high selenium-based particulate pollution has led to various forms of periodontal and dental diseases with large-scale carcinogenic effects (Xie et al. 2006). Leaching of selenium and its bioaccumulation in the aquatic habitat of Rocky Mountain landscape in Alberta, Canada was demonstrated by various field studies owing to open cast coal mining (Lemly 2019). In a study conducted at Xuzhou City, in the Northwest of China, samples of soil near coal mining and power plant areas had high selenium content compared to farming soils and irrigational water, indicating mining and power plants as a major selenium pollution source (Huang et al. 2009). Oil exploration and gas flare phenomena in Nigeria have resulted in severe selenium contamination of its regional surface and groundwater reservoirs (Fayiga et al. 2018). Mollusks from Mannering Bay, Australia were observed to have high selenium content in their tissues. The presence of high selenium levels is due to bioaccumulation by macrobenthos along with bacteria and bioturbation from nearby power plant-based industrial areas (Peters et al. 1999). The dozens of deformed bird embryos observed at the Kesterson Reservoir tragedy event at San Francisco, USA were also caused due to deposition of leached selenium-enriched soil from irrigated soil of the San Joaquin Valley (Deborah et al. 2017). Groundwater of Dorno-Gobi aimag in the south of Mongolia are severely contaminated by high selenium levels due to high rate of urbanization, mining, industrial development, large-scale livestock production, and ever-growing production of cereals and vegetables (Golubkina et al. 2018). The Nyle river system of South Africa had selenium levels higher than the standard recommended value due to nearby intense agricultural activities using pesticides, increased use of hydrocarbon fuels, and leaching from weathered rocks of surrounding geology. Increased selenium levels had led to changes in feeding behavior, skeletal abnormalities in newborn, skewed growth rate, and eventually death of living organisms (Greenfield et al. 2012).

Freshwater of the Elk River was observed to have 12 times greater selenium concentration than the recommended value of 2 μg/l due to surface runoff from five open cast coal mines of southeastern British Columbia located in the Elk River basin (Lussier et al. 2003).

The adverse impact of selenium pollution is observed by its bioconcentration and bioaccumulation effect in the food web. From the Chiapas coast, Mexico, the sample of Silky shark (*Carcharhinus falciformis*) showed a high level of selenium concentration. High levels of selenium also occurred in its prey *Coryphaena hippurus*, indicating clear signs of food chain transfer and biomagnification in *C. falciformis* (Rodríguez-Gutiérrez et al. 2020). A study conducted in Australia on tadpoles of Australian native frogs *Limnodynastes peronii* revealed severe bioaccumulation of Se (IV) in digestive and excretory organs. Of this absorbed selenium, 19% was retained even after 12 days of depuration in clean water (Lanctôt et al. 2017). High discharge and bioaccumulation of selenium are found in various aquatic organisms from poultry feed production in Bangladesh (Islam et al. 2007). Chlorella species, which form half of the global microalgal biomass production was found to accumulate as high as 7.5 μM selenate, which is a highly toxic form of selenium (Gojkovic et al. 2015). Glassfish, *Ambassis jacksoniensis*, found in the estuarine belt in the eastern coast of Australia, was found to accumulate as high as 23 ± 2.2% of selenium. The source of its contamination is the discharges from the towns and industrial units containing high selenium levels (Creighton and Twining 2010).

The above-stated facts and studies highlight the point that though selenium pollution was not given due concerns at early era, it is adversely affecting the ecosystems. In addition, its bioaccumulation is posing a severe threat to aquatic and terrestrial organisms, including humans.

17.4.4 Selenium Pollution of Aquifers in India

In developing countries like India, selenium has been largely unrecognized as pollutant for many years due to the preoccupation with other high-priority pollutants (Lemly 2004). The first survey report for selenium pollution was reported from Gujarat in 1970 owing to a mysterious disease outbreak (Sharma 1984).

The surface runoff from seleniferous soil is considered one of the most important sources of selenium pollution. Consumption of crops grown on seleniferous soils lead to human and animal toxicity (Dhillon and Dhillon 2019). Seleniferous soils are characterized by selenium levels of 0.5 mg Se/kg or more, thereby producing toxic effects on the produce (Dhillon and Dhillon 2003). Seleniferous soils are unevenly distributed in India, mainly concentrating on the northwestern belt (Dhillon and Dhillon 2003). Seleniferous soils were observed to be distributed in the areas of Haryana, Punjab, Meghalaya, West Bengal, and Assam (Dhillon and Dhillon 2003). Runoff from seleniferous soil rich areas illustrated as Figure 17.2. is a leading cause of selenium pollution of aquifers in North India. For instance, the majority of the tube wells sampled in Amritsar, Gurdaspur, and Tarn Taran districts lying in the Majha belt of Punjab have selenium level higher than the BIS standard, i.e. 0.01 mg/l (Virk 2018). The Barwa and Jainpur villages in Punjab also had contaminated concentration of Se (45–341 μg/l) in their groundwater due to leaching from their seleniferous soils. The soils have 2.3–11.6 mg/kg and 3.1 mg/kg of selenium at Jainpur and Barwa respectively (Bajaj et al. 2011). The aquifers found in Bist-Doab area of Punjab have high

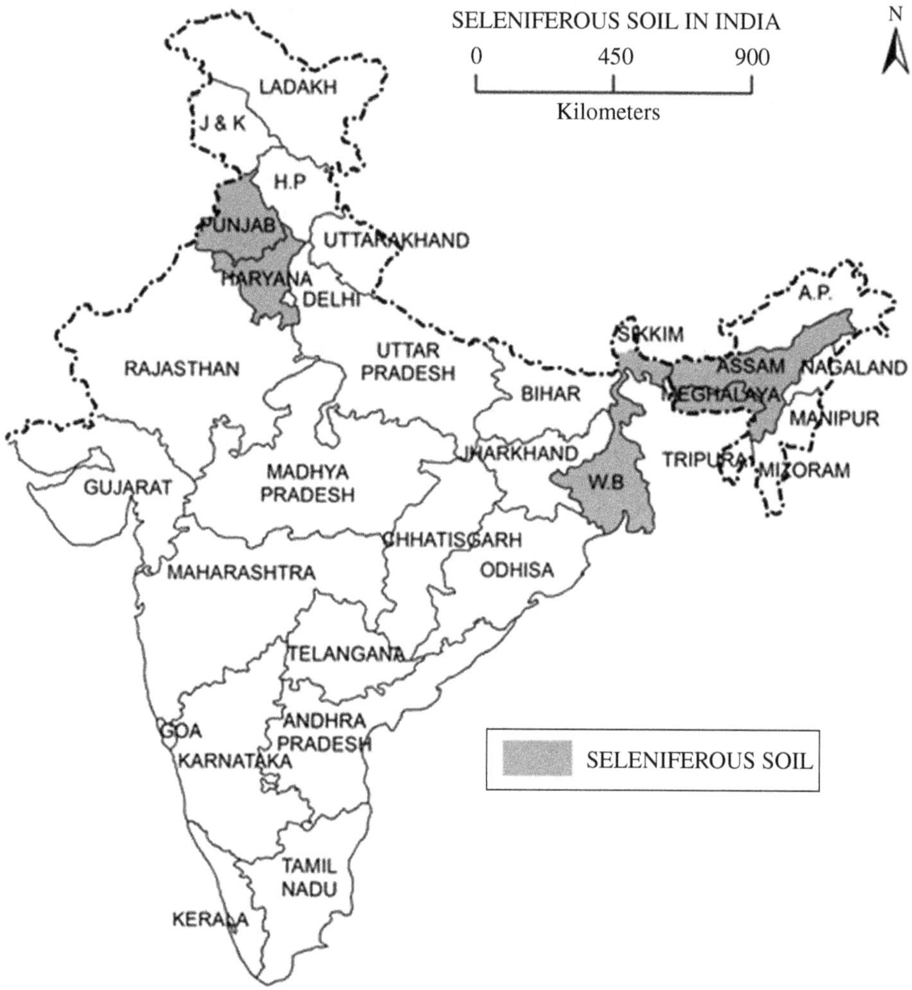

Figure 17.2 Map showing distribution of Seleniferous soil in India.

selenium concentration, i.e. from 10 to 40 μg/l (Lapworth et al. 2017). Groundwater of Jhajjar district, Haryana are contaminated with selenium due to geogenic contamination from seleniferous soils (Gupta and Misra 2018). The Siwalik Foothill Zone in Northeast has higher than average concentration of Se owing to runoff from seleniferous soils of Siwalik region (Singh et al. 2020a; Dhillon and Dhillon 2014).

Industries, mining, coal burns, power plants, agricultural, and surface runoff from shale rocks form other important sources of selenium pollution (Jyothi et al. 2020). Damodar river sediment in the Raniganj-Durgapur industrial belt, West Bengal was observed to have selenium levels much above the recommended standards (De et al. 1985). Groundwater of Guwahati was highly contaminated, with selenium ranging from 0.4 to 3.85 mg/l because of the solid waste generated from different selenium-based industries (Lakhimi 2011). A high level of selenium was found in the water of River Ganges at Kanpur due to industrial discharge and agricultural runoff (Singh et al. 2020b). In Ahmedabad, two lakes, i.e. Ambli and Makarba Lake, has high selenium concentration, i.e. 395.4 and 641.4 μg/l respectively due to different selenium-based industrial discharge (Paikaray 2016; Vediya and Srivastava 2008). Selenium content of more than 4.48 μg/g was observed around Sagar Island in West Bengal, polluted by industrial emission (Paikaray 2016). Elevated selenium levels were found in the Alwar and Jaipur districts of Rajasthan due to industrial discharge (Yadav et al. 2005). Water samples collected from oil drilling sites at Assam had high selenium content compared to other sites (Dey et al. 2015). Coal mining sites in the Assam coal fields were observed to release up to 0.24 mg/kg of selenium (Mukherjee and Srivastava 2005). Selenium contamination was found in certain districts of Maharashtra due to surface runoff from shale rocks (Paikaray 2016). Effluent from Jharia coal field, Jharkhand was observed as the main environmental concern for selenium contamination of aquifers (Finkelman and Stracher 2011). Industrial effluents and emissions from the Barauni industrial area, Bihar has led to selenium contamination of its groundwater aquifers (Sharma and Kumar 2006). Untreated industrial waste dumped in low-lying ponds and groundwater aquifers has led to severe selenium pollution in Chennai city, Tamil Nadu (Kumar and Riyazuddin 2011). Figure 17.3 illustrates the selenium-polluted areas based on the literature mentioned above.

17.5 Identification of Vulnerable Areas

Based on the information obtained from literature, an attempt was made to identify vulnerable areas for selenium pollution. On the basis of the number of sources, gradient map vulnerability for different districts was prepared (Figure 17.4). Single-source states were considered the least vulnerable. States having two sources were put into the moderately vulnerable category while those having more than two sources were identified as highly vulnerable (Figure 17.4).

17.5.1 Least Vulnerable Areas

States having a single source of selenium pollution were classified as least vulnerable. In many districts around the country industrial discharge was observed as the prominent source of selenium pollution. The states classified in this category were Assam, Gujarat, Karnataka, Kerala, Maharashtra, and Tamil Nadu. The districts relating to each state were:

1) **Assam**: Tinsukia district is a prime coal mining area and hence it is one of the least vulnerable areas to selenium pollution (Baruah et al. 2016).
2) **Gujarat**: The majority of its districts, namely Ahmedabad, Anand, Bharuch, Panch-Mahal, Bhavnagar, Patan, Mahesana, Gandhinagar, Kheda, and Sabar-Kantha, are dedicated to industry (Rathi 2003). These industrial districts can serve as potential sources of selenium discharge into nearby aquifers. Furthermore since districts are related to a single source of selenium they are the least vulnerable to pollution from it.
3) **Karnataka**: This state is an industrial hub, with many of its districts dedicated to industrial setup. Districts where large number of industries located are Bangaluru, Chamrajnagar, Hassan, Mandya, Mysuru, Ramnagarana, and Tumkur (Chertow

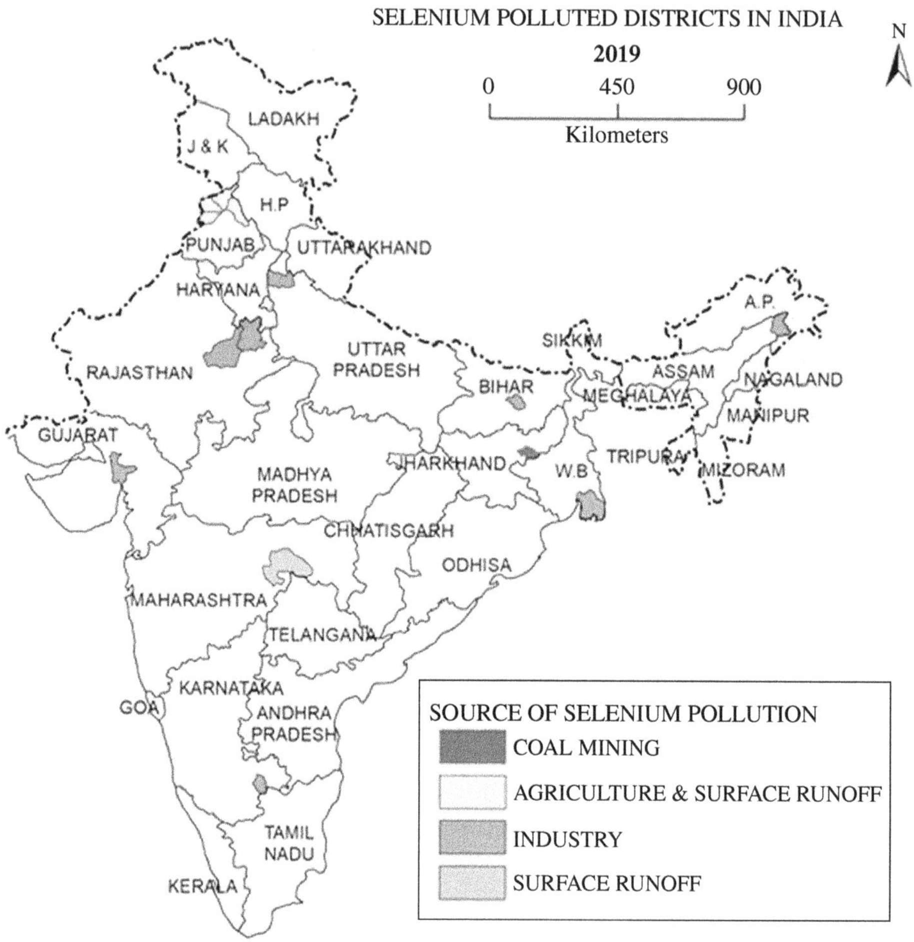

Figure 17.3 Map showing selenium-polluted districts in India with their source of pollution.

et al. 2019; Kambalagere 2020; Kerutagi et al. 2020; Kuzhiparambil 2020; Prabhavathi and Ramakrishna Parama 2019; Shanshak et al. 2020). These districts have industrial discharge serving as the potential source of selenium pollution. Since selenium discharge is observed only through a single source they are therefore categorized as least vulnerable.

4) **Kerala**: Certain districts of Kerala are dedicated to heavy industries releasing large amounts of heavy metals, contaminating nearby aquifers. In Alappuha, Kollam, Kottayam, Pathanamthitta, and Tiruanathpuram districts, large clusters of industries exist (Meera and Anand 2019). These clusters form potential sources for selenium discharge into water bodies. Since industries were observed as the only source of selenium pollution, these districts therefore fall into the least vulnerable group.
5) **Maharashtra**: Two districts namely Ahmednagar and Bid were categorized as least vulnerable to selenium pollution as they have a single selenium source of effluent affecting water reservoirs. These districts have large clusters of chemical industries (Jadhav

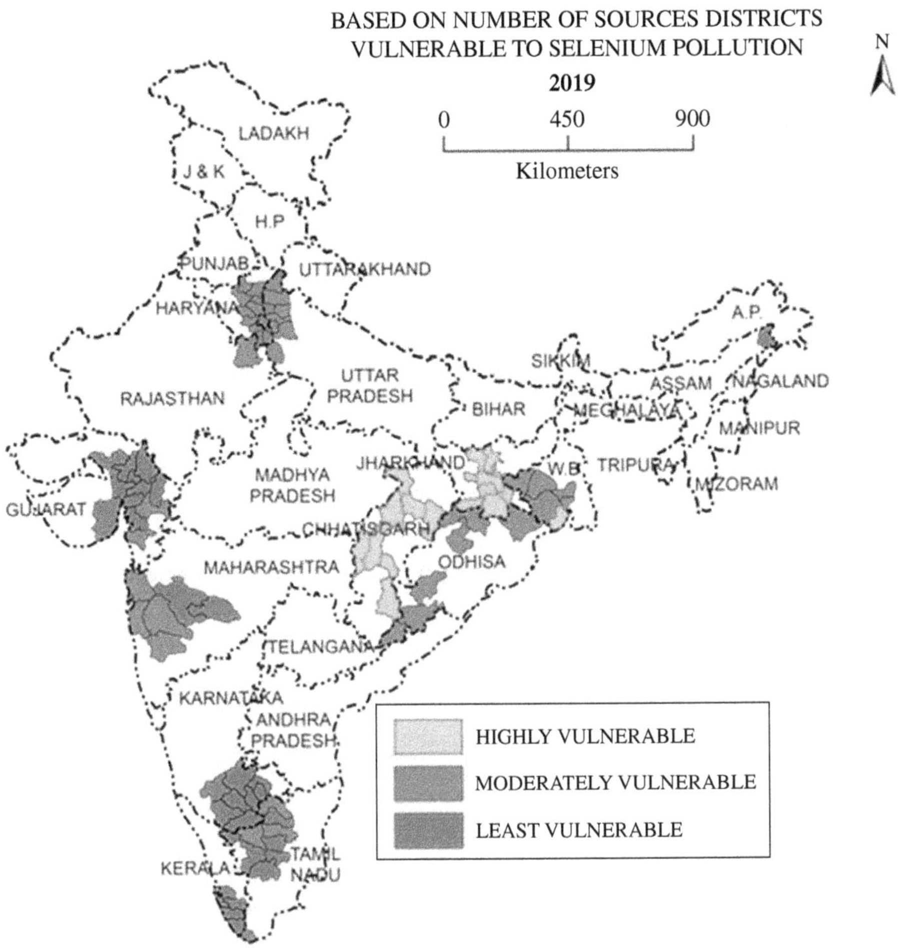

Figure 17.4 Map of potential vulnerable areas to selenium pollution based on number of sources.

and Bharambe 2020; Sawant 2019) which may serve as potential discharge points of selenium into nearby aquifers.

6) **Tamil Nadu**: Aquifers of certain districts were frequently reported to be contaminated by chemicals released from chemical industries. In Namakkal, Nilgiri, Dharmapuri, Tiruppur, and Dindigul districts there are large clusters of heavy industries (Jeyaraj et al. 2019; Rajeswari et al. 2019; Sasmitha et al. 2019; Tamilchelvan and Muralimohan 2019). Since there are clusters of chemical industries there is a enormous possibility of release of selenium. However, since no other sources of selenium exist, these districts are clustered as least vulnerable to selenium pollution.

17.5.2 Moderately Vulnerable Areas

States having two different selenium pollution sources were classified as moderately vulnerable. Pairs of sources mainly identified in these states for discharge of selenium were industry and power plant, industry and surface runoff, industry and petroleum-based shale rocks, industry and agricultural runoff, industry, and coal mining. The states classified in this category were Gujarat, Haryana, West Bengal, Maharashtra, Tamil Nadu, Odisha, and Uttar Pradesh. Districts vulnerable in these states are discussed below:

1) **Gujarat:** Surat and Vadodara are two districts moderately vulnerable to selenium pollution as they have chemical industries as well as power plants (Fujita et al. 2019; Nihalani and Mishra 2020). Since these districts have two sources, the aquifers of these districts are more vulnerable to get contaminated with selenium pollution.
2) **Haryana**: This state has seleniferous soils prone to release of selenium in nearby aquifers as surface runoff (Dhillon and Dhillon 2019). Further large numbers of chemical industries are present in different districts, having the potential to release large excess of selenium into water bodies (Kumar et al. 2019; Ravish et al. 2019, 2020; Singh et al. 2019). Based on these two facts, districts namely Faridabad, Kaithal, Jhajjar, Palwal, Panipat, Karnal, Jind, Gurgaon, Mewat, Rohtak, and Sonipat may serve as moderately vulnerable sites for selenium pollution. This is due to the presence of two potential sources of selenium discharges into aquifers.
3) **Maharashtra**: Industries and runoff from petroleum origin shale rocks are major potential sources for selenium discharge. Mumbai, being situated in the coastal areas, receives runoff from petroleum origin shale rocks and from clusters of many chemical industries into its aquifers (Dasgupta and Mukherjee 2020; Kumar et al. 2019; Manaschintan and Nalini 2020). Therefore Mumbai becomes a district moderately vulnerable to selenium pollution. Apart from this, agricultural runoff also becomes an important potential source of selenium in other districts of Maharashtra. Other districts, namely Mumbai-suburban, Pune, Raigarh, Satara, and Thane serve as moderately vulnerable spots for selenium pollution. The aquifers of these districts receive agriculture runoff as well as discharge from the chemical industry (Gaikwad and Pawar 2020; Jagtap 2019; Khamkar and Mhaske 2019; Prajapati et al. 2020; Shree and Nishikant 2019; Sonawane 2019). As these districts receive discharges from two potential selenium sources, they were categorized as moderately vulnerable areas for selenium pollution.
4) **Tamil Nadu**: Industries and agricultural runoff forms two potential sources of selenium pollution. These potential sources were observed at large scale in four identified districts, namely Erode, Kanyakumari, Salem, and Krishnagiri (Divahar et al. 2019; Esakkimuthu et al. 2015; Jehamalar et al. 2010; Mohanakavitha et al. 2019; Satyanarayanan et al. 2017; Sudha and Ambujam 2011; Thilagavathi et al. 2015). Since two potential selenium sources are found at these four identified districts, they are categorized as moderately vulnerable for selenium pollution.
5) **Odisha**: Industries and coal mining forms two potential selenium releasing sources for this state. Certain districts have higher potential of releasing selenium owing to the presence of numerous industries and coal mining activity. Sambalpur, Kalahandi, Koraput, Malkangiri, Mayurbhunj, and Sundargarh districts are moderately vulnerable for selenium pollution of aquifers owing to large-scale runoff from chemical industries and mining activities (Basti et al. 2015; Mishra et al. 2012; Nayak 2012; Sahoo et al. 2014; Sahu 2016).

6) **West Bengal**: Industries and agriculture are two important potential sources of selenium pollution in this state. Districts having potential exposure of selenium contamination from these two sources are Howrah, Purulia, Hooghly, West Medinipur, and Bankura (Aktar et al. 2011; Chakraborty et al. 2016; Chakravorty et al. 2014; Chowdhury et al. 2009; Ghosh and Panigrahi 2018; Kiran and Srivastava 2014; Mandal et al. 2018; Masanta and Mallik 2009; Mondal et al. 2018). Hence these districts are mapped as moderately vulnerable areas for selenium pollution.
7) **Uttar Pradesh**: This state has high population density, numerous industries having potential for releasing selenium. In addition, certain districts also have small or large power plants, further increasing their potential to release selenium in the nearby water bodies. Districts having both power plants and industries are Baghpat, Gautambhudhnagar, Bulandshahr, Ghaziabad, Mathura, Saharanpur, Muzaffarnagar, Delhi, and Kanpur (Agrawal et al. 2017; Chand et al. 2009; Gopal and Chauhan 2007; Mittal et al. 2013; Sindhwani et al. 2015; Singh et al. 2018; Yadav and Yadav 2014). As two potential selenium releasing sources are identified in these districts, they are hence categorized as moderately vulnerable areas for selenium pollution.

17.5.3 Highly Vulnerable Areas

States having districts with presence of more than two potential selenium-discharging sources are categorized as highly vulnerable areas to selenium pollution. The states falling into this category are West Bengal, Chhattisgarh, and Jharkhand. The potential sources present in combination are coal mining, power plants, and industries. Districts in each state are:

1) **Chhattisgarh**: Districts, namely Bastar, Bilaspur, Dhamtari, Durg, Korba, Korea, Raigarh, and Rajnand village, have power plants, industries, agricultural runoff as well as coal mining activities, thereby having the potential for a huge amount of selenium discharge (Chakraborty et al. 2016; Jain et al. 2007; Sahu and Prasad 2018; Shrivastava and Shrivastava 2012; Tiwari et al. 2015; Vaishnav and Sahu 2006; Yadav and Haleem 2014). With the presence of more than two potential selenium sources, these districts can be classified as highly vulnerable areas for selenium pollution.
2) **Jharkhand:** Power plants, coal mining, and industries form potential selenium sources in this state. Districts having all the three potential sources are Chatra, Hazaribarg, Khunti, Loharbaga, Ramgarh, Ranchi, Saraikela kharsawan, Simdega, and West Singbhum (Hoelscher et al. 2017; Mondal et al. 2013; Pandit et al. 2014; Shubhankar and Ambade 2016). Since more than two potential selenium-discharging sources are present, these districts can be categorized as highly vulnerable to selenium pollution.
3) **West Bengal**: In the state of West Bengal, East Medinipur district has three potential sources of selenium pollution, i.e. industries, agricultural runoff, and power plants (Dasgupta and Paul 2011; Shamba et al. 2013). Since there are more than two potential sources of selenium discharge in this district it can be considered as a highly vulnerable area for selenium pollution in West Bengal.

The vulnerability map thus developed (Figure 17.4) may help identify and prioritize areas for management actions. This may also help in designing better management strategies for pollution abatement and conservation of water resources. Selenium for long has been an pollutant that little is known about, but the era now demands a thorough study of selenium-affected areas for their better identification and understanding. This will help efficient and sustainable administration of various types of water resources, conservation of their biodiversity, as well as reducing lethality to humans from selenium toxicity, thereby leading to societal wellbeing.

References

Abdel-Hamid, M.I. and Skulberg, O.M. (1995). Effect of selenium on the growth of some selected green and blue-green algae. *Lakes and Reservoirs: Research and Management* 1 (3): 205–211.

Adani, G., Filippini, T., Michalke, B., and Vinceti, M. (2020). Selenium and other trace elements in the etiology of Parkinson's disease: a systematic review and meta-analysis of case-control studies. *Neuroepidemiology* 54 (1): 1–23.

Adedara, I.A., Adebowale, A.A., Atanda, O.E. et al. (2019). Selenium abates reproductive dysfunction via attenuation of biometal accumulation, oxido-inflammatory stress and caspase-3 activation in male rats exposed to arsenic. *Environmental Pollution* 254: 113079.

Agrawal, V., Agrawal, A.M., and Mohanty, R.P. (2017). Application of fuzzy MCDM in supplier selection of fertiliser manufacturing industry. *International Journal of Business Performance and Supply Chain Modelling* 9 (2): 133–159.

Aguilar, F., Autrup, H., Barlow, S. et al. (2008). Selenium-enriched yeast as source for selenium added for nutritional purposes in foods for particular nutritional uses and foods (including food supplements) for the general population scientific opinion of the panel on food additives, flavourings, processing aids and materials in contact with food. *EFSA Journal* 766: 1–42.

Aktar, M.W., Sengupta, D., and Chowdhury, A. (2011). Occurrence of heavy metals in fish: a study for impact assessment in industry prone aquatic environment around Kolkata in India. *Environmental Monitoring and Assessment* 181 (1–4): 51–61.

Alexander, J. (2015). Selenium. In: *Handbook on the Toxicology of Metals*, 4e (eds. G. Nordberg, B. Fowler and M. Nordberg), 1175–1208. Academic Press Publications.

Arteel, G.E., Briviba, K., and Sies, H. (1999). Protection against peroxynitrite. *FEBS Letters* 445 (2–3): 226–230.

Bajaj, M., Eiche, E., Neumann, T. et al. (2011). Hazardous concentrations of selenium in soil and groundwater in North-West India. *Journal of Hazardous Materials* 189 (3): 640–646.

Baruah, J., Baruah, B.K., Kalita, S., and Choudhury, S.K. (2016). Impact analysis of open cast coal mining on land use/land cover using remote sensing and GIS technique in Ledo Margherita region of Assam, India. *Imperial Journal of Interdisciplinary Research (IJIR)* 2 (8): 671–676.

Basti, S., Sahu, C., and Sahu, S.K. (2015). Seasonal variation in physicochemical and biological parameters of pond waters in Bhawanipatna town, Kalahandi, Odisha. *International Journal of Emerging Research in Management and Technology* 4: 44–52.

Belzile, N., Chen, Y.W., and Xu, R. (2000). Early diageneticbehaviour of selenium in freshwater sediments. *Applied Geochemistry* 15 (10): 1439–1454.

Beyzi, S.B., Konca, Y., Köknur, Ö. et al. (2019). Effects of vitamin and mineral premix withdrawal from diets on carcass and meat quality of feedlot steers. *Tropical Animal Health and Production* 51 (7): 1919–1925.

Blanco, J.A. (2019). Suitability of Totora (Schoenoplectus californicus (CA Mey.)Soják) for its use in constructed wetlands in areas polluted with heavy metals. *Sustainability* 11 (1): 19.

Boyd, R. (2011). Selenium stories. *Nature Chemistry* 3 (7): 570–570.

Brown, K.M. and Arthur, J.R. (2001). Selenium, selenoproteins and human health: a review. *Public Health Nutrition* 4 (2b): 593–599.

Butler, J.A., Beilstein, M.A., and Whanger, P.D. (1989). Influence of dietary methionine on the metabolism of selenomethionine in rats. *The Journal of Nutrition* 119 (7): 1001–1009.

Chakraborty, P., Khuman, S.N., Selvaraj, S. et al. (2016). Polychlorinated biphenyls and organochlorine pesticides in River Brahmaputra from the outer Himalayan range and River Hooghly emptying into the Bay of Bengal: occurrence, sources and ecotoxicological risk assessment. *Environmental Pollution* 219: 998–1006.

Chakravorty, P.P., Sinha, M., and Chakraborty, S.K. (2014). Impact of industrial effluent on water quality and benthic macro invertebrate diversity in fresh water ponds in Midnapore district of West Bengal, India. *Indian Journal of Entomology and Zoology Studies* 2: 93–101.

Chand, P., Kumar, A., Gaur, A., and Mahna, S.K. (2009). Elemental analysis of ash using X-ray fluorescence technique. *Asian Journal of Chemistry* 21 (10): 220–224.

Chawla, R., Filippini, T., Loomba, R. et al. (2019). Exposure to a high selenium environment in Punjab, India: biomarkers and health conditions. *Science of The Total Environment* 719: 134541.

Chertow, M., Gordon, M., Hirsch, P., and Ramaswami, A. (2019). Industrial symbiosis potential and urban infrastructure capacity in Mysuru, India. *Environmental Research Letters* 14 (7): 075003.

Chowdhury, A., Jha, M.K., Chowdary, V.M., and Mal, B.C. (2009). Integrated remote sensing and GIS-based approach for assessing groundwater potential in West Medinipur district, West Bengal, India. *International Journal of Remote Sensing* 30 (1): 231–250.

Creighton, N. and Twining, J. (2010). Bioaccumulation from food and water of cadmium, selenium and zinc in an estuarine fish, Ambassis jacksoniensis. *Marine Pollution Bulletin* 60 (10): 1815–1821.

Dai, Z., Rizwan, M., Gao, F. et al. (2020). Nitric oxide alleviates selenium toxicity in rice by regulating antioxidation, selenium uptake, speciation and gene expression. *Environmental Pollution* 257: 113540.

Dasgupta, T. and Mukherjee, S. (2020). *Sediment Compaction and Applications in Petroleum Geoscience*. Springer.

Dasgupta, A. and Paul, S. (2011). Fly ash and its impact on land: a case study of Kolaghat thermal power plant, Purba Medinipur, West Bengal. *Indian Journal of Spatial Science* 11 (2): 1–11.

De, A.K., Sen, A.K., Karim, M.R. et al. (1985). Pollution profile of Damodar river sediment in Raniganj-Durgapur industrial belt, West Bengal, India. *Environment International* 11 (5): 453–458. https://doi.org/10.1016/0160-4120(85)90228-4.

Deborah, G., Hladun, K.R., Vindiola, B.G., and Trumble, J.T. (2017). Arthropod communities in a selenium-contaminated habitat with a focus on ant species. *Environmental Pollution* 220: 234–241.

Dey, T., Gogoi, K., Unni, B. et al. (2015). Role of environmental pollutants in liver physiology: special references to peoples living in the oil drilling sites of Assam. *PLoS One* 10 (4): e0123370.

Dhillon, K.S. and Dhillon, S.K. (2003). Distribution and management of seleniferous soils. *Advances in Agronomy* 79 (1): 119–184.

Dhillon, K.S. and Dhillon, S.K. (2014). Development and mapping of seleniferous soils in northwestern India. *Chemosphere* 99: 56–63.

Dhillon, K.S. and Dhillon, S.K. (2019). Genesis of seleniferous soils and associated animal and human health problems. In: *Advances in Agronomy*, vol. 154 (ed. D.L. Sparks), 1–80. Academic Press.

Divahar, R., Raj, P.S., Sangeetha, S.P., and Mohanakavitha, T. (2019). Impact of industrial wastewater disposal on surface water bodies in Kalingarayan Canal, Erode district. *Indian Journal of Ecology* 46 (4): 823–827.

Doblin, M.A., Baines, S.B., Cutter, L.S., and Cutter, G.A. (2006). Sources and biogeochemical cycling of particulate selenium in the San Francisco Bay estuary. *Estuarine, Coastal and Shelf Science* 67 (4): 681–694.

Donner, M.W., Cuss, C.W., Poesch, M. et al. (2018). Selenium in surface waters of the lower Athabasca River watershed: chemical speciation and implications for aquatic life. *Environmental Pollution* 243: 1343–1351.

dos Reis, A.R., Boleta, E.H.M., Alves, C.Z. et al. (2020). Selenium toxicity in upland field-grown rice: seed physiology responses and nutrient distribution using the μ-XRF technique. *Ecotoxicology and Environmental Safety* 190: 110147.

Du, H., Jiao, L., Wang, Q. et al. (2011). Selenium: another metalloid beneficial for the electrochemical performance of co electrode in alkaline rechargeable batteries. *Journal of Power Sources* 196 (24): 10748–10752.

Ellis, A.S., Johnson, T.M., Herbel, M.J., and Bullen, T.D. (2003). Stable isotope fractionation of selenium by natural microbial consortia. *Chemical Geology* 195 (1–4): 119–129.

El-Ramady, H., Abdalla, N., Alshaal, T. et al. (2014). Selenium in soils under climate change, implication for human health. *Environmental Chemistry Letters* 13 (1): 1–19.

Esakkimuthu, K., Kumar, K.V., and Ponram, P. (2015). Assessment of water-polluting sources by multivariate statistical methods in Putheri Lake, Kanyakumari, Tamil Nadu, India [08° 12′ 27.1 ″N 77° 25′ 54.7 ″E]. *Sustainable Water Resources Management* 1 (4): 349–353.

Etteieb, S., Magdouli, S., Zolfaghari, M., and Brar, S. (2020). Monitoring and analysis of selenium as an emerging contaminant in mining industry: a critical review. *Science of the Total Environment* 698: 134339.

Fayiga, A.O., Ipinmoroti, M.O., and Chirenje, T. (2018). Environmental pollution in Africa. *Environment, Development and Sustainability* 20 (1): 41–73.

FDA (1994). Food additives permitted in feed and drinking water of animals. Selenium.21CFR573 .4-1-11.

Filippini, T., Michalke, B., Mandrioli, J. et al. (2018). Selenium neurotoxicity and amyotrophic lateral sclerosis: an epidemiologic perspective. In: *Selenium* (ed. B. Michalke), 231–248. Cham: Springer.

Finkelman, R.B. and Stracher, G.B. (2011). Environmental and health impacts of coal fires. In: *Coal and Peat Fires: A Global Perspective: Coa – Geology and Combustion*, vol. 1 (eds. G.B. Stracher, A. Prakash and E.V. Sokol), 115–125. Elsevier.

Finley, J.W., Davis, C.D., and Feng, Y. (2000). Selenium from high selenium broccoli protects rats from colon cancer. *The Journal of Nutrition* 130 (9): 2384–2389.

Fishbein, L. (1983). Environmental selenium and its significance. *Fundamental and Applied Toxicology* 3 (5): 411–419.

Fordyce, F. (2005). Selenium deficiency and toxicity in the environment. In: *Essentials of Medical Geology* (eds. O. Selinus, B. Alloway, J.A. Centeno, et al.), 373–415. Amsterdam: Elsevier.

Fujita, H., Nakano, K., Chaisomphob, T., and Hambali, E. (2019). Greenhouse gas emission of electricity generation and field abandonment scenarios using empty fruit bunches at a palm oil mill, Surat Thani Province, Thailand. *Journal of the Japan Institute of Energy* 98 (5): 119–123.

Gaikwad, V.P. and Pawar, S.N. (2020). A case study of rooftop rainwater harvesting potentiality in rural settlement in drought prone area of Satara District, Maharashtra. *Studies in Indian Place Names* 40 (60): 6443–6456.

Gallignani, M., Valero, M., Brunetto, M.R. et al. (2000). Sequential determination of Se (IV) and Se (VI) by flow injection-hydride generation-atomic absorption spectrometry with HCl/HBr microwave aided pre-reduction of Se (VI) to Se (IV). *Talanta* 52 (6): 1015–1024.

Ghosh, P. and Panigrahi, A.K. (2018). A comprehensive study on correlation of gastropod diversity with some hydro-environmental parameters of selected waterbodies of lower Damodar basin, West Bengal, India. *Journal of Applied and Natural Science* 10 (4): 1259–1265.

Goehring, T.B., Palmer, I.S., Olson, O.E. et al. (1984). Effects of seleniferous grains and inorganic selenium on tissue and blood composition and growth performance of rats and swine. *Journal of Animal Science* 59 (3): 725–732.

Gojkovic, Ž., Garbayo, I., Ariza, J.L.G. et al. (2015). Selenium bioaccumulation and toxicity in cultures of green microalgae. *Algal Research* 7: 106–116.

Golubkina, N., Erdenetsogt, E., Tarmaeva, I. et al. (2018). Selenium and drinking water quality indicators in Mongolia. *Environmental Science and Pollution Research* 25 (28): 28619–28627.

Gopal, B. and Chauhan, M. (2007). River Yamuna from source to Delhi: human impacts and approaches to conservation. In: *Restoring River Yamuna* (eds. P. Martin, B. Gopal and C. Southey), 45–69. New Delhi: National Institute of Ecology.

Greenfield, R., Van Vuren, J.H.J., and Wepener, V. (2012). Heavy metal concentrations in the water of the Nyl River system, South Africa. *African Journal of Aquatic Science* 37 (2): 219–224.

Gunnarsson, L.G. and Bodin, L. (2018). Amyotrophic lateral sclerosis and occupational exposures: a systematic literature review and meta-analyses. *International Journal of Environmental Research and Public Health* 15 (11): 2371.

Gupta, M. and Gupta, S. (2017). An overview of selenium uptake, metabolism, and toxicity in plants. *Frontiers in Plant Science* 7: 2074.

Gupta, R. and Misra, A.K. (2018). Groundwater quality analysis of quaternary aquifers in Jhajjar District, Haryana, India: focus on groundwater fluoride and health implications. *Alexandria Engineering Journal* 57 (1): 375–381.

Hadrup, N. and Ravn-Haren, G. (2020). Acute human toxicity and mortality after selenium ingestion: a review. *Journal of Trace Elements in Medicine and Biology* 58: 126435.

Hamilton, S.J. (1999). Hypothesis of historical effects from selenium on endangered fish in the Colorado River basin. *Human and Ecological Risk Assessment: An International Journal* 5 (6): 1153–1180.

Hansson, E. and Jacobsson, S.O. (1966). Uptake of [75Se] selenomethionine in the tissues of the mouse studied by whole-body autoradiography. *Biochimicaet Biophysica Acta (BBA)-General Subjects* 115 (2): 285–293.

Haygarth, P.M. (1994). Global importance and global cycling of selenium. In: *Selenium in the Environment* (eds. W.T. Frankenberger Jr. and S. Benson), 1–27. New York: Marcel Dekker Inc.

He, Y., Xiang, Y., Zhou, Y. et al. (2018). Selenium contamination, consequences and remediation techniques in water and soils: a review. *Environmental Research* 164: 288–301.

Hoelscher, K., Miklian, J., and Vadlamannati, K.C. (2017). Hearts and mines: a district level empirical analysis of the Maoist conflict in India. *International Area Studies Review* 15 (2): 2012.

Hoffman, D.J. (2002). Role of selenium toxicity and oxidative stress in aquatic birds. *Aquatic Toxicology* 57 (1–2): 11–26.

Hopkins, W.A., Staub, B.P., Baionno, J.A. et al. (2005). Transfer of selenium from prey to predators in a simulated terrestrial food chain. *Environmental Pollution* 134 (3): 447–456.

Hu, Y., Han, Y., Wang, L. et al. (2019). Toxicity effects of different dietary selenium forms on sea cucumber, *Apostichopus japonicus*. *Aquaculture Reports* 15: 100209.

Huang, S., Ming, H.U.A., Jinshun, F.E.N.G. et al. (2009). Assessment of selenium pollution in agricultural soils in the Xuzhou District, Northwest Jiangsu, China. *Journal of Environmental Sciences* 21 (4): 481–487.

Huang, X., Liu, X., Luo, Q. et al. (2011). Artificial selenoenzymes: designed and redesigned. *Chemical Society Reviews* 40 (3): 1171–1184.

Huang, L.Q., Bi, Y.F., Mu, L.L., and Zhang, H.B. (2014). The process and mechanism of electrolytic manganese anode slime lead removal. *Advanced Materials Research* 878: 163–170.

IFA (2011). *The Role of Fertilizers in Nutrition Security. Three Examples of Micronutrient Fertilization*. Paris, France: IFA (International Fertilizer Industry Association).

Islam, M.S., Kazi, M.A.I., Hossain, M.M. et al. (2007). Propagation of heavy metals in poultry feed production in Bangladesh. *Bangladesh Journal of Scientific and Industrial Research* 42 (4): 465–474.

Ivory, K., Prieto, E., Spinks, C. et al. (2017). Selenium supplementation has beneficial and detrimental effects on immunity to influenza vaccine in older adults. *Clinical Nutrition* 36 (2): 407–415.

Jadhav, S.S. and Bharambe, S.N. (2020). The assessment of land use land cover (LULC) change detection in Ahmednagar District, Maharashtra: using GIS techniques. *Studies in Indian Place Names* 40 (71): 1692–1699.

Jagtap, M.M. (2019). Assessment of suitability of irrigation canal water for agriculture: a study based on new Mutha right Bank Canal waters from Pune, Maharashtra, India. *Proceedings of the International Academy of Ecology and Environmental Sciences* 9 (3): 89–97.

Jain, S.K., Agarwal, P.K., and Singh, V.P. (2007). Mahanadi, Subernarekha and Brahmani Basins. In: *Hydrology and Water Resources of India*, 597–639. Dordrecht: Springer.

Janz, D.M., DeForest, D.K., Brooks, M.L. et al. (2010). Selenium toxicity to aquatic organisms. In: *Ecological Assessment of Selenium in the Aquatic Environment* (eds. P.M. Chapman et al.), 141–231. CRC Press.

Jehamalar, E.E., Benitha Golda, D., and Sam Manohar Das, S. (2010). Water quality index and its seasonal variation of Thamiraparani River at KAnyakumari District of Tamil Nadu, South India. *Journal of Basic and Applied Biology* 4 (3): 110–116.

Jeyaraj, M., Mahalingam, V., Indhuleka, A. et al. (2019). Chemical analysis of surface water quality of river Noyyal connected tank in Tirupur District, Tamil Nadu, India. *IWRA (India) Journal (Half Yearly Technical Journal of Indian Geographical Committee of IWRA)* 8 (2): 23–28.

Jyothi, M.N., Ramaiah, B.J., and Maliyekkal, S.M. (2020). Occurrence, contamination, speciation and analysis of selenium in the environment. In: *Measurement, Analysis and Remediation of Environmental Pollutants* (eds. T. Gupta, S.P. Singh, P. Rajput and A.K. Agarwal), 245–269. Singapore: Springer.

Kambalagere, Y. (2020). A study on air quality index (AQI) of Bengaluru, Karnataka during lockdown period to combat coronavirus disease (Covid-19): air quality turns 'Better' from 'Hazardous'. *Studies in Indian Place Names* 40 (69): 59–66.

Karthikeyan, P., Marigoudar, S.R., Nagarjuna, A., and Sharma, K.V. (2019). Toxicity assessment of cobalt and selenium on marine diatoms and copepods. *Environmental Chemistry and Ecotoxicology* 1: 36–42.

Kavlak, G. and Graedel, T.E. (2013). Global anthropogenic selenium cycles for 1940–2010. *Resources, Conservation and Recycling* 73: 17–22.

Kaya, S., Bilgili, A., Baydan, E. et al. (1998). *Toxicology in Veterinary Medicine*. Food and Agricultural Organization of United Nation.

Kerutagi, M.G., Pavithra, A.S., and Gollagi, S.G. (2020). Performance of palm industry in Karnataka: a case study in Tumkur District. *Economic Affairs* 65 (1): 85–96.

Kessi, J., Ramuz, M., Wehrli, E. et al. (1999). Reduction of selenite and detoxification of elemental selenium by the Phototrophic Bacterium Rhodospirillum rubrum. *Applied Environmental Microbiology* 65 (11): 4734–4740.

Khamkar, D.J. and Mhaske, S.Y. (2019). Identification of landslide susceptible settlements using geographical information system of Yelwandi river basin, Maharashtra (India). *Natural Hazards* 96 (3): 1263–1287.

Khamkhash, A., Srivastava, V., Ghosh, T. et al. (2017). Mining-related selenium contamination in Alaska, and the state of current knowledge. *Minerals* 7 (3): 46.

Kiełczykowska, M., Kopciał, E., Kocot, J. et al. (2018). Lithium disturbs homeostasis of essential microelements in erythrocytes of rats: selenium as a protective agent? *Pharmacological Reports* 70 (6): 1168–1172.

Kiran, V.S.S. and Srivastava, Y.K. (2014). Micro watershed level water resource management based on three years runoff estimation using remote sensing and GIS techniques for Simlapal block, Bankura, West Bengal, India. *International Journal of Earth Science and Engineering* 7 (1): 80–92.

Kolbert, Z., Molnár, Á., Feigl, G., and Van Hoewyk, D. (2019). Plant selenium toxicity: proteome in the crosshairs. *Journal of Plant Physiology* 232: 291–300.

Kumar, A.R. and Riyazuddin, P. (2011). Speciation of selenium in groundwater: seasonal variations and redox transformations. *Journal of Hazardous Materials* 192 (1): 263–269.

Kumar, A., Patil, R.S., Dikshit, A.K., and Kumar, R. (2019). Assessment of spatial ambient concentration of NH_3 and its health impact for Mumbai City. *Asian Journal of Atmospheric Environment (AJAE)* 13 (1): 11–19.

Kuzhiparambil, A. (2020). Skills development initiatives and labour migration in a secondary circuit of globalized production: evidence from the garment industry in India. In: *Mobilities of Labour and Capital in Asia* (eds. P.S. Aulakh and P.F. Kelly), 153–171. Cambridge University Press.

Lacourciere, G.M. and Stadtman, T.C. (2001). Utilization of selenocysteine as a source of selenium for selenophosphate biosynthesis. *Biofactors* 14 (1–4): 69–74.

Lakhimi, G. (2011). Solid waste disposal and its health implications in guwahati city a study in medical geography (*Doctoral dissertation*), Faculty of Science, Gauhati University.

Lanctôt, C.M., Melvin, S.D., and Cresswell, T. (2017). Selenium speciation influences bioaccumulation in Limnodynastes peronii tadpoles. *Aquatic Toxicology* 187: 1–8.

Lapworth, D.J., Krishan, G., MacDonald, A.M., and Rao, M.S. (2017). Groundwater quality in the alluvial aquifer system of northwest India: new evidence of the extent of anthropogenic and geogenic contamination. *Science of the Total Environment* 599: 1433–1444.

Lemly, A.D. (1997). Environmental implications of excessive selenium: a review. *Biomedical and Environmenta Sciences* 10: 415–435.

Lemly, A.D. (2002). Selenium pollution around the world. In: *Selenium Assessment in Aquatic Ecosystems* (ed. A.D. Lemly), 3–17. New York;, NY: Springer.

Lemly, A.D. (2004). Aquatic selenium pollution is a global environmental safety issue. *Ecotoxicology and Environmental Safety* 59 (1): 44–56.

Lemly, A.D. (2019). Environmental hazard assessment of Benga Mining's proposed Grassy Mountain coal project. *Environmental Science and Policy* 96: 105–113.

Li, J., Liang, D., Qin, S. et al. (2015). Effects of selenite and selenate application on growth and shoot selenium accumulation of pakchoi (Brassica chinensis L.) during successive planting conditions. *Environmental Science and Pollution Research* 22 (14): 11076–11086.

Lima, L.W., Checchio, M.V., dos Reis, A.R. et al. (2019). Selenium restricts cadmium uptake and improve micronutrients and proline concentration in tomato fruits. *Biocatalysis and Agricultural Biotechnology* 18: 101057.

Liu, K., Cai, M., Hu, C. et al. (2019). Selenium (Se) reduces Sclerotinia stem rot disease incidence of oilseed rape by increasing plant Se concentration and shifting soil microbial community and functional profiles. *Environmental Pollution* 254: 113051.

Luoma, S.N. and Presser, T.S. (2009). Emerging opportunities in management of selenium contamination. *Environmental Science and Technology* 43 (22): 8483–8487.

Luoma, S.N. and Rainbow, P.S. (2008). *Metal Contamination in Aquatic Environments: Science and Lateral Management*. Cambridge University Press.

Lussier, C., Veiga, V., and Baldwin, S. (2003). The geochemistry of selenium associated with coal waste in the Elk River Valley, Canada. *Environmental Geology* 44 (8): 905–913.

Maier, K.J., Ogle, R.S., and Knight, A.W. (1988). The selenium problem in lentic ecosystems. *Lake and Reservoir Management* 4 (2): 155–163.

Maier, K.J., Foe, C.G., and Knight, A.W. (1993). Comparative toxicity of selenate, selenite, seleno-DL-methionine and seleno-DL-cystine to Daphnia magna. *Environmental Toxicology and Chemistry: An International Journal* 12 (4): 755–763.

Manaschintan, S.B. and Nalini, S.S. (2020). Effect of land use–land cover change on runoff characteristics in Mumbai City. In: *Applications of Geomatics in Civil Engineering* (eds. J.K. Ghosh and I. da Silva), 183–192. Singapore: Springer.

Mandal, B., Dolui, G., and Satpathy, S. (2018). Land suitability assessment for potential surface irrigation of river catchment for irrigation development in Kansai watershed, Purulia, West Bengal, India. *Sustainable Water Resources Management* 4 (4): 699–714.

Martens, D.A. (2003). Selenium. In: *Encyclopedia of Water Science* (eds. B.A. Stewart and T.A. Howel), 840–842. New York;: Dekker https://doi.org/10.1081/E-EWS 120010212.

Masanta, S. and Mallik, S. (2009). Effect of mulch and irrigation on yield and water use efficiency of wheat under Patloi Nala micro-watershed in Purulia district of West Bengal. *Journal of Crop and Weed* 5 (2): 22–24.

Meera, D. and Anand, S.P. (2019). Bioaccumulation of selected heavy metals in a freshwater fish Anabas testudineus collected from a pond in an industrial area in Kollam District of Kerala. *Advances in Applied Research* 11 (1): 59–60.

Mehdi, Y., Hornick, J.L., Istasse, L., and Dufrasne, I. (2013). Selenium in the environment, metabolism and involvement in body functions. *Molecules* 18 (3): 3292–3311.

Meseck, S.L. and Cutter, G.A. (2006). Evaluating the biogeochemical cycle of selenium in San Francisco Bay through modeling. *Limnology and Oceanography* 51 (5): 2018–2032.

Mézes, M. and Balogh, K. (2009). Prooxidant mechanisms of selenium toxicity–a review. *Acta Biologica Szegediensis* 53 (suppl): 15–18.

Mishra, M.E., Patel, A.K., and Behera, N.I. (2012). An assessment of coliform bacteria in the river Mahanadi system of Sambalpur. *Bioscan* 7: 463–467.

Mittal, P.K., Nanda, N., Singh, O.P. et al. (2013). Establishment of a focus on Anopheles fluviatilis, an important malaria vector near the National Thermal Power Corporation Project in Dadri CHC area in District Gautam Budh Nagar, Uttar Pradesh, India: a case study. *Journal of Vector Borne Diseases* 50 (4): 307.

Mo, A., Wang, J., Yuan, M. et al. (2019). Effect of sub-chronic dietary L-selenomethionine exposure on reproductive performance of red swamp crayfish, (*Procambarus clarkii*). *Environmental Pollution* 253: 749–758.

Mohanakavitha, T., Shankar, K., Divahar, R. et al. (2019). Impact of industrial wastewater disposal on surface water bodies in Kalingarayan canal, Erode district, Tamil Nadu, India. *Archives of Agriculture and Environmental Science* 4 (4): 379–387.

Mohtashami, R., Dehnavi, M.M., Balouchi, H., and Faraji, H. (2020). Improving yield, oil content and water productivity of dryland canola by supplementary irrigation and selenium spraying. *Agricultural Water Management* 232: 106046.

Mondal, G.C., Singh, A.K., Singh, T.B. et al. (2013). Hydrogeochemistry and quality assessment of mine water of West Bokaro coalfields, Hazaribag, Jharkhand, India. *Journal of Materials Science and Engineering. A* 3 (8A): 540–549.

Mondal, R., Mukherjee, A., Biswas, S., and Kole, R.K. (2018). GC-MS/MS determination and ecological risk assessment of pesticides in aquatic system: a case study in Hooghly River basin in West Bengal, India. *Chemosphere* 206: 217–230.

Mugesh, G. and Singh, H.B. (2000). Synthetic organoselenium compounds as antioxidants: glutathione peroxidase activity. *Chemical Society Reviews* 29 (5): 347–357.

Mukherjee, S. and Srivastava, S.K. (2005). Trace elements in high-sulfur Assam coals from the Makum coalfield in the northeastern region of India. *Energy and Fuels* 19 (3): 882–891.

Muzembo, B.A., Ngatu, N.R., Januka, K. et al. (2019). Selenium supplementation in HIV-infected individuals: a systematic review of randomized controlled trials. *Clinical Nutrition ESPEN* 34: 1–7.

Nancharaiah, Y.V. and Lens, P.N. (2015). Selenium biomineralization for biotechnological applications. *Trends in Biotechnology* 33 (6): 323–330.

Nayak, S. (2012). Sustainable mineral-intensive growth in Odisha, India. *Journal of the Institution of Engineers (India): Series D* 93 (1): 43–51.

Naz, F.S., Yusuf, M., Khan, T.A. et al. (2015). Low level of selenium increases the efficacy of 24-epibrassinolide through altered physiological and biochemical traits of Brassica juncea plants. *Food Chemistry* 185: 441–448.

Nihalani, S.A. and Mishra, Y.D. (2020). Water consumption management for thermal power plant. In: *Environmental Processes and Management* (eds. R.M. Singh, P. Shukla and P. Singh), 135–153. Cham: Springer.

Ogle, R.S., Maier, K.J., Kiffney, P. et al. (1988). Bioaccumulation of selenium in aquatic ecosystems. *Lake and Reservoir Management* 4 (2): 165–173.

Paikaray, S. (2016). Origin, mobilization and distribution of selenium in a soil/water/air system: a global perspective with special reference to the Indian scenario. *Clean–Soil, Air, Water* 44 (5): 474–487.

Pandit, A.P., Sinha, V.S., Kumar, N., and Choudhary, U. (2014). Assessment of ground water quality of Noamundi, Jharkhand with special reference to iron. *Journal of Chemistry and Chemical Sciences* 4 (3): 118–124.

Peters, G.M., Maher, W.A., Jolley, D. et al. (1999). Selenium contamination, redistribution and remobilisation in sediments of Lake Macquarie, NSW. *Organic Geochemistry* 30 (10): 1287–1300.

Pokhrel, G.R., Wang, K.T., Zhuang, H. et al. (2020). Effect of selenium in soil on the toxicity and uptake of arsenic in rice plant. *Chemosphere* 239: 124712.

Ponce, M., Giraldez, I., Calero, S. et al. (2018). Toxicity and biochemical transformation of selenium species in rotifer (Brachionusplicatilis) enrichments. *Aquaculture* 484: 105–111.

Prabhavathi, N. and Ramakrishna Parama, V.R. (2019). Effect of sugar industry solid waste pressmud and bio compost on soil physical and chemical properties at different intervals during finger millet crop. *Journal of Pharmacognosy and Phytochemistry* 8 (3): 3038–3042.

Prajapati, M., Jariwala, N., and Agnihotri, P. (2020). Evaluation of groundwater quality with special emphasis on fluoride contamination using multivariate statistical analysis in rural parts of Surat district, Gujarat. *ISH Journal of Hydraulic Engineering* 26 (2): 179–186.

Presser, T.S., Piper, D.Z., Bird, K.J. et al. (2004). The Phosphoria formation: a model for forecasting global selenium sources to the environment. In: *Handbook of Exploration and Environmental Geochemistry*, vol. 8 (ed. J.R. Hein), 299–319. Elsevier Science BV.

Pyrzyńska, K. (2002). Determination of selenium species in environmental samples. *Microchimica Acta* 140 (1–2): 55–62.

Rad, E.Y., Falahi, E., Saboori, S. et al. (2019). Effect of selenium supplementation on lipid profile levels: an updated systematic review and meta-analysis of randomized controlled clinical trials. *Obesity Medicine* 15: 100113.

Rajeswari, A., Christy, E.J.S., Gopi, S. et al. (2019). Department of Chemistry, The Gandhigram Rural Institute Deemed to be University, Dindigul, Tamil Nadu, India. *Interfaces in Particle and Fibre Reinforced Composites: Current Perspectives on Polymer, Ceramic, Metal and Extracellular Matrices*, 219.

Rathi, A.K.A. (2003). Promotion of cleaner production for industrial pollution abatement in Gujarat (India). *Journal of Cleaner Production* 11 (5): 583–590.

Ravish, S., Setia, B., and Deswal, S. (2019). Monitoring of pre-and post-monsoon groundwater quality of north-eastern Haryana region using GIS. *Environmental Technology* 41: 1–27.

Ravish, S., Setia, B., and Deswal, S. (2020). Hydro-chemistry for the analysis of sub-surface water quality in North-Eastern Haryana: a fast-urbanizing region. In: *Numerical Optimization in Engineering and Sciences* (eds. D. Dutta and B. Mahanty), 1–13. Singapore: Springer.

Reich, H.J. and Hondal, R.J. (2016). Why nature chose selenium. *ACS Chemical Biology* 11 (4): 821–841.

Rodríguez-Gutiérrez, J., Galván-Magaña, F., Jacobo-Estrada, T. et al. (2020). Mercury–selenium concentrations in silky sharks (Carcharhinus falciformis) and their toxicological concerns in the southern Mexican Pacific. *Marine Pollution Bulletin* 153: 111011.

Sahoo, S., Mahesh, A., and Panigrahi, A. (2014). Impact of contamination by bamboo decomposition and lubricant oil leak and deterioration of Balimela reservoir water standard at Malkangiri of Odisha. *Natural Journal of Life Science* 11 (2): 01–06.

Sahu, P.C. (2016). Groundwater conditions in a tribal dominated area of Mayurbhanj district, Odisha. *International Journal of Advance Research and Review* 1 (5): 75–84.

Sahu, I. and Prasad, A.D. (2018). Assessment of hydro potential using integrated tool in QGIS. *ISPRS Annals of Photogrammetry, Remote Sensing and Spatial Information Sciences* 4 (5): 115–119.

Salbe, A.D. and Levander, O.A. (1990). Comparative toxicity and tissue retention of selenium in methionine-deficient rats fed sodium selenate or L-selenomethionine. *The Journal of Nutrition* 120 (2): 207–212.

Sandholm, M., Oksanen, H.E., and Pesonen, L. (1973). Uptake of selenium by aquatic organisms. *Limnology and Oceanography* 18 (3): 496–499.

Sasmitha, R., Arunachalam, R., and Muhammad Iqshanullah, A. (2019). Documentation of soil related environmental issues and its contributing factors: a study among the hilly tribes of the Nilgiri District. *Madras Agricultural Journal* 106: 284–287.

Satyanarayanan, M., Eswaramoorthi, S., Subramanian, S., and Periakali, P. (2017). Factor analysis of rock, soil and water geochemical data from Salem magnesite mines and surrounding area, Salem, southern India. *Applied Water Science* 7 (5): 2607–2616.

Sawant, P. (2019). Study of human resource management practices in chemical industries with special reference to Navi Mumbai (*Doctoral dissertation*, Tilak Maharashtra Vidyapeeth).

Schiavon, M., Ertani, A., Parrasia, S., and DallaVecchia, F. (2017). Selenium accumulation and metabolism in algae. *Aquatic Toxicology* 189: 1–8.

Schrauzer, G.N. (2000). Selenomethionine: a review of its nutritional significance, metabolism and toxicity. *The Journal of Nutrition* 130 (7): 1653–1656.

Schrauzer, G.N. (2001). Nutritional selenium supplements: product types, quality, and safety. *Journal of the American College of Nutrition* 20 (1): 1–4.

Shahid, M., Niazi, N.K., Khalid, S. et al. (2018). A critical review of selenium biogeochemical behavior in soil-plant system with an inference to human health. *Environmental Pollution* 234: 915–934.

Shamba, C., Soumik, B., Samrat, G., and Mathur, A.K. (2013). Isolation of pyridine degrading bacteria from soils contaminated with petrochemical industry effluents in Purba Medinipur. *Journal of Biological and Environmental Sciences* 7 (20): 109–119.

Shamberger, R.J. (1981). Selenium in the environment. *Science of the Total Environment* 17 (1): 59–74.

Shanshak, M., Budagumpi, S., Małecki, J.G., and Keri, R.S. (2020). Green synthesis of 3, 4-disubstituted isoxazol-5 (4H)-ones using ZnO@ Fe_3O_4 core–shell nanocatalyst in water. *Applied Organometallic Chemistry* 34 (4): e5544.

Sharma, S. (1984). Selenium research in India. *International Journal of Environmental Studies* 22 (3–4): 231–239.

Sharma, M.C. and Kumar, P. (2006). Impact of pollution on livestocks of adjoining areas of barauni industrial area in Northern India. *Toxicology Letters* 164: S178–S179.

Sharma, S. and Vance, G.F. (2007). Dissolution chemistry of inorganic selenium in alkaline mine soils. *Developments in Environmental Science* 5: 363–380.

Sharma, V.K., McDonald, T.J., Sohn, M. et al. (2015). Biogeochemistry of selenium. a review. *Environmental Chemistry Letters* 13 (1): 49–58.

Shree, B.V. and Nishikant, G. (2019). Examining the heavy metal contents of an estuarine ecosystem: case study from Maharashtra, India. *Journal of Coastal Conservation* 23 (6): 977–984.

Shrift, A. (1973). Metabolism of selenium in plants and microorganisms. In: *Organic Selenium Compounds in Chemistry and Biology* (eds. D.E. Klayman and W.H.H. Gunther), 763–814. New York;: Wiley-Interscience.

Shrivastava, S. and Shrivastava, L. (2012). Studies on fly ash and animals. *Discovery Science* 2 (5): 48–54.

Shu, Y., Wu, M., Yang, S. et al. (2020). Association of dietary selenium intake with telomere length in middle-aged and older adults. *Clinical Nutrition* 39 (10): 3086–3091.

Shubhankar, B. and Ambade, B. (2016). Chemical characterization of carbonaceous carbon from industrial and semi urban site of eastern India. *SpringerPlus* 5 (1): 837.

Silva, V.M., Boleta, E.H.M., Lanza, M.G.D.B. et al. (2018). Physiological, biochemical, and ultrastructural characterization of selenium toxicity in cowpea plants. *Environmental and Experimental Botany* 150: 172–182.

Sindhwani, R., Goyal, P., Kumar, S., and Kumar, A. (2015). Anthropogenic emission inventory of criteria air pollutants of an urban agglomeration-National Capital Region (NCR), Delhi. *Aerosol and Air Quality Research* 15 (4): 1681–1697.

Singh, U., Singh, S., Tiwari, R.K., and Pandey, R.S. (2018). Water pollution due to discharge of industrial effluents with special reference to Uttar Pradesh, India–a review. *International Archive of Applied Sciences and Technology* 9 (4): 111–121.

Singh, H., Alam, M.S., Ingle, S.R., and Raizada, S. (2019). Effect of sewage and industrial effluents application on soil physical properties and nutrients uptake by plants under Faridabad district of Haryana, India. *Journal of Pharmacognosy and Phytochemistry* 8 (1): 835–839.

Singh, P., Singh, R.P., and Srivastava, V. (eds.) (2020a). *Contemporary Environmental Issues and Challenges in Era of Climate Change*. Springer.

Singh, V., Ngpoore, N.K., Chand, J., and Lehri, A. (2020b). Monitoring and assessment of pollution load in surface water of River Ganga around Kanpur, India: a study for suitability of this water for different uses. *Environmental Technology and Innovation* 9: 100676.

Sonawane, D.V. (2019). Study of the impact of heavy metals and physico-chemical parameters on the quality of the wells water around Yedgaon Dam area Junner, Dist: Pune. *Current Pharma Research* 9 (3): 2867–2877.

Spallholz, J.E. (1997). Free radical generation by selenium compounds and their prooxidant toxicity. *Biomedical and Environmental Sciences: BES* 10 (2–3): 260–270.

Spallholz, J.E. and Hoffman, D.J. (2002). Selenium toxicity: cause and effects in aquatic birds. *Aquatic Toxicology* 57 (1–2): 27–37.

Staicu, L.C., Van Hullebusch, E.D., and Lens, P.N. (2015). Production, recovery and reuse of biogenic elemental selenium. *Environmental Chemistry Letters* 13 (1): 89–96.

Staicu, L.C., Hullebusch, E.D.V., Rittmann, B.E., and Lens, P.N.L. (2017). Industrial selenium pollution: sources and biological treatment technologies. In: *Bioremediation of Selenium Contaminated Wastewater* (ed. E.D.V. Hullebusch), 75–101. Cham: Springer International Publishing.

Sudha, V. and Ambujam, N.K. (2011). Agricultural non point source pollution effects in the freshwater reservoir, Krishnagiri, Tamilnadu, India. *Journal of Environmental Research and Development* 6 (2): 222–231.

Sullivan, D., Zywicki, R., and Yancey, M. (2013). Method for the determination of total selenium in a wide variety of foods using inductively coupled plasma/mass spectrometry. *Journal of AOAC International* 96 (4): 786–794.

Sun, H.J., Rathinasabapathi, B., Wu, B. et al. (2014). Arsenic and selenium toxicity and their interactive effects in humans. *Environment International* 69: 148–158.

Tamilchelvan, P. and Muralimohan, N. (2019). A case study on Assessment of Groundwater Quality in and Around Namakkal District in Tamil Nadu. *International Research Journal of Multidisciplinary Technovation* 1 (6): 662–667.

Tan, L.C., Nancharaiah, Y.V., van Hullebusch, E.D., and Lens, P.N. (2016). Selenium: environmental significance, pollution, and biological treatment technologies. *Biotechnology Advances* 34 (5): 886–907.

Thilagavathi, N., Subramani, T., Suresh, M., and Karunanidhi, D. (2015). Mapping of groundwater potential zones in Salem Chalk Hills, Tamil Nadu, India, using remote sensing and GIS techniques. *Environmental Monitoring and Assessment* 187 (4): 164.

Tiwari, S.N., Mazumder, S.M., Pathak, B.N., and Agrawal, R. (2015). Assessment of application potential of blast furnace flue dust. *Indian Journal of Environmental Protection* 35 (1): 58–65.

Toppo, S., Vanin, S., Bosello, V., and Tosatto, S.C. (2008). Evolutionary and structural insights into the multifaceted glutathione peroxidase (Gpx) superfamily. *Antioxidants and Redox Signaling* 10 (9): 1501–1514.

Ulusoy, H.İ., Yılmaz, Ö., and Gürkan, R. (2013). A micellar improved method for trace levels selenium quantification in food samples, alcoholic and nonalcoholic beverages through CPE/FAAS. *Food Chemistry* 139 (1–4): 1008–1014.

Vaishnav, M.M. and Sahu, D. (2006). Study of some physico-chemical characteristics of Hasdeo river water at Korba (India). *Journal of Environmental Research and Development* 1 (2): 140–142.

Vediya, S.D. and Srivastava, A.K. (2008). Pollution status of Ambli lake and Makarba lake situated at Ahmedabad, Gujarat with reference to heavy metals. In: *Proceedings of the 12th World Lake Conference*, 307–311.

Velez, M.E.V., da Luz, J.M.R., da Silva, M.D.C.S. et al. (2019). Production of bioactive compounds by the mycelial growth of *Pleurotus djamor* in whey powder enriched with selenium. *LWT* 114: 108376.

Virk, H.S. (2018). Selenium contamination of groundwater of Majha Belt of Punjab (India). *Research and Reviews: A Journal of Toxicology* 8 (2): 1–7.

Wang, Z. and Gao, Y. (2001). Biogeochemical cycling of selenium in Chinese environments. *Applied Geochemistry* 16 (11–12): 1345–1351.

Welna, M., Szymczycha-Madeja, A., and Pohl, P. (2018). Selenium and other beneficial elements in fruit juices. In: *Fruit Juices* (eds. G. Rajauria and B.K. Tiwari), 75–93. Academic Press.

Wen, H. and Carignan, J. (2007). Reviews on atmospheric selenium: emissions, speciation and fate. *Atmospheric Environment* 41 (34): 7151–7165.

Winkel, L.H., Johnson, C.A., Lenz, M. et al. (2012). Environmental selenium research: from microscopic processes to global understanding. *Environmental Science and Technology* 46 (2): 571–579.

Wu, C.Y., Wong, C.S., Chung, C.J. et al. (2019). The association between plasma selenium and chronic kidney disease related to lead, cadmium and arsenic exposure in a Taiwanese population. *Journal of Hazardous Materials* 375: 224–232.

Xie, R., Seip, H.M., Wibetoe, G. et al. (2006). Heavy coal combustion as the dominant source of particulate pollution in Taiyuan, China, corroborated by high concentrations of arsenic and selenium in PM10. *Science of the Total Environment* 370 (2–3): 409–415.

Yadav, K.C. and Haleem, A. (2014). Thermal Power Plant Performance Centered Operational Recommendations. In: *National Seminar on Thermal Power Plant Performance Management*, 36–40.

Yadav, M.K. and Yadav, S.K. (2014). Operational power plants based impacts on biodiversity to assess ecosystem degradation in UP (India). *International Journal of Environmental Research and Development* 4 (3): 219–226.

Yadav, S.K., Singh, I., Singh, D., and Han, S.D. (2005). Selenium status in soils of northern districts of India. *Journal of Environmental Management* 75 (2): 129–132.

Yang, X., Tian, Y., Ha, P., and Gu, L. (1997). Determination of the selenomethionine content in grain and human blood. *Wei sheng yanjiu: Journal of Hygiene Research* 26 (2): 113–116.

Young, T.F., Finley, K., Adams, W.J. et al. (2010). Appendix A: selected case studies of ecosystem contamination by se. *Ecological Assessment of Selenium in the Aquatic Environment*: 257–292.

Yücel, U.M., Başbuğan, Y., Uyar, A. et al. (2020). Use of an antiarrhythmic drug against acute selenium toxicity. *Journal of Trace Elements in Medicine and Biology* 59: 126471.

Zahedi, S.M., Abdelrahman, M., Hosseini, M.S. et al. (2019). Alleviation of the effect of salinity on growth and yield of strawberry by foliar spray of selenium-nanoparticles. *Environmental Pollution* 253: 246–258.

Zang, H., Qian, S., Li, J. et al. (2020). The effect of selenium on the autophagy of macrophage infected by Staphylococcus aureus. *International Immuno Pharmacology* 83: 106406.

Zhang, L., Zeng, H., and Cheng, W.H. (2018). Beneficial and paradoxical roles of selenium at nutritional levels of intake in healthspan and longevity. *Free Radical Biology and Medicine* 127: 3–13.

Zhang, Y., Li, H., Lin, T. et al. (2019). Plasma selenium levels and risk of new-onset diabetes in hypertensive adults. *Journal of Trace Elements in Medicine and Biology* 56: 6–12.

Index

Selenium Contamination in Water, First Edition. Edited by Pooja Devi, Pardeep Singh, Arindam Malakar, and Daniel Snow.

b

g

h

i

t

www.ingramcontent.com/pod-product-compliance
Lightning Source LLC
LaVergne TN
LVHW062204200926
848484LV00006B/1317
* 9 7 8 1 1 1 9 6 9 3 4 5 1 *